Equidistribution of Dynamical Systems

Time-Quantitative Second Law

Fractals and Dynamics in Mathematics, Science, and the Arts: Theory and Applications

ISSN: 2382-6320

Published

Vol. 7 *Equidistribution of Dynamical Systems: Time-Quantitative Second Law*
by Jozsef Beck (Rutgers University, USA)

Vol. 6 *Exploring Scale Symmetry*
by Thomas Lowe (Commonwealth Scientific and Industrial Research Organisation (CSIRO), Australia)

Vol. 5 *Analysis, Probability and Mathematical Physics on Fractals*
edited by Patricia Alonso Ruiz (Texas A&M University, USA), Joe P. Chen (Colgate University, USA), Luke G. Rogers (University of Connecticut, USA), Robert S. Strichartz (Cornell University, USA) and Alexander Teplyaev (University of Connecticut, USA)

For the complete list of volumes in this series, please visit
www.worldscientific.com/series/fds

Fractals and Dynamics in Mathematics, Science, and the Arts: Vol. 7 Theory and Applications

Equidistribution of Dynamical Systems

Time-Quantitative Second Law

Jozsef Beck
Rutgers University, USA

World Scientific

NEW JERSEY • LONDON • SINGAPORE • BEIJING • SHANGHAI • HONG KONG • TAIPEI • CHENNAI • TOKYO

Published by

World Scientific Publishing Co. Pte. Ltd.
5 Toh Tuck Link, Singapore 596224
USA office: 27 Warren Street, Suite 401-402, Hackensack, NJ 07601
UK office: 57 Shelton Street, Covent Garden, London WC2H 9HE

British Library Cataloguing-in-Publication Data
A catalogue record for this book is available from the British Library.

Fractals and Dynamics in Mathematics, Science, and the Arts: Theory and Applications — Vol. 7
EQUIDISTRIBUTION OF DYNAMICAL SYSTEMS
Time-Quantitative Second Law

ISBN 978-981-122-555-0 (hardcover)
ISBN 978-981-122-556-7 (ebook for institutions)
ISBN 978-981-122-557-4 (ebook for individuals)

For any available supplementary material, please visit
https://www.worldscientific.com/worldscibooks/10.1142/11974#t=suppl

Desk Editor: Liu Yumeng

Contents

Preface

We wrote an unusually long Preface. It is partly an introduction written in an essay style and partly a brief informal summary of the whole book in a dozen pages. We felt the subject of "Second Law and the mathematical rigor" was quite controversial, along the line with the well known love-and-hate relation between mathematicians and physicists, which deserved such a "very talkative", perhaps provocative, Preface. After reading the long Preface, the reader should have a pretty good idea whether or not he/she wants to read the rest of the book.

In classical mechanics the two-body problem has a well known explicit solution, but already the general three-body problem has no explicit solution (we cannot solve the corresponding system of differential equations, and computer simulation indicates hopelessly "chaotic" behavior). It is no surprise, therefore, that we know very little about the time-evolution of many-particle closed dynamical systems, the subject of the book. For example, what can we say about the typical time evolution of a many-particle *far-from-equilibrium* closed system on a *realistic* time scale? There is hardly any nontrivial result in the literature.

We emphasize that in this book we study global questions about the whole system, instead of focusing on the motion of a single particle, or a few particles, as a small part of a large system — like the Lorentz gas models, and other scattering billiards. These "local" questions have a large literature; see e.g. the book of Chernov and Makarian [Ch-M2006] and the conference volume [Sz2000].

The most important global issue about the time evolution of a large closed system is the famous Second Law (of thermodynamics). This postulate of physics says in a nutshell that the time evolution of closed systems with many degrees of freedom, starting from nonequilibrium, approaches equilibrium (usually very rapidly), and exhibits entropy increase.

Unfortunately, there are plenty of notorious problems surrounding the Second Law. So, what is exactly the "trouble" with the Second Law? Here is a list in form of some questions.

(1) How to define entropy? How to convert the well known intuitions (like "disorder" and "energy spreading") into a precise mathematical definition? Boltzmann's well known formula entropy $= k \log W$ is a great guiding intuition for physicists, but it is far too vague for mathematicians. Physicists define the Boltzmann entropy by dividing the phase-space $\mathcal{P}$ into sub-regions ("small boxes") — called a "course-graining" of $\mathcal{P}$ — where the points represent microstates that are "macroscopically indistinguishable". Now the Boltzmann-entropy of a "macrostate" is $k \log V$, where the macrostate (i.e., small box) has volume V in $\mathcal{P}$, and k is Boltzmann's constant. Unfortunately the concept of "coarse-graining" seems to be a trap. For instance, how small are these "small boxes"? There is no precise answer. In the last 100 years nobody could precisely define *macrostates*, and turn this approach into rigorous mathematics.

(We will discuss the "trouble" with the alternative information-theoretic approach "progressive loss of information by the observer", based on the Shannon's entropy, later.)

(2) How to express the Second Law in terms of a rigorous mathematical theorem that eliminates the well known paradoxes (like the Loschmidt-paradox/irreversibility-paradox)?

(3) How to prove such a precise theorem? (Boltzmann's so-called H-theorem does not have a mathematical proof.)

(4) The Second Law is a "soft" qualitative statement about entropy increase that does not say anything about the absolutely crucial time factor. How to formulate a time-quantitative Second Law that describes the necessary time to reach equilibrium, and the size of the entropy increase as an *explicit* function of time?

For more about entropy in physics, see Appendix 1, where we elaborate on these questions.

The good news is that in this book we rigorously prove such a time-quantitative Second Law, which works on a realistic time scale. As a byproduct, we can rigorously clarify the Loschmidt-paradox/irreversibility-paradox. In fact, we give positive answer to all questions (1)–(4).

The bad news is that it takes a whole book to formulate and prove our version of the Second Law (in models of implicitly interacting particles — to be explained below). Physics provides the motivation, but what we really do in the book is mathematics with precise proofs.

A *dynamical system* in the broadest/abstract sense is an arbitrary "rule" (= mapping) which describes the time evolution of a "point" (representing the state of the whole system) in its ambient space (called phase space or state space). In this book we are studying concrete dynamical systems: we examine the time evolution of explicit many-particle idealized mechanical models, like many-particle billiard systems and generalized billiards.

We are going to rigorously prove approach to spatial equilibrium — i.e., equidistribution of the system in the "gas container" — and entropy increase. We do it by using an intrinsic entropy concept of our own. Intuitively, it is an entropy concept defined on the microscopic level — i.e., expressed in terms of the individual particles — so, we call it **micro-entropy**.

Following Boltzmann's ingenious insight that entropy increase is a statistical phenomenon, we study the *typical* time evolution of a large system in a closed environment, where the particles are in a "vessel" or "container" (it can be a cube, a sphere, etc.), starting from an arbitrary explicit off-equilibrium initial configuration. What we rigorously mean by "equilibrium" is an intrinsic equilibrium concept of our own that we call **micro-equilibrium**. Intuitively, it is a spatial equilibrium in the container, defined on the microscopic level ("optimal equidistribution in the container"). We study the time evolution: we prove fast approach to micro-equilibrium, and also prove that it exhibits the stability of the micro-equilibrium for an extremely long time. We allow *arbitrary off-equilibrium* initial condition, including such extreme cases where all particles start from the same point — we call it a "Big Bang"; or anything similar to "Big Bang", e.g., a configuration where all particles are in the (say) left half of a box. The number N of the particles can be arbitrary, but our results become really interesting for large N, e.g., $N = 10^3$ or $N = 10^6$ or larger like the range of the Avogadro number. In fact, our favorite size for N in the illustrations is in the range of 10^{27}, which is the approximate number of gas molecules in a cubic gas container with side length one meter.

(It is not an accident that we cautiously say *approach to equilibrium* instead of the direct/straightforward term of *convergence to equilibrium.* The reason is very simple: convergence to (snapshot) micro-equilibrium *fails* for an infinite sequence of time instances tending to infinity. We return to this interesting technical detail later; see e.g. Lemma 1.2 below. What we call snapshot micro-equilbrium is arguably the most natural equilibrium concept.)

Here is a brief summary of the book in one paragraph. The Second Law is an axiom of thermodynamics. It is fair to say that most physicists are pragmatists, and are willing to accept the Second Law without any precise mathematical proof. But not Boltzmann: he felt it was necessary, and possible, to prove the Second Law in terms of the motion of individual particles in Newtonian mechanics. Boltzmann's kinetic theory (Boltzmann equation and his H-theorem) was an attempt to prove that the entropy increase comes from particle-particle collision. But Boltzmann failed to give a mathematical proof. The rigorous analysis of Boltzmann's argument shows that the entropy increase comes from the *ad hoc* initial condition of "precollision molecular chaos", which cannot be proved on a mechanical basis — or at least nobody succeeded to do so in the last 100 years. (It is the assumption that molecules are uncorrelated before collision, but not after the collision. So Boltzmann introduced a fundamental asymmetry in time. We return to this in Appendix 1.)

Here we still work with individual particles and deterministic orbits (Laplacian setup), but we have a completely different approach. We rigorously prove fast entropy increase *without* any direct reference to collision (i.e., particle-particle collision). Instead we allow (basically) arbitrary non-linear orbits for the individual particles. The idea is that arbitrary orbits and arbitrary motions on these orbits imitate the particle-particle collision, and also imitate the effect of the external and internal forces. We give some justification for this *implicit interaction* approach very soon in Section 2, where we elaborate on this critical issue.

We study *implicit interaction* models, because, similarly to Boltzman, we cannot handle the mathematical difficulties of the direct form of particle-particle collision either.

The physicist's view on the Second Law in a nutshell is that large closed systems exhibit entropy increase, and *this* entropy increase forces the system to approach equilibrium. In the rigorous proof of our time-quantitative Second Law we reverse this order: we prove the approach to equilibrium *first*. More precisely,
(1) first we show that in our implicit interaction models the typical time evolution of the large closed system approaches a well-defined microscopic equilibrium; (2) then we define a new entropy concept on a microscopic level that measures the "discrepancy" from the microscopic equilibrium (in fact, our Micro-Entropy is equal to an absolute constant minus the "discrepancy", so "entropy increase" means "discrepancy decrease"); and finally, (3) we prove that this Micro-Entropy satisfies a time-quantitative Second Law, i.e., the typical time evolution of the system

exhibits Micro-Entropy increase in an explicit time-dependent form such that reaching the stage of maximum Micro-Entropy coincides with the state of microscopic equilibrium.

We emphasize that we do not need any *ad hoc* initial condition like the "molecular chaos".

Many physicists — perhaps the majority — prefer Gibbs to Boltzmann. (For example, the classical book of Landau and Lifshitz (see La-Li40]), arguably the best book on the subject, calls the Boltzmann entropy an "outdated" concept that "cannot be rigorously justified", and the book exclusively uses the Gibbs entropy.) Because the Gibbs entropy remains constant (by Liouville's theorem), the current explanation of the Second Law by the "Gibbs school" (representing the majority of physicists) is to refer to "chaos" in the phase space, without any mathematically precise definition of "chaos". This kind of "proof" means drawing chaotic pictures with many "tentacles", instead of providing any mathematical theorem. The philosophy behind it is that the "fine-grained entropy" (= Gibbs entropy) remains constant, but the "course-grained" entropy — which reflects the limitations of the human observer — does increase.

Another, somewhat similar, way to explain the Second Law is the information-theoretic approach: it refers to the "progressive loss of information by the observer". This approach usually includes a discussion about the Maxwell demon, and some reference to Shannon entropy and Information Theory in mathematics, but again they never mention any concrete relevant mathematical theorem justifying the entropy increase. What the information-theoretic approach really means is that we are describing the state of our knowledge rather than the state of the system (like gas in a container, say). And if that is so, is not *entropy increase* a subjective phenomenon caused by our severely limited capacity for handling information? Is the Second Law a product of the human condition? The information-theoretic approach cannot give a convincing answer to these rather embarrassing questions.

The mainstream view of physicists about entropy and the Second Law is well illustrated by what Leonard Susskind said in his video lecture series on Statistical Mechanics at Stanford University (2013), widely available on the internet:

"The *microscopic* entropy remains constant (due to Liouville's theorem), but the *coarse-grained* entropy does increase. This is the Second Law. The reason behind it is our limited resolution powers, explaining why we cannot keep track of the fine structure.

In other words, entropy is not really a property of the system: it is the property of the system *and* what we know about the system. If we could follow in infinite detail every point in the phase space, then entropy would be conserved.

The resolution of the appareant contradiction between the Second Law and the reversibility of classical mechanics lies in the lack of precision of our observations, and in chaos (= undectable differences in initial conditions lead to large changes in results)."

Notice that Susskind did not mention any mathematical theorem. Instead he used "chaos" as a magic word that completely clarifies the issue.

In sharp contrast with the mainstream view of physicists expressed by Susskind, we define a meaningful microscopic entropy that is not constant. What is more, we prove that this micro-entropy does increase: it satisfies a time-quantitative Second Law!

The overwhelming majority of physics textbooks imply that the Second Law is well understood. Occasionally they give an intuitive argument that falls short of mathematical rigor, or discuss a non-mechanical toy model like the probabilistic Ehrenfest urn model ("Markov chain") or the Kac ring model (pure combinatorics; not even close to a mechanical model of particles moving on individual trajectories). Neither one is particularly convincing. Or they say the mathematical proof of the Second Law is beyond the scope of the book. Just a few articles and lectures tend to admit that the origins of the Second Law remain mysterious. What we have here is a diversity of intuitive explanations. Besides (1) Boltzmann's problematic H-theorem, or (2) using vaguely defined chaos without any theorem, or (3) "increasing loss of information" (that we already mentioned), books and articles attribute the validity of the Second Law to (4) unknown constrains on the initial conditions, (5) some appeal to cosmology (Big Bang), or (6) to unknown features of quantum mechanics.

Such uncertainty/dubiety by the physicists about the Second Law clearly indicates that the basic questions have not been completely answered yet. This is particularly troubling, since many physics books call the Second Law the "most universal law in physics", and many physicists consider Boltzmann's formula $Entropy = k \log W$ (engraved on his tombstone) arguably the second most important formula in physics (perhaps after Einstein's $E = mc^2$).

It just adds to the confusion that cosmologists freely talk about the "entropy of the universe", but they hardly ever address the following basic questions. Why is it fair to extrapolate from the "observable universe" to the whole universe? Is the universe a closed system? Is the universe in equilibrium, and if it is not, which definition of entropy can we use? (For example, the meaning of Boltzmann's formula $Entropy = k \log W$ is problematic in off-equilibrium, and it becomes totally meaningless for a possibly infinite universe.)

Even many introductory physics books prefer to express the Second Law in the unnecessarily, and rather misleadingly, general form of involving the "entropy of the universe" like this: the entropy of the universe never decreases. (Of course these books never provide a precise definition of the concept.) These books make the false impression to the young students that "entropy of the universe" is a well-established concept. But nothing is further from the truth.

Being a mathematician, in this book we study the following six basic mathematical questions:

(1) *How to define micro-equilibrium — a spatial equilibrium concept, an optimal form of equidistribution in the "container" — in a large (= many-particle) system?*

(2) *Suppose we are given a concrete initial velocity distribution (e.g., Gaussian, or photon-like isotropic distribution with constant speed) and have an arbitrary concrete initial point configuration (i.e., the starting points of the particles) with a typical set of initial velocities (typical in the given fixed velocity distribution). How long does it take to reach micro-equilibrium?*

(3) *How to define our intrinsic micro-entropy as an "inherent" property/parameter of the initial point configuration and the given fixed velocity distribution?*

(4) *Does micro-entropy really increase, and if it does, what explicit time-quantitative function describes it?*

(5) *What can we say about the long-term stability of the micro-equilibrium?*

and last but not least,

(6) *What can we say about (2)–(5) on a* **realistic** *short time scale?*

We answer questions (1) and (2) in Chapter 1, and supply the necessary mathematical proofs in Chapter 2. We start answering question (3) at the end of Chapter 1, and finish it in Chapter 3. We answer questions (4), (5), and (6) in Chapters 3–4. We discuss question (5) in Chapter 5. And we elaborate on question (6) in all five chapters. (It may help the reader that at the beginning of the sections or the chapters we give a short summary of the content in a couple of sentences.)

We give a detailed analysis of concrete many-particle models, including time-quantitative results like "starting from this initial configuration... it takes between 3 and 5 milliseconds for the typical time evolution of the closed system to reach micro-equilibrium" or "starting from that initial configuration...it takes between 100 and 200 years for the typical time evolution of the system to reach micro-equilibrium" (where in both cases "typical" means 99.999 ... 9 percent).

This book is about mathematics: it is at the crossroads of number theory, Fourier analysis and dynamical systems. But of course the motivation and the guiding intuition come from the Second Law of thermodynamics. The reader does not need any expertise in thermodynamics beyond the naive "high-school level".

This is a research monograph, where the overwhelming majority of the results are due to the author and are published here for the first time. It is intended to mathematicians, including Ph.D. students, also to mathematical physicists, and, last but not least, to physicists who are interested in the mathematical foundation of the Second Law.

Equilibrium statistical mechanics is basically applied probability theory, and it works with probabilistic models, where the particles are represented either by independent random variables, or the time evolution is described by a well-known stochastic process (e.g., a Markov chain). It means that statistical mechanics completely ignores the individual trajectories of particles. (For a rigorous treatment of statistical mechanics; see e.g. D. Ruelle [Ru69] or O. Penrose [PeO70], the latter is based on the "Markovian Postulate".)

In sharp contrast, here we study large closed dynamical systems that are defined in terms of the individual trajectories of the particles. What we study are hybrid models: they are *basically deterministic* with some inevitable random part. It means the following. The initial configuration is deterministic — because we are interested in the nonequilibrium case! — and the trajectories of the particles are deterministic. In this respect our setup is completely different from the setup of (equilibrium) statistical mechanics.

On the other hand, we study *typical* initial velocities (where *typical* means "overwhelming majority with respect to some natural probability measure", e.g., Gauss velocity distribution). This reflects Boltzmann's discovery about the statistical nature of the Second Law. We point out that it is impossible to upgrade "typical initial velocities" to "all possible initial velocities", since as we demonstrate on several examples, the Second Law clearly fails for pathological initial velocities! In other words, the study of *typical* initial velocities is an inevitable necessity to prove any kind of positive result. *Typical behavior* is really the best that we can hope for.

To define *typical* velocities, we have to fix a velocity distribution (say, the 3-dimensional normal distribution, also called Gaussian), and assume that the initial velocities of the particles are independent random variables with the given distribution. Independence implies a product measure, which makes our (basically combinatorial) calculations "doable".

This is what we cannot do. In our models the particles move on deterministic trajectories, so, we cannot apply probability theory the way as Statistical Mechanics does. The time evolution of our closed systems defines a measure-preserving flow in the phase space, and this time-flow is often ergodic. Nevertheless, we cannot apply (measure-theoretic) ergodic theory either, because the initial configurations are *not* typical (ergodic theory ignores zero-sets). In fact, what we do is the complete opposite: we work with *explicit* far-from-equilibrium initial configurations (e.g., Big Bang), and they of course represent a zero measure set in the phase space. We do not use abstract dynamical systems theory; in particular, and do not use hyperbolic systems. We do not use differential equations at all (for example, the Boltzmann equation has a large literature, but it does not seem to help us). Similarly, we cannot use the well known entropy concept of measure-preserving dynamical systems: the Kolmogorov–Sinai entropy (motivated by Shannon's entropy). The reason is that the Kolmogorov–Sinai entropy is a *constant*, defined as an *infinite* time limit. Constant entropy (i.e., the Kolmogorov–Sinai entropy) does not help us, because our goal is to prove entropy increase in a realistic short time scale.

In this book we develop our tools from scratch, mostly using Fourier analysis and combinatorics. We define our own entropy as a "microstate functions" (= "configuration function"), see later.

The existing entropy concepts in physics are all "macrostate functions" ignoring the particles: Boltzmann entropy and Gibbs entropy (and we can add to the list the similar mathematical concepts of the Shannon entropy and the Kolmogorov–Sinai measure-theoretic entropy). "Macrostate" is an intuitive concept meaning "observable level, far above the microscopic level, where one can already measure the basic physical quantities like temperature, pressure, and so on".

(By the way, another sign of "trouble with entropy" is that there are too many entropy concepts; see e.g. the paper of Lebowitz and Maes [L-M2003]. Note that we did not even mention the original Clausius thermodynamic entropy, the von Neumann entropy (= quantum version of the Gibbs entropy) and the topological entropy.)

To distinguish our entropy from the existing "macrostate" entropy concepts, we call it *micro-entropy*. The name micro-entropy emphasizes the fact that our entropy concept is a "microstate function", i.e., it is expressed in terms of the location and the velocity of the individual particles.

The usual entropy concepts of statistical mechanics (Boltzmann entropy and Gibbs entropy) are designed for equilibrium, and they are meaningless in our off-equilibrium (dynamical) system models. Nevertheless, at the beginning of Section 11 we recall the Boltzmann entropy to point out an interesting *formal* analogy between Boltzmann's famous formula (Entropy = $k \log W$, where W is the number of microstates consistent with the given macrostate) and our micro-entropy. This formal similarity is a second reason — beside our Second Law — that justifies the name *micro-entropy*. (Another remarkable property of our micro-entropy is its "robustness"; see the explanation in Section 6 after the Metatheorem.)

As we said already, physics defines entropy as a macrostate function, and this is exactly where the trouble with mathematical rigour begins. *Macrostate* is a good intuition, but it is not clear how to give a precise and meaningful mathematical definition of "macrostate" when the system is far from equilibrium; see Appendix 1. Our basic idea was to completely avoid this trap: our entropy concept is a microstate function (in fact, it is a configuration function). Microstates — unlike macrostates — have a perfectly straightforward rigorous definition — expressed in terms of Newtonian mechanics — in both equilibrium and off-equilibrium. Microstate simply means that in every instant we know the location and velocity of every individual particle (i.e., we stay in the frameworks of classical mechanics).

Note that our book is completely different from the nearly 100-page long paper of Lieb and Yngvason titled *The Physics and Mathematics of the Second Law of Thermodynamics*; see [L-Y99]. Lieb and Yngvason develop a new axiomatic approach for the foundations of classical thermodynamics. Classical thermodynamics, a subject that was developed on a phenomenological basis, by definition gives up on the Newtonian view of particles moving on individual trajectories (e.g., it considers gases as an infinitely divisible fluid without any substructure). Moreover, Lieb and Yngvason deal only with states of systems in equilibrium and do not attempt to define "entropy" for systems not in equilibrium.

What we do is the complete opposite. We study the time evolution of closed many-particle mechanical systems, starting from far-from-equilibrium, and focus on the entropy of states that are *not* in equilibrium. Despite our diametrically opposite ways of approaching the Second Law and entropy, we share a common goal with Lieb and Yngvason. We quote the last sentence in the abstract of their paper: "Our formulation offers a certain clarity and rigor that goes beyond most textbook discussions of the second law." The same applies to our work: our goal was to formulate precise theorems, and give rigorous mathematical proofs.

The goal of our book is to give a mathematical justification of the basic postulate of nonequilibrium thermodynamics. This postulate of physics says in a nutshell that the time evolution of closed systems with many degrees of freedom, starting from nonequilibrium, approaches equilibrium (usually very rapidly), and exhibits entropy increase. Moreover, we prove that equilibrium is characterized by *long-term stability*. It requires fairly subtle mathematics to make the concepts of *micro-equilibrium*, *micro-entropy*, and *long-term stability* precise, and to eliminate the well known paradoxes. It explains the considerable length of the book.

We already mentioned the fact that the entropy concepts used in statistical mechanics (the Boltzmann entropy and the Gibbs entropy) are non-intrinsic macrostate functions. Nevertheless, here we briefly elaborate on this critical issue. The Boltzmann entropy is defined in terms of the given macrostates ("coarse-grained description" of the system in the phase space), and "macrostate" is already a rather problematic — subjective! — concept in the equilibrium case. And it is not even well-defined in the non-equilibrium case! Different sets of macrostates — a partition of the phase space — lead to different Boltzmann entropies. This fact is the source of some long-standing controversy. For example, if we refine the macrostates, in the limit (tending to infinity) we obtain some absurd value for the Boltzmann entropy (due to $\log 0 = -\infty$ and $\log 1 = 0$). So, what is the "proper coarse-graining"? What are the "good" macrostates? If the choice of the macrostates is a subjective reflection of our ignorance of a system, then how can we make an objective statement about the entropy? The "Boltzmann school" does not give a straight answer to these annoying questions.

The most extreme view is represented by Jaynes and his school (see [Ja89]), who have vigorously asserted that entropy is purely subjective, a quantification of one's lack of knowledge of the microstate (= molecular state) of the system. The following quote is attributed to Wigner: "entropy is an anthropomorphic concept". Jaynes often refers to Wigner, telling how much he was motivated by this bite-the-bullet type statement.

This kind of absurd — or at least highly problematic — view by leading physicists is very troubling for us mathematicians. It fits to recall here the well known story of what von Neumann said to Shannon when Shannon derived his (now famous) formula for "uncertainty" of information. Shannon asked von Neumann (who was at home in both mathematics and physics) what he should call this "uncertainty". Von Neumann replied: "You should call it entropy, because nobody really knows what entropy is, and whenever you use the term you will always be at an advantage." Reading the old and new physics books about entropy, we must conclude that von Neumann's sarcastic remark/joke remains just as true even now, 70 years later!

(Of course von Neumann admired the creative intuition of physicists; his best friends were all physicists. What von Neumann criticized was the disappointing ambiguity; the lack of precise definitions, theorems, proofs about entropy and the second law, i.e., the low level of mathematics in general surrounding this important subject.)

Since Shannon's entropy is the basis for a well-established mathematical theory — information theory — many physicists "solve" the dilemma of entropy by claiming that *entropy = lack of information.*

It is telling that there are more than 20 different formulations of the Second Law in physics. This led a well-known science-historian and phylosopher to remark (rather sarcastically) that "every physicist knows what the second law is, the problem is that no two agree on what it is" (see [Sk93]).

As we said, we study many-particle closed systems motivated by classical mechanics; mainly the (non-equilibrium) kinetic theory of gases. This automatically assumes the Newtonian deterministic/mechanical viewpoint. It means that every individual point-particle has a well-defined trajectory, and the initial condition uniquely determines the time evolution of the system forever. The simplest warm-up model is a large system of point-billiards (Bernoulli model for the non-equilibrium ideal gas), where every trajectory is linear or piecewise linear, depending on the cases of torus or box (= "container"). We quickly move far beyond point billiards in a box, and study far-reaching generalizations where the particles move on (almost) arbitrary nonlinear parametrized curves (as trajectories) in a closed environment. Note that arbitrary nonlinear curves may serve as a model for the complicated thermodynamical process of "energy spreading" in liquids and solids, far beyond the case of gases. This is how we mimick particle-particle collision, and the effects of external and internal forces. We will elaborate on this in Section 2. We may also replace the box with other natural containers; for example, the sphere.

So, what we study is the time evolution of large systems of particles in a container ("closed system"), where the interaction and the forces are expressed *implicitly* in the form of arbitrary non-linear individual orbits.

We have to say a few words about the issue of *entropy increase and implicit interaction.* In real life phenomena the interaction of particles (e.g., collision of gas molecules) is a crucial ingredient in the precess of approaching equilibrium. For example, Maxwell gave a very convincing intuitive explanation of why particle-particle collision of the gas molecules is the reason behind the quick approach to the Gaussian velocity distribution, representing equilibrium. (This is why physicists also call it the Maxwellian velocity distribution). Unfortunately the rigorous mathematical discussion of particle-particle interaction in a direct form leads to excruciating technical difficulties. For illustration note that Boltzmann claimed a "mathematical proof" of the Second Law (i.e., entropy increase) via his H-theorem. This proof is at the center of a century old controversy. The first minor problem is that the H-theorem — unlike what we do in this book — is *not* a statement about the individual trajectory of the whole system. Instead the H-theorem is a statistical statement about the *distribution function* of particle velocity $\mathbf{v}$ in position $\mathbf{x}$ (infinitesimal element in the 6-dimensional space) at time t. Second, and this is the main issue, there is a notorious problem with mathematical rigor. In fact, there are two results here: the first result is that the typical time evolution of dilute gas, starting from off-equilibrium, satisfies the Boltzmann equation. The second result is that a solution of the Boltzmann equation defines the so-called H-functional (a negative entropy type quantity), which is a decreasing function of the time. There is no problem with the proof of the second result (which is just a couple of lines, and it is based on the simple fact that the function $f(x, y) = (x-y)\log(x/y)$ is always non-negative). It is the proof of the first result that is problematic; the justification of the Boltzmann equation is the source of a century old controversy. The key step in the proof is an analysis of particle-particle collision, and, according to Boltzmann's argument, this is the ultimate source of the entropy increase. Unfortunately, from the viewpoint of rigorous mathematics, Boltzmann's argument is faulty, because he (indirectly) assumed that molecules are uncorrelated before collision, but not after the collision, so he introduced a fundamental asymmetry in time ("one-sided or pre-collision molecular chaos"). This is a logical fallacy in an argument whose goal is to justify entropy-increase/irreversibility/arrow-of-time. Even now, 130 years after the discovery of the Boltzmann equation, we are nowhere near proving Boltzmann's assumption on the "molecular chaos". Rigorously proving the long time propagation of the molecular chaos remains a wide open mathematical problem. We conclude, therefore, that — strictly speaking — Boltzmann's H-theorem is not a "theorem". It is a conditional result based on the unproved hypothesis of mathematical chaos. We will give a more detailed discussion of this and other related issues about the Second Law later in Appendix 1.

Another attempt to do rigorous mathematics is to use Markov chain models (initiated by P. and T. Ehrenfest who introduced the Ehrenfest urn model). Then of course we face the fundamental problem of what justifies the giant leap from deterministic Newtonian mechanics to a stochastic process. But even if we ignore this fundamental problem, we still face the above-mentioned logical problem of introducing a fundamental asymmetry in time. More precisely, Markov chains do not reflect "reversibility", a basic symmetry of Newtonian mechanics. Indeed, it is a well known result in probability theory that, if a Markov chain is time-reversible (in the usual sense that the reversed process has a time-independent transition matrix), then the process is necessarily in equilibrium, meaning the stationary distribution. (As usual, by a Markov chain we mean a time homogeneous Markov chain.) The proof is just one line: consider a finite Markov chain with states $1, 2, \ldots, n$, let p_i, $1 \le i \le n$ be a probability distribution on the states, let $P(i \to j)$ denote the transition probabilities, and assume time-reversibility

$$p_i P(i \to j) = p_j P(j \to i) \quad \text{for all} \quad 1 \le i, j \le n,$$

called "local balance" or "detailed balance". Then

$$\sum_{i=1}^{n} p_i P(i \to j) = \sum_{i=1}^{n} p_j P(j \to i) = p_j \left(\sum_{i=1}^{n} P(j \to i) \right) = p_j,$$

which proves that p_i, $1 \le i \le n$ *is* the stationary (= equilibrium) distribution of the Markov chain.

Perhaps the reader is wondering why the "detailed balance" condition

$$p_i P(i \to j) = p_j P(j \to i) \quad \text{for all} \quad 1 \le i, j \le n,$$

means time-reversibility. Well, the message of "detailed balance" is that the rate at which the forward process makes a transition from i to j is always equal to the rate at which the reverse process makes a transition from j to i. If the process is time reversible, the two rates must be equal, explaining why "local balance" represents a natural form of time-reversibility.

So, a time-reversible Markov chain exhibits a trivial case of the Second Law where the system is already in equilibrium, i.e., there is no entropy increase. We conclude, therefore, that the use of Markov chain models (e.g., the Ehrenfest urn model) to explain the Second Law, and to resolve the "reversibility vs. irreversibility paradox", is highly problematic — to say the least.

Our (oversimplified) mechanical models are of course time-reversible, and we prove typical entropy increase (i.e., Second Law) in both directions of time.

We prove the Second Law by focusing on spatial equilibrium, i.e., an optimal form of *equidistribution* of the system in the container, and for *implicit interaction models* (where the particle-particle collision, and the external and internal forces, are represented in an indirect way by allowing arbitrary non-linear orbits).

Focusing on spatial equilibrium seems like a very one-sided approach, which probably makes physicists rather disappointed. However, we can defend this one-sided approach with the following intuition. In large closed physical systems the equilibrium in the velocities of the molecules gets established very, very quickly, due to frequent particle-particle collisions. Then comes the longer second stage, when (relatively) slowly the spatial equilibrium gets established. **Our models represent the slower second stage, when the the velocities of the molecules are already in equilibrium, and we study the approach to spatial equilibrium, i.e., the approach to equidistribution of the system in the container.**

We prove that, under surprisingly general conditions, a large closed system of particles, moving on (almost) arbitrary non-linear orbits, and starting from an arbitrary fixed initial configuration (e.g., Big Bang), approaches spatial equilibrium, exhibits entropy increase (and exhibits long-term stability of equilibrium). More precisely, this holds for a typical time evolution. The approach to equilibrium can be very rapid or very slow, heavily depending on the "initial velocity distribution". For example, it is very fast for the Gaussian velocity distribution (a couple of milliseconds in some "realistic" models), and it is very, very slow in the case where the particles have the same constant speed (for motivation note that photons move with the same speed). It remains true even if the system starts from Big Bang, or any other extreme initial configuration. (When we compare two different initial velocity distributions, then of course we always assume that they have the same mean speed.)

On one hand, our work is about large (dynamical) systems. On the other hand, one may say that our work belongs to Uniform Distribution and its counter-part called Irregularities of Distribution (= Discrepancy

Theory), a particular chapter of number theory. (For the interested reader we recommend the books [Dr-Ti97] and [Be-Ch87], [Mat99].) By using Fourier analysis and combinatorics, we develop a time-quantitative theory of *square-root fluctuation type spatial equilibrium* — a new chapter in (continuous) Uniform Distribution. It is based on our micro-entropy and on our time-quantitative Second Law(s).

We prove "entropy increase" in the form of "discrepancy decrease" in the spatial distribution. This holds for the typical time evolution of a closed many-particle system.

For illustration consider the simplest warm-up model.

Our rigorous Second Law for the simplest nonequilibrium ideal gas model: a nutshell summary. *Consider a system of N point billiards in a box $[0,1]^3$ (N is large) starting from an arbitrary fixed initial N-element point configuration. Fix a basically arbitrary initial velocity distribution (for example, the Gaussian = 3-dimensional normal distribution), and for every particle independently choose a concrete initial velocity according to the given probability distribution (independence implies a product measure). Then the time evolution of the system is completely determined, where each particle moves with constant speed on a piecewise linear billiard path. We show that, with the typical choice of the initial velocities (according to the product measure), the system exhibits entropy increase, which is explicitly described by the Time Evolution Function — a computable function that depends only on the given initial configuration (and the given fixed velocity distribution).*

For illustration note that, in the torus model where the particles move in the 3-dimensional unit torus $[0,1)^3$, and every particle has unit mass, the Micro-Entropy of an N-element point set $\mathcal{Z} = \{\mathbf{z}_1, \ldots, \mathbf{z}_N\} \subset [0,1)^3$ is an explicit function of the following infinitely positive terms (each one is a square-modulus)

$$s(\mathcal{Z};\mathbf{n}) = \left| \sum_{k=1}^{N} e^{2\pi i \mathbf{n}\cdot\mathbf{z}_k} \right|^2, \quad \mathbf{n} \in \mathbb{Z}^3 \setminus \mathbf{0}. \tag{0.1}$$

Here we insert a key insight that we will clarify later. If the terms $s(\mathcal{Z};\mathbf{n})$, $\mathbf{n} \in \mathbb{Z}^3 \setminus \mathbf{0}$ in (0.1) are "small" — namely, much smaller than the trivial upper bound N^2 — then the sum of the N complex numbers inside (0.1) mostly "cancels out". This is equivalent to the fact that the N-element point set $\mathcal{Z} \subset [0,1)^3$ exhibits a quantitative form of spatial equidistribution (depending on this "smallness"). And vice versa: a quantitative form of *spatial equidistribution of the N-element point set is equivalent to the fact that the terms $s(\mathcal{Z};\mathbf{n})$, $\mathbf{n} \in \mathbb{Z}^3 \setminus \mathbf{0}$ in (0.1) are all "small". This explains the relevance of the terms in (0.1) in the definition of our Micro-Entropy.*

Beside (0.1) the Micro-Entropy depends only on the given fixed velocity distribution. More precisely, we define the Micro-Entropy by putting together the critical terms $s(\mathcal{Z};\mathbf{n})$, $\mathbf{n} \in \mathbb{Z}^3 \setminus \mathbf{0}$ in a particular way that depends only on the given velocity distribution. For example, see (0.2)–(0.3) below for the special case of the Gaussian distribution; for more special cases, see Section 6.

A more detailed description goes as follows. We study the necessary time — or the corresponding "drifting distance" traveled by a typical particle for the general case where the particles move on general nonlinear curves — for a concrete evolution of the closed system to reach (Spatial) Equilibrium (= micro-equilibrium). As we show, this amount of time in the simplest model (or drifting distance in the general case of arbitrary orbits) is basically the same for the majority of the initial velocities (chosen independently from the given velocity distribution). We call it the Relaxation Time in the simplest billiard/torus model (or Relaxation Drifting Distance in the general case) of the given initial configuration (and the given velocity distribution).

It means that in the simplest billiards-in-a-box model we can associate an intrinsic Relaxation Time (= typical relaxation time) with every point in the $3N$-dimensional configuration space (under the condition of a fixed initial velocity distribution). A far from (Spatial) Equilibrium initial point configuration is characterized with the property that it has a long Relaxation Time; an initial point configuration is in (Spatial) Equilibrium if and only if its Relaxation Time is $O(1)$, i.e., bounded by a constant independent of N.

We can prove that, in a typical time evolution of the closed system, represented by a parametrized curve in the configuration space (= trajectory of the whole system), we have monotonicity in the following sense: the parametrized curve keeps moving from a point with longer Relaxation Time to a point with shorter Relaxation Time.

Warning: Just because the time evolution of a system reached a point in the configuration space with short Relaxation Time, it does not mean that the system will shortly reach Equilibrium. The actual remaining time to reach Equilibrium can be much longer, but it can also be shorter — it depends on the Relaxation Time of the initial configuration, and it is explicitly determined by a "Time Evolution Function". We refer to this as the "entropy paradox" (see Section 17). This "entropy paradox", and the related Loschmidt-paradox/irreversibility-paradox, explain why the rigorous proof of the Second Law is such a subtle mathematical problem (where the devil is in the details).

In the simplest model our intrinsic Micro-Entropy (= an "inherent" property of the given initial configuration and the given initial velocity distribution) is defined in terms of the Relaxation Time. Micro-Entropy is a renormalized Relaxation Time — it is renormalized in terms of the initial velocity distribution in such a way that Micro-Entropy becomes nearly independent of the initial velocity distribution. (We refer to this fact as follows: "Micro-Entropy is almost equal to Reduced Micro-Entropy"; see Section 6.) Since Micro-Entropy is defined as a monotone decreasing function of the Relaxation Time, we obtain the key property of the Second Law: **entropy increase**.

For example, if the given fixed velocity distribution is the 3-dimensional Gaussian (= normal), and $\mathcal{Z} = \{\mathbf{z}_1, \ldots, \mathbf{z}_N\}$ is an arbitrary set of N particles in the torus $[0,1)^3$, then the micro-entropy of $\mathcal{Z}$ is defined as

$$\text{MicroEntropy}_{Gauss}(\mathcal{Z}) = N^2 - \text{Imbalance}_{Gauss}(\mathcal{Z}), \tag{0.2}$$

where

$$\text{Imbalance}_{Gauss}(\mathcal{Z}) = \max_{\mathbf{n}\in\mathbb{Z}^3\setminus\mathbf{0}} \left(\frac{\left|\sum_{k=1}^{N} e^{2\pi i \mathbf{n}\cdot\mathbf{z}_k}\right|^2}{N} \text{ or } 1 \right)_{\max}^{1/|\mathbf{n}|^2}, \tag{0.3}$$

with $\mathbf{n} = (n_1, n_2, n_3)$ and $|\mathbf{n}| = (n_1^2 + n_2^2 + n_3^2)^{1/2} \geq 1$ ("length"). Note that

$$\text{Imbalance}_{Gauss}(\mathcal{Z}) = \exp\left(2\pi \cdot \text{RelaxationTime}_{Gauss}(\mathcal{Z})\right)^2.$$

As we said above, we prove "entropy increase" in the form of "discrepancy decrease", and (0.2) reflects this relation.

Since the Micro-Entropy depends on all individual particles, it is not surprising at all that the definition is complicated. What is simple here is the simple behavior of entropy increase (that we can precisely prove for this complicated entropy in a time-quantitative form). Unfortunately, the proof is again rather complicated.

In the more interesting general models, where each particle moves on its own (basically) arbitrary (possibly nonlinear) parametrized curve in the 3-space (in the torus models we take everything modulo one, i.e., we take the fractional parts of the coordinates), we can prove similar results. A minor (necessary) technical difference is that we have to replace the Relaxation Time with concept called Relaxation Drifting Distance.

Unfortunately here in the Preface we cannot give a more convincing short intuition to motivate these at first sight quite mysterious and complicated terms and formulas, like "Micro-Entropy" and "Imbalance". We give the detailed clarification/justification later in the book.

The first place to read more about Micro-Entropy is Section 6. A reader in big rush may jump ahead to Section 6 right now. And after finishing Section 6, such a reader may skip the proofs and jump to the "Illustrations": Sections 15–17.

What happens *after* a large system — starting from off-equilibrium — already had reached equilibrium? Well, in Chapter 5 we prove results demonstrating long-term stability of the square-root fluctuation type equilibrium.

Beside stability, equilibrium is also associated with "randomness". That problem is beyond the scope of this book. We devoted another book to that subject titled *Strong Uniformity and Large Dynamical Systems*; see [Be2018]. There is hardly any overlapping between the two books, except Appendix 2 here, which is Section 3 in [Be2018]. We decided to include Appendix 2 for the convenience of the reader.

For other mathematical approaches to define entropy of non-equilibrium systems, completely different from our approach, see e.g. [Ru99] and [Go-Le2004].

Finally note that the notation we use in the book is rather standard in mathematics. The only exception is that we use Gamma(z) to denote Euler's Gamma function. The reason is that we save $\Gamma(t)$ for the trajectory of the whole system in the configuration space, which is a key concept in the book (and of course t denotes *time*). We use log for the natural (base e) logarithm, use $\log_2$ for the binary (base 2) logarithm, and for a (usually finite) set A we use $|A|$ to denote the number of elements of A. For simplicity we usually refer to the d-dimensional Lebesgue measure in the familiar low dimensions $d = 1$, 2, and 3 — respectively — as *length*, *area*, and *volume*. The d-dimensional Lebesgue measure with $d > 3$ is usually denoted by vol_d. In the book we use all three standard notations for the Lebesgue integral of f over a (measurable) subset $B \subset \mathbf{R}^d$ of the d-space

$$\int_B f\, d\text{V} = \int_B f\, d\text{vol}_d = \int_B f(\mathbf{x})\, d\mathbf{x}$$

(where of course V indicates volume).

We use $\lfloor x \rfloor$, $\lceil x \rceil$, and $\{x\}$ for the lower integral part, the upper integral part, and the fractional part of a real number x in this order (so $x = \lfloor x \rfloor + \{x\}$), and we frequently use $\|x\|$ to denote the distance of x from the nearest integer. We often say "x modulo one", which means the fractional part $0 \leq \{x\} < 1$, unless it is explicitly stated otherwise (like say between $-1/2$ and $1/2$).

Finally, given a vector $\mathbf{v} = (v_1, \ldots, v_d)$, for simplicity the ordinary length (= magnitude) of the vector is denoted by $|\mathbf{v}| = (v_1^2 + \cdots + v_d^2)^{1/2}$.

Chapter 1

Formulating a Time-Quantitative Second Law for Large Systems

1. Defining our implicit interaction models; Unrealistic versus realistic time scale

What is in Sections 1–2? *As a motivation, let us start with a little bit of thermodynamics, since the original entropy concept is the Clausius thermodynamic entropy. Using the terminology of thermodynamics, entropy (an extensive non-conserved macrostate variable) is the counterpart of (the intensive macrostate variable) temperature the same way as (the extensive macrostate variable) volume is the counterpart of (the intensive macrostate variable) pressure. These are called "conjugate variables". The key fact is that the internal energy — the main variable of the system — is expressed in terms of these conjugate variables (volume and pressure, entropy and temperature). For example, pressure differences force a change in the volume, and their product is the energy lost by the system in the form of mechanical work. In a similar way, temperature differences drive changes in entropy, and their product is energy transferred by heat flow. The mystery of entropy is that the other three — temperature, volume and pressure — are well-known perfectly ordinary everyday concepts, but the intuitive meaning of entropy is far less obvious.*

So, what is the informal intuitive meaning of entropy? A physicist would say it is a measure of disorder; a chemist would say it is a measure of energy dispersal, a quality of energy that measures its capacity to perform mechanical work; for more, see Appendix 1: The Second Law in physics. These informal definitions describe entropy from two different viewpoints. Our approach represents a third, purely mathematical viewpoint: our micro-entropy (a microstate variable!) measures how far the dynamical system is from micro-equilibrium. The key point is that for large dynamical systems we can make this third intuitive meaning mathematically precise! We can provide a rigorous definition of micro-equilibrium (which is, in fact, a "snapshot square-root fluctuation type equilibrium", an optimal form of equidistribution in the container) — we do this in Section 2. In Sections 1–2 we introduce the models for which we can define micro-equilibrium, and later define our micro-entropy, and provide a rigorous proof for a Second Law. We also elaborate on the practical issue of realistic time versus unrealistic time.

Here come the mathematical details. As a warm-up we begin with the simplest model of linear orbits. We will introduce a more interesting general model — with not necessarily linear orbits — at the end of Section 1 (see also Section 2 for the general torus model).

Simplest case: Bernoulli billiard model. Consider a large system of point billiards in the unit cube $[0, 1]^3$ (we call it the container). Motivated by the kinetic theory of gases, a more or less realistic special case is where the unit cube $[0, 1]^3$ contains $N = 10^{27}$ point billiards. (Note that for a typical gas, a cubic container of side length 1 meter contains around $10^{26} - 10^{27}$ gas molecules, and the molecules move with a speed between a few hundred to a few thousand meter per second.) The system may start from an *arbitrary* initial point configuration, allowing extreme cases such as all particles are in the (say) left-hand side of the cube, or even that all particles start from the same point ("Big Bang"). In this simplest model each particle (= point billiard) exhibits constant speed piecewise linear motion on its own billiard path. We refer to this simplest model as the (linear) Bernoulli model, or the billiards-in-a-box model.

We investigate the following questions: Does the system approach any kind of "equilibrium"? How to give a precise definition of "equilibrium"?

In this simplest many-particle model (point billiards) the classical Kronecker–Weyl equidistribution theorem — in fact, the continuous version about torus lines — describes the *long-term* time evolution of the system (we also need the geometric trick of *unfolding*; see Figures 1–2 in Section 1). Unfortunately, the famous Kronecker–Weyl theorem (see [We16] or the book [Dr-Ti97]) gives "soft" qualitative results, where long-term means practically infinite time. Despite this fundamental weakness, we still care about these "soft" long-term results, because they give us the first hint about what we are *really* interested in: the realistic short-time approach to equilibrium. This is why we start with the Kronecker–Weyl theorem.

The continuous version of the Kronecker–Weyl equidistribution theorem says that an infinite torus line $t\mathbf{a}$ modulo one, $0 \le t < \infty$ is uniformly distributed (= equidistributed) in the d-dimensional unit cube $[0,1]^d$ if and only if the d coordinates $a_1, a_2, \ldots, a_d$ of the direction vector $\mathbf{a} = (a_1, a_2, \ldots, a_d)$ are linearly independent over the rationals. Here *uniformity* refers to the traditional definition (usually associated with the name of Weyl) that, given an arbitrary Jordan measurable test set $S \subset [0,1]^d$, the relative time the torus line spends in the test set S converges to the d-dimensional volume of S as the total time tends to infinity. Or equivalently, for an arbitrary Riemann integrable test function f defined on the d-dimensional unit torus $[0,1)^d$ (i.e., f is periodic),

$$\lim_{T\to\infty} \frac{1}{T} \int_0^T f(t\mathbf{a})\, dt = \int_{[0,1)^d} f\, d\mathrm{V},$$

where V refers to the d-dimensional Lebesgue measure ("volume").

(Note that for the class of bounded sets in the d-space $\mathbf{R}^d$ the following 3 properties are equivalent: (1) S is Jordan measurable, (2) the characteristic function of S, $f = \chi_s$, is Riemann integrable, (3) the boundary of S has d-dimensional Lebesgue measure zero.)

It is interesting to note that the original reason for studying these questions was their relation to applied mathematics: secular perturbations in astronomy (see Bohl [Bo09]) and the kinetic theory of gases (see Weyl [We14]). We may say that this book is a far-reaching extension of the line of research started in the paper of Weyl [We14].

Many consider the Kronecker–Weyl theorem the first "ergodic" theorem, despite the fact that it is restricted to Jordan measurable sets, which is not a σ-algebra. (The ergodic theorems are formulated in terms of a σ-algebra like the family of Borel sets or Lebesgue measurable sets.) The Kronecker–Weyl theorem was the first result with the message that the "long-term time average of an orbit equals the space average". (In fact, it holds for *every* starting point, which is not true for Birkhoff's individual ergodic theorem.) We emphasize that in this book we do not use ergodic theory/theorems at all. The well-known (measure-theoretic) Birkhoff's individual ergodic theorem is not useful for us for two reasons. (1) We study far-from-equilibrium initial configurations that represent a zero-set, (the traditional measure-theoretic) ergodic theory ignores zero-sets). (2) Since we are interested in the individual trajectory of the system, Birkhoff's individual ergodic theorem seems to be a very promising tool to be used. But, unfortunately, it does not help, because it is silent about the speed of convergence to the limit. So it does not say anything about the realistic time evolution. We cannot use the alternative version of the ergodic theorem — called operator form (e.g. "transfer operator") — either. Neither form seems to help describing the time-quantitative evolution of our micro-entropy. We use completely different tools.

To have a chance for *any* kind of reasonable "approach to equilibrium", we certainly need some assumption about the initial velocities of the point billiards in the box. (For example, if the initial point configuration is a crowded cluster, and every point-billiard has the *same velocity*, then the same cluster moves around, from wall to wall, back and forth, and the system does not approach any "equilibrium".) Since *same velocity* does not lead anywhere, to guarantee any kind of meaningful approach to equilibrium, we need some condition with the message that the velocity vectors are not the same in a "stronger quantitative sense". We make it precise by using the concept of linear independence over the rationals — exactly the condition of the Kronecker–Weyl equidistribution theorem.

Formally, assume that there are N point billiards in the unit cube $[0,1]^3$ (say, N is around 10^{27}, or larger), and the motion of the k-th point billiard in the unit cube $[0,1]^3$ is described by $\mathbf{x}_k(t)$, $0 \le t < \infty$; so the starting point is $\mathbf{x}_k(0) = \mathbf{y}_k$, and the initial velocity is

$$\frac{d}{dt}\mathbf{x}_k(t)|_{t=0} = \mathbf{x}_k'(0) = \mathbf{v}_k = (v_{k,1}, v_{k,2}, v_{k,3}). \qquad (1.1).$$

The initial point configuration of the system is given by a $3N$-dimensional vector

$$\Gamma(0) = (\mathbf{x}_1(0), \mathbf{x}_2(0), \ldots, \mathbf{x}_N(0)) = (\mathbf{y}_1, \mathbf{y}_2, \ldots, \mathbf{y}_N), \qquad (1.2)$$

and the time-evolution of the system is represented by the parametrized curve ("trajectory of the system")

$$\Gamma(t) = (\mathbf{x}_1(t), \mathbf{x}_2(t), \ldots, \mathbf{x}_N(t)), \quad 0 \le t < \infty \qquad (1.3)$$

in the $3N$-dimensional unit cube $[0,1]^{3N}$ (= configuration space); t is the time parameter.

Approach to equilibrium means decreasing "disparity". We detect "disparity" with test sets, or test functions in general. For example, consider an arrangement where at some moment all particles are in the left half of the unit cube: it represents a *macroscopic disparity* in that moment. Let S denote the left half of the unit cube; then the *macroscopic disparity* detected by S is the huge discrepancy $N/2$. Indeed, S contains N points, but the volume of S is $1/2$, so S is expected to contain roughly $N/2$ points, which gives the huge discrepancy $N - N/2 = N/2$.

Note that microscopic disparities are represented by more complicated test sets S.

Assume that the $3N$ real numbers (see (1.1))

$$\{v_{k,1}, v_{k,2}, v_{k,3} :\ 1 \le k \le N\} \tag{1.4}$$

are linearly independent over the rationals. Let $S \subset [0,1]^3$ be an arbitrary but fixed Jordan measurable subset of the unit cube with volume $0 < \mathrm{vol}(S) < 1$. Let $N_S(t)$ denote the point-counting function:

$$N_S(t) = \sum_{\substack{1 \le k \le N: \\ \mathbf{x}_k(t) \in S}} 1. \tag{1.5}$$

It is natural to expect that the system reaches spatial equilibrium, i.e., equidistribution of the system in the container, when the point-counting function $N_S(t)$ gets very close to the "expectation" $\mathrm{vol}(S)N$ in the precise sense that the discrepancy exhibits standard deviation size fluctuations, i.e., constant times

$$\sqrt{\mathrm{vol}(S)(1-\mathrm{vol}(S))N},$$

motivated by probability theory ("time-lapse randomness").

Applying the (continuous version of the) Kronecker–Weyl equidistribution theorem in the very high dimensional configuration space ($\dim = 3N$), and combining it with the geometric idea of *unfolding* (see Figures 1–2 in Section 1), it is a routine exercise to prove the following asymptotic result.

Lemma 1.1. (CLT-Corollary of the Kronecker–Weyl theorem) *If the $3N$ initial velocity coordinates of the point-billiard system (see (1.1)–(1.5)) are linearly independent over the rationals, then for every initial point configuration, for every Jordan measurable set $S \subset [0,1]^3$, and for every real number $-\infty < y < \infty$,*

$$\lim_{T\to\infty} \frac{1}{T} \mathrm{length}\left\{ t \in [0,T] :\ \frac{N_S(t) - pN}{\sqrt{pqN}} \le y \right\}$$

$$= \sum_{n \le pN + y\sqrt{pqN}} \binom{N}{n} p^n q^{N-n}$$

$$= \frac{1}{\sqrt{2\pi}} \int_{-\infty}^{y} e^{-u^2/2}\, du + O(1/\sqrt{pqN}), \tag{1.6}$$

where $p = \mathrm{vol}(S)$ and $q = 1 - p$.

Here *length* in (1.6) means the 1-dimensional Lebesgue measure, and CLT stands for the central limit theorem.

Lemma 1.1 is a trivial consequence of the following result: under the condition of Lemma 1.1,

$$\lim_{T\to\infty}\frac{1}{T}\text{length}\,\{t\in[0,T]:\ N_S(t)=n\}=\binom{N}{n}p^nq^{N-n},\tag{1.7}$$

where $p=\text{vol}(S)$ and $q=1-p$. Notice that the right-hand side of (1.7) is the general term in the binomial distribution with parameters N and $0<p<1$.

In fact, (1.7) is a consequence of the following even simpler result. Let $I\subset\{1,2,\ldots,N\}$ be an arbitrary subset; then, under the condition of Lemma 1.1,

$$\lim_{T\to\infty}\frac{1}{T}\text{length}\,\{t\in[0,T]:\ \mathbf{x}_k(t)\in S\iff k\in I\}=p^{|I|}q^{N-|I|},\tag{1.7$'$}$$

where again $p=\text{vol}(S)$ and $q=1-p$. We give the full proof in Appendix 2.

Note that there are a few other related corollaries of the Kronecker–Weyl theorem, e.g. a Poisson limit theorem; we discuss them in Appendix 2.

Lemma 1.1 is an asymptotic result: it is about the (very!) long-term time evolution of a large system of N point-billiards in a cube $[0,1]^3$, starting from an arbitrary initial point configuration (e.g., Big Bang). We assume that the $3N$ velocity coordinates are linearly independent over the rationals; we pick an arbitrary Jordan measurable test set $S\subset[0,1]^3$ with $p=\text{vol}(S)\le 1/2$ (otherwise we take the complement of S, i.e., replace p with q), and study the distribution of the point-counting function $N_S(t)$ where t runs in a very, very long time interval $0\le t\le T$. If Np is "large" then we obtain a central limit theorem (if Np is "in the constant range" then we obtain a Poisson limit theorem). The message is that *equilibrium = time-lapse randomness*, at least in a weak asymptotic sense as $t\to\infty$.

The central limit theorem (and the Poisson limit theorem; see Appendix 2) with negligible error term is just a simple consequence of (1.7) (via the De Moivre–Laplace theorem) due to the fact that the number of particles N is very large. In fact, the error term $O(1/\sqrt{pqN})$ in (1.6) is the inevitable discrepancy between the (continuous) standard normal distribution and the (discrete) normalized binomial distribution with parameters N and $0<p<1$.

The requirement of linear independence of the $3N$ velocity coordinates is easily satisfied by choosing a *typical* (also called generic) set of N velocities: almost all choices satisfy linear independence.

Lemma 1.1 is a very elegant and interesting result, but, unfortunately, it has a fundamental shortcoming: it does not say anything about the time scale. What we are really interested in is the question of **how long does it take to reach equilibrium?** and **what happens in realistic time?** In most real life experiments, the transition from nonequilibrium to equilibrium is fast, and this is not reflected in (1.6)–(1.7).

Next we mention some "absurd" corollaries of the Kronecker–Weyl theorem, which inevitably require ridiculously large "unrealistic time scale". Again we study the time evolution of a large system of N point-billiards in a cube $[0,1]^3$, and assume that the $3N$ velocity coordinates are linearly independent over the rationals. Let S be the right half $[\frac{1}{2},1] \times [0,1] \times [0,1]$ of the unit cube $[0,1]^3$. Whatever initial point configuration we had at the start, if we wait long enough, it will happen that all N particles will be in S at the same time. And what is more, the system will return to this "absurd" state infinitely many times. (We call it "absurd", because in real life the cubic gas container would collapse, due to the outside pressure.)

It is easy to upgrade this to an even more absurd result. Under the same condition ("linear independence") it will happen that, given any *rational* sub-box (Cartesian product of closed intervals that have rational endpoints, and the volume is positive), all N particles will be in this sub-box at the same time, and the system will return to each rational box infinitely many times. (The set of rational boxes is countable, and every box of positive volume contains a rational box.)

The most general case is the following.

Lemma 1.2. (Superdiscrepancy) *Assume that the $3N$ initial velocity coordinates of the point-billiard system in the unit cube $[0,1]^3$ are linearly independent over the rationals. Then the time evolution of the system exhibits the following property that we call superdiscrepancy. Given any sequence $B_1, B_2, \dots, B_N$ of N rational sub-boxes and any ordering of the N particles, starting from an arbitrary initial point configuration, it will happen that the first particle will be in the first rational box B_1, the second particle will be in the second rational box B_2, and so on, the N-th particle will be in the N-th rational box B_N, all at the same time, and the system will return to each such $(B_1, \dots, B_N)$-configuration infinitely many times.*

This (number-theoretic) superdiscrepancy result — a kind of paradox — has the same message as the well known Zermelo–Poincare paradox in ergodic theory and statistical mechanics (i.e., a combination of Poincare's recurrence theorem and Liouville's volume-preserving theorem). Both "paradoxes" say, very roughly speaking, that, given any many-particle closed system, if one waits long enough, "crazy things will happen". (Of course, "long enough" is really very, very, very long; a totally unrealistic time frame.)

Lemma 1.2 explains why we cannot expect *convergence to "snapshot" equilibrium* as $t \to \infty$ for individual time evolution. We have to downgrade it to *approach to "snapshot" equilibrium*, emphasizing the temporary nature of (snapshot) equilibrium on the individual level. (On the other hand, in view of Lemma 1.1 we can talk about convergence to "time-lapse" equilibrium.)

Notice that superdiscrepancy also demonstrates some kind of "randomness" (similarly to the CLT-Corollary). Indeed, superdiscrepancy can be interpreted as an analogue of the Large Deviation Theorem in probability theory (see e.g. the excellent book [Fel71]), representing the case of "extremely large deviation". Note that the proof of Lemma 1.2 is a simple by-product of the proof of Lemma 1.1; see Appendix 2.

So far we have been studying the (extremely!) long-term time evolution of large system of point billiards in a cube. Point billiard systems represent an oversimplified gas-in-a-box model, a many-particle dynamical system mimicking the non-equilibrium ideal gas (Bernoulli model). Now we switch to what we are *really* interested in: we study the approach to equilibrium in realistic short time. Our goal is to prove "hard" time-quantitative theorems (Lemma 1.1 is a "soft" qualitative theorem). The good news is that we can prove surprisingly good "hard" theorems for *typical* initial velocities in the Bernoulli model. It means we can prove approach to equilibrium in realistic short time for the overwhelming majority of the initial velocities in a given velocity distribution. (The bad news is that we cannot prove anything interesting or new for any explicit set of velocities.)

We have to clarify what a Bernoulli model with a "given velocity distribution" means. First note that we restrict ourselves to the *isotropic* case: we assume uniformity in all directions. To give an isotropic velocity distribution, it suffices to give a speed distribution.

It is natural to start with what physicists call the "velocity equilibrium for gases": the Gaussian (= Maxwellian) velocity distribution. It reflects physical reality for gas particles in thermal equilibrium. The Gaussian velocity distribution is isotropic, and it is determined by the following density function for the speed:

$$\frac{1}{\sigma^3}\sqrt{\frac{2}{\pi}}y^2 e^{-y^2/(2\sigma^2)} \tag{1.8}$$

for all $y \geq 0$, where the physical meaning of parameter $\sigma > 0$ is $\sigma = \sqrt{k_0 T/m}$; here m is the mass, T is the absolute temperature, and k_0 is the Boltzmann's constant. (Physicists call (1.8) the Maxwellian speed distribution; we simply call it the Gaussian speed distribution.)

Using the substitution $x = y/\sigma$ in (1.8) we have

$$\frac{1}{\sigma^3}\int_0^u y^2 e^{-y^2/(2\sigma^2)}\,dy = \int_0^{u/\sigma} x^2 e^{-x^2/2}\,dx,$$

which implies that the Gaussian speed distributions in (1.8) are basically the same. So, in the rest we assume that $\sigma = 1$:

$$\sqrt{\frac{2}{\pi}}y^2 e^{-y^2/2} \quad \text{for all } y \geq 0. \tag{1.9}$$

Perhaps the reader is wandering: if we want to understand the transition from nonequilibrium to equilibrium, then why do we assume "velocity equilibrium"? A good motivation comes from the following intuition. Nonequilibrium thermodynamics is far less developed than its equilibrium counterpart — large part of it is unproved speculations. G.E. Uhlenbeck, an expert physicist, speculated that the transition from nonequilibrtium to equilibrium in realistic irreversible processes has, roughly speaking, two very different stages. Uhlenbeck predicted that the first stage is shorter: very quickly, after a few particle-particle collisions, the equilibrium in the velocities of the molecules gets established (Gaussian distribution, at least with good approximation). Then comes the longer second stage, when slowly the spatial equilibrium gets established, parallel and interacting with the further convergence of the velocities to the Gaussian distribution (see e.g. [U-F63] and [Th72]). This intuition justifies our assumption of the Gaussian distribution: what we study is the longer second stage, when the spatial equilibrium gets established.

Case 1: Gaussian distribution

We rewrite (1.9) in a more formal way: assume that ρ_j (representing "radius"), $1 \leq j \leq N$ are N independent, identically distributed positive random variables, with common distribution

$$\Pr[\rho_j \leq u] = \sqrt{\frac{2}{\pi}}\int_0^u y^2 e^{-y^2/2}\,dy \tag{1.9'}$$

for all positive $u > 0$ (here the function $e^{-y^2/2}$ comes from the standard normal distribution, and y^2 is motivated by the surface area $4\pi y^2$ of a sphere of radius $y > 0$). We also assume that $\mathbf{u}_j$ (= direction), $1 \leq j \leq N$

are N identically distributed random variables, where each $\mathbf{u}_j$ is uniformly distributed on the unit sphere $\mathbf{S}^2$, and the $2N$ random variables $\mathbf{u}_j$, ρ_j, $1 \leq j \leq N$ are mutually independent. Formally, we consider the product measure on the Cartesian product set

$$(\rho_1, \mathbf{u}_1, \ldots, \rho_N, \mathbf{u}_N) \in [0,\infty) \times \mathbf{S}^2 \times \cdots \times [0,\infty) \times \mathbf{S}^2 = ([0,\infty) \times \mathbf{S}^2)^N$$

$$= \Omega_1(Bernoulli),$$

where the factor measure is a probability measure (we divide the surface area on the unit sphere by 4π). So, the product measure, denoted by ProdMeas_1, is also a probability measure on $([0,\infty) \times \mathbf{S}^2)^N = \Omega_1(Bernoulli)$.

The actual speed of the j-th particle is a constant (depending on the particle) $|\mathbf{x}_j'(t)| = v_j(t) = \rho_j, \quad 0 \leq t \leq T$, and the whole billiard-path (piecewise linear zig-zag) is determined by the initial velocity $\mathbf{x}_j'(0) = \mathbf{v}_j(0) = \rho_j \mathbf{u}_j$. (In fact, the piecewise linear path is determined by the unit vector $\mathbf{u}_j$ alone. Note we denote the length of a vector $\mathbf{x}$ with $|\mathbf{x}|$.)

Let us return to (1.9); using integration by parts, we have

$$\sqrt{\frac{2}{\pi}} \int_0^\infty y \cdot y^2 e^{-y^2/2}\, dy = 2\sqrt{\frac{2}{\pi}} \int_0^\infty y \cdot e^{-y^2/2}\, dy = \sqrt{\frac{8}{\pi}},$$

that is,

$$\sqrt{8/\pi} \ \text{ is the mean speed.} \tag{1.10}$$

Typical Gaussian distribution precisely means $(1-\varepsilon)$-part with some small butfixed $\varepsilon > 0$ (say, $\varepsilon = 10^{-2}$ means 99 percent) of the product set $([0,\infty) \times \mathbf{S}^2)^N$, measured with the product measure ProdMeas_1. Formally,

$$(\rho_1, \mathbf{u}_1, \ldots, \rho_N, \mathbf{u}_N) \in \mathcal{A}_1 \ \text{ where } \ \mathcal{A}_1 \subset ([0,\infty) \times \mathbf{S}^2)^N = \Omega_1(Bernoulli)$$

$$\text{with } \ \text{ProdMeas}_1(\mathcal{A}_1) = 1 - \varepsilon. \tag{1.11}$$

We refer to this as the (simplest) *Bernoulli model of the first kind.*

A case completely different from the Gaussian distribution is where all particles have the same constant speed. It is unrealistic for gas molecules; nevertheless, it is still very interesting in physics, since photons, or elementary particles with zero rest mass, move with the speed of light. For simplicity assume that the common speed is one.

Case 2 ("like photons"): We change (1.9′) to $\rho_1 = \rho_2 = \ldots = \rho_N = 1$

Formally, we consider the product measure on the Cartesian product set

$$(\mathbf{u}_1, \ldots, \mathbf{u}_N) \in \mathbf{S}^2 \times \cdots \times \mathbf{S}^2 = (\mathbf{S}^2)^N = \Omega_2(Bernoulli),$$

where the factor measure is the normed surface area (i.e., we divide by $4\pi =$ the surface area of $\mathbf{S}^2$). So, the product measure, denoted by ProdMeas_2, is a probability measure on $(\mathbf{S}^2)^N = \Omega_2(Bernoulli)$.

Typical means $(1-\varepsilon)$-part, with some small but fixed $\varepsilon > 0$, of the product set $(\mathbf{S}^2)^N$, measured with the product measure ProdMeas_2. Formally,

$$(\mathbf{u}_1, \ldots, \mathbf{u}_N) \in \mathcal{A}_2 \quad \text{where} \quad \mathcal{A}_2 \subset (\mathbf{S}^2)^N = \Omega_2(Bernoulli)$$

$$\text{with} \quad \text{ProdMeas}_2(\mathcal{A}_2) = 1 - \varepsilon. \tag{1.12}$$

We refer to this as the (simplest) *Bernoulli model of the second kind.*

In the third case we replace the uniform distribution on the unit sphere $|\mathbf{u}| = 1$ with uniform distribution in the unit ball $|\mathbf{x}| \leq 1$. We may say that Case 3 is "between" Case 1 and Case 2. Case 3 is still quite interesting and natural from the viewpoint of pure mathematics.

Case 3 ("uniform in the unit ball"): We change (1.9′) to

$$\Pr[\rho_j \leq u] = u^3 = \int_0^u 3y^2\, dy \quad \text{for } 0 \leq u \leq 1$$

(and of course we have 0 for $u < 0$ and 1 for $u > 1$). We have

$$\text{mean speed} = \int_0^1 y \cdot 3y^2\, dy = 3/4.$$

An equivalent definition is to assume that $\mathbf{v}_j = \rho_j \mathbf{u}_j$, $\mathbf{u}_j \in \mathbf{S}^2$, $1 \leq j \leq N$ are N independent, identically distributed random variables, where each $\mathbf{v}_j$ is uniformly distributed in the unit 3-ball $B^3 = \{|\mathbf{x}| \leq 1\} \subset \mathsf{R}^3$. Formally, we consider the product measure on the Cartesian product set

$$(\mathbf{v}_1, \ldots, \mathbf{v}_N) \in B^3 \times \cdots \times B^3 = (B^3)^N = \Omega_3(Bernoulli),$$

where the factor measure is the normed volume (i.e., we divide by $4\pi/3$). So, the product measure, denoted by ProdMeas_3, is a probability measure on $(B^3)^N = \Omega_3(Bernoulli)$.

Typical means $(1-\varepsilon)$-part, with some small but fixed $\varepsilon > 0$, of the product set $(B^3)^N$, measured with the product measure ProdMeas_3. Formally,

$$(\mathbf{v}_1, \ldots, \mathbf{v}_N) \in \mathcal{A}_3 \quad \text{where} \quad \mathcal{A}_3 \subset (B^3)^N = \Omega_3(Bernoulli)$$

$$\text{with} \quad \text{ProdMeas}_3(\mathcal{A}_3) = 1 - \varepsilon. \tag{1.13}$$

We refer to this as the (simplest) *Bernoulli model of the third kind.*

We refer to the huge product spaces $\Omega_1(Bernoulli)$ in Case 1, $\Omega_2(Bernoulli)$ in Case 2, and $\Omega_3(Bernoulli)$ in Case 3 as the corresponding (initial) *velocity space* of the whole system.

So far we mentioned three isotropic velocity distributions: in Case 1 the speed density is

$$\sqrt{\frac{2}{\pi}}y^2e^{-y^2/2} \quad \text{for } y > 0 \quad \text{and zero for } y \leq 0,$$

in Case 3 the speed density is $3y^2$ for $0 \leq y \leq 1$ and 0 otherwise, and in Case 2 the speed density is the Dirac delta function concentrated on $y = 1$. Of course, there are many more interesting speed distributions. In the general isotropic case we have an arbitrary speed density function $g(y)$ — i.e., $\Pr[\rho_j \leq u] = \int_0^u g(y)\,dy$ — such that

$$\infty > g(y) \geq 0 \text{ for } 0 \leq y < \infty, \quad \int_0^\infty g(y)\,dy = 1 \quad \text{and} \quad \int_0^\infty ug(u)\,du < \infty, \tag{1.14}$$

where the last equation means that the "mean speed is finite". (We allow g to be a generalized function like the Dirac delta function concentrated on 1 representing the second kind, but we exclude the trivial case of Dirac delta concentrated on 0, i.e., the mean speed is strictly positive.)

We are going to compare how fast systems approach equilibrium with different speed density functions. It is natural to assume, therefore, that they have the same mean speed, preferably one. But, mainly for notational convenience, we are not so rigid, and study speed density functions for which the mean speed is "close to one". For example, the mean speed for the first kind is $\sqrt{8/\pi}$, and for the third kind it is 3/4.

Temporarily we focus on Cases 1–3, and study the general case (1.14) later.

We study the following question: Given an arbitrary N-element initial point configuration

$$\mathcal{Y} = \{\mathbf{y}_1, \mathbf{y}_2, \ldots, \mathbf{y}_N\} \subset [0,1]^3$$

and an initial velocity distribution — i.e., Bernoulli model of the first, or second, or third kind — how long does it take for the typical time evolution, starting from $\mathcal{Y}$, to reach "equilibrium"? We want to define "equilibrium" in terms of "maximum uniformity": the system reaches "equilibrium" when its uniformity is close to the theoretical maximum.

Warning: We often use the term *initial (point) configuration* for both the N-element subset

$$\{\mathbf{y}_1, \mathbf{y}_2, \ldots, \mathbf{y}_N\} \subset [0,1]^3$$

of the container (that holds the particles) and the $3N$-dimensional vector

$$(\mathbf{y}_1, \mathbf{y}_2, \ldots, \mathbf{y}_N) \in \left([0,1]^3\right)^N$$

representing a point in the configuration space. We hope that this slight ambiguity will not confuse the reader.

To define the "equilibrium" precisely, first we employ the switch from billiard paths to torus lines via *unfolding* (see Figures 1–2 below).

From billiard path to torus line: unfolding. By using the well known geometric trick of *unfolding* one can convert every motion on a billiard path in the h-dimensional cube $[0,1]^h$ into a motion on the torus line in the torus $[0,1)^h$.

For simplicity, we consider first the 2-dimensional case. Let $\mathbf{x}(t) = (x_1(t), x_2(t))$, $0 < t < \infty$ represent an infinite billiard path with starting point $\mathbf{x}(0) = (x_1(0), x_2(0)) = \mathbf{y} \in [0,1]^2$ and initial angle $\theta \in [0, 2\pi)$.

The intuitively plausible concept of "billiard path" precisely means that a point-mass (representing a tiny billiard ball) moves freely along a straight line inside the unit square with constant speed until it hits the boundary (i.e., one of the four sides of the square; in the pathological case when the point billiard hits the corner we assume that it returns along its former path — this convention is suggested by considerations of continuity). The reflection off the boundary is elastic, meaning the familiar law of reflection: the angle of incidence equals the angle of reflection. After the reflection, the point-billiard continues its linear motion with the new velocity (but of course the speed remains the same) until it hits the boundary again, and so forth (we ignore friction, air resistance, etc.). The initial condition, i.e., the starting point $\mathbf{y} \in [0,1]^2$ of the billiard path and the initial angle $\theta \in [0, 2\pi)$, uniquely determine an infinite piecewise linear billiard path $\mathbf{x}(t) = (x_1(t), x_2(t))$, $0 < t < \infty$ in the unit square.

The figure below illustrates the geometric trick of *unfolding* the billiard path inside the unit square to a straight line in the entire plane. The idea of *unfolding* is very simple and elegant: we keep reflecting the square itself in the respective side and unfold the piecewise linear billiard path to a straight line.

Two (straight) lines in the plane correspond to the same billiard path if and only if they differ by a translation through an integral vector where both coordinates are even, i.e., where the vector is from the "double" square lattice $2\mathbb{Z} \times 2\mathbb{Z}$. In other words, the problem of the long-term distribution of a billiard path in the unit square is equivalent to the distribution of the corresponding torus-line in the 2×2 square.

The law of reflection implies that there are at most four different directions along the billiard path (the initial direction is preserved modulo $\pi/2$, which is one-fourth of the whole angle 2π; the same holds for any rectangle). If we use unit speed, then of course *arclength* and *time* are the same.

Formally, a billiard path in the unit square $[0,1]^2$ has the form

$$\mathbf{x}(t) = (x_1(t), x_2(t)),\ 0 < t < \infty \text{ with } x_j(t) = 2\,\|(y_j + t\alpha_j)/2\|\,,\ j = 1, 2, \tag{1.15}$$

where $\mathbf{a} = (\alpha_1, \alpha_2)$ is a nonzero vector, and $\|z\|$ denotes the distance of a real number z from the nearest integer. Here $\mathbf{y} = (y_1, y_2)$ is the starting point, and $\mathbf{a}$ gives the initial direction; so $\theta = \arctan \frac{\alpha_2}{\alpha_1}$ ("inverse tangent") is the initial angle.

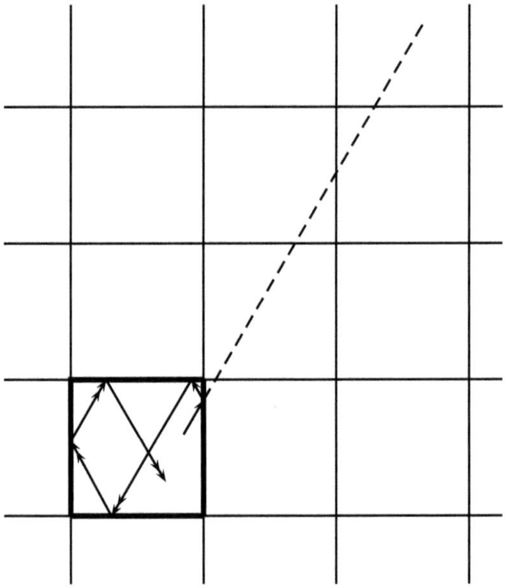

Figure 1

It follows from *unfolding* that a billiard path in the unit square intersects a given test set S at time t exactly when the corresponding torus-line in the 2×2 square intersects the union set $S_1 \cup S_2 \cup S_3 \cup S_4$ (see the second figure below).

Indeed, unfolding defines 4 reflected copies S_1, S_2, S_3, S_4 of S (each one of the 4 unit squares on the second figure below contains a reflected copy of the given test set S). In the last step we shrink the underlying 2×2 square to the unit square $I^2 = [0,1)^2$.

Of course, the test set S can be upgraded to any "not too ugly" periodic function f, i.e., $f(x,y)$ is a Riemann integrable periodic complex-valued function with period one in both variables. In the case of the figure, f means the 0-1 valued characteristic function of the union $S_1 \cup S_2 \cup S_3 \cup S_4$ of the four half-size reflected copies of the given subset $S \subset I^2 = [0,1)^2$.

Note that the unfolding of the billiard path in the h-dimensional unit cube $[0,1]^h$ with $h \geq 3$ can be defined in the analog way. Formally, unfolding means the map

$$2\|x/2\| \to \{x/2\} \text{ applied to each coordinate,} \tag{1.16}$$

where $0 \leq \{x\} < 1$ denotes the fractional part of a real number x (i.e., $x = \lfloor x \rfloor + \{x\}$ where $\lfloor x \rfloor$ is the largest integer $\leq x$).

We use the fact that for every set $S \subset [0,1)$,

$$2\|x/2\| \in S \iff \{x/2\} \in \frac{1}{2}S \cup \left(1 - \frac{1}{2}S\right), \tag{1.17}$$

where

$$\frac{1}{2}S = \{y/2 : \ y \in S\} \text{ and } 1 - \frac{1}{2}S = \left\{1 - \frac{y}{2} : \ y \in S\right\}. \tag{1.18}$$

It follows from *unfolding* that a billiard path in the h-dimensional unit cube $[0,1]^h$ intersects a given test set $S \subset [0,1]^h$ at time t exactly when the corresponding torus-line in the h-dimensional $2 \times \cdots \times 2$ (hyper)cube intersects the union of the 2^h reflected copies of S. (Note that each one of the 2^h unit (hyper)cubes contains a reflected copy of the given test set S.) In the last step we shrink the h-dimensional $2 \times \cdots \times 2$ (hyper)cube to the unit (hyper)cube $I^h = [0,1)^h$.

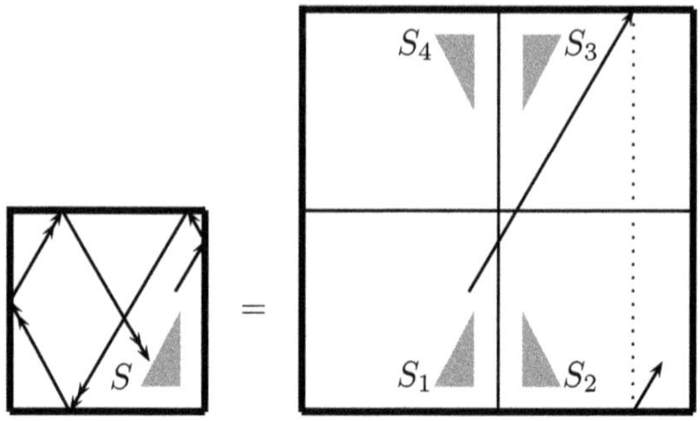

Figure 2

An alternative — but basically equivalent — way to do the switch "billiard $\to$ torus" is to reduce the 2^h reflected half-copies of the test set S to a single half-copy $\frac{1}{2}S$ (see (1.18)), and to work with 2^h reflected torus lines. The formal execution of this idea goes as follows. Let

$$\mathbf{x}(t) = (x_1(t), \ldots, x_h(t)),\ 0 < t < \infty \text{ with } x_i(t) = 2\,\|(y_i + t\alpha_i)/2\|\,,$$
$$1 \leq i \leq h, \tag{1.19}$$

be a billiard path in $[0,1)^h$ with starting point $\mathbf{y} = (y_1, \ldots, y_h)$ and initial direction $\mathbf{a} = (\alpha_1, \ldots, \alpha_k)$ (nonzero vector; as usual, $\|z\|$ denotes the distance of a real number from the nearest integer); and let $S \subset [0,1)^h$ be a (measurable) test set. Given a ± 1-vector $(\varepsilon_1, \ldots, \varepsilon_h)$ with $\varepsilon_i = 1$ or -1, $1 \leq i \leq h$, we define a torus line $\mathbf{x}^*(\varepsilon_1, \ldots, \varepsilon_h; t)$ by specifying its i-th coordinate ($1 \leq i \leq h$):

$$\left\{\frac{y_i + t\alpha_i}{2}\right\} \text{ if } \varepsilon_i = 1, \text{ and } \left\{1 - \frac{y_i + t\alpha_i}{2}\right\} = \left\{-\frac{y_i + t\alpha_i}{2}\right\} \text{ if } \varepsilon_i = -1,$$

that is,

$$\mathbf{x}^*(\varepsilon_1, \ldots, \varepsilon_h; t) = (\{\varepsilon_1(y_1 + t\alpha_1)\}, \ldots, \{\varepsilon_h(y_h + t\alpha_h)\}) \tag{1.20}$$

(as usual $\{z\}$ denotes the fractional part). Combining (1.17)–(1.20) we obtain

$$\mathbf{x}(t) \in S \iff \text{there is a } \pm 1\text{−vector } (\varepsilon_1, \ldots, \varepsilon_h)$$
$$\text{such that } \mathbf{x}^*(\varepsilon_1, \ldots, \varepsilon_k; t) \in \frac{1}{2}S. \tag{1.21}$$

What is more, we can clearly upgrade "there is a" in (1.21) to "there is exactly one" (a plausible geometric explanation is that the four sets S_1, S_2, S_3, S_4 in the figure above are disjoint).

A far-reaching generalization of the linear Bernoulli model with general orbits. Note that (1.19) represents a constant speed piecewise linear point billiard motion. As a far-reaching generalization, let $\mathbf{c}(t) = (c_1(t), \ldots, c_h(t))$ be an arbitrary continuous parametrized space curve in the h-dimensional Euclidean space $\mathbf{R}^h$; then

$$\mathbf{x}(t) = (x_1(t), \ldots, x_h(t)),\ 0 < t < \infty \text{ with } x_i(t) = 2\left\|c_i(t)/2\right\|,\ 1 \le i \le h \tag{1.22}$$

defines a general orbit in $[0,1)^h$. Again let $S \subset [0,1)^h$ be a (measurable) test set. Given a ± 1-vector $(\varepsilon_1, \ldots, \varepsilon_h)$ with $\varepsilon_i = 1$ or -1, $1 \le i \le h$, we define the corresponding torus curve $\mathbf{x}^*(\varepsilon_1, \ldots, \varepsilon_h; t)$ as follows:

$$\mathbf{x}^*(\varepsilon_1, \ldots, \varepsilon_h; t) = (\{\varepsilon_1 c_1(t)\}, \ldots, \{\varepsilon_h c_h(t)\}). \tag{1.23}$$

What we obtain is analogous to (1.21):

$$\begin{aligned}\mathbf{x}(t) \in S \iff &\text{ there is a } \pm 1\text{-vector } (\varepsilon_1, \ldots, \varepsilon_h)\\ &\text{such that } \mathbf{x}^*(\varepsilon_1, \ldots, \varepsilon_h; t) \in \frac{1}{2}S.\end{aligned} \tag{1.24}$$

And again, we can clearly upgrade "there is a" to "there is exactly one".

Consider now an N-element **generalized (non-linear) particles-in-a-box model** in $[0,1)^h$:

$$\begin{aligned}\mathbf{x}_k(t) = (x_{k,1}(t), \ldots, x_{k,h}(t)),\ 0 < t < \infty \text{ with } x_{k,i}(t) = 2\left\|c_{k,i}(t)/2\right\|,\\ 1 \le i \le h,\end{aligned} \tag{1.25}$$

where

$$\mathbf{c}_k(t) = (c_{k,1}(t), \ldots, c_{k,h}(t)),\ 1 \le k \le N \tag{1.26}$$

are N continuous parametrized space curves in the h-dimensional space $\mathbf{R}^h$. And again let $S \subset [0,1)^h$ be a (measurable) test set. The point-counting function $N_S(t)$ of this generalized particle model equals the point-counting function of the corresponding torus model of $2^h N$ torus curves defined in (1.23):

$$\sum_{\substack{1 \le h \le N:\\ \mathbf{x}_k(t) \in S}} 1 = \sum_{1 \le k \le N} \sum_{\substack{(\varepsilon_1, \ldots, \varepsilon_h):\ \delta_i = \pm 1\\ \mathbf{x}_k^*(\varepsilon_1, \ldots, \varepsilon_h; t) \in \frac{1}{2}S}} 1. \tag{1.27}$$

Notice that (1.27) follows from (1.24) (and from the upgrading of "there is a" to "there is exactly one").

The message of (1.27) is that any particles-in-a-box model — including the simplest linear orbit with constant speed case in (1.19), and also the general (not necessarily linear) case in (1.22) — is equivalent to a torus model (see (1.23)). We refer to this as the **from-box-to-torus via unfolding model.**

Remark. As far as we know, the first appearance of the geometric trick of unfolding is in the paper of D. König and A. Szűcs [Kö-Sz13], and it became widely known after Hardy and Wright included it in their well known book on number theory. König and Szűcs combined the trick of unfolding with the Kronecker–Weyl theorem, and thus they obtained the following elegant result about the billiard path in the unit square. If the slope of the initial angle is rational, then the billiard path is periodic, and if the slope of the initial angle is irrational, then the billiard path becomes uniformly distributed (= equidistributed) in the unit square. Since unfolding works in every dimension ≥ 2, there is an analog result in all higher dimensions.

2. The first step is the hardest: How to define a quantitative form of microscopic equilibrium in large systems?

From the preface we recall the one-sentence summary of the book:
(1) first we show that in our implicit interaction models the typical time evolution of the large closed system approaches a well-defined microscopic equilibrium; (2) then we define a new entropy concept on a microscopic level that measures the distance from this microscopic equilibrium (in fact, our Micro-Entropy = absolute constant minus this "distance"); and finally, (3) we prove that this Micro-Entropy satisfies a time-quantitative Second Law, i.e., the typical time evolution of the system exhibits Micro-Entropy increase in an explicit time-dependent form such that reaching maximum Micro-Entropy coincides with the state of microscopic equilibrium.

The clarification of this single sentence is the subject of the book. It is far from easy — due to, among others, the "entropy paradox" and the Loschmidt-paradox/irreversibility-paradox; see Section 17. The following diagram illustrates the key steps of the clarification. It summarizes the (unfortunately!) long chain of definitions needed, starting from a precise definition of microscopic equilibrium and eventually leading to a quantitative time-dependent form of the Second Law:

Square Root Fluctuation Type (Snapshot) Equilibrium

$\longrightarrow$ Relative Disparity (a Boltzmann entropy like quantity)

$\longrightarrow$ Disparity Function $\longrightarrow$ Unevenness Function

$\longrightarrow$ Relaxation Time and Relaxation Drifting Distance

$\longrightarrow$ Imbalance $\longrightarrow$ MicroEntropy

$\longrightarrow$ MicroEntropy increase described by the Time Evolution Function

= Time – quantitative Second Law.

Note that all of these concepts are defined by explicit computable "microstate" functions.

The arrow $\longrightarrow$ above means that the definition of Equilibrium (= micro-equilibrium) motivates the definition of Relative Disparity, which motivates the definition of Disparity Function, and so on.

In the book we give precise definitions of Equilibrium, Relaxation Time and Drifting Distance, Time Evolution Function, Micro-Entropy, which at the end lead to a proof of the Second Law.

We may say, with some exaggeration, that Chapters 1–4 are all about how to turn this diagram into precise theorems. It would be nice to find a shortcut in this very long diagram, but we do not know how to do that.

Let us go back to the first step: it is to define equilibrium on a microscopic level. Unfortunately **the innocent-looking first step already contains large part of the conceptual difficulties** we face in any attempt to rigorously define entropy and prove entropy increase. It takes long preparation to formulate the precise definition. Here we just tell in advance that our definition of "microscopic equilibrium" — in a class of some natural many-particle models, i.e., large dynamical systems — will turn out to be a spatial equilibrium concept — a form of equidistribution of the system in the container.

Of course for a small dynamical system it is trivial to define equilibrium: it is simply a "fixed point", representing stability. But the concept of "fixed point" is far too vague — almost meaningless — for the large systems that we are interested in.

For a useful quantitative definition we turn to thermodynamics, at least to get some motivation and driving intuition. Thermodynamics tells us that a large system is in equilibrium if its (micro)state can be described by a few well-defined macrovariables, which are basically constant everywhere (= defining a macrostate), allowing the relatively negligible — but inevitable! — "white noise" fluctuations coming from the microscopic level. "White noise" in more precise quantitative terms usually means square-root size fluctuations. This is another great intuition. We use the part "inevitable square-root size fluctuations" from this intuition to motivate our precise definition that we call "square-root equilibrium"; see Definitions 2.3 and 2.4 below. We call Definitions 2.3–2.4 "Lebesgue Equilibrium" and "Lebesgue Average Equilibrium", and use the short notation L-equilibrium and LA-equilibrium.

Our rigorous definition of equilibrium does not use the physical concept of macrostate. In the book a "state" always means a microstate. So, our equilibrium is a perfectly well-defined concept on the microscopic level. We should call it "micro-equilibrium", but for simplicity in the rest we just say "square-root equilibrium" or "equilibrium". (Note that we never simplify our "micro-entropy" to "entropy", because we want to avoid the confusion with the traditional entropy concepts.)

For our rigorous definition of (microscopic) equilibrium we turn to the mathematical subject of Uniform Distribution; in particular, to "strong uniformity".

Note that the main model of this book is a general torus model that we define below. It is a (slight) generalization of the "from-box-to-torus via unfolding" model defined in (1.19)–(1.27). (The advantage of the torus is that Fourier analysis works particularly well on the torus. For comparison, note that Fourier analysis is not nearly as convenient on the sphere.)

We visualize "absolute (spatial) equilibrium" in the torus as a state of "maximum balance" — a "best possible state" that cannot be improved on. (Here we talk about "snapshot" equilibrium, i.e., the time $t = t_0$ is fixed. It is completely different from "time-lapse" equilibrium, where the time runs in a long interval, say $0 \leq t \leq T$ with a large T, possibly $T = \infty$; see e.g. Lemma 1.1.) In fact, there are three natural ways to talk about "balance":

Case 1: particle balance,

Case 2: mass balance (assuming different particles may have different masses), and finally,

Case 3: energy balance (meaning kinetic energy, and assuming different particles move with different speeds).

Note that in terms of physics mass balance means the system has a well-defined density, i.e., density becomes a "macrovariable" of the system.

Similarly, kinetic energy balance means that the system has a well-defined "temperature", i.e., the kinetic energy is uniform through the "gas container". It means that temperature becomes a "macrovariable" of the system.

From the viewpoint of physics energy balance represents the most interesting case. However, since we are going to face considerable mathematical difficulties, it is best to start with the simplest case of particle balance. We will discuss the case of energy balance later in Section 24.

Our precise equilibrium is going to be a "snapshot square-root type equilibrium" (see Definitions 2.3 and 2.4), which requires some preparations.

Note that the snapshot of the system at time $t = t_0$ is an N-element point set (explaining how Uniform Distribution enters the story). This raises the question: is there a state of "maximum particle balance" or "maximum uniformity", and how to define it? To answer this question, we introduce the concept of *LA-uniformity* of a continuous family of N-element point sets in the unit torus $[0, 1)^h$; see Definition 2.1 below. Later we use the same concept to precisely define "absolute (spatial) equilibrium". Note that L stands for Lebesgue, and A stands for average (in fact, a quadratic average).

The main idea. In our study of uniformity (= equidistribution) we make a bold approach: *we take maximum over all possible Lebesgue measurable test sets*, which is the largest family we can work with. (We will use the same idea to define our micro-entropy; see later.)

Note that most physics books give the intuitive interpretation of the Second Law as the law of "increasing disorder", but they never give a mathematically precise definition of what "disorder" actually means. In our models "increasing disorder" represents "increasing uniformity", and our rigorous Second Law states that the typical time evolution exhibits increasing uniformity with respect to all Lebesgue measurable test sets. A usual criticism of the Boltzmann and Gibbs entropies is that they are "subjective" and "anthropomorphic". In the definition of our micro-entropy we take maximum over all possible Lebesgue measurable test sets (in a rather subtle way to be explained below). Since "measurable" is the most general ultimate setup, this way we automatically eliminate any kind of criticism of being "subjective" or "anthropomorphic".

Perhaps the reader finds the idea of *taking maximum over all possible Lebesgue measurable test sets* hopelessly general, too naive to be feasible. To convince the reader that this is in fact the *right idea*, here we insert a short detour that the reader may skip (but we recommend not to skip).

An Instructive Detour. Here we briefly discuss a recent result of ours which is exactly about *how* to take maximum over all possible Lebesgue measurable test sets in a subtle way. Actually, we go beyond test sets, and work with test functions. In the following illustration the natural function space is $f \in L^2([0,1)^2)$, i.e., the space of all complex-valued Lebesgue square integrable functions defined on the unit square $[0,1)^2$. We study the uniformity of "typical" 2-dimensional torus lines with arbitrary fixed starting point $\mathbf{y} \in [0,1)^2$. So we just need a measure on the set of all initial directions (angles) $\theta \in [0,2\pi)$. We choose the most natural one: the normalized 1-dimensional Lebesgue measure in the interval $[0,2\pi)$. This way the term "$1-\varepsilon$ part of all angles" in Theorem A below becomes perfectly precise.

Theorem A ([Be2015]). *Let $f \in L^2([0,1)^2)$, and let $T > 100$ be an arbitrarily large but fixed real number. Let $\mathbf{x}(t) = \mathbf{x}(\mathbf{y};\theta;t) = (x_1(t), x_2(t))$, $0 \le t \le T$ with*

$$x_1(t) = \{y_1 + t\cos\theta\} \quad \text{and} \quad x_2(t) = \{y_2 + t\sin\theta\}$$

(where $\{\ldots\}$ is the fractional part) denote a torus line of length T in the unit square; the starting point $\mathbf{y} = (y_1, y_2) \in [0,1)^2$ is arbitrary but fixed, and the angle is $\theta \in [0,2\pi)$.

Let

$$\Delta_f(\mathbf{y};\theta;T) = \int_0^T f(\mathbf{x}(\mathbf{y};\theta;t))\,dt$$

denote the integral of f along the torus line of length T. Let $0 < \varepsilon < 1/2$ be arbitrary. Then for $1-\varepsilon$ part of all angles $\theta \in [0,2\pi)$, the discrepancy of the corresponding torus line

$$\text{discrepancy}_f(\mathbf{y};\theta;T) = \Delta_f(\mathbf{y};\theta;T) - T\int_{I^2} f(\mathbf{z})\,d\mathbf{z}$$

is estimated from above by

$$|\text{discrepancy}_f(\mathbf{y};\theta;T)| < \frac{C^\star \sigma_0(f)}{\varepsilon}(\log T)^3(\log\log T)^2 + C^{\star\star}(f),$$

where

$$\sigma_0^2(f) = \int_{I^2} \left| f(\mathbf{y}) - \int_{I^2} f(\mathbf{z})\,d\mathbf{z} \right|^2 d\mathbf{y},$$

$C^\star$ is an effectively computable absolute constant, and $C^{\star\star}$ is an effectively computable constant that depends only on the given test function f.

Note that log stands for the natural (base e) logarithm (we do not use ln in the book).

We can rewrite Theorem A in the following form.

First Inequality:

$$\sup_{\substack{f\in L^2:\\ \sigma_0^2(f)>0}} \sup_{\mathbf{y}\in[0,1)^2} \inf_{\substack{\mathcal{A}\subset[0,2\pi):\\ \text{length}(\mathcal{A})=(1-\varepsilon)2\pi}} \sup_{\theta\in\mathcal{A}} \frac{1}{\sigma_0(f)} \left| \Delta_f(\mathbf{y};\theta;T) - T\int_{I^2} f(\mathbf{z})\,d\mathbf{z} \right|$$

$$\leq \frac{C^\star \sigma_0(f)}{\varepsilon}(\log T)^3(\log\log T)^2 + C^{\star\star}(f),$$

where for simplicity length stands for the 1-dimensional Lebesgue measure, and

$$\sigma_0(f) = \left(\int_{I^2} \left| f(\mathbf{y}) - \int_{I^2} f(\mathbf{z})\,d\mathbf{z} \right|^2 d\mathbf{y} \right)^{1/2}$$

is the natural norming factor.

For comparison, we state the following very simple fact.

Second Inequality:

$$\sup_{\substack{f\in L^2:\\ \sigma_0^2(f)>0}} \inf_{(\mathbf{y};\theta)\in[0,1)^2\times[0,2\pi)} \frac{1}{\sigma_0(f)} \left| \Delta_f(\mathbf{y};\theta;T) - T\int_{I^2} f(\mathbf{z})\,d\mathbf{z} \right| \geq T.$$

Comparing the two inequalities, we see a striking contrast between the polylogarithmic order of magnitude

$$(\log T)^{3+\varepsilon} \text{ and the linear } T$$

as T tends to infinity.

The First Inequality has the interpretation that for every Lebesgue measurable test set in the unit square, and even for every test function $f \in L^2$, the overwhelming majority of the torus lines of length T are "superuniform" with maximum error $(\log T)^{3+\varepsilon}$, which is a shockingly small polylogarithmic error term. On the other hand, the Second Inequality is completely trivial. It follows from the obvious fact that for every torus line of length T there is a test set — for example, the complement of the torus line in the unit square — for which the torus line is "totally anti-uniform" (since a set does not intersect its own complement!).

The Second Inequality is a useless, trivial way to take maximum over all possible test functions $f \in L^2$ (with $\sigma_0^2(f) > 0$). The First Inequality, on the other hand, is a *subtle* way to take maximum over all possible test functions $f \in L^2$ (with $\sigma_0^2(f) > 0$), providing a shockingly good polylogarithmic upper bound about the maximum error. We will use the same subtle idea in the definition of "absolute equilibrium", and also later in the definition of our "micro-entropy".

We stress the key fact that in Theorem A the starting point $\mathbf{y}$ is *arbitrary* (and not just generic). Similarly, when in later sections we study the time evolution of a large closed non-equilibrium system, the initial configuration is always *arbitrary* (and not just generic). This is absolutely crucial, since our goal is to describe the time evolution from an arbitrary off-equilibrium state to equilibrium.

For comparison we mention the following result.

Theorem B ([Be2010]). *Similarly to Theorem A, we study the discrepancy of the same torus line*

$$\text{discrepancy}_f(\mathbf{y};\theta;T) = \Delta_f(\mathbf{y};\theta;T) - T\int_{I^2} f(\mathbf{z})\, d\mathbf{z},$$

where $f \in L^2([0,1)^2)$ and $T > 100$ are arbitrary. Let $0 < \varepsilon < 1/2$ be arbitrary. Then for $1-\varepsilon$ part of all **initial conditions** $(\mathbf{y},\theta) \in [0,1)^2 \times [0,2\pi)$, *the discrepancy is estimated from above by*

$$|\text{discrepancy}_f(\mathbf{y};\theta;T)| < \frac{30\sigma_0(f)}{\varepsilon}\sqrt{\log T}.$$

Notice that the condition "$1-\varepsilon$ part of all" has two different meanings in Theorem A and Theorem B. The initial condition $(\mathbf{y},\theta) \in [0,1)^2 \times [0,2\pi)$ in Theorem B includes both the starting point and the angle, so Theorem B does not say anything about a given fixed starting point $\mathbf{y}$. Theorem B has a substantially smaller upper bound, but it has the wrong message. Theorem A has a substantially weaker upper bound, but it has the right message: it works for *every starting point.*

By the way, in the product space $[0,1)^2 \times [0,2\pi)$ we take the (natural) product measure: the (Cartesian) product of the 2-dimensional Lebesgue measure in the unit square and the 1-dimensional Lebesgue measure in $[0,2\pi)$. This makes the term "$1-\varepsilon$ part of all initial conditions" in Theorem B perfectly precise.

Note that Theorem A was just an illustration; we do not need this result in the rest. (We do not need Theorem B either.) The interested reader can find the proof of Theorem A and many more related "strong uniformity and Khinchin's conjecture" type results in [Be2015] (see also [Mar70], and for applications of strong uniformity; see the book [Be2018]). *This completes the Detour.*

As we said above, to study uniformity, it is natural — and beneficial — to go beyond test sets, and work with the larger class of test functions. In dimension h a test function is a (complex valued) function $f = f(x_1, \ldots, x_h) \in L^2(I^h)$ defined on the unit torus $I^h = [0,1)^h$, i.e., $f(x_1, \ldots, x_h)$ is periodic in all h variables with period 1. An important special case is where $f = \chi_S$ is a 0-1 valued characteristic function of a test set $S \subset [0,1)^h$; here S is an arbitrary measurable subset with $0 < \mathrm{vol}(S) < 1$ (for simplicity vol stands for the h-dimensional Lebesgue measure vol_h, and we extend $f = \chi_S$ periodically). We study the uniformity of an N-element point set

$$\mathcal{Z} = \{\mathbf{z}_1, \ldots, \mathbf{z}_N\} \subset [0,1)^h$$

with respect to a given test set $S \subset [0,1)^h$, and with respect to a test function $f \in L^2(I^h)$ in general.

Let

$$\sigma_0^2(f) = \int_{I^h} \left| f - \int_{I^h} f \, d\mathrm{V} \right|^2 d\mathrm{V} = \int_{I^h} |f|^2 \, d\mathrm{V} - \left| \int_{I^h} f \, d\mathrm{V} \right|^2 \tag{2.1}$$

denote the "variance" of the test function f. We always assume that $\sigma_0^2(f) > 0$. In the special case $f = \chi_S$, we have $\sigma_0^2(f) = pq$ with $p = \mathrm{vol}(S)$ and $q = 1 - p$.

Uniformity means that $\sum_{\mathbf{z} \in \mathcal{Z}} f(\mathbf{z})$ and $N \int_{I^h} f \, d\mathrm{V}$ are close to each other, i.e., the Riemann sum is a good approximation of the integral. Since the torus I^h is translation invariant, any translated copy $S + \mathbf{w}$ (mod 1), $\mathbf{w} \in \mathbf{R}^h$ is just as good of a test set as S itself. Similarly, any translation of f by a vector $\mathbf{w}$ modulo one is just as good of a test function as f itself. Here "translation of f by $\mathbf{w}$ modulo one" means the function

$$f_{\mathbf{w}}(\mathbf{x}) = f(\mathbf{x} - \mathbf{w})$$

(note that f is periodic with period one; this is why we can restrict $\mathbf{w} \in \mathbf{R}^h$ to $\mathbf{w} \in I^h$). This motivates why we study the *average of the square-discrepancy* over all translations of the torus I^h:

$$\text{AvSqDi}(\mathcal{Z}; f) = \int_{I^h} \left| \sum_{\mathbf{z} \in \mathcal{Z}} f(\mathbf{z} - \mathbf{w}) - N \int_{I^h} f \, d\text{V} \right|^2 d\mathbf{w}.$$

The following simple identity will play a key role in the rest of the book:

$$\text{AvSqDi}(\mathcal{Z}; f) = \int_{I^h} \left| \sum_{\mathbf{z} \in \mathcal{Z}} f(\mathbf{z} - \mathbf{w}) - N \int_{I^h} f \, d\text{V} \right|^2 d\mathbf{w}$$

$$= \sum_{\substack{\mathbf{n} \in \mathbb{Z}^h: \\ \mathbf{n} \neq \mathbf{0}}} |a_{\mathbf{n}}|^2 \left| \sum_{k=1}^{N} e^{2\pi i \mathbf{n} \cdot \mathbf{z}_k} \right|^2, \tag{2.2}$$

where $a_{\mathbf{n}}$, $\mathbf{n} \in \mathbb{Z}^h$ are the Fourier coefficients of the given test function $f \in L^2$. Note that (2.2) is based on the L^2-theory of Fourier series and Parseval's formula (for the details, see the argument of (7.6′)).

We may say, with some gross exaggeration, that the whole book is based on the simple identity (2.2). As an illustration, we briefly introduce the somewhat related concept of Reduced Micro-Entropy of a point set $\mathcal{Z} = \{\mathbf{z}_1, \ldots, \mathbf{z}_N\} \subset [0,1)^h$:

$$\text{ReducedMicroEntropy}(\mathcal{Z}) = \max_{\substack{\mathbf{n} \in \mathbb{Z}^h: \\ |\mathbf{n}|=1}} \left| \sum_{k=1}^{N} e^{2\pi i \mathbf{n} \cdot \mathbf{z}_k} \right|^2.$$

We call it "reduced" because the maximum is restricted to the shortest nonzero lattice points (i.e., we restrict it to the $2h$ lattice points $(\pm 1, 0, \ldots, 0), \ldots, (0, \ldots, 0, \pm 1)$). Reduced Micro-Entropy has an important meaning; we explain it later (including why this concept deserves the name "entropy") in Section 6 after the Metatheorem.

Here we just briefly note that for every nonzero lattice point $\mathbf{n} \in \mathbb{Z}^h \setminus \mathbf{0}$, the corresponding term

$$\left| \sum_{k=1}^{N} e^{2\pi i \mathbf{n} \cdot \mathbf{z}_k} \right|^2$$

exhibits two basic properties of entropy (see Section 11):

(1) "entropy is additive in thermodynamical equilibrium", and

(2) a formal analogy with Boltzmann's formula "entropy is (some absolute constant times) $\log W$, where W is the number of microstates consistent with the given macrostate".

Our Micro-Entropy (the key concept in our Second Law) is more complicated than the Reduced Micro-Entropy. It will be defined as a function of the infinitely many terms

$$\left|\sum_{k=1}^{N} e^{2\pi i \mathbf{n}\cdot\mathbf{z}_k}\right|^2, \quad \mathbf{n} \in \mathbb{Z}^h \setminus \mathbf{0},$$

where the function depends on the initial velocity distribution (this represents the special case where every particle has unit mass). Despite the formal dependens on the initial velocity distribution, Micro-Entropy is "almost the same" as the Reduced Micro-Entropy, where the latter does *not* depend on the initial velocity distribution.

What makes Micro-Entropy special is that, beyond properties (1) and (2), it also describes the Relaxation Time (i.e., the necessary time of the typical time evolution of the system, starting from the initial configuration $\mathbf{z}_1, \ldots, \mathbf{z}_N$, to reach "square-root equilibrium"), and last but not least, it satisfies a Second Law ("entropy increase"). For the details; see Sections 6 and Chapters 3–4.

We are now ready to define *LA-uniformity.*

Definition 2.1. Let $\mathcal{Z}_\omega$, $\omega \in \Omega$ be a family of N-element point sets in the h-dimensional unit cube $[0,1)^h$ ($h \geq 1$), i.e., $\mathcal{Z}_\omega \subset [0,1)^h$ with $|\mathcal{Z}_\omega| = N$ for every $\omega \in \Omega$; moreover, assume that there is a probability space $(\Omega, \mathcal{B}, \text{meas})$ on Ω: $\mathcal{B}$ is a σ-algebra of the subsets of Ω and meas is a (positive) measure defined on $\mathcal{B}$ such that $\text{meas}(\Omega) = 1$. Let $0 < \varepsilon < 1$ and $C > 0$. We say that the family $\mathcal{Z}_\omega$, $\omega \in \Omega$ exhibits **LA-uniformity** with parameters (ε, C) if for every square-integrable test function $f \in L^2(I^h)$ there exists a subset $\Omega(\varepsilon; C; f) \in \mathcal{B}$ of Ω with $\text{meas}(\Omega(\varepsilon; C; f)) \geq 1 - \varepsilon$ such that, for every $\omega \in \Omega(\varepsilon; C; f)$,

$$\sqrt{\text{AvSqDi}(\mathcal{Z}_\omega; f)} = \sqrt{\int_{I^h} \left|\sum_{\mathbf{z}\in\mathcal{Z}_\omega} f(\mathbf{z}-\mathbf{w}) - N\int_{I^h} f\, d\text{V}\right|^2 d\mathbf{w}} \leq C\sigma_0(f)\sqrt{N}. \tag{2.3}$$

Remarks. The key point in the definition is that the parameters ε and C are independent of N and $f \in L^2$. Of course in the applications we want "reasonable" values for ε and C; for example, $\varepsilon = 10^{-2}$ and $C = 10$

are reasonable values. Notice that $\sqrt{\text{AvSqDi}(\mathcal{Z}_\omega; f)}$ is a quadratic mean of the discrepancy, the term $\sigma_0^2(f)N$ represents the variance, and $\sigma_0(f)\sqrt{N}$ represents the standard deviation of N independent components, so the requirement (2.3) can be interpreted as "close to random" apart from a constant factor C.

The following technical lemma has the important message that for the whole class of Lebesgue measurable test sets discrepancy of size $\sqrt{N}$ is *inevitable.*

Lemma 2.1. *For every integer $N > 1$ and every dimension $h \geq 1$ there exists an $m = m(N)$ such that we can find m Lebesgue measurable subsets $S_1, \ldots, S_m$ of the h-dimensional unit cube $[0,1)^h$ (the dimension $h \geq 1$ is arbitrary) with the following property: given any N-element point set*

$$\mathcal{Z} = \{\mathbf{z}_1, \ldots, \mathbf{z}_N\} \subset [0,1)^h,$$

for at least two-thirds of the m sets $S_1, \ldots, S_m$,

$$||\mathcal{Z} \cap S_j| - \text{vol}(S_j)N| > \frac{\sqrt{2}}{20}\sqrt{N}.$$

We postpone the proof to the end of Section 2.

(Perhaps the reader is wondering why we have the "large" square-root size discrepancy in Lemma 2.1 and the much smaller polylogarithmic discrepancy in Theorem A. The reason is that there is a surprising difference between the discrete and continuous versions of the so-called "Khinchin's conjecture on strong uniformity" in Uniform Distribution. For more about it; see [Be2015].)

We use Lemma 2.1 to explain why (2.3) is the best possible result (apart from a constant factor). So LA-uniformity represents "absolute" in the double sense of *absolute minimum error*, and also it represents the *absolute maximum* for the family of test sets (since the class of Lebesgue measurable sets is the largest class for which measure is defined).

Corollary of Lemma 2.1. *Inequality (2.3) in Definition 2.1 is best possible apart from a constant factor. More precisely, (2.3) cannot hold with $\varepsilon = 1/3$ and $C = 1/10$.*

Proof of Lemma 2.1 $\Rightarrow$ Corollary. Assume that (2.3) *does* hold with $\varepsilon = 1/3$ and $C = 1/10$, and we derive a contradiction. By hypothesis, for some $N > 1$ there exists a continuous family of N-element point sets $\mathcal{Z}_\omega \subset [0,1)^h$, $\omega \in \Omega$ such that there is a probability space $(\Omega, \mathcal{B}, \text{meas})$ on

Ω (i.e., $\mathcal{B}$ is a σ-algebra of the subsets of Ω and meas is a measure defined on $\mathcal{B}$ such that $\text{meas}(\Omega) = 1$), and for every test function $f \in L^2(I^h)$ there exists a subset $\Omega(f) \in \mathcal{B}$ of Ω with $\text{meas}(\Omega(f)) \geq 2/3$ and

$$\sqrt{\text{AvSqDi}(\mathcal{Z}_\omega; f)} \leq \frac{1}{10}\sigma_0(f)\sqrt{N} \tag{2.4}$$

for every $\omega \in \Omega(f)$. We show that (2.4) contradicts Lemma 2.1.

To obtain a contradiction, consider the $m(N)$ sets $S_1, \ldots, S_m$ whose existence is guaranteed by Lemma 2.1, and let $f_j = \chi_{S_j}$, $1 \leq j \leq m$ denote their 0-1 valued characteristic functions (extended periodically). Applying (2.4) with $f = f_j$, $1 \leq j \leq m$ we obtain m $\mathcal{B}$-measurable subsets $\Omega_j \subset \Omega$ with $\text{meas}(\Omega_j) \geq 2/3$ such that

$$\sqrt{\text{AvSqDi}(\mathcal{Z}_\omega; f_j)} \leq \frac{1}{10}\sigma_0(f_j)\sqrt{N}$$

holds for every $\omega \in \Omega_j$ $(1 \leq j \leq m)$. The fact

$$\frac{1}{m}\sum_{j=1}^{m} \text{meas}(\Omega_j) \geq 2/3$$

implies that there must exist an $\omega_0 \in \Omega$ which is contained by at least $2m/3$ sets Ω_j. Then the corresponding N-set $\mathcal{Z}_{\omega_0}$ has the property

$$\sum_{j\in J_1} \text{AvSqDi}(\mathcal{Z}_{\omega_0}; f_j)$$

$$= \sum_{j\in J_1} \int_{I^h} \left| \sum_{\mathbf{z}\in\mathcal{Z}_{\omega_0}} f_j(\mathbf{z}-\mathbf{w}) - N\int_{I^h} f_j\, dV \right|^2 d\mathbf{w} \leq \frac{N}{100}\sum_{j\in J_1} \sigma_0^2(f_j) \tag{2.5}$$

for some index set $J_1 \subset \{1, 2, \ldots, m\}$ with $|J_1| \geq 2m/3$.

Since (2.5) represents an "average", and of course "minimum $\leq$ average", (2.5) implies the existence of a translation $\mathbf{w}_0 \in [0,1)^h$ such that

$$\sum_{j\in J_1} \left(|(\mathcal{Z}_{\omega_0} + \mathbf{w}_0) \cap S_j| - \text{vol}(S_j)N\right)^2 \leq \frac{N}{100}\sum_{j\in J_1} \sigma_0^2(f_j) = \frac{N}{100}\sum_{j\in J_1} p_j q_j, \tag{2.6}$$

where $p_j = \text{vol}(S_j)$ and $q_j = 1 - p_j$, and $\mathcal{Z}_{\omega_0} + \mathbf{w}_0$ is the translated copy of $\mathcal{Z}_{\omega_0}$ modulo one (we are in the torus).

On the other hand, applying Lemma 2.1 with $\mathcal{Z} = \mathcal{Z}_{\omega_0} + \mathbf{w}_0$, we obtain that there is another index set $J_2 \subset \{1, 2, \ldots, m\}$ with $|J_2| \geq 2m/3$ such that

$$\left||(\mathcal{Z}_{\omega_0} + \mathbf{w}_0) \cap S_j| - \text{vol}(S_j)N\right| > \frac{\sqrt{2N}}{20} \tag{2.7}$$

for every $j \in J_2$. Let $J_3 = J_1 \cap J_2$ be the intersection. Combining (2.6) and (2.7), we have

$$\frac{|J_3|N}{200} < \sum_{j\in J_3} (|(\mathcal{Z}_{\omega_0} + \mathbf{w}_0) \cap S_j| - \text{vol}(S_j)N)^2$$

$$\leq \sum_{j\in J_1} (|(\mathcal{Z}_{\omega_0} + \mathbf{w}_0) \cap S_j| - \text{vol}(S_j)N)^2 \leq \frac{N}{100} \sum_{j\in J_1} p_j q_j \leq \frac{|J_1|N}{400},$$

which implies

$$|J_3| < |J_1|/2. \tag{2.8}$$

Clearly

$$|J_3| \geq |J_1| + |J_2| - m \geq |J_1| + (2m/3) - m = |J_1| - (m/3). \tag{2.9}$$

Combining (2.8) and (2.9), we conclude

$$|J_1| - \frac{m}{3} < |J_1|/2,$$

which contradicts the hypothesis $|J_1| \geq 2m/3$. This contradiction proves that (2.4) is false, implying that (2.3) is the best possible result apart from a constant factor. This completes the proof of the Corollary. □

The message of the Corollary can be summarized as follows: the order of $\sqrt{N}$ is the *absolute minimum error* for the whole class of Lebesgue measurable test sets.

We just proved that LA-uniformity exhibits the absolute minimum error. Nevertheless, we can slightly refine the concept by getting rid of the average in (2.3). This gives us the concept of L-uniformity (where again L refers to Lebesgue, i.e., Lebsgue measurable test sets and functions).

Definition 2.2. Similarly to Definition 2.1, let $\mathcal{Z}_\omega$, $\omega \in \Omega$ be a family of N-element point sets in the h-dimensional unit cube $[0,1)^h$ ($k \geq 1$), i.e., $\mathcal{Z}_\omega \subset [0,1)^h$ with $|\mathcal{Z}_\omega| = N$ for every $\omega \in \Omega$; moreover, assume that there is a probability space $(\Omega, \mathcal{B}, \text{meas})$ on Ω: $\mathcal{B}$ is a σ-algebra of the subsets of Ω and meas is a (positive) measure defined on $\mathcal{B}$ such that $\text{meas}(\Omega) = 1$. Let $0 < \varepsilon < 1$ and $C > 0$. We say that the family $\mathcal{Z}_\omega$, $\omega \in \Omega$ exhibits **L-uniformity** with parameters (ε, C) if for every square-integrable test function $f \in L^2(I^h)$ there exists a subset $\Omega(\varepsilon; C; f) \in \mathcal{B}$ of Ω with $\text{meas}(\Omega(\varepsilon; C; f)) \geq 1 - \varepsilon$ such that, for every $\omega \in \Omega(\varepsilon; C; f)$

$$\left| \sum_{\mathbf{z}\in\mathcal{Z}_\omega} f(\mathbf{z}) - N \int_{I^h} f \, dV \right| \leq C\sigma_0(f)\sqrt{N}. \tag{2.10}$$

Remarks. Comparing the two definitions it is obvious that L-uniformity is the simpler concept. Despite these advantages, nevertheless, it is the more complicated LA-uniformity that we start working with. The reason is that LA-uniformity is the natural concept from the viewpoint of Fourier analysis, due to a key application of Parseval's formula (see (2.2)). Since Fourier analysis is our basic tool, the concept of LA-uniformity motivates/explains the definition of our *micro-entropy*; see Section 6.

(It does not mean that we ignore L-uniformity in the rest. For example, in the second half of Section 24 we show how to define a similar intrinsic entropy based on the concept of L-uniformity.)

At the end of Section 1 we already mentioned a far-reaching generalization allowing the particles to travel on (almost) arbitrary space curves, not just on straight lines like point billiards. This is exactly how we attempt to involve — indirectly — particle-particle collision and external and internal forces in our models. We elaborate on this. To give a real life motivation for arbitrary space curves, note that in a constant magnetic field $\mathbf{B}$, there is a force on a moving charged particle, say a proton; the force is $\mathbf{F} = \mathbf{v} \times \mathbf{B}$ (cross product), where $\mathbf{v}$ is the velocity of the proton. This is why the proton travels on a helix like $\mathbf{r}(t) = (\cos t, \sin t, t)$, instead of a straight line. (An uncharged particle travels on a straight line.)

A second motivation is that we try to imitate particle-particle collision by allowing random-walk-like chaotic zig-zag trajectories. If the system is homogeneous, consisting of the same type of particles only, then it is just a minor oversimplification/cheating to assume that two colliding particles, P_1 and P_2 (say), simply switch directions, that is, a collision means that P_1 follows the line of P_2 and P_2 follows the line of P_1 (i.e., P_1 and P_2 trade places). Since all what we care about is the number of particles in a given test set in a given time t, the particle-switch is quite acceptable, giving up the identity of the particles is not a big problem.

However, if the system is *not* homogeneous, i.e., it consists of at least two different types of particles, then we cannot give up the identity of the particles. Then we have to replace the straight line trajectory with a random-walk-like chaotic trajectory, due to the effect of the frequent particle-particle collisions. Of course a random-walk like trajectory is totally different from a straight line. For example, we know that a random walk is "slow": its expected distance from the start is the square-root of the time, which is much slower than the constant speed motion along a straight line, for which the distance from the start is a linear function of the time.

We also know that the 3-dimensional random walk "rarely returns". The same holds for the 2-dimensional random walk. (The 1-dimensional random walk, on the other hand, is totally different.)

A third real life motivation for arbitrary space curves is that the time evolution of such a general many-particle system mimicks a key aspect of the Second Law: "energy dispersal" on a microscopic level. We witness "energy dispersal" in our every day life. Beyond the case of gases, thermodynamics also applies for fluids and solids, and then "spreading of energy" is a more complicated process than the collision of gas molecules (which is well approximated by linear billiard paths). Using arbitrary space curves instead of billiard paths is a better model. (It is interesting to point out that to intuitively describe the Second Law most chemistry books prefer to use "energy dispersal" instead of "increasing disorder"; the latter is used in most physics books.)

Let us mention in advance a surprising corollary of our results. As it turns out, the complexity of the curves of the particles is irrelevant; what really matters is the "drifting distance"; see (2.11) below.

Now we define the main model of our book: the **general torus model**. In the general torus model (intuitively an "energy dispersal model") every particle has its own *type* of trajectory described by a parametrized space curve: the trajectory of the k-th particle is $\mathbf{r}_k(t)$ modulo one, where $\mathbf{r}_k(0) = \mathbf{0}$, $1 \le k \le N$.

We of course assume that each parametrized space curve is continuous and has arc length (rectifiable), but beyond that they are almost arbitrary (straight line, bending line, twisted line, any kind of general helix, spiral, etc.). We require, however, that the parametrized space curves $\mathbf{r}_k(t)$ have a tendency of "drifting away from the start $\mathbf{0}$". A parametrized curve modulo one describes the motion of a particle in the torus, and $|\mathbf{r}_k(t)|$ denotes the euclidean distance in the whole space (not in the torus!) from the start $\mathbf{r}_k(0) = \mathbf{0}$ at time t. We emphasize that here we do *not* take $\mathbf{r}_k(t)$ modulo one yet; for example, on a straight line $t\mathbf{e}$, where $\mathbf{e}$ is a unit vector, the distance of $t\mathbf{e}$ from the start $\mathbf{0}$ is t.

In the rest it should be always clear from the context of the sentence that the *distance* between two points $\mathbf{x}, \mathbf{y} \in \mathsf{R}^d$ is the usual R^d-distance (= euclidean distance) $\left(\sum_{j=1}^d (x_j - y_j)^2\right)^{1/2}$ or the "unit torus distance" $\left(\sum_{j=1}^d \|x_j - y_j\|^2\right)^{1/2}$, where $\|z\|$ is the distance of z from the nearest integer.

Our object is to prove a (spatial) Second Law with the following message. The time evolution of a large closed system of particles, in this case the general torus model, starting from an arbitrary initial point configuration (including such an extreme case as the "Big Bang"), exhibits rapidly increasing uniformity, assuming the particles start with typical initial velocities.

We use $\mathbf{x}_k(t)$ to denote the trajectory of the k-th particle on the torus $I^3 = [0,1)^3$ $(1 \le k \le N)$. We assume arbitrary initial point configuration: let $\mathbf{x}_k(0) = \mathbf{y}_k \in I^3$, $1 \le k \le N$ be arbitrary but fixed. Let $\mathbf{r}_k(t) \subset \mathsf{R}^3$, $1 \le k \le N$ be an arbitrary family of N continuous parametrized space curves with $\mathbf{r}_k(0) = \mathbf{0}$, and write

$$\text{drifting distance at time } t = \min_{1 \le k \le N} |\mathbf{r}_k(t)|. \tag{2.11}$$

We know that $\mathbf{r}_k(t)$ describes the *type* of the trajectory of the k-th particle (of course for the torus we take $\mathbf{r}_k(t)$ modulo one), and $\mathbf{x}_k(0) = \mathbf{y}_k$ is the starting point, but we also need the initial velocity. Similarly to the Bernoulli model of the first kind (Gaussian velocity distribution), we choose a scale ρ_k in $0 < \rho_k < \infty$ with distribution (1.9′), and also choose a "direction", which means more than simply choosing a point on the unit sphere $\mathbf{S}^2$. Indeed, here "direction" actually means a *rotation* of the 3-space. The reason is that a general space curve (say a helix) is a proper 3-dimensional object (unlike a straight line), so we must choose a rotation of the 3-space $\vartheta_k \in SO(3)$.

Note that the 2-dimensional case is simpler: the rotation of the plane simply means to choose a point on the unit circle $\mathbf{S}^1$, or equivalently, to choose an angle $0 \le \theta < 2\pi$.

The choice of the scale ρ_k and the rotation ϑ_k in the 3-space make the initial condition complete $(1 \le k \le N)$. So, the *real* trajectory of the k-th particle in the torus model is

$$\mathbf{x}_k(t) = \mathbf{y}_k + \vartheta_k(\rho_k \mathbf{r}_k(t)) \quad \text{modulo one}, \quad 1 \le k \le N, \tag{2.12}$$

which describes the time evolution of the whole N-point system.

To define *typical*, we need a measure. As usual, we assume that ρ_1, ϑ_1, ρ_2, $\vartheta_2, \ldots$, ρ_N, ϑ_N are $2N$ independent random variables, where ρ_1, $\rho_2, \ldots, \rho_N$ have the same Gaussian distribution described by (1.9′), and ϑ_1, $\vartheta_2, \ldots, \vartheta_N$ are all uniformly distributed in the rotation group $SO(3)$ (= special orthogonal group of all 3×3 real matrices) with respect to the Haar measure. This means we use the corresponding product measure, denoted by ProdMeas_1, on the product space

$$\omega = (\rho_1, \vartheta_1, \rho_2, \vartheta_2, \ldots, \rho_N, \vartheta_N) \in \Omega_1 = \left([0,\infty) \times SO(3)\right)^N. \tag{2.13}$$

We have $\int_{\Omega_1} 1\,d\omega = 1$, where for notational simplicity $d\omega = d\mathrm{ProdMeas}_1(\omega)$.

We refer to this as the (general) 3-dimensional torus model of the first kind (i.e., Gaussian distribution). We call the huge product space Ω_1 the "initial velocity space" of the first kind.

Note that the Haar measure on $SO(3)$ is equivalent to a simple product measure: the product of the surface area on the unit sphere $\mathbf{S}^2$ and the arc-length on the unit circle $\mathbf{S}^1$. This follows from the well known fact that every nontrivial rotation in the 3-space can be specified by an axis of rotation together with an angle of rotation about this axis (more precisely, one also needs to specify orientations), which leads to a measure-theoretic isomorphism between $SO(3)$ and $\mathbf{S}^2 \times \mathbf{S}^1$.

Let

$$\mathcal{Y} = \{\mathbf{y}_1, \mathbf{y}_2, \ldots, \mathbf{y}_N\} \quad \text{and} \quad \mathcal{R} = (\mathbf{r}_1, \mathbf{r}_2, \ldots, \mathbf{r}_N)$$

be, respectively, an initial point configuration and a family of parametrized curves. (We always start at time $t = 0$, and study the time evolution "forward" $t > 0$, but of course the same results hold going "backward" in time.) Given

$$\omega = (\rho_1, \vartheta_1, \rho_2, \vartheta_2, \ldots, \rho_N, \vartheta_N) \in \Omega_1 = ([0,\infty) \times SO(3))^N,$$

write

$$\mathcal{Y}_\omega(t) = \mathcal{Y}_{\mathcal{R};\omega;t}$$

$$= \{\mathbf{y}_1 + \vartheta_1(\rho_1\mathbf{r}_1(t)) \text{ modulo one}, \ldots, \mathbf{y}_N + \vartheta_N(\rho_N\mathbf{r}_N(t)) \text{ modulo one}\}. \tag{2.14}$$

Note that (2.14) represents the time evolution of the system via snapshots: (2.14) gives a snapshot of the system at a fixed time t. We say that the time evolution of the general torus model (2.12)–(2.14) reaches **LA-equilibrium** (**L-equilibrium**) with parameters (ε, C) at time t when for every $\tau \geq t$ the continuous family $\mathcal{Y}_\omega(\tau)$, $\omega \in \Omega_1$ of N-element sets, defined in (2.14), exhibits **LA-uniformity** (**L-uniformity**) with parameters (ε, C).

It is easy to extend the definitions of LA-equilibrium and L-equilibrium for the more general case of arbitrary mass distribution; see Definitions 2.3 and 2.4 below. It makes the notation a little bit more complicated; the details go as follows. As usual, we fix a (complex valued) test function $f = f(x,y,z) \in L^2$ defined in the unit torus I^3 (so $f(x,y,z)$ is periodic in all three variables with period 1). An important special case is $f = \chi_S$, where $S \subset I^3 = [0,1)^3$ is a measurable subset with $0 < \mathrm{vol}(S) < 1$. We study the uniformity of the N-particle system at time t with respect to the given set S, and with respect to $f \in L^2$ in general.

As usual, let

$$\mathcal{Y} = \{\mathbf{y}_1, \mathbf{y}_2, \ldots, \mathbf{y}_N\} \tag{2.15}$$

denote the initial point configuration, and write

$$\mathcal{M} = \{\mu_1, \mu_2, \ldots, \mu_N\} \tag{2.16}$$

for a given mass distribution. That is, $\mathbf{y}_k \in I^3$ is the starting point of the k-th particle, and $\mu_k > 0$ denotes its mass.

We combine (2.15) and (2.16) to a single vector

$$(\mathcal{Y}, \mathcal{M}) = (\mathbf{y}_1, \mu_1; \mathbf{y}_2, \mu_2; \ldots; \mathbf{y}_N, \mu_N), \tag{2.17}$$

and call it the "initial (point) configuration with given mass distribution".

As a natural norming, we assume that the average mass is 1, that is,

$$\frac{1}{N}\sum_{k=1}^{N} \mu_k = 1, \quad \text{or equivalently,} \quad \sum_{k=1}^{N} \mu_k = N, \tag{2.18}$$

where the most important special case is $1 = \mu_1 = \mu_2 = \ldots = \mu_N$.

Again write

$$\mathcal{R} = (\mathbf{r}_1, \mathbf{r}_2, \ldots, \mathbf{r}_N) \tag{2.19}$$

for a family of N continuous rectifiable parametrized space curves $\mathbf{r}_k = \mathbf{r}_k(t)$ with $\mathbf{r}_k(0) = \mathbf{0}$, and write (see (2.11))

$$\mathcal{R}(t) = \min_{1 \le k \le N} |\mathbf{r}_k(t)| \tag{2.20}$$

for the "drifting distance". The family of curves $\mathcal{R}$ describes the types of the trajectories of the particles.

We recall (2.13):

$$\omega = (\rho_1, \vartheta_1, \rho_2, \vartheta_2, \ldots, \rho_N, \vartheta_N) \in \Omega_1 = \left([0, \infty) \times SO(3)\right)^N. \tag{2.21}$$

We may say that ω determines the initial-velocities of the system (this is why we call Ω_1 the initial velocity space). Note that the triple $(\mathcal{Y}; \mathcal{R}; \omega)$, or the quadruple $(\mathcal{Y}, \mathcal{M}; \mathcal{R}; \omega)$ for general mass distribution, is the complete *initial condition* of the particle-system: it determines the trajectory of the whole system in the phase space for all time t.

For a given initial condition $(\mathcal{Y}, \mathcal{M}; \mathcal{R}; \omega)$, let $N(\mathcal{Y}, \mathcal{M}; \mathcal{R}; \omega; S; t)$ denote the mass-counting function in S at time t:

$$\text{MassCounting} = N(\mathcal{Y}, \mathcal{M}; \mathcal{R}; \omega; S; t) = \sum_{\substack{1 \le k \le N: \\ \mathbf{x}_k(t) \in S}} \mu_k, \tag{2.22}$$

where $\mathbf{x}_k(t) = \mathbf{y}_k + \vartheta_k(\rho_k \mathbf{r}_k(t))$ modulo one is the trajectory of the k-th particle; see (2.12).

In the special case $1 = \mu_1 = \mu_2 = \ldots = \mu_N$ we simply write

$$\text{PointCounting} = N(\mathcal{Y}; \mathcal{R}; \omega; S; t) = \sum_{\substack{1 \le k \le N: \\ \mathbf{x}_k(t) \in S}} 1, \tag{2.23}$$

omitting $\mathcal{M}$.

In view of (2.18) uniformity means that $N(\mathcal{Y}, \mathcal{M}; \mathcal{R}; \omega; S; t)$ and $\text{vol}(S) \sum_{k=1}^{N} \mu_k = N \cdot \text{vol}(S)$ are close to each other, so we study the discrepancy

$$N(\mathcal{Y}, \mathcal{M}; \mathcal{R}; \omega; S; t) - N \cdot \text{vol}(S). \tag{2.24}$$

In general, for an arbitrary test function $f \in L^2$, we have the perfect analogues of (2.22)–(2.24). Write

$$N(\mathcal{Y}, \mathcal{M}; \mathcal{R}; \omega; f; t) = \sum_{1 \le k \le N} f(\mathbf{x}_k(t))\mu_k, \tag{2.25}$$

and in the special case $1 = \mu_1 = \mu_2 = \ldots = \mu_N$ we simply write

$$N(\mathcal{Y}; \mathcal{R}; \omega; f; t) = \sum_{1 \le k \le N} f(\mathbf{x}_k(t)), \tag{2.26}$$

omitting $\mathcal{M}$.

We recall (2.1):

$$\sigma_0^2(f) = \int_{I^3} \left| f - \int_{I^3} f \, dV \right|^2 dV = \int_{I^3} |f|^2 \, dV - \left| \int_{I^3} f \, dV \right|^2,$$

noting that we always assume $\sigma_0^2(f) > 0$.

Uniformity means

$$N(\mathcal{Y}, \mathcal{M}; \mathcal{R}; \omega; f; t) = N \int_{I^3} f \, dV \; + \; \text{negligible},$$

where I^3 is the 3-dimensional unit torus ($N = \sum_{k=1}^{N} \mu_k$ is the total mass and V indicates volume). So, we study the discrepancy

$$N(\mathcal{Y}, \mathcal{M}; \mathcal{R}; \omega; f; t) - N \int_{I^3} f \, dV, \tag{2.27}$$

in fact, we study its absolute value

$$\mathbf{D}(\mathcal{Y}, \mathcal{M}; \mathcal{R}; \omega; f; t) = \left| N(\mathcal{Y}, \mathcal{M}; \mathcal{R}; \omega; f; t) - N \int_{I^3} f \, dV \right|. \tag{2.27$'$}$$

Definition 2.3. We study the time evolution of an N-element torus model of the first kind $(\mathcal{Y}, \mathcal{M}; \mathcal{R}; \Omega_1)$. We say that the system is in **L-equilibrium** with parameters $0 < \varepsilon < 1$ and $C > 0$ at time t, if for every test function $f \in L^2$ and for every real number $\tau \geq t$ there exists a subset $\Omega_1(\varepsilon; C; f; \tau) \subset \Omega_1$ of the initial velocity space with measure at least $1 - \varepsilon$ (i.e., $\text{ProdMeas}_1(\Omega_1(\varepsilon; C; f; \tau)) \geq 1 - \varepsilon$) such that, for every $\omega \in \Omega_1(\varepsilon; C; f; \tau)$

$$\mathbf{D}(\mathcal{Y}, \mathcal{M}; \mathcal{R}; \omega; f; \tau) = \left| N(\mathcal{Y}, \mathcal{M}; \mathcal{R}; \omega; f; \tau) - N\int_{I^3} f\, dV \right| \leq C\sigma_0(f)\sqrt{\sum_{k=1}^{N} \mu_k^2}.$$

Integrating the square

$$\mathbf{D}^2(\mathcal{Y}, \mathcal{M}; \mathcal{R}; \omega; f; t) = \left| N(\mathcal{Y}, \mathcal{M}; \mathcal{R}; \omega; f; t) - N\int_{I^3} f\, dV \right|^2 \tag{2.28}$$

over the huge initial velocity space Ω_1, we obtain

$$\mathbf{D}^2_{\Omega_1}(\mathcal{Y}, \mathcal{M}; \mathcal{R}; f; t) = \int_{\Omega_1} \mathbf{D}^2(\mathcal{Y}, \mathcal{M}; \mathcal{R}; \omega; f; t)\, d\omega. \tag{2.29}$$

We refer to $\mathbf{D}^2(\mathcal{Y}, \mathcal{M}; \mathcal{R}; f; t)$ as the *square-discrepancy* (relative to f), and refer to $\mathbf{D}^2_{\Omega_1}(\mathcal{Y}, \mathcal{M}; \mathcal{R}; f; t)$ as the *Big Square-Discrepancy* of the first kind (relative to f). We call it "Big" because it is a big average over the huge initial velocity space Ω_1; it is the average of the square-discrepancy with respect to the given test function $f \in L^2$ at time t (with initial point configuration $\mathcal{Y}$ and mass distribution $\mathcal{M}$).

To obtain the concept of **LA-equilibrium**, we define an extension of (2.28). We use the concept of "translation of f by $\mathbf{w}$ modulo one"

$$f_{\mathbf{w}}(\mathbf{x}) = f(\mathbf{x} - \mathbf{w}) \tag{2.30}$$

(note that f is periodic; by periodicity we can restrict $\mathbf{w} \in \mathsf{R}^3$ to $\mathbf{w} \in I^3$). We study the *average* of (2.28) over all translations of the torus I^3: let

$$\overline{\mathbf{D}^2}(\mathcal{Y}, \mathcal{M}; \mathcal{R}; \omega; f; t) = \int_{I^3} \mathbf{D}^2(\mathcal{Y}, \mathcal{M}; \mathcal{R}; \omega; f_{\mathbf{w}}; t)\, d\mathbf{w}$$

$$= \int_{I^3} \left| N(\mathcal{Y}, \mathcal{M}; \mathcal{R}; \omega; f_{\mathbf{w}}; t) - N\int_{I^3} f\, dV \right|^2 d\mathbf{w}, \tag{2.31}$$

where $f_{\mathbf{w}}$ is defined in (2.30). We call $\overline{\mathbf{D}^2}(\mathcal{Y}, \mathcal{M}; \mathcal{R}; \omega; f; t)$ the *average square-discrepancy* (relative to f).

It is interesting (and useful in later applications) to point out that the average square-discrepancy relative to f

$$\overline{\mathbf{D}^2}(\mathcal{Y}, \mathcal{M}; \mathcal{R}; \omega; f; t)$$

coincides with the square-discrepancy relative to f

$$\mathbf{D}^2(\mathcal{Y}, \mathcal{M}; \mathcal{R}; \omega; f; t)$$

if f is any complex exponential function of the form

$$f(\mathbf{x}) = e^{2\pi i \mathbf{n}\cdot\mathbf{x}}, \quad \mathbf{n} \in \mathbb{Z}^3 \setminus \mathbf{0}.$$

Next we introduce an analog of Definition 2.3.

Definition 2.4. Similarly to Definition 2.3, we study the time evolution of an N-element torus model of the first kind $(\mathcal{Y}, \mathcal{M}; \mathcal{R}; \Omega_1)$. We say that the system is in **LA-equilibrium** with parameters $0 < \varepsilon < 1$ and $C > 0$ at time t, if for every test function $f \in L^2$ and for every real number $\tau \geq t$ there exists a subset $\Omega_1(\varepsilon; C; f; \tau) \subset \Omega_1$ of the initial velocity space with measure at least $1 - \varepsilon$ (i.e., $\text{ProdMeas}_1(\Omega_1(\varepsilon; C; f; \tau)) \geq 1 - \varepsilon$) such that, for every $\omega \in \Omega_1(\varepsilon; C; f; \tau)$

$$\sqrt{\overline{\mathbf{D}^2}(\mathcal{Y}, \mathcal{M}; \mathcal{R}; \omega; f; \tau)} \leq C\sigma_0(f)\sqrt{\sum_{k=1}^{N} \mu_k^2}.$$

Note in advance that both Definitions 2.3 and 2.4 can be easily extended to the torus model of the second kind, the third kind, and for any other initial velocity distribution.

Definitions 2.3 and 2.4 have the common message that once the system reaches equilibrium (at time t, say), it stays in equilibrium forever ("for every time τ with $\tau \geq t$") — well, at least it holds for the *collective* time evolution in some well-defined sense.

Note, however, that the statement "it stays in equilibrium forever" fails(!) for the *individual* time evolution; see Lemma 1.2. For a detailed discussion, see the beginning of Section 25.

Let us return to (2.31): integrating (2.31) over the huge initial velocity space Ω_1, we define

$$\overline{\mathbf{D}^2}_{\Omega_1}(\mathcal{Y}, \mathcal{M}; \mathcal{R}; f; t) = \int_{\Omega_1} \overline{\mathbf{D}^2}(\mathcal{Y}, \mathcal{M}; \mathcal{R}; \omega; f; t)\, d\omega. \tag{2.32}$$

We refer to $\overline{\mathbf{D}^2}_{\Omega_1}(\mathcal{Y}, \mathcal{M}; \mathcal{R}; f; t)$ as the *Big Average Square-Discrepancy* of the first kind (relative to f).

Motivated by physics, so far we have focused on the 3-dimensional case. We emphasize that our definitions clearly work in every dimension $h \geq 1$. Thus we obtain the h-dimensional torus model of the first kind, and the corresponding concepts of LA-equilibrium and L-equilibrium.

Next we switch from the general torus model to the special "from-box-to-torus via unfolding" model defined in (1.19)–(1.27) at the end of Section 1. For simplicity we just call it the "torus-via-unfolding" (or "TorViaUnfold") model. It is a technical nuisance that the "torus-via-unfolding" model is formally *not* a special case of the general torus model. This is why we have to define the corresponding LA-equilibrium and L-equilibrium separately.

Let

$$\mathcal{R} = (\mathbf{r}_1(t), \mathbf{r}_2(t), \ldots, \mathbf{r}_N(t)) \tag{2.33}$$

be a family of N continuous rectifiable parametrized space curves with $\mathbf{r}_k(0) = \mathbf{0}$. The family of curves $\mathcal{R}$ in (2.33) describes the types of the trajectories of the point billiards.

For a given

$$\omega = (\rho_1, \vartheta_1, \rho_2, \vartheta_2, \ldots, \rho_N, \vartheta_N) \in \Omega_1 = ([0, \infty) \times SO(3))^N,$$

we rewrite (1.25)–(1.26) as follows:

$$\mathbf{c}_k(t) = \mathbf{y}_k + \vartheta_k(\rho_k \mathbf{r}_k(t), \ 1 \leq k \leq N. \tag{2.34}$$

Motivated by (1.27), we have the following analog of (2.23) (where $S \subset I^h = [0,1)^h$ is a test set):

$$\text{MassCounting(TorViaUnfold)} = N(\text{TorViaUnfold}; \mathcal{Y}, \mathcal{M}; \mathcal{R}; \omega; S; t)$$

$$= \sum_{1 \leq k \leq N} \sum_{\substack{(\varepsilon_1, \ldots, \varepsilon_h):\ \varepsilon_i = \pm 1 \\ \mathbf{x}_k^*(\varepsilon_1, \ldots, \varepsilon_h; t) \in \frac{1}{2} S}} \mu_k, \tag{2.35}$$

which implies that the discrepancy

$$N(\mathcal{Y}, \mathcal{M}; \mathcal{R}; \omega; S; t) - N\text{vol}(S)$$

is replaced by the analog

$$N(\text{TorViaUnfold}; \mathcal{Y}, \mathcal{M}; \mathcal{R}; \omega; S; t) - N 2^h \text{vol}\left(\frac{1}{2} S\right)$$

$$= N(\text{TorViaUnfold}; \mathcal{Y}, \mathcal{M}; \mathcal{R}; \omega; S; t) - N\text{vol}(S). \tag{2.36}$$

We define easily the "TorViaUnfold analog" of (2.25) in a similar way for every $f \in L^2(I^h)$: let

$$\begin{aligned}\text{MassCounting(TorViaUnfold)} &= N(\text{TorViaUnfold}; \mathcal{Y}, \mathcal{M}; \mathcal{R}; \omega; f; t)\\ &= \sum_{1 \le k \le N} \sum_{\substack{(\varepsilon_1, \ldots, \varepsilon_h):\\ \varepsilon_i = \pm 1 \ (1 \le i \le h)}} f(\mathbf{x}_k^*(\varepsilon_1, \ldots, \varepsilon_h; t))\mu_k. \end{aligned} \tag{2.37}$$

Note that (2.37) gives back (2.35) in the special case where $f \in L^2(I^h)$ is the characteristic function of the "half set" $\frac{1}{2}S$.

Again the discrepancy

$$N(\mathcal{Y}, \mathcal{M}; \mathcal{R}; \omega; f; t) - N \int_{I^3} f \, d\text{V}$$

is replaced by the analog

$$N(\text{TorViaUnfold}; \mathcal{Y}, \mathcal{M}; \mathcal{R}; \omega; f; t) - N2^h \int_{I^h} f \, d\text{V}. \tag{2.38}$$

Then we just repeat (2.27)–(2.32) and Definitions 2.3–2.4. This is how we define the LA-equilibrium and L-equilibrium for torus-via-unfolding model.

The concept of L-equilibrium can be characterized as a "square-root equilibrium", and similarly, LA-equilibrium can be characterized as an "average square-root equilibrium". The main reason why we define "equilibrium" as a "square-root equilibrium" is the *optimum property*: the order of $\sqrt{N}$ discrepancy is the best what one can expect for the class of test functions $f \in L^2$; see the Remarks after Definition 2.1, including Lemma 2.1. A second motivation comes from the CLT-Corollary of the Kronecker–Weyl Theorem (see Lemma 1.1 in Section 1), where the norming factor is const$\sqrt{N}$ (in fact, $\sqrt{pqN}$ with $p = \text{vol}(S)$ and $q = 1 - p$). This implies that the typical long-term fluctuation (i.e., $t \to \infty$) is in the range of $\sqrt{N}$.

Finally, as we promised, we conclude Section 2 with a proof of Lemma 2.1. (A reader in rush may skip it, and jump ahead to Section 3.) We actually prove the following 1-dimensional result.

Lemma 2.2. *Let $N \ge 1$ be an arbitrary integer, and let $r = r(N)$ denote the unique integer satisfying $2N \le 2^r < 4N$. We show that there exist $m = 2^{2^r}$ Lebesgue measurable subsets $S_1, S_2, \ldots, S_m$ of the unit interval with the following property: given any N-element subset $\mathcal{X} = \{x_1, x_2, \ldots, x_N\}$ in the unit interval $[0, 1]$, for at least two-thirds of the m subsets $S_1, S_2, S_3, \ldots, S_m$,*

$$||\mathcal{X} \cap S_j| - N \cdot \text{length}(S_j)| > \frac{1}{20}\sqrt{2^r} \ge \frac{\sqrt{2}}{20}\sqrt{N},$$

where length stands for the 1-dimensional Lebesgue measure.

The deduction of the h-dimensional Lemma 2.1 from the 1-dimensional Lemma 2.2 is trivial. Indeed, for every $S_j \subset [0,1)$, $1 \le j \le m$ guaranteed by Lemma 2.2 we define the h-dimensional set $S'_j \subset [0,1)^h$ as the Cartesian product $S'_j = S_j \times [0,1)^{h-1}$.

Proof of Lemma 2.2. We combine the ideas of *trivial error* and the *probabilistic method.* Let $N \ge 1$ and $2N \le 2^r < 4N$; we divide the unit interval into 2^r congruent subintervals

$$[(i-1)2^{-r}, i2^{-r}), \quad i = 1, 2, 3, \ldots, 2^r. \tag{2.39}$$

Trivial error means that, given any N-element subset $\mathcal{X} = \{x_1, x_2, \ldots, x_N\}$ in the unit interval $[0,1]$, at least half of the subintervals (2.39) do not contain any element of $\mathcal{X}$. We call them *empty* subintervals.

Applying the probabilistic method means that we make a *random construction*: we take the union of a *random subset* of intervals (2.39). More precisely, let $Y_1, Y_2, \ldots, Y_{2^r}$ be a sequence of 2^r independent and identically distributed Bernoulli random variables with $\Pr[Y_i = 0] = \Pr[Y_i = 1] = 1/2$ (one may call it a sequence of Heads and Tails). Independence means that the underlying probability space Ω_0 is a Cartesian product:

$$\Omega_0 = \{\omega = (\delta_1, \ldots, \delta_{2^r}) : \quad \delta_i \in \{0,1\},\ i = 1, 2, \ldots, 2^r\},$$

and $Y_i(\omega)$ is simply the i-th coordinate δ_i of ω. Note that $|\Omega_0| = 2^{2^r}$.

Every $\omega \in \Omega_0$ defines a measurable subset $B(\omega)$ (in fact, a union of some intervals (2.39)) of the unit interval in the following natural way:

$$B(\omega) = \bigcup_{i:\ Y_i(\omega)=1} [(i-1)2^{-r}, i2^{-r}). \tag{2.40}$$

Now let $\mathcal{X} = \{x_1, x_2, \ldots, x_N\} \subset [0,1]$ be arbitrary. Since $2N \le 2^r < 4N$, at least half of the subintervals (2.39) are empty ("trivial error"): let $I = I(\mathcal{X})$ denote the set of indices i such that the interval $[(i-1)2^{-r}, i2^{-r})$ contains no element of $\mathcal{X}$. Clearly $|I(\mathcal{X})| \ge 2^r - N \ge 2^{r-1} \ge N$.

Consider the sum of independent (and identically distributed) random variables

$$\sum_{i \in I(\mathcal{X})} Y_i$$

with expectation (denoted by $\mathbf{E}$) and variance

$$\mathbf{E}\left(\sum_{i \in I(\mathcal{X})} Y_i\right) = \frac{1}{2}|I(\mathcal{X})| \quad \text{and} \quad \text{Variance}\left(\sum_{i \in I(\mathcal{X})} Y_i\right) = \frac{1}{4}|I(\mathcal{X})|.$$

The central limit theorem says that, for any fixed real number $\lambda > 0$,

$$\Pr\left[\sum_{i\in I(\mathcal{X})} Y_i - \frac{1}{2}|I(\mathcal{X})| > \frac{\lambda}{2}\sqrt{|I(\mathcal{X})|}\right] = \frac{1}{\sqrt{2\pi}}\int_\lambda^\infty e^{-u^2/2}\,du + O(N^{-1/2}), \tag{2.41}$$

where "Pr" means equiprobability in the discrete probability space Ω_0.

For any (measurable) subset $S \subset [0,1]$, consider the "error"

$$\text{error}(\mathcal{X}, S) = |\mathcal{X} \cap S| - |\mathcal{X}|\text{length}(S).$$

Let

$$J(\mathcal{X}) = \{1, 2, \ldots, 2^r\} \setminus I(\mathcal{X})$$

$$= \{i : \ [(i-1)2^{-r}, i2^{-r}) \text{ does contain an element of } \mathcal{X}\},$$

and $|J(\mathcal{X})| = \ell$. Then there are 2^ℓ possible vectors of the form

$$\mathbf{v} = (Y_j(\omega) : \ j \in J(\mathcal{X})) \in \{0,1\}^\ell,$$

where the elements $j \in J(\mathcal{X})$ are written in increasing order.

Let $\mathbf{v} \in \{0,1\}^\ell$ be arbitrary but fixed, and consider the subset

$$\Omega_{\mathbf{v}} = \Omega_{\mathbf{v}}(\mathcal{X}) = \{\omega \in \Omega_0 : \ (Y_j(\omega) : \ j \in J(\mathcal{X})) = \mathbf{v}\}$$

of the probability space Ω_0. We study the fluctuation of the "error"

$$\text{error}(\mathcal{X}, S) = |\mathcal{X} \cap S| - |\mathcal{X}|\text{length}(S)$$

as S runs through the sets $B(\omega)$, $\omega \in \Omega_{\mathbf{v}}$.

By definition, the cardinality $|\mathcal{X} \cap B(\omega)|$ is the *same* for every $\omega \in \Omega_{\mathbf{v}}(\mathcal{X})$, and

$$|\mathcal{X}|\text{length}(B(\omega)) = \frac{N}{2^r}\sum_{i\in I(\mathcal{X})} Y_i + \frac{N}{2^r}(\text{sum of the coordinates of } \mathbf{v})$$

for every $\omega \in \Omega_{\mathbf{v}}(\mathcal{X})$. Thus for every $\omega \in \Omega_{\mathbf{v}}(\mathcal{X})$,

$$\text{error}(\mathcal{X}, \omega) = \text{error}(\mathcal{X}, B(\omega)) = |\mathcal{X} \cap B(\omega)| - |\mathcal{X}|\text{length}(B(\omega))$$

$$= C(\mathcal{X}, \mathbf{v}) - \frac{N}{2^r}\left(\sum_{i\in I(\mathcal{X})} Y_i - \frac{1}{2}|I(\mathcal{X})|\right), \tag{2.42}$$

where $C(\mathcal{X}, \mathbf{v})$ is a constant depending only on $\mathcal{X}$ and $\mathbf{v}$.

Applying the central limit theorem (2.41) in (2.42), we obtain that, independently of the value of the constant $C(\mathcal{X}, \mathbf{v})$ in (2.42), we can estimate the following conditional probability:

$$\Pr\left[|\text{error}(\mathcal{X},\omega)| \le \frac{1}{20}\sqrt{2^r} \mid \omega \in \Omega_{\mathbf{v}}(\mathcal{X})\right]$$

$$\le \Pr\left[|\text{error}(\mathcal{X},\omega)| \le \frac{\sqrt{2}}{20}\sqrt{|I(\mathcal{X})|} \mid \omega \in \Omega_{\mathbf{v}}(\mathcal{X})\right] \le \frac{1}{3}, \tag{2.43}$$

where we used the lower bound bound $|I(\mathcal{X})| \ge 2^{r-1}$ and $\Pr[E|F]$ denotes the probability of event E under the condition F. Since (2.43) is independent of the choice of the vector $\mathbf{v}$, we can switch to the unconditional probability, and obtain

$$\Pr\left[|\text{error}(\mathcal{X},\omega)| \le \frac{1}{20}\sqrt{2^r}\right] \le \frac{1}{3}. \tag{2.44}$$

Since $2N \le 2^r < 4N$, by (2.44), for at least two-thirds of the 2^{2^r} subsets $B(\omega) \subset [0,1]$,

$$||\mathcal{X} \cap B(\omega)| - N\text{length}(B(\omega))| > \frac{1}{20}\sqrt{2^r} \ge \frac{\sqrt{2}}{20}\sqrt{N},$$

completing the proof of Lemma 2.2. □

Lemma 2.2 represents the special case of equal masses $\mu_1 = \cdots = \mu_N = 1$. It is easy to see that the proof of Lemma 2.2 extends to the general case where different particles may have different masses. The message remains the same: square-root equilibrium is *optimal*.

Finally note that Definitions 2.1 and 2.2 define *particle equilibrium*, and similarly, Definitions 2.3 and 2.4 define *mass equilibrium*. Of course, we can also define *energy equilibrium* along the same lines. It formally means to modify the meaning of μ_i: let $\mu_i = m_i v_i^2/2$ $(1 \le i \le N)$, where m_i and v_i are the mass and the speed of the i-th particle.

For simplicity in Sections 3–23 we focus on *mass equilibrium* (and of course *particle equilibrium* shows up as an interesting special case). In Section 24 we explain what kind of minor changes are needed to cover the case of *energy equilibrium*.

3. First surprise: Shockingly fast approach to micro-equilibrium in the Gaussian case

What is in Section 3? *We recall the following part from the one-sentence summary of the book:* **(1) first we show that in our implicit interaction models the typical time evolution of the large closed system approaches a well-defined microscopic equilibrium.** *Our first theorem, Theorem 3.1 below, implies extremely fast — in fact, superexponentially fast — approach to square-root equilibrium (= micro-equilibrium). In other words, the Relaxation Time is extremely short.*

Our ultimate goal is to formulate and prove a time-quantitative Second Law. It simply means that the time is explicitly involved in the statement. Theorem 3.1 is the first step in this direction.

The formal statement of Theorem 3.1 is quite long, but it has a simple meaning to be explained below.

Theorem 3.1. *Consider a general 3-dimensional N-particle torus model with Gaussian initial velocity distribution $(\mathcal{Y}, \mathcal{M}; \mathcal{R}; \Omega_1)$, i.e., the torus model of the first kind. Write*

$$\mathcal{R}(t) = \min_{1 \le k \le N} |\mathbf{r}_k(t)|$$

for the drifting distance. So, the trajectory of the k-th particle is

$$\mathbf{x}_k(t) = \mathbf{y}_k + \vartheta_k(\rho_k \mathbf{r}_k(t)) \quad \text{modulo one},$$

where

$$\omega = (\rho_1, \vartheta_1, \rho_2, \vartheta_2, \ldots, \rho_N, \vartheta_N) \in \Omega_1 = \left([0, \infty) \times SO(3)\right)^N$$

is the initial velocity distribution (it describes the time evolution of the system; $\vartheta_k \in SO(3)$, and $0 < \rho_k < \infty$ has the Gaussian distribution, $1 \le k \le N$).

Let $f \in L^2$ be an arbitrary complex valued square-integrable function defined on the torus I^3; f is our test function to measure uniformity. Then, starting at $t = 0$, for every $t > 0$ we have the following inequality for the Big Average Square-Discrepancy (see (2.31)–(2.32)):

$$\left| \overline{\mathbf{D}^2}_{\Omega_1}(\mathcal{Y}, \mathcal{M}; \mathcal{R}; f; t) - \sigma_0^2(f) \sum_{k=1}^{N} \mu_k^2 \right| \le e^{-(2\pi \mathcal{R}(t))^2} \sigma_0^2(f) N^2. \tag{3.1}$$

The inequality for the Big Square-Discrepancy (see (2.28)–(2.29)) is more complicated:

$$\left| \mathbf{D}^2_{\Omega_1}(\mathcal{Y}, \mathcal{M}; \mathcal{R}; f; t) - \sigma_0^2(f) \sum_{k=1}^{N} \mu_k^2 \right| \le e^{-(2\pi \mathcal{R}(t))^2} \sigma_0^2(f) N^2$$

$$+\sigma_0^2(f)N^2\frac{2\exp\left(-4(\pi\mathcal{R}(t))^2\right)}{1-\exp\left(-2(\pi\mathcal{R}(t))^2\right)}\left(1+\frac{2\exp\left(-2(\pi\mathcal{R}(t))^2\right)}{1-\exp\left(-2(\pi\mathcal{R}(t))^2\right)}\right)^2$$

$$+\sigma_0^2(f)\left(\sum_{1\le k\le N}\mu_k^2\right)\frac{2\exp\left(-2(\pi\mathcal{R}(t))^2\right)}{1-\exp\left(-2(\pi\mathcal{R}(t))^2\right)}\left(1+\frac{2\exp\left(-2(\pi\mathcal{R}(t))^2\right)}{1-\exp\left(-2(\pi\mathcal{R}(t))^2\right)}\right)^2, \tag{3.2}$$

where, as usual, exp *stands for the exponential function.*

Next consider the special case $\mathbf{r}_1(t)=\mathbf{r}_2(t)=\ldots=\mathbf{r}_N(t)=\mathbf{r}(t)$. *Then* $\mathcal{R}(t)=\mathbf{r}(t)$, *and writing* $\mathcal{R}=\mathbf{r}$, *we have*

$$\left|\overline{\mathbf{D}^2}_{\Omega_1}(\mathcal{Y},\mathcal{M};\mathbf{r};f;t)-\sigma_0^2(f)\sum_{k=1}^{N}\mu_k^2\right|$$

$$\le e^{-(2\pi|\mathbf{r}(t)|)^2}\max\left\{\overline{\mathbf{D}^2}_{\Omega_1}(\mathcal{Y},\mathcal{M};\mathbf{r};f;0),\sigma_0^2(f)\sum_{k=1}^{N}\mu_k^2\right\}. \tag{3.3}$$

In this special case we have the explicit formula

$$\overline{\mathbf{D}^2}_{\Omega_1}(\mathcal{Y},\mathcal{M};\mathbf{r};f;t)=\sum_{k=1}^{N}\mu_k^2\sum_{\substack{\mathbf{n}\in\mathbb{Z}^3:\\ \mathbf{n}\neq\mathbf{0}}}|a_{\mathbf{n}}|^2\left(1-\exp\left(-(2\pi|\mathbf{r}(t)|\,|\mathbf{n}|)^2\right)\right)$$

$$+\sum_{\substack{\mathbf{n}\in\mathbb{Z}^3:\\ \mathbf{n}\neq\mathbf{0}}}|a_{\mathbf{n}}|^2\left|\sum_{k=1}^{N}\mu_k e^{2\pi i\mathbf{n}\cdot\mathbf{y}_k}\right|^2\exp\left(-(2\pi|\mathbf{r}(t)|\,|\mathbf{n}|)^2\right), \tag{3.4}$$

where $a_{\mathbf{n}}$, $\mathbf{n}\in\mathbb{Z}^3$ *are the Fourier coefficients of the Lebesgue square-integrable test function* f.

Finally, the results above on the 3-dimensional unit torus $I^3=[0,1)^3$ *have a straightforward generalization for the d-dimensional unit torus* $I^d=[0,1)^d$ *in every dimension* $d\ge 1$ *(of course* $\mathbf{n}\in\mathbb{Z}^3$ *is replaced by* $\mathbf{n}\in\mathbb{Z}^d$*).*

For the formal definition of the Gaussian speed distribution in arbitrary dimension $d\neq 3$; see the end of this section.

The proof of Theorem 3.1 is based on Fourier analysis. We postpone it to Section 7. Nevertheless, as a warm-up, we formulate a key lemma (Lemma 3.1) at the very end of Section 3, including the short Fourier analysis proof. This lemma will be repeatedly used: first in the proof of Theorem 3.1, and later in the proofs of Theorem 4.1 and Theorem 4.2.

To explain the simple meaning of the complicated Theorem 3.1, we rewrite (3.4) in the alternative form (note that it works in every dimension $d \geq 1$)

$$\overline{\mathbf{D}^2}_{\Omega_1}(\mathcal{Y}, \mathcal{M}; \mathbf{r}; f; t) = \mathrm{Eqlbrm}(\mathcal{M}; f)$$

$$+ \overline{\mathbf{D}^2}_{1,negl}(\mathcal{Y}, \mathcal{M}; \mathbf{r}; f; t) + \overline{\mathbf{D}^2}_{1,uneven}(\mathcal{Y}, \mathcal{M}; \mathbf{r}; f; t), \tag{3.5}$$

where the "equilibrium part", the "negligible part" and the part of "unevenness" are defined as follows. We have

$$\mathrm{Eqlbrm}(\mathcal{M}; f) = \sigma_0^2(f) \sum_{k=1}^{N} \mu_k^2 = \left(\sum_{\substack{\mathbf{n} \in \mathbb{Z}^d: \\ \mathbf{n} \neq \mathbf{0}}} |a_{\mathbf{n}}|^2 \right) \left(\sum_{k=1}^{N} \mu_k^2 \right) \tag{3.6}$$

is called the *equilibrium part*, because it is independent of both the time and the distribution of the point set $\mathcal{Y}$. Notice that at the end of (3.6) we used Parseval's formula. Moreover,

$$\overline{\mathbf{D}^2}_{1,negl}(\mathcal{Y}, \mathcal{M}; \mathbf{r}; f; t) = - \sum_{k=1}^{N} \mu_k^2 \sum_{\substack{\mathbf{n} \in \mathbb{Z}^d: \\ \mathbf{n} \neq \mathbf{0}}} |a_{\mathbf{n}}|^2 \exp\left(-(2\pi |\mathbf{r}(t)|\ |\mathbf{n}|)^2\right) \tag{3.7}$$

is called the *negligible part* of the first kind, because it tends to zero as $|\mathbf{r}(t)| \to \infty$, and also we have the inequality

$$0 \leq \mathrm{Eqlbrm}(\mathcal{M}; f) + \overline{\mathbf{D}^2}_{1,negl}(\mathcal{Y}, \mathcal{M}; \mathbf{r}; f; t) \leq \mathrm{Eqlbrm}(\mathcal{M}; f) = \sigma_0^2(f) \sum_{k=1}^{N} \mu_k^2. \tag{3.8}$$

Finally, we have the part describing how *uneven* the configuration is at time t; we call it the *unevenness* of the first kind:

$$\overline{\mathbf{D}^2}_{1,uneven}(\mathcal{Y}, \mathcal{M}; \mathbf{r}; f; t) = \sum_{\substack{\mathbf{n} \in \mathbb{Z}^d: \\ \mathbf{n} \neq \mathbf{0}}} |a_{\mathbf{n}}|^2 \left| \sum_{k=1}^{N} \mu_k e^{2\pi i \mathbf{n} \cdot \mathbf{y}_k} \right|^2 \exp\left(-(2\pi |\mathbf{r}(t)|\ |\mathbf{n}|)^2\right). \tag{3.9}$$

Note that (3.9) represents the dominating part in (3.4) as long as the system stays in off-equilibrium. The choice of the word *uneven* is motivated by our main goal: we want to prove that, in a typical time evolution, a large system becomes less and less "uneven" (Second Law).

Note that in the special case $f = \chi_S$ we have $\sigma_0^2(f) = pq$ with $p = \mathrm{vol}(S)$ and $q = 1 - p$, which is the variance of a single particle (with respect to the given test set S) in the standard *probabilistic model* of equilibrium thermodynamics in Statistical Mechanics. Indeed, then the particles are represented by independent random variables uniformly distributed in the unit cube, the probability that the k-th particle is in the given test set S is $\text{vol}(S) = p$, and

$$\text{Variance}(Z_1 + \ldots + Z_N) = Npq,$$

where $Z_1, \ldots, Z_N$ are N independent 0-1 valued random variables with common distribution

$$\Pr[Z_k = 1] = p \quad \text{and} \quad \Pr[Z_k = 0] = q, \quad 1 \le k \le N.$$

We refer to Npq as the *spatial equilibrium variance* (with respect to the given test set S).

Remark. We just briefly point out the possibility of an alternative way to define equilibrium. Let $(S_1, S_2, \ldots, S_\ell)$ be an arbitrary partition of the box $I^3 = [0,1)^3$; we assume that S_i are all measurable. By (3.6) and $p_i = \text{vol}(S_i)$, we have

$$\sum_{i=1}^{\ell} \text{Eqlbrm}(\mathcal{M}; \chi_{S_i}) = \left(\sum_{k=1}^{N} \mu_k^2\right) \sum_{i=1}^{\ell} p_i(1 - p_i)$$

$$\le \left(\sum_{k=1}^{N} \mu_k^2\right) \sum_{i=1}^{\ell} p_i = \sum_{k=1}^{N} \mu_k^2,$$

that is, the upper bound is independent of the partition of the box $I^3 = [0,1)^3$. This simple observation suggests the possibility of a slightly different alternative way to define equilibrium: we could take maximum over all measurable partitions of the box instead of just taking maximum over all measurable test sets. We stop here; we leave it to the reader to work out the details of this idea (if he/she finds it interesting).

What we are really interested in is the time evolution of an individual system. Given an arbitrary but fixed initial point configuration $\mathcal{Y}$, mass distribution $\mathcal{M}$, family of trajectory types $\mathcal{R}$, and test function $f \in L^2$, we study

$$\mathbf{D}_f^2(\omega; t) = \mathbf{D}^2(\mathcal{Y}, \mathcal{M}; \mathcal{R}; \omega; f; t) \quad \text{and} \quad \overline{\mathbf{D}^2}_f(\omega; t) = \overline{\mathbf{D}^2}(\mathcal{Y}, \mathcal{M}; \mathcal{R}; \omega; f; t),$$

as t runs in $0 < t < \infty$, for a typical initial velocity distribution $\omega \in \Omega_1$.

The main message of Theorem 3.1 is that, as the time t increases, both the Big Square-Discrepancy and the Big Average Square-Discrepancy of the first kind

$$\mathbf{D}_f^2(1;t) = \mathbf{D}^2_{\Omega_1}(\mathcal{Y},\mathcal{M};\mathcal{R};f;t) \quad \text{and} \quad \overline{\mathbf{D}^2}_f(1;t) = \overline{\mathbf{D}^2}_{\Omega_1}(\mathcal{Y},\mathcal{M};\mathcal{R};f;t)$$

converge to the equilibrium part (= spatial equilibrium variance)

$$\sigma_0^2(f)\sum_{k=1}^{N}\mu_k^2 = \mathrm{Eqlbrm}(\mathcal{M};f)$$

superexponentially fast in terms of the drifting distance $\mathcal{R}(t)$.

Assume the system starts from off-equilibrium; as the time t increases, the dominating part of

$$\overline{\mathbf{D}^2}_f(1;t) = \overline{\mathbf{D}^2}_{\Omega_1}(\mathcal{Y},\mathcal{M};\mathbf{r};f;t)$$

turns out to be the "unevenness"

$$\overline{\mathbf{D}^2}_{1,uneven}(\mathcal{Y},\mathcal{M};\mathbf{r};f;t)$$

(here for simplicity we consider the special case $\mathcal{R}=\mathbf{r}$). We emphasize that "unevenness" is a **strictly decreasing** function of the drifting distance $|\mathbf{r}(t)|$. This fact immediately follows from (3.9): $\overline{\mathbf{D}^2}_{1,uneven}(\mathcal{Y},\mathcal{M};\mathbf{r};f;t)$ equals the sum

$$\sum_{\substack{\mathbf{n}\in\mathbb{Z}^d:\\ \mathbf{n}\neq\mathbf{0}}} |a_{\mathbf{n}}|^2 \left|\sum_{k=1}^{N}\mu_k e^{2\pi i\mathbf{n}\cdot\mathbf{y}_k}\right|^2 \exp\left(-(2\pi|\mathbf{r}(t)|\,|\mathbf{n}|)^2\right),$$

and every term in this sum is decreasing as $|\mathbf{r}(t)|$ is increasing.

Notice that "unevenness" decreases superexponentially fast to zero in terms of the drifting distance. So, "unevenness" loses its dominant role when it goes below the equilibrium part $\mathrm{Eqlbrm}(\mathcal{M};f)$ (which is time-independent). In fact, "unevenness" loses its dominant role exactly when the system reaches LA-equilibrium (see Definition 2.4).

Next we give some time-quantitative illustrations. First note that we can measure time in seconds or in years, and of course these different choices of units lead to very different answers. Also, the speed of the particles and the size of the box make an obvious difference in describing the approach to equilibrium. What we really want is an absolute number. This is what we do: for a large point billiard system in a box we simply count the *average number of bounces* per particle to reach (spatial) equilibrium.

Unfolding converts a billiard path into a torus line, and for a torus line the *number of jumps* has a similar clear-cut meaning. It means the number of times an average particle hits the walls of the imaginary boundary of the torus, and jumps to the opposite wall, until the system reaches (spatial) equilibrium (we visualize the torus as a box with permeable boundary).

It is easy to see that in a point billiard system *unfolding* eliminates half of the bounces, so the number of jumps in the derived torus model is half of the number of bounces.

For simplicity we refer to

$$\inf_{\tau \geq t} \mathcal{R}(\tau)$$

as the *necessary number of jumps*, or the *relaxation distance* (it is in fact a lower bound on the actual number of necessary jumps).

Focusing on the *necessary number of jumps* means that we prefer to replace the relaxation time with the analog concept of relaxation distance.

(Note that the dimension makes some minor difference here. For example, in the 3-dimensional case it would be preferable to use $\sqrt{8/\pi}\mathcal{R}(t)$ instead of $\mathcal{R}(t)$, due to the fact that the Gaussian speed distribution with density function $\sqrt{2/\pi} y^2 e^{-y^2/2}$ has mean speed $\sqrt{8/\pi}$. However, the accidental constant factor $\sqrt{8/\pi}$ is close to 1, so for simplicity we just ignore it.)

To reach (spatial) equilibrium, we clearly need at least one jump

$$\mathcal{R}(t) \geq 1. \tag{3.10}$$

Indeed, the diameter of the unit cube $[0,1]^3$ is $\sqrt{3}$, and, if the billiard system starts from a Big Bang in a corner (or starts from close to Big Bang), the particles need to reach the diagonally opposite corner of the box, which is a trivial necessary condition for spatial equilibrium. (For the unit torus $[0,1)^3$ the maximum torus-distance is attained between $(0,0,0)$ and $(1/2,1/2,1/2)$; it equals $\sqrt{3}/2$.)

For simplicity assume that the particles have unit mass: $\mu_k = 1$ for all $1 \leq k \leq N$. Then by (3.1),

$$\left| \overline{\mathbf{D}^2}_{\Omega_1}(\mathcal{Y}; \mathcal{R}; f; t) - N\sigma_0^2(f) \right| \leq e^{-4\pi^2 \mathcal{R}^2(t)} N^2 \sigma_0^2(f), \tag{3.11}$$

which implies that

$$\left| \overline{\mathbf{D}^2}_{\Omega_1}(\mathcal{Y}; \mathcal{R}; f; t) - N\sigma_0^2(f) \right| \leq N\sigma_0^2(f) \tag{3.12}$$

if

$$e^{4\pi^2 \mathcal{R}^2(t)} \geq N \iff \mathcal{R}(t) \geq \frac{\sqrt{\log N}}{2\pi}. \tag{3.13}$$

Therefore, if (3.13) holds, by (3.12) we have

$$\overline{\mathbf{D}^2}_{\Omega_1}(\mathcal{Y}; \mathcal{R}; f; t) = \int_{\Omega_1} \mathbf{D}^2(\mathcal{Y}; \mathcal{R}; \omega; f; t)\, d\omega \leq 2N\sigma_0^2(f). \tag{3.14}$$

Applying the (trivial) Markov inequality in (3.14), we obtain the following result.

Corollary 3.1. *Consider a general N-particle torus model of the first kind $(\mathcal{Y};\mathcal{R};\Omega_1)$ in the 3-dimensional unit torus $I^3=[0,1)^3$ such that every particle has mass one. Let t be a time satisfying*

$$\inf_{\tau\geq t}\mathcal{R}(\tau)\geq\frac{\sqrt{\log N}}{2\pi}.$$

Then, for every $0<\varepsilon<1$, the system $(\mathcal{Y};\mathcal{R};\Omega_1)$ is in LA-equilibrium with parameters ε and $C=2/\varepsilon$ at time t.

Switching from the 3-dimensional unit torus I^3 to the general d-dimensional unit torus I^d $(d\geq 1)$ the same result holds. □

In the range of the Avogadro number $N\approx 10^{24}$ we have

$$2>\frac{\sqrt{\log N}}{2\pi},$$

so by (3.13)

$$\inf_{\tau\geq t}\mathcal{R}(\tau)\geq 2$$

suffices to reach LA-equilibrium independently of the initial point configuration. In other words, for a typical time evolution **two jumps suffice** to reach LA-equilibrium even in the worst case scenario (four bounces in the billiard model).

Note that the LA-equilibrium here can be easily upgraded to L-equilibrium. Indeed, we just replace (3.1) with (3.2), and a slightly more complicated calculation gives the perfect analog of (3.14): if (3.13) holds, then

$$\int_{\Omega_1}\mathbf{D}^2(\mathcal{Y};\mathcal{R};\omega;f;t)\,d\omega\leq 2N\sigma_0^2(f). \tag{3.15}$$

Applying the trivial Markov inequality in (3.15), we obtain the following result.

Corollary 3.2. *Consider a general N-particle torus model of the first kind $(\mathcal{Y};\mathcal{R};\Omega_1)$ in the 3-dimensional unit torus $I^3=[0,1)^3$ such that every particle has mass one. Let t be a time satisfying*

$$\inf_{\tau\geq t}\mathcal{R}(\tau)\geq\frac{\sqrt{\log N}}{2\pi}.$$

Then, for every $0<\varepsilon<1$, the system $(\mathcal{Y};\mathcal{R};\Omega_1)$ is in L-equilibrium with parameters ε and $C=2/\varepsilon$ at time t.

The same holds in any dimension $d\geq 1$. □

So, for a typical time evolution **two jumps suffice** to reach L-equilibrium even in the worst case scenario (four bounces in the billiard model).

Two jumps is a shockingly small number (since starting from close to Big Bang, we need at least one jump to go from one end to the other end of the torus).

What happens if we increase the number N of particles? Physicists predict that the number of atoms in the observable universe is less than $N = 10^{100}$. Writing $N = 10^{100}$ in (3.13), the same argument shows that for a typical time evolution **four jumps suffice** to reach L-equilibrium, independently of the initial point configuration.

These astonishingly small numbers — 2 jumps for $N = 10^{24}$ and 4 jumps for $N = 10^{100}$ — are consequences of the fact that $\sqrt{\log N}$ in (3.13) is a very slowly increasing function.

These results prove extremely fast approach to square-root equilibrium for the majority of time evolutions, both forward and backward! On the other hand, we do not know *any* explicit set of initial velocities for which the approach to square-root equilibrium happens in realistic short time.

Note that "both forward and backward" may seem like a paradox at first sight — a fact that seemingly contradicts the Second Law. It is basically the well known Loschmidt paradox. For a resolution of this paradox; see Paradoxes in Section 17.

It is quite surprising that the "ugliness" (= "complexity") of the test function f is entirely irrelevant. The only relevant parameter is the integral $\sigma_0^2(f)$, and in the integral it does not make any difference how complicated f is.

As we mentioned in Section 2, we can imitate particle-particle collision by working with random-walk-like chaotic zig-zag trajectories $\mathbf{r}_k(t)$, $1 \leq k \leq N$. Of course a random-walk like trajectory is totally different from a straight line. It is "slow": its expected distance from the start is the *square-root* of the time, which is much slower than the constant speed motion along a straight line (for which the distance from the start is a linear function of the time). It follows that, for random walk like trajectories, the requirement

$$\inf_{\tau \geq t} \mathcal{R}(\tau) \geq \frac{\sqrt{\log N}}{2\pi}$$

in Corollary 3.2 actually means logarithmicly small time (by squaring $\sqrt{\log N}$). That is, the case of the Gaussian initial velocity distribution exhibits a very fast — exponentially fast — approach to equilibrium even when the trajectories are slowly drifting random-walk-like chaotic zig-zags.

We can say, as a rule, that the complexity of the parametrized curves of the particles is irrelevant; what matters is the "drifting distance".

Since the Gaussian initial velocity distribution exhibits a very fast approach to equilibrium, it is natural to ask: Is it true that the Gaussian case represents the *fastest* possible approach to equilibrium? The answer is *no*: there exist "super-duper fast" speed distributions (note that we always assume isotropic velocity distributions with mean speed one) which exhibit *instant equilibrium*. We give a detailed discussion at the end of Section 5.

The results in the next section demonstrate a dramatic change in the number of necessary jumps to reach equilibrium. If we replace the 3-dimensional *Bernoulli model of the first kind* with the *Bernoulli model of the second kind* or the *Bernoulli model of the third kind*, then, starting from extreme off-equilibrium (e.g., Big Bang, or where the particles are all in the right half of the box), the number of necessary jumps switches from the order $\sqrt{\log N}$ in (3.13) to the much larger $\sqrt{N}$ or $N^{1/4}$, respectively (see later). Since N is very large (say $N = 10^{24}$), $\sqrt{N}$ and $N^{1/4}$ are still large. It means that, starting from off-equilibrium, the 3-dimensional Bernoulli model of the second — or third — kind exhibits a much slower approach to equilibrium.

The 2-dimensional case is even slower. The number of necessary jumps switches from the order $\sqrt{\log N}$ to N or $N^{1/3}$, respectively.

In the 4-dimensional case the number of necessary jumps switches from the order $\sqrt{\log N}$ to $N^{1/3}$ or $N^{1/5}$, respectively.

In the d-dimensional case the number of jumps switches from the order $\sqrt{\log N}$ to $N^{1/(d-1)}$ or $N^{1/(d+1)}$, respectively (of course here we assume that N is much larger than the dimension d).

Notice that $N^{1/(d-1)}$ becomes formally meaningless for the one-dimensional Bernoulli model of the second kind. This is perfectly natural, since the 1-dimensional Bernoulli model of the second kind is trivial: it is periodic with period 2, which excludes any kind of approach to equilibrium.

Here is an almost trivial *elementary* lower bound for the speed of reaching L-equilibrium. Suppose that we modify the speed density function (1.9′) to

$$\Pr[\rho_j \le u] = \sqrt{\frac{2}{\pi}} \int_0^u e^{-y^2/2}\, dy \quad \text{for all positive } u \tag{3.16}$$

by dropping the factor y^2; we refer to this as the torus model of the *fourth kind* (see Section 18). Because we dropped the quadratic factor, the new density function $g(y) = \sqrt{2/\pi} e^{-y^2/2}$ is nonzero at zero: $g(0) = \sqrt{2/\pi}$. It means that for typical velocities (in the sense of (3.16)) too many particles move "very slowly", so it takes a long time to reach L-equilibrium. Indeed, consider the simplest case where

$$\mathbf{r}_1(t) = \ldots = \mathbf{r}_N(t) = (t, 0, 0),$$

and the initial point configuration is a Big Bang: the system starts from a single point $\mathbf{c}$. Consider the ball B of radius $N^{-1/6}$ centered at $\mathbf{c}$ (we are in the 3-dimensional unit torus). Choosing typical velocities according to (3.16), there are more than $100\sqrt{N}$ particles with speed less than $200N^{-1/2}$. So, these slow particles all stay in B for a long time interval

$$0 \le t \le T = \frac{N^{-1/6}}{200N^{-1/2}} = N^{1/3}/200.$$

Since the volume of B is $4\pi N^{-1/2}/3$, so at time $t = N^{1/3}/200$ the ball B contains more than $100\sqrt{N}$ particles, and the expected number is $N \cdot 4\pi N^{-1/2}/3$, which is around $5\sqrt{N}$. The discrepancy between $100\sqrt{N}$ and $5\sqrt{N}$ shows that the system is still not in L-equilibrium. In other words, to reach L-equilibrium even for such a simple test set as a ball, in the case of the simplest fourth kind (see (3.16)), the typical time evolution needs more time than $t = N^{1/3}/200$. (If N is very large than of course $N^{1/3}/200$ is still large.)

This argument works in the general isotropic case with any speed density function $g(y) \geq 0$ for $0 \leq y < \infty$ and $\int_0^\infty g(y)\,dy = 1$ such that $g(0) > 0$ and $g(y)$ is continuous at 0^+.

This is the only elementary lower bound argument that we know. The rest of our lower bound results are nonelementary: they are all based on nontrivial Fourier analysis.

The next result is about the time evolution of the Average Square-Discrepancy $\overline{\mathbf{D}^2}(\mathcal{Y}; \mathbf{r}; \omega; f; t)$ for a typical $\omega \in \Omega_1$ when the system is already in LA-equilibrium. The choice (say)

$$|\mathbf{r}(t)| \geq \sqrt{\log N} \tag{3.17}$$

guarantees that Corollary 3.1 applies, meaning that for a typical $\omega \in \Omega_1$ the system is in LA-equilibrium:

$$\overline{\mathbf{D}^2}(\mathcal{Y}; \mathbf{r}; \omega; f; t) \text{ is in the range of constant times } N\sigma_0^2(f) = \text{Eqlbrm}(f) \tag{3.18}$$

for the overwhelming majority of $\omega \in \Omega_1$.

Combining (3.11) with (3.17) we obtain that the "expectation" is practically equal to the equilibrium part $N\sigma_0^2(f)$:

$$\overline{\mathbf{D}^2}_{\Omega_1}(\mathcal{Y}; \mathbf{r}; f; t) = \int_{\Omega_1} \overline{\mathbf{D}^2}(\mathcal{Y}; \mathbf{r}; \omega; f; t)\,d\omega = N\sigma_0^2(f) + O(1). \tag{3.19}$$

The negligible additive constant error $O(1)$ makes (4.35) very sharp. So it is natural to ask whether or not (3.18) can be upgraded to

$$\overline{\mathbf{D}^2}(\mathcal{Y};\mathbf{r};\omega;f;t) = (1+o(1))N\sigma_0^2(f) \tag{3.20}$$

for the overwhelming majority of $\omega \in \Omega_1$, where $o(1) \to 0$ as $t \to \infty$.

The following result shows that the answer is *no*.

Theorem 3.2. *Consider the general torus model of the first kind in the special case $\mathcal{R} = \mathbf{r}$; assume that every mass is one: $1 = \mu_1 = \ldots = \mu_N$; and also assume that the test function f is the complex exponential function $f(\mathbf{x}) = e^{2\pi i \mathbf{n}\cdot\mathbf{x}}$ for some fixed lattice point $\mathbf{n} \in \mathbb{Z}^d \setminus \mathbf{0}$. If*

$$|\mathbf{r}(t)| \geq \sqrt{\log N}, \tag{3.21}$$

then

$$\text{ProdMeas}_1\left\{\omega \in \Omega_1 : \left|\overline{\mathbf{D}^2}(\mathcal{Y};\mathbf{r};\omega;f;t) - \overline{\mathbf{D}^2}_{\Omega_1}(\mathcal{Y};\mathbf{r};f;t)\right| \geq \frac{N}{2} + O(1)\right\}$$

$$\geq \frac{3}{32}\cdot\frac{1}{1+\delta_1} + \delta_2, \tag{3.22}$$

where

$$\delta_1 = \frac{6N^2+O(1)}{N^2(N-1)^2} = \frac{6+o(1)}{N^2} \quad \text{and} \quad \delta_2 = \frac{O(1)}{N^2}. \tag{3.23}$$

Note that (3.22)–(3.23) is a special case of the following general result:

$$\text{ProdMeas}_1\left\{\omega \in \Omega_1 : \left(\overline{\mathbf{D}^2}(\mathcal{Y},\mathcal{M};\mathbf{r};\omega;f;t) - \overline{\mathbf{D}^2}_{\Omega_1}(\mathcal{Y},\mathcal{M};\mathbf{r};f;t)\right)^2\right.$$

$$\left.\geq \frac{1}{4}\left(\left(\sum_{j=1}^N \mu_j^2\right)^2 - \sum_{j=1}^N \mu_j^4\right)\kappa_0^4(f) + O(1)\right\} \geq \frac{3}{32}\cdot\frac{1}{1+\delta_1} + \delta_2, \tag{3.24}$$

where

$$\delta_1 = \frac{6\left(\sum_{j=1}^N \mu_j^4\right)^2\sigma_0^8(f) + O(1)}{\left(\left(\left(\sum_{j=1}^N \mu_j^2\right)^2 - \sum_{j=1}^N \mu_j^4\right)\kappa_0^4(f)\right)^2}, \tag{3.25}$$

$$\delta_2 = \frac{O(1)}{\left(\left(\left(\sum_{j=1}^N \mu_j^2\right)^2 - \sum_{j=1}^N \mu_j^4\right)\kappa_0^4(f)\right)^2 + \left(\sum_{j=1}^N \mu_j^4\right)^2\sigma_0^8(f) + O(1)} \tag{3.26}$$

and

$$\kappa_0^4(f) = \sum_{\substack{\mathbf{n}\in\mathbb{Z}^d:\\ \mathbf{n}\neq\mathbf{0}}} |a_{\mathbf{n}}|^4.$$

We postpone the proof of Theorem 3.2 to Section 10.

Theorem 3.2 proves that (3.20) is *false*. That is, if the system is already in LA-equilibrium (see (3.21)), we cannot expect that for $1-\varepsilon$ part (= overwhelming majority) of the initial velocities $\omega \in \Omega_1$,

$$\overline{\mathbf{D}^2}(\mathcal{Y};\mathbf{r};\omega;f;t) = (1+o(1))N\sigma_0^2(f). \tag{3.27}$$

Indeed, since $3/32$ is more than 9 percent, in view of (3.19) and (3.25)–(3.26), for at least 9 percent of the initial velocities the discrepancy is "too large":

$$\left|\overline{\mathbf{D}^2}(\mathcal{Y};\mathbf{r};\omega;f;t) - N\sigma_0^2(f)\right|$$

$$= \left|\overline{\mathbf{D}^2}(\mathcal{Y};\mathbf{r};\omega;f;t) - \overline{\mathbf{D}^2}_{\Omega_1}(\mathcal{Y};\mathbf{r};f;t)\right| + O(1) \geq \frac{N}{2} + O(1),$$

disproving any hope for discrepancy $o(N)$ for the overwhelming majority of the initial velocities (i.e., the ε in $1-\varepsilon$ cannot be less than 9 percent).

The message of Theorem 3.2 is that in LA-equilibrium the typical time evolution of $\overline{\mathbf{D}^2}(\mathcal{Y};\mathbf{r};\omega;f;t)$ shows inevitable fluctuations of size const$\cdot N\sigma_0^2(f)$ around the constant $N\sigma_0^2(f)$ (= equilibrium part). We refer to the fluctuations of size const$\cdot N\sigma_0^2(f)$ as the unavoidable "white noise". (Here we assumed that the test function $f \in L^2$ is "reasonable".)

Next consider the case of general mass distribution. We start with two simple observations: the condition

$$\max_{1\leq j\leq N} \mu_j = o(N) \tag{3.28}$$

clearly implies

$$\frac{\sum_{j=1}^N \mu_j^4}{\left(\sum_{j=1}^N \mu_j^2\right)^2} = o(1).$$

Moreover, for a "reasonable" test function $f \in L^2$ the ratio

$$\frac{\kappa_0^4(f)}{\sigma_0^4(f)} = \frac{\sum_{\mathbf{n}\in\mathbb{Z}^d\setminus\mathbf{0}} |a_{\mathbf{n}}|^4}{\left(\sum_{\mathbf{n}\in\mathbb{Z}^d\setminus\mathbf{0}} |a_{\mathbf{n}}|^2\right)^2} \tag{3.29}$$

is not too small. (Here and in the rest the slightly ambiguous notation $\mathbb{Z}^d \backslash \mathbf{0}$ more precisely means $\mathbb{Z}^d \setminus \{\mathbf{0}\}$, i.e., the zero lattice point is excluded.)

It follows that in the general case the message of Theorem 3.2 remains the same: in LA-equilibrium the typical time evolution of $\overline{\mathbf{D}^2}(\mathcal{Y}, \mathcal{M}; \mathbf{r}; \omega; f; t)$ shows inevitable fluctuations of size const·Eqlbrm($\mathcal{M}; f$) around the constant Eqlbrm($\mathcal{M}; f$) (equilibrium part). Here we assume that the test function $f \in L^2$ is "reasonable" in the sense that ratio (3.29) is not too small, and also assume (3.28).

As we promised after Theorem 3.1, we include the definition of the Gaussian speed distribution for arbitrary dimension $d \geq 1$. In the 1-dimensional case we have the density function

$$\sqrt{\frac{2}{\pi}} e^{-y^2/2} \text{ for } y > 0,$$

and 0 otherwise; and in general, for every odd dimension $d \geq 3$ we have the density function

$$\sqrt{\frac{2}{\pi}} \frac{y^{d-1} e^{-y^2/2}}{(d-2)!!} \quad \text{for } y > 0, \tag{3.30$'$}$$

and 0 otherwise (here we use the well known notation $k!! = k(k-2)(k-4)\cdots$, where the last factor is 1 or 2 depending on the parity of k). Finally, for every even $d \geq 2$, we have the density function

$$\frac{y^{d-1} e^{-y^2/2}}{(d-2)!!} = \frac{y^{d-1} e^{-y^2/2}}{2^{\frac{d}{2}-1}(\frac{d}{2}-1)!} \quad \text{for } y > 0, \tag{3.30$''$}$$

and 0 otherwise.

The disadvantage of definition (3.30′)–(3.30″) (describing the speed distribution) is that it comes without any a priory motivation for the "weird" constant factors in (3.30′)–(3.30″). An equivalent alternative definition — which is certainly more instructive — is to describe the Gaussian velocity distribution $\mathbf{v} = (v_1, \ldots, v_d)$ with its multivariate density function

$$f(x_1, \ldots, x_d) = \prod_{j=1}^{d} \frac{1}{\sqrt{2\pi}} e^{-x_j^2/2}. \tag{3.31$'$}$$

Since (3.31′) is the product of the factors $\frac{1}{\sqrt{2\pi}} e^{-x_j^2/2}$, the coordinates $v_1, \ldots, v_d$ of the velocity $\mathbf{v}$ are independent random variables with standard normal distribution each. This is why (3.31′) is called the density function of the d-dimensional standard normal distribution.

We can rewrite (3.31′) in the form

$$f(\mathbf{x}) = (2\pi)^{-d/2} e^{-|\mathbf{x}|^2/2}, \tag{3.31$''$}$$

where $|\mathbf{x}| = \sqrt{x_1^2 + \ldots + x_d^2}$ is the usual Euclidean distance. So (3.31″) implies that the distribution of the velocity $\mathbf{v}$ — i.e., the distribution of the d-dimensional standard normal distribution — is isotropic. Since $|\mathbf{v}|$ is the speed, by (3.31″) the density function of the speed distribution is

$$g(y) = (2\pi)^{-d/2}\text{SurfArea}(\mathbf{S}^{d-1}(y))e^{-y^2/2}, \tag{3.32}$$

where $\mathbf{S}^{d-1}(y) = \{\mathbf{x} \in \mathsf{R}^d : |\mathbf{x}| = y\}$ is the sphere in the d-space of radius y, and SurfArea $(\mathbf{S}^{d-1}(y))$ stands for its surface area.

The next step is to show that (3.32) gives back (3.30′)–(3.30″), and thus it explains, in retrospect, the geometric meaning of the *ad hoc* constant factors in (3.30′)–(3.30″). Since a solid sphere (= ball) is the union of concentric spheres, we can express SurfArea $(\mathbf{S}^{d-1}(y))$ in terms of the derivative of the volume (warning: $\frac{\mathrm{d}}{\mathrm{d}y}$ indicates derivative and d is the dimension)

$$\text{SurfArea}(\mathbf{S}^{d-1}(y)) = \frac{\mathrm{d}}{\mathrm{d}y}\text{Volume(solid sphere } \mathbf{S}^{d-1}(y)) = dy^{d-1}C^*(d), \tag{3.33}$$

where $C^*(d)$ is the (hyper)volume of the d-dimensional unit ball. A well known recurrence formula on trigonometric integrals gives that $C^*(d)$ is equal to the ratio of $\pi^{d/2}$ and the Gamma function at $\frac{d}{2}+1$, i.e.,

$$C^*(d) = \text{volume of the } d\text{–dimensional unit ball}$$

$$= \frac{\pi^{d/2}}{(d/2)!} \quad \text{or} \quad \frac{2^{\lceil d/2\rceil}\pi^{\lfloor d/2\rfloor}}{1\cdot 3\cdot 5\cdot 7\cdots d} \tag{3.34}$$

if d is even or odd (in the odd case $\lceil x\rceil$ and $\lfloor x\rfloor$ denote the upper and lower integral parts of a real number x). Combining (3.32), (3.33) and (3.34), we obtain, respectively, (3.30′) and (3.30″). This proves the equivalence of the two definitions of the multi-dimensional Gaussian distribution.

Note that the initial velocity space $\Omega_1(\dim = d)$ is the straightforward generalization of the case $d = 3$ (see (2.13)). That is,

$$\omega = (\rho_1, \vartheta_1, \rho_2, \vartheta_2, \ldots, \rho_N, \vartheta_N) \in \Omega_1(\dim = d) = ([0,\infty) \times SO(d))^N,$$

where $\Omega_1(\dim = d)$ is equipped with the product measure: the normed Haar measure on the group $SO(d)$ and the following probability measure on $[0,\infty)$:

$$\Pr[\rho_j \le u] = \frac{1}{2^{\frac{d}{2}-1}(\frac{d}{2}-1)!}\int_0^u y^{d-1}e^{-y^2/2}\,dy$$

for every even $d \geq 2$,

$$\Pr[\rho_j \leq u] = \sqrt{\frac{2}{\pi}} \frac{1}{(d-2)!!} \int_0^u y^{d-1} e^{-y^2/2}\, dy$$

for every odd $d \geq 3$, and

$$\Pr[\rho_j \leq u] = \sqrt{\frac{2}{\pi}} \int_0^u e^{-y^2/2}\, dy$$

for $d = 1$.

We conclude Section 3 with the following lemma, which plays a crucial role in the proof of Theorem 3.1 (and the proofs of Theorems 4.1 and 4.2; see the next section).

Lemma 3.1. *Let* $\mathbf{w}$ *be an arbitrary 3-dimensional vector. Then*

$$\frac{1}{4\pi} \int_{\mathbf{v} \in \mathbf{S}^2} e^{i\mathbf{w} \cdot \mathbf{v}}\, d\mathrm{SA}(\mathbf{v}) = \frac{\sin(|\mathbf{w}|)}{|\mathbf{w}|}, \tag{3.35}$$

where $d\mathrm{SA}(\mathbf{v})$ *represents the surface area.*

Moreover,

$$(2\pi)^{-3/2} \int_{\mathbf{v} \in R^3} e^{i\mathbf{w} \cdot \mathbf{v}} e^{-|\mathbf{v}|^2/2}\, d\mathbf{v} = e^{-|\mathbf{w}|^2/2}, \tag{3.36}$$

where $d\mathbf{v}$ *represents the volume.*

Remark. Notice that the right-hand side in (3.36) tends to zero as $|\mathbf{w}| \to \infty$ much faster than the right-hand side in (3.35). Since $\mathbf{v}$ is the initial velocity, this dramatic difference in the asymptotic behaviors of (3.35) and (3.36) gives some indication why the case of Gaussian initial velocity distribution reaches equilibrium so much faster than the case of isotropic constant unit speed.

Proof of Lemma 3.1. We start with (3.35). The special case $\mathbf{w} = \mathbf{0}$ is trivial: the left side is equal to 1, and the right side tends to 1 as $\mathbf{w}$ tends to $\mathbf{0}$; so we may assume that $\mathbf{w} \neq \mathbf{0}$. We need a well known geometric fact about the surface area of some special spherical domains. Let $\mathbf{e} \in \mathbf{S}^2$ be an arbitrary but fixed unit vector, and let $0 < \delta < 1$ be an arbitrary real number. Consider the "spherical belt" (often called the "spherical zone"):

$$\mathbf{S}^2(\mathbf{e}; \delta) = \left\{ \mathbf{v} \in \mathbf{S}^2 : \ |\mathbf{e} \cdot \mathbf{v}| \leq \delta \right\}.$$

On one hand, it is trivial that the surface area of the "belt" $\mathbf{S}^2(\mathbf{e}; \delta)$ is independent of the choice of the unit vector $\mathbf{e}$; it is a very surprising fact, on the other hand, that the dependence on δ is just plain linear (discovered by Archimedes)

$$\text{SurfaceArea}\left(\mathbf{S}^2(\mathbf{e};\delta)\right) = \text{SurfArea}\left(\mathbf{S}^2(\mathbf{e};\delta)\right) = 4\pi\delta. \tag{3.37}$$

(3.37) implies that, for any nonzero 3-dimensional vector $\mathbf{w}$ and any real number ϱ with $0 < \varrho < |\mathbf{w}|$,

$$\text{SurfArea}\left(\left\{\mathbf{v} \in \mathbf{S}^2 : \; |\mathbf{w}\cdot\mathbf{v}| \le \varrho\right\}\right) = 4\pi \cdot \frac{\varrho}{|\mathbf{w}|}. \tag{3.38}$$

By using (3.38), we have

$$\frac{1}{4\pi}\int_{\mathbf{S}^2} e^{\mathrm{i}\mathbf{w}\cdot\mathbf{v}}\, d\mathrm{SA}(\mathbf{v}) = \frac{1}{2}\int_{-1}^{1} e^{\mathrm{i}|\mathbf{w}|u}\, du = \frac{\sin(|\mathbf{w}|)}{|\mathbf{w}|}, \tag{3.39}$$

proving (3.35).

For later application we prove (3.36) in arbitrary dimension $d \ge 1$. Write $\mathbf{w} = (w_1, \ldots, w_d)$ and $\mathbf{v} = (v_1, \ldots, v_d)$, and factorize the integral over the d-space coordinatewise. The proof of is a straightforward corollary of the well known fact that the Fourier transform of $e^{-x^2/2}$ is itself. Indeed, we have

$$\begin{aligned}
(2\pi)^{-d/2}\int_{\mathbf{v}\in\mathbf{R}^d} e^{\mathrm{i}\mathbf{w}\cdot\mathbf{v}} e^{-|\mathbf{v}|^2/2}\, d\mathbf{v} &= \prod_{j=1}^{d} \frac{1}{\sqrt{2\pi}}\int_{-\infty}^{\infty} e^{\mathrm{i}w_j v_j} e^{-v_j^2/2}\, dv_j \\
&= \prod_{j=1}^{d}\left(\text{Fourier transform of } e^{-v^2/2} \text{ at } w_j\right) \\
&= \prod_{j=1}^{d}\left(e^{-w_j^2/2}\right) = e^{-(w_1^2+\ldots+w_d^2)} = e^{-|\mathbf{w}|^2/2}.
\end{aligned} \tag{3.40}$$

The special case $d = 3$ in (3.40) gives (3.36). Thus the proof of Lemma 3.1 is complete. □

4. Second surprise: Shockingly slow approach to micro-equilibrium in the "photon-like" constant speed case

What is in Section 4? *Section 3 was devoted to the general torus model of the first kind, i.e., the case of the Gaussian initial velocity distribution. Here we study what happens if we switch to the second kind and the third kind.*

We start with the 3-dimensional case, i.e., we replace the Gaussian speed distribution

$$\Pr[\rho \le u] = \sqrt{\frac{2}{\pi}} \int_0^u y^2 e^{-y^2/2}\, dy \quad \text{for all } u > 0$$

with "speed one" $\rho = 1$. This is the general torus analog of the 3-dimensional Bernoulli model of the second kind (see the end of Section 1). In this case the trajectory of the k-th particle is

$$\mathbf{x}_k(t) = \mathbf{y}_k + \vartheta_k \mathbf{r}_k(t) \quad \text{modulo one,} \tag{4.1}$$

where

$$\omega = (\vartheta_1, \vartheta_2, \ldots, \vartheta_N) \in \Omega_2 = (SO(3))^N. \tag{4.2}$$

This describes the time evolution of the whole N-point system. (Formally we just removed the factor ρ_k in the Gaussian case.)

(4.2) means that $\vartheta_1, \vartheta_2, \ldots, \vartheta_N$ are all uniformly distributed in the rotation group $SO(3)$ (= special orthogonal group of all 3×3 real matrices) with respect to the Haar measure. We use the corresponding product measure ProdMeas_2 on the product space, and, for simplicity, let $d\omega = d\text{ProdMeas}_2(\omega)$ denote the normalized version of this product measure on the set Ω_2: $\int_{\Omega_2} 1\, d\omega = 1$.

We refer to this as the (general) *torus model of the second kind.* We call the huge product space Ω_2 the "initial velocity space" for the *torus model of the second kind.*

The switch to the *torus model of the second kind* means that we study the Average Square-Discrepancy (relative to f; see (2.31))

$$\overline{\mathbf{D}^2}(\mathcal{Y}, \mathcal{M}; \mathcal{R}; \omega; f; t) \quad \text{for a typical } \omega \in \Omega_2. \tag{4.3}$$

Let

$$\overline{\mathbf{D}^2}_{\Omega_2}(\mathcal{Y}, \mathcal{M}; \mathcal{R}; f; t) = \int_{\Omega_2} \overline{\mathbf{D}^2}(\mathcal{Y}, \mathcal{M}; \mathcal{R}; \omega; f; t)\, d\omega, \tag{4.4}$$

which is the analog of (2.32). Again we refer to $\overline{\mathbf{D}^2}_{\Omega_2}(\mathcal{Y}, \mathcal{M}; \mathcal{R}; f; t)$ as the *Big Average Square-Discrepancy* (relative to f).

Of course these definitions work in every dimension $d \ge 2$; we refer to it as the d-dimensional torus model of the second kind.

The following result is an analog of Theorem 3.1. The surprising novelty here is that the short Relaxation Time in Theorem 3.1 is replaced by long Relaxation Time.

Theorem 4.1. *Consider a general 3-dimensional N-particle torus model of the second kind $(\mathcal{Y}, \mathcal{M}; \mathcal{R}; \Omega_2)$. Similarly to Theorem 3.1, write*

$$\mathcal{R}(t) = \min_{1 \le k \le N} |\mathbf{r}_k(t)|.$$

Let $f \in L^2$ be an arbitrary complex test function defined on the torus I^3. Then, starting at $t = 0$, for every $t > 0$ we have the following inequality for the Big Average Square-Discrepancy:

$$\left| \overline{\mathbf{D}^2}_{\Omega_2}(\mathcal{Y}, \mathcal{M}; \mathcal{R}; f; t) - \sigma_0^2(f) \sum_{k=1}^{N} \mu_k^2 \right| \le \min\left\{ \frac{1}{(2\pi \mathcal{R}(t))^2}, 1 \right\} \sigma_0^2(f) N^2. \tag{4.5}$$

Next consider the special case $\mathbf{r}_1(t) = \mathbf{r}_2(t) = \ldots = \mathbf{r}_N(t) = \mathbf{r}(t)$. Then the drifting distance $\mathcal{R}(t) = |\mathbf{r}(t)|$, and writing $\mathcal{R} = \mathbf{r}$, we have

$$\left| \overline{\mathbf{D}^2}_{\Omega_2}(\mathcal{Y}, \mathcal{M}; \mathbf{r}; f; t) - \sigma_0^2(f) \sum_{k=1}^{N} \mu_k^2 \right|$$

$$\le \min\left\{ \frac{1}{(2\pi |\mathbf{r}(t)|)^2}, 1 \right\} \max\left\{ \overline{\mathbf{D}^2}_{\Omega_2}(\mathcal{Y}, \mathcal{M}; \mathbf{r}; f; 0), \sigma_0^2(f) \sum_{k=1}^{N} \mu_k^2 \right\}. \tag{4.6}$$

In this special case we have the equality

$$\overline{\mathbf{D}^2}_{\Omega_2}(\mathcal{Y}, \mathcal{M}; \mathbf{r}; f; t) = \sum_{k=1}^{N} \mu_k^2 \sum_{\substack{\mathbf{n} \in \mathbb{Z}^3: \\ \mathbf{n} \ne \mathbf{0}}} |a_{\mathbf{n}}|^2 \left(1 - \left(\frac{\sin(2\pi |\mathbf{r}(t)|\ |\mathbf{n}|)}{2\pi |\mathbf{r}(t)|\ |\mathbf{n}|} \right)^2 \right)$$

$$+ \sum_{\substack{\mathbf{n} \in \mathbb{Z}^3: \\ \mathbf{n} \ne \mathbf{0}}} |a_{\mathbf{n}}|^2 \left| \sum_{k=1}^{N} \mu_k e^{2\pi i \mathbf{n} \cdot \mathbf{y}_k} \right|^2 \left(\frac{\sin(2\pi |\mathbf{r}(t)|\ |\mathbf{n}|)}{2\pi |\mathbf{r}(t)|\ |\mathbf{n}|} \right)^2 \tag{4.7}$$

where $a_{\mathbf{n}}$, $\mathbf{n} \in \mathbb{Z}^3$ are the Fourier coefficients of the test function $f \in L^2$.

Finally, the results above on the 3-dimensional unit torus $I^3 = [0,1)^3$ extend for the d-dimensional unit torus $I^d = [0,1)^d$ in arbitrary dimension $d \ge 2$ with the following modification: the critical function $\frac{\sin z}{z}$ in (4.7) is replaced by

$$(d-2)!! \frac{J_{(d-2)/2}(z)}{z^{(d-2)/2}} \cdot \left\{ 1 \text{ or } \sqrt{\frac{\pi}{2}} \right\},$$

depending on whether $d \ge 2$ is even or odd, accordingly. Here $J_\nu(z)$ is the well known Bessel function.

Remarks. As usual $k!! = k(k-2)(k-4)\cdots$, where the last factor is 1 or 2 depending on the parity of k, and $0!! = 1$.

Note that the Bessel functions arise as solutions of some important differential equations, but here we define the Bessel function (of the first kind) $J_\nu(z)$ (where ν is called the order) in a shortcut way via power series

$$J_\nu(z) = \sum_{m=0}^{\infty} \frac{(-1)^m}{m!\text{Gamma}(m+\nu+1)} \left(\frac{z}{2}\right)^{2m+\nu},$$

where Gamma(z) is Euler's Gamma function.

If $\nu \geq 0$ is a half-integer, then $J_\nu(z)$ has a finite form, e.g.,

$$J_{\frac{1}{2}}(z) = \sqrt{\frac{2}{\pi z}} \sin z \text{ and } J_{\frac{3}{2}}(z) = \sqrt{\frac{2}{\pi z}} \left(-\cos z + \frac{\sin z}{z}\right),$$

but for $\nu \geq 0$ integer the Bessel function $J_\nu(z)$ does not have a finite expression. What we can still use is an asymptotic expression — called Hankel's formula — that works for all $\nu \geq 0$; see later.

The unusual notation Gamma(s) for the Gamma function is explained by our preference to save letter Γ for the trajectory of the whole system in the configuration space.

Again we write "$\mathbf{r}$" instead of "$\mathcal{R} = \mathbf{r}$". We rewrite (4.7):

$$\overline{\mathbf{D}^2}_{\Omega_2}(\mathcal{Y}, \mathcal{M}; \mathbf{r}; f; t) = \text{Eqlbrm}(\mathcal{M}; f)$$

$$+\overline{\mathbf{D}^2}_{2,negl}(\mathcal{Y}, \mathcal{M}; \mathbf{r}; f; t) + \overline{\mathbf{D}^2}_{2,uneven}(\mathcal{Y}, \mathcal{M}; \mathbf{r}; f; t), \tag{4.8}$$

where

$$\text{Eqlbrm}(\mathcal{M}; f) = \sigma_0^2(f) \sum_{k=1}^{N} \mu_k^2 = \left(\sum_{\substack{\mathbf{n}\in\mathbb{Z}^3:\\ \mathbf{n}\neq\mathbf{0}}} |a_{\mathbf{n}}|^2 \right) \left(\sum_{k=1}^{N} \mu_k^2 \right)$$

is the *equilibrium part* defined in (3.6), and

$$\overline{\mathbf{D}^2}_{2,negl}(\mathcal{Y}, \mathcal{M}; \mathbf{r}; f; t) = -\sum_{k=1}^{N} \mu_k^2 \sum_{\substack{\mathbf{n}\in\mathbb{Z}^3:\\ \mathbf{n}\neq\mathbf{0}}} |a_{\mathbf{n}}|^2 \left(\frac{\sin(2\pi|\mathbf{r}(t)|\,|\mathbf{n}|)}{2\pi|\mathbf{r}(t)|\,|\mathbf{n}|} \right)^2 \tag{4.9}$$

is the *negligible part*: it tends to zero as $|\mathbf{r}(t)| \to \infty$, and we also have the inequality

$$0 \leq \text{Eqlbrm}(\mathcal{M}; f) + \overline{\mathbf{D}^2}_{2,negl}(\mathcal{Y}, \mathcal{M}; \mathbf{r}; f; t) \leq \text{Eqlbrm}(\mathcal{M}; f) = \sigma_0^2(f) \sum_{k=1}^{N} \mu_k^2, \tag{4.10}$$

which is the perfect analog of (3.8). Finally, we have the *unevenness*

$$\overline{\mathbf{D}^2}_{2,uneven}(\mathcal{Y},\mathcal{M};\mathbf{r};f;t)=\sum_{\substack{\mathbf{n}\in\mathbb{Z}^3:\\ \mathbf{n}\neq\mathbf{0}}}|a_{\mathbf{n}}|^2\left|\sum_{k=1}^{N}\mu_k e^{2\pi i\mathbf{n}\cdot\mathbf{y}_k}\right|^2\left(\frac{\sin(2\pi|\mathbf{r}(t)|\ |\mathbf{n}|)}{2\pi|\mathbf{r}(t)|\ |\mathbf{n}|}\right)^2, \tag{4.11}$$

which is the dominating part in (4.8) as long as the system stays in off-equilibrium.

(4.8)–(4.11) represent the 3-dimensional case. The d-dimensional analog is obtained by replacing $\frac{\sin z}{z}$ with

$$(d-2)!!\frac{J_{(d-2)/2}(z)}{z^{(d-2)/2}}\cdot\left\{1 \text{ or } \sqrt{\frac{\pi}{2}}\right\},$$

depending on whether $d\geq 2$ is even or odd, accordingly.

For the torus model of the second kind $(\mathcal{Y},\mathcal{M};\mathcal{R};\Omega_2)$ we can define LA-equilibrium (L-equilibrium) by simply replacing Ω_1 and ProdMeas_1 in Definition 2.4 (Definition 2.3) with Ω_2 and ProdMeas_2.

Next we derive an analog of Corollary 3.1 from Theorem 4.1 exactly the same way as we derived Corollary 3.1 from Theorem 3.1. Again for simplicity assume that the particles have unit mass: $\mu_k=1$ for all $1\leq k\leq N$. Then by (4.5),

$$\left|\overline{\mathbf{D}^2}_{\Omega_2}(\mathcal{Y};\mathcal{R};f;t)-N\sigma_0^2(f)\right|\leq\min\left\{\frac{1}{(2\pi\mathcal{R}(t))^2},1\right\}N^2\sigma_0^2(f)N^2,$$

which implies that

$$\left|\overline{\mathbf{D}^2}_{\Omega_2}(\mathcal{Y};\mathcal{R};f;t)-N\sigma_0^2(f)\right|\leq N\sigma_0^2(f)$$

if

$$4\pi^2\mathcal{R}^2(t)\geq N \iff \mathcal{R}(t)\geq\frac{\sqrt{N}}{2\pi}. \tag{4.12}$$

Therefore, if (4.12) holds, we have

$$\int_{\Omega_2}\overline{\mathbf{D}^2}(\mathcal{Y};\mathcal{R};\omega;f;t)\,d\omega\leq 2N\sigma_0^2(f).$$

Applying the (trivial) Markov inequality, we obtain the following result.

Corollary 4.1. *Consider a general 3-dimensional N-particle torus model of the second kind $(\mathcal{Y};\mathcal{R};\Omega_2)$ in the 3-dimensional unit torus $I^3=[0,1)^3$ such that every particle has mass one. Let t be a time satisfying*

$$\inf_{\tau\geq t}\mathcal{R}(\tau)\geq\frac{\sqrt{N}}{2\pi}.$$

Then, for every $0 < \varepsilon < 1$, the system $(\mathcal{Y};\mathcal{R};\Omega_2)$ is in LA-equilibrium with parameters ε and $C = 2/\varepsilon$ at time t.

Switching from the 3-dimensional unit torus I^3 to the general d-dimensional unit torus I^d with $d \geq 2$, $\sqrt{N}$ is replaced by $N^{1/(d-1)}$. □

Motivated by the *Bernoulli model of the third kind* (see the end of Section 1), next we study the 3-dimensional torus model with the speed density function $3y^2$ for $0 \leq y \leq 1$, and 0 otherwise. In this case the trajectory of the k-th particle is

$$\mathbf{x}_k(t) = \mathbf{y}_k + \vartheta_k(\rho_k \mathbf{r}_k(t)) \quad \text{modulo one}, \quad 1 \leq k \leq N, \tag{4.13}$$

which describes the time evolution of the whole N-point system. Notice that (4.13) is an analog of the Gaussian case, but the distribution of the scale ρ_k ($1 \leq k \leq N$) is different. Similarly to the Gaussian distribution, we assume that $\rho_1, \vartheta_1, \rho_2, \vartheta_2, \ldots, \rho_N, \vartheta_N$ are $2N$ independent random variables, where $\rho_1, \rho_2, \ldots, \rho_N$ have the same distribution described by

$$\Pr[\rho_k \leq y] = 3\int_0^y u^2\, du = y^3 \quad \text{for } 0 \leq y \leq 1, \tag{4.14}$$

and 0 otherwise, and $\vartheta_1, \vartheta_2, \ldots, \vartheta_N$ are all uniformly distributed in the rotation group $SO(3)$ (= special orthogonal group of all 3×3 real matrices) with respect to the Haar measure. This means we use the corresponding product measure ProdMeas_3 on the product space

$$\omega = (\rho_1, \vartheta_1, \rho_2, \vartheta_2, \ldots, \rho_N, \vartheta_N) \in \Omega_3 = ([0,1] \times SO(3))^N. \tag{4.15}$$

For simplicity, let $d\omega = d\text{ProdMeas}_3(\omega)$ denote the normalized version of this product measure on the set Ω_3; i.e., $\int_{\Omega_3} 1\, d\omega = 1$.

We refer to this as the (general) *torus model of the third kind.* We call the huge product space Ω_3 the "initial velocity space" for the *torus model of the third kind.*

The switch to the *torus model of the third kind* means that we study the Average Square-Discrepancy (relative to f)

$$\overline{\mathbf{D}^2}(\mathcal{Y},\mathcal{M};\mathcal{R};\omega;f;t) \quad \text{for a typical } \omega \in \Omega_3. \tag{4.16}$$

Let

$$\overline{\mathbf{D}^2}_{\Omega_3}(\mathcal{Y},\mathcal{M};\mathcal{R};f;t) = \int_{\Omega_3} \overline{\mathbf{D}^2}(\mathcal{Y},\mathcal{M};\mathcal{R};\omega;f;t)\, d\omega, \tag{4.16$'$}$$

and again we refer to (4.16′) as the *Big Average Square-Discrepancy* (relative to f; see (2.31)).

Of course these definitions work in every dimension $d \geq 1$; we refer to it as the d-dimensional torus model of the third kind.

We have the following analog of Theorems 3.1 and 4.1.

Theorem 4.2. *(a) Consider a general 3-dimensional N-particle torus model of the third kind $(\mathcal{Y}, \mathcal{M}; \mathcal{R}; \Omega_3)$.*

Let $f \in L^2$ be an arbitrary complex valued square-integrable test function on the torus I^3. Then, starting at $t = 0$, for every $t > 0$ we have the following inequality for the Big Average Square-Discrepancy (see (4.16′)):

$$\left| \overline{\mathbf{D}^2}_{\Omega_3}(\mathcal{Y}, \mathcal{M}; \mathcal{R}; f; t) - \sigma_0^2(f) \sum_{k=1}^{N} \mu_k^2 \right|$$

$$\leq \min\left\{ \frac{18}{(2\pi\mathcal{R}(t))^4} + \frac{18}{(2\pi\mathcal{R}(t))^6}, 1 \right\} \sigma_0^2(f) N^2. \tag{4.17}$$

Next consider the special case $\mathbf{r}_1(t) = \mathbf{r}_2(t) = \ldots = \mathbf{r}_N(t) = \mathbf{r}(t)$. Then $\mathcal{R}(t) = |\mathbf{r}(t)|$, and writing $\mathcal{R} = \mathbf{r}$, we have

$$\left| \overline{\mathbf{D}^2}_{\Omega_3}(\mathcal{Y}, \mathcal{M}; \mathbf{r}; f; t) - \sigma_0^2(f) \sum_{k=1}^{N} \mu_k^2 \right| \leq \min\left\{ \frac{18}{(2\pi|\mathbf{r}(t)|)^4} + \frac{18}{(2\pi|\mathbf{r}(t)|)^6}, 1 \right\} \cdot$$

$$\max\left\{ \overline{\mathbf{D}^2}_{\Omega_3}(\mathcal{Y}, \mathcal{M}; \mathbf{r}; f; 0), \sigma_0^2(f) \sum_{k=1}^{N} \mu_k^2 \right\}. \tag{4.18}$$

In this special case we have the equality

$$\overline{\mathbf{D}^2}_{\Omega_3}(\mathcal{Y}, \mathcal{M}; \mathbf{r}; f; t) = \sum_{k=1}^{N} \mu_k^2 \sum_{\substack{\mathbf{n} \in \mathbb{Z}^3: \\ \mathbf{n} \neq \mathbf{0}}} |a_{\mathbf{n}}|^2$$

$$\cdot \left(1 - 9 \left(\frac{\cos(2\pi|\mathbf{r}(t)|\ |\mathbf{n}|)}{(2\pi|\mathbf{r}(t)|\ |\mathbf{n}|)^2} - \frac{\sin(2\pi|\mathbf{r}(t)|\ |\mathbf{n}|)}{(2\pi|\mathbf{r}(t)|\ |\mathbf{n}|)^3} \right)^2 \right)$$

$$+ \sum_{\substack{\mathbf{n} \in \mathbb{Z}^3: \\ \mathbf{n} \neq \mathbf{0}}} |a_{\mathbf{n}}|^2 \left| \sum_{k=1}^{N} \mu_k e^{2\pi i \mathbf{n} \cdot \mathbf{y}_k} \right|^2 \cdot 9 \left(\frac{\cos(2\pi|\mathbf{r}(t)|\ |\mathbf{n}|)}{(2\pi|\mathbf{r}(t)|\ |\mathbf{n}|)^2} - \frac{\sin(2\pi|\mathbf{r}(t)|\ |\mathbf{n}|)}{(2\pi|\mathbf{r}(t)|\ |\mathbf{n}|)^3} \right)^2 \tag{4.19}$$

where $a_{\mathbf{n}}$, $\mathbf{n} \in \mathbb{Z}^3$ are the Fourier coefficients of the test function $f \in L^2$.

The results above on the 3-dimensional unit torus I^3 extend for the d-dimensional unit torus I^d in arbitrary dimension $d \geq 1$ with the following modification: the critical function

$$-3\left(\frac{\cos z}{z^2} - \frac{\sin z}{z^3}\right)$$

is replaced by $\frac{\sin z}{z}$ *for* $d = 1$, *and replaced by*

$$\frac{d!!}{w^{d/2}} J_{d/2}(w) \cdot \left\{1 \text{ or } \sqrt{\frac{\pi}{2}}\right\}$$

for the rest of the dimensions, depending on whether $d \geq 2$ *is even or odd, accordingly.*

(b) Finally, consider the general isotropic case in the standard 3-space: again the trajectory of the k-th particle is

$$\mathbf{x}_k(t) = \mathbf{y}_k + \vartheta_k(\rho_k \mathbf{r}_k(t)) \quad \text{modulo one}, \quad 1 \leq k \leq N,$$

but the distribution of the scale ρ_k $(1 \leq k \leq N)$ *is basically arbitrary.*

For simplicity we just consider the case where there is an initial speed density function $g(y)$ *— i.e.,* $\Pr[\rho_j \leq u] = \int_0^u g(y)\, dy$ *— such that*

$$\infty > g(u) \geq 0 \text{ for } 0 \leq u < \infty, \quad \int_0^\infty g(u)\, du = 1, \quad \text{and} \quad \int_0^\infty u g(u)\, du < \infty,$$

where the last equation means that the "mean speed is finite". We allow g *to be a generalized function like the Dirac delta function concentrated on 1 representing the second kind, but we exclude the trivial case of Dirac delta concentrated on 0, i.e., the mean speed is strictly positive. The corresponding initial velocity space is*

$$\omega = (\rho_1, \vartheta_1, \rho_2, \vartheta_2, \ldots, \rho_N, \vartheta_N) \in \Omega_g = ([0,\infty) \times SO(3))^N$$

equipped with the product measure (Haar measure for $SO(3)$ *and the probability density function* g *for* $[0,\infty)$*).*

Let

$$\overline{\mathbf{D}^2}_{\Omega_g}(\mathcal{Y}, \mathcal{M}; \mathcal{R}; f; t) = \int_{\Omega_g} \overline{\mathbf{D}^2}(\mathcal{Y}, \mathcal{M}; \mathcal{R}; \omega; f; t)\, d\omega, \tag{4.16''}$$

and again we refer to (4.16″) as the Big Average Square-Discrepancy (relative to f*; see (2.31)). Notice that (4.16″) is a straightforward generalization of (4.16′) (and (4.4) and (2.32)).*

Then the key functions

$$e^{-w^2/2}, \quad \frac{\sin w}{w} \quad \text{and} \quad -3\frac{\cos w}{w^2} + 3\frac{\sin w}{w^3}$$

in Theorems 3.1, 4.1 and 4.2(a) are replaced by the corresponding function

$$\mathbf{F}_g(\mathbf{w}) = \int_0^\infty \frac{1}{4\pi} \int_{\mathbf{v}\in\mathbf{S}^2} e^{iu\mathbf{v}\cdot\mathbf{w}} g(u)\, du\, d\mathrm{SA}(\mathbf{v})$$

$$= \int_0^\infty \frac{1}{4\pi} \int_{\mathbf{v}\in\mathbf{S}^2} \cos(u\mathbf{v}\cdot\mathbf{w}) g(u)\, du\, d\mathrm{SA}(\mathbf{v})$$

where $\mathbf{w}$ *is an arbitrary 3-dimensional vector and* $d\mathrm{SA}(\mathbf{v})$ *represents the surface area. We refer to* $\mathbf{F}_g$ *as the "isotropic Fourier transform" of* g*. Due to the integration over the unit sphere,* $\mathbf{F}_g(\mathbf{w})$ *defines a real-valued function depending only on the length* $w = |\mathbf{w}|$ *of the input vector, and we have the simple but important property*

$$\mathbf{F}_g(0) = \mathbf{F}_g(\mathbf{0}) = 1 \quad \text{and} \quad |\mathbf{F}_g(w)| = |\mathbf{F}_g(\mathbf{w})| < 1 \quad \text{for} \quad w = |\mathbf{w}| > 0.$$

The analogs of (4.17)–(4.19) in the general case go as follows:

$$\left| \overline{\mathbf{D}^2}_{\Omega_g}(\mathcal{Y}, \mathcal{M}; \mathcal{R}; f; t) - \sigma_0^2(f) \sum_{k=1}^N \mu_k^2 \right|$$

$$\leq \min\left\{ \sup_{u \geq 2\pi\mathcal{R}(t)} \mathbf{F}_g^2(u), 1 \right\} \sigma_0^2(f) N^2, \tag{4.17$'$}$$

moreover, in the special case $\mathbf{r}_1(t) = \mathbf{r}_2(t) = \ldots = \mathbf{r}_N(t) = \mathbf{r}(t)$ *we have (writing* $\mathcal{R} = \mathbf{r}$*)*

$$\left| \overline{\mathbf{D}^2}_{\Omega_g}(\mathcal{Y}, \mathcal{M}; \mathbf{r}; f; t) - \sigma_0^2(f) \sum_{k=1}^N \mu_k^2 \right| \leq \min\left\{ \sup_{u \geq 2\pi|\mathbf{r}(t)|} \mathbf{F}_g^2(u), 1 \right\} \sigma_0^2(f) N^2, \tag{4.18$'$}$$

and in this special case we have the equality

$$\overline{\mathbf{D}^2}_{\Omega_g}(\mathcal{Y}, \mathcal{M}; \mathbf{r}; f; t) = \sum_{k=1}^N \mu_k^2 \sum_{\substack{\mathbf{n}\in\mathbb{Z}^3:\\ \mathbf{n}\neq\mathbf{0}}} |a_{\mathbf{n}}|^2 \left(1 - \mathbf{F}_g^2(2\pi|\mathbf{r}(t)|\,|\mathbf{n}|)\right)$$

$$+ \sum_{\substack{\mathbf{n}\in\mathbb{Z}^3:\\ \mathbf{n}\neq\mathbf{0}}} |a_{\mathbf{n}}|^2 \left| \sum_{k=1}^N \mu_k e^{2\pi i \mathbf{n}\cdot\mathbf{y}_k} \right|^2 \cdot \mathbf{F}_g^2(2\pi|\mathbf{r}(t)|\,|\mathbf{n}|), \tag{4.19$'$}$$

where $a_{\mathbf{n}}$*,* $\mathbf{n} \in \mathbb{Z}^3$ *are the Fourier coefficients of the test function* $f \in L^2$*.*

Due to the crucial role of the isotropic Fourier transform $\mathbf{F}_g(\mathbf{w})$*, we give a very useful alternative formula for* $\mathbf{F}_g(\mathbf{w})$*. First we extend the probability density function* g *defined in* $[0, \infty)$ *over the whole real line as an even* *function* $g(-u) = g(u)$, *and introduce the (ordinary) Fourier transform* $\widehat{g}(x)$ *of* g. *Then we can express the "isotropic Fourier transform"* $\mathbf{F}_g(\mathbf{w})$ *in terms of the (ordinary) Fourier transform* $\widehat{g}(x)$ *as follows:*

$$\mathbf{F}_g(w) = \mathbf{F}_g(\mathbf{w}) = \frac{\sqrt{\pi}}{2\sqrt{2}w}\int_{-w}^{w}\widehat{g}(x)\,dx = \sqrt{\frac{\pi}{2}}\frac{1}{w}\int_0^w \widehat{g}(x)\,dx,$$

where by definition

$$\widehat{g}(x) = \frac{1}{\sqrt{2\pi}}\int_{-\infty}^{\infty} e^{-ixu}g(u)\,du.$$

If g *is an (ordinary) Lebesgue measurable function then by the Riemann–Lebesgue theorem* $\widehat{g}(x) \to 0$ *as* $x \to \infty$, *and so the last formula implies* $\mathbf{F}_g(w) \to 0$ *as* $w \to \infty$.

Moreover, if $\int_{-\infty}^{\infty}|\widehat{g}(x)|\,dx < \infty$, *then by the Fourier inversion formula*

$$g(0) = \frac{1}{\sqrt{2\pi}}\int_{-\infty}^{\infty}\widehat{g}(x)\,dx,$$

and thus we can rewrite $\mathbf{F}_g(w)$ *as*

$$\mathbf{F}_g(w) = \frac{\sqrt{\pi}}{2\sqrt{2}w}\left(g(0)\sqrt{2\pi} - \int_{|x|\geq w}\widehat{g}(x)\,dx\right).$$

Finally, we point out the following simple but important equality, which means to switch from spherical coordinates to Cartesian coordinates:

$$\mathbf{F}_g(\mathbf{w}) = \int_0^{\infty}\frac{1}{4\pi}\int_{\mathbf{v}\in\mathbf{S}^2} e^{iu\mathbf{v}\cdot\mathbf{w}}g(u)\,du\,d\mathrm{SA}(\mathbf{v})$$

$$= \int_{R^3} e^{i\mathbf{x}\cdot\mathbf{w}}\frac{g(|\mathbf{x}|)}{|\mathbf{x}|^2}\,d\mathbf{x} = (2\pi)^{3/2}\widehat{\mathbf{g}_3}(-\mathbf{w}),$$

where

$$\mathbf{g}_3(\mathbf{x}) = \frac{g(|\mathbf{x}|)}{|\mathbf{x}|^2}$$

is a rotation-invariant function in 3 variables, and $\widehat{\mathbf{g}_3}(\mathbf{w})$ *denotes its Fourier transform (which is also rotation invariant; so* $\widehat{\mathbf{g}_3}(-\mathbf{w}) = \widehat{\mathbf{g}_3}(\mathbf{w})$*).*

This 3-dimensional fact has a straightforward generalization in every dimension $d \geq 2$:

$$\mathbf{F}_g(\mathbf{w}) = \frac{1}{\mathrm{SuArea}(\mathbf{S}^{d-1})}\int_{\mathbf{v}\in\mathbf{S}^{d-1}} e^{iu\mathbf{v}\cdot\mathbf{w}}g(u)\,du\,d\mathrm{SA}(\mathbf{v})$$

$$= \int_{R^d} e^{i\mathbf{x}\cdot\mathbf{w}} \frac{g(|\mathbf{x}|)}{|\mathbf{x}|^{d-1}}\, d\mathbf{x} = (2\pi)^{d/2}\widehat{\mathbf{g}_d}(-\mathbf{w}),$$

where

$$\mathbf{g}_d(\mathbf{x}) = \frac{g(|\mathbf{x}|)}{|\mathbf{x}|^{d-1}}$$

is a rotation-invariant function in d variables, $\widehat{\mathbf{g}_d}(\mathbf{w})$ *denotes its Fourier transform (which is also rotation invariant; so* $\widehat{\mathbf{g}_d}(-\mathbf{w}) = \widehat{\mathbf{g}_d}(\mathbf{w})$*), and* $\mathrm{SuArea}(\mathbf{S}^{d-1})$ *denotes the surface area of the unit sphere* $\mathbf{S}^{d-1} = \left\{\mathbf{u} \in \mathsf{R}^d : \ |\mathbf{u}| = 1\right\}$ *in the d-dimensional space.*

Remarks. The message of the last fact is that the isotropic Fourier transform $\mathbf{F}_g(w) = \mathbf{F}_g(\mathbf{w})$, $w = |\mathbf{w}|$ of the speed density function g is in fact an ordinary multivariable Fourier transform (multiplied by a constant).

The other formula

$$\mathbf{F}_g(w) = \frac{\sqrt{\pi}}{2\sqrt{2}w}\left(g(0)\sqrt{2\pi} - \int_{|x|\geq w} \widehat{g}(x)\, dx\right)$$

will be repeatedly used later, and it raises the question of how to satisfy the condition $\widehat{g} \in L^1(\mathsf{R})$. That is, under what conditions about the probability density function g can we guarantee that the Fourier transform of the symmetric extension of g is integrable? A simple necessary condition — a special case of Lemma 19.1 later — goes as follows. If g is twice continuously differentiable such that $g', g'' \in L^1(\mathsf{R})$ and $\lim_{x\to\infty} g'(x) = 0$, $\lim_{x\to\infty} g''(x) = 0$, then

$$|\widehat{g}(x)| \leq \frac{\text{const}}{(1+|x|)^2},$$

which implies $\widehat{g} \in L^1(\mathsf{R})$ (i.e., $\int_{-\infty}^{\infty} |\widehat{g}(x)|\, dx < \infty$).

Notice that formula (4.19') — which includes (3.4), (4.7) and (4.19) in the special cases of the first, second and third kinds, respectively — is a generalization of the simple Parseval type identity (2.2). Note in advance that (4.19') will be the key to define the Time Evolution Function, which will describe the typical time evolution of the Micro-Entropy ("Second Law"); see Section 6 and Chapter 3.

Again we write "$\mathbf{r}$" instead of "$\mathcal{R} = \mathbf{r}$". As usual, we rewrite (4.19):

$$\overline{\mathbf{D}^2}_{\Omega_3}(\mathcal{Y}, \mathcal{M}; \mathbf{r}; f; t) = \mathrm{Eqlbrm}(\mathcal{M}; f)$$

$$+ \overline{\mathbf{D}^2}_{3,negl}(\mathcal{Y}, \mathcal{M}; \mathbf{r}; f; t) + \overline{\mathbf{D}^2}_{3,uneven}(\mathcal{Y}, \mathcal{M}; \mathbf{r}; f; t), \tag{4.20}$$

where

$$\text{Eqlbrm}(\mathcal{M}; f) = \sigma_0^2(f) \sum_{k=1}^{N} \mu_k^2 = \left(\sum_{\substack{\mathbf{n} \in \mathbb{Z}^3: \\ \mathbf{n} \neq \mathbf{0}}} |a_{\mathbf{n}}|^2 \right) \left(\sum_{k=1}^{N} \mu_k^2 \right)$$

is the familiar *equilibrium part*, and

$$\overline{\mathbf{D}^2}_{3,negl}(\mathcal{Y}, \mathcal{M}; \mathbf{r}; f; t)$$

$$= - \sum_{k=1}^{N} \mu_k^2 \sum_{\substack{\mathbf{n} \in \mathbb{Z}^3: \\ \mathbf{n} \neq \mathbf{0}}} |a_{\mathbf{n}}|^2 \, 9 \left(\frac{\cos(2\pi |\mathbf{r}(t)| \, |\mathbf{n}|)}{(2\pi |\mathbf{r}(t)| \, |\mathbf{n}|)^2} - \frac{\sin(2\pi |\mathbf{r}(t)| \, |\mathbf{n}|)}{(2\pi |\mathbf{r}(t)| \, |\mathbf{n}|)^3} \right)^2 \tag{4.21}$$

is the *negligible part*: it tends to zero as $|\mathbf{r}(t)| \to \infty$, and we also have the inequality

$$0 \leq \text{Eqlbrm}(\mathcal{M}; f) + \overline{\mathbf{D}^2}_{3,negl}(\mathcal{Y}, \mathcal{M}; \mathbf{r}; f; t) \leq \text{Eqlbrm}(\mathcal{M}; f) = \sigma_0^2(f) \sum_{k=1}^{N} \mu_k^2. \tag{4.22}$$

Finally, we have the *unevenness*

$$\overline{\mathbf{D}^2}_{3,uneven}(\mathcal{Y}, \mathcal{M}; \mathbf{r}; f; t) = \sum_{\substack{\mathbf{n} \in \mathbb{Z}^3: \\ \mathbf{n} \neq \mathbf{0}}} |a_{\mathbf{n}}|^2 \left| \sum_{k=1}^{N} \mu_k e^{2\pi i \mathbf{n} \cdot \mathbf{y}_k} \right|^2$$

$$\cdot 9 \left(\frac{\cos(2\pi |\mathbf{r}(t)| \, |\mathbf{n}|)}{(2\pi |\mathbf{r}(t)| \, |\mathbf{n}|)^2} - \frac{\sin(2\pi |\mathbf{r}(t)| \, |\mathbf{n}|)}{(2\pi |\mathbf{r}(t)| \, |\mathbf{n}|)^3} \right)^2, \tag{4.23}$$

which is the dominating part as long as the system stays in off-equilibrium.

For the torus model of the third kind $(\mathcal{Y}, \mathcal{M}; \mathcal{R}; \Omega_3)$ we can define LA-equilibrium (L-equilibrium) by simply replacing Ω_1 and ProdMeas_1 in Definition 2.4 (Definition 2.3) with Ω_3 and ProdMeas_3.

Next we derive an analog of Corollary 4.1 from Theorem 4.2(a) exactly the same way as we derived Corollary 3.1 from Theorem 3.1 (or derived Corollary 4.1 from Theorem 4.1). Again for simplicity assume that the particles have unit mass: $\mu_k = 1$ for all $1 \leq k \leq N$. Then by (4.17),

$$\left| \overline{\mathbf{D}^2}_{\Omega_3}(\mathcal{Y}; \mathcal{R}; f; t) - N\sigma_0^2(f) \right|$$

$$\leq \min \left\{ \frac{18}{(2\pi \mathcal{R}(t))^4} + \frac{18}{(2\pi \mathcal{R}(t))^6}, 1 \right\} N^2 \sigma_0^2(f),$$

which implies that

$$\left|\overline{\mathbf{D}}_3(\mathcal{Y};\mathcal{R};f;t) - N\sigma_0^2(f)\right| \le N\sigma_0^2(f)$$

if $\mathcal{R}(t) \ge N^{1/4}/2$. Therefore, if $\mathcal{R}(t) \ge N^{1/4}/2$ holds, we have

$$\int_{\Omega_3} \overline{\mathbf{D}^2}(\mathcal{Y};\mathcal{R};\omega;f;t)\,d\omega \le 2N\sigma_0^2(f).$$

Applying the (trivial) Markov inequality, we obtain the following result.

Corollary 4.2. *Consider a general 3-dimensional N-particle torus model of the third kind* $(\mathcal{Y};\mathcal{R};\Omega_3)$ *in the 3-dimensional unit torus* $I^3 = [0,1)^3$ *such that every particle has mass one. Let t be a time satisfying*

$$\inf_{\tau\ge t} \mathcal{R}(\tau) \ge N^{1/4}/2.$$

Then, for every $0<\varepsilon<1$, *the system* $(\mathcal{Y};\mathcal{R};\Omega_3)$ *is in LA-equilibrium with parameters* ε *and* $C = 2/\varepsilon$ *at time t.*

Switching from the 3-dimensional unit torus I^3 *to the general d-dimensional unit torus* I^d *with* $d\ge 1$, $N^{1/4}$ *is replaced by* $N^{1/(d+1)}$. □

One can easily obtain the d-dimensional analogs of these results by making the modification described at the end of Theorem 4.2.

Corollaries 3.1, 4.1 and 4.2(a) indicate very different upper bounds for the required time to reach LA-equilibrium. Starting from an arbitrary N-element initial point configuration (including Big Bang and any other extreme case), it requires $\le \sqrt{\log N}$, $\le \sqrt{N}$ and $\le N^{1/4}$ jumps (= relaxation distance) in the torus model of the first, second and third kind, respectively. Next we show that these upper bounds are best possible (apart from some absolute constant). For simplicity we focus on the 3-dimensional case. We study the special case where $\mathcal{R} = \mathbf{r}$, the particles have unit mass: $\mu_k = 1$ for all $1\le k\le N$, and the initial point configuration is a Big Bang $\mathbf{y}_1 = \mathbf{y}_2 = \ldots = \mathbf{y}_N$: formally $\mathcal{Y} = \mathcal{Y}_{BB}$. Since this is an extreme off-equilibrium, the corresponding "unevenness" overwhelmingly dominates the Big Average Square-Discrepancy

$$\overline{\mathbf{D}^2}_{\Omega_1}(\mathcal{Y}_{BB};\mathbf{r};f;t),\ \overline{\mathbf{D}^2}_{\Omega_2}(\mathcal{Y}_{BB};\mathbf{r};f;t)\ \text{ and }\ \overline{\mathbf{D}^2}_{\Omega_3}(\mathcal{Y}_{BB};\mathbf{r};f;t).$$

In this special case the first kind "unevenness" simplifies to

$$\sup_{\tau\ge t}\overline{\mathbf{D}^2}_{1,uneven}(\mathcal{Y}_{BB};\mathbf{r};f;\tau) = N^2 \sup_{\tau\ge t} \sum_{\substack{\mathbf{n}\in\mathbb{Z}^3:\\ \mathbf{n}\ne\mathbf{0}}} |a_{\mathbf{n}}|^2 \exp\left(-(2\pi|\mathbf{r}(\tau)|\,|\mathbf{n}|)^2\right)$$

$$\approx \sigma_0^2(f) N^2 \sup_{\tau\ge t} \exp\left(-(2\pi|\mathbf{r}(\tau)|)^2\right) \tag{4.24}$$

if the test function f is "reasonable".

Similarly, for the general torus model of the second kind we have

$$\sup_{\tau\ge t}\overline{\mathbf{D}^2}_{2,uneven}(\mathcal{Y}_{BB};\mathbf{r};f;\tau) = N^2\sup_{\tau\ge t}\sum_{\substack{\mathbf{n}\in\mathbb{Z}^3:\\ \mathbf{n}\ne\mathbf{0}}}|a_{\mathbf{n}}|^2\left(\frac{\sin(2\pi|\mathbf{r}(\tau)|\,|\mathbf{n}|)}{2\pi|\mathbf{r}(\tau)|\,|\mathbf{n}|}\right)^2$$

$$\approx \sigma_0^2(f)N^2\max\left\{\sup_{\tau\ge t}\frac{1}{(2\pi|\mathbf{r}(\tau)|)^2},1\right\}, \tag{4.25}$$

if the test function f is "reasonable".

Finally, for the general torus model of the third kind,

$$\sup_{\tau\ge t}\overline{\mathbf{D}^2}_{3,uneven}(\mathcal{Y}_{BB};\mathbf{r};f;\tau)$$

$$= N^2\sup_{\tau\ge t}\sum_{\substack{\mathbf{n}\in\mathbb{Z}^3:\\ \mathbf{n}\ne\mathbf{0}}}|a_{\mathbf{n}}|^2\, 9\left(\frac{\cos(2\pi|\mathbf{r}(\tau)|\,|\mathbf{n}|)}{(2\pi|\mathbf{r}(\tau)|\,|\mathbf{n}|)^2}-\frac{\sin(2\pi|\mathbf{r}(\tau)|\,|\mathbf{n}|)}{(2\pi|\mathbf{r}(\tau)|\,|\mathbf{n}|)^3}\right)^2$$

$$\approx \sigma_0^2(f)N^2\max\left\{\sup_{\tau\ge t}\frac{9}{(2\pi|\mathbf{r}(\tau)|)^4},1\right\}, \tag{4.26}$$

if the test function f is "reasonable".

In this special case the equilibrium part simplifies to

$$\text{Eqlbrm}(f) = \sigma_0^2(f)N. \tag{4.27}$$

Reaching LA-equilibrium means that the "unevenness" goes down to the size of the equilibrium part. By (4.24)–(4.27) this happens when

$$N^2\exp\left(-(2\pi|\mathbf{r}(t)|)^2\right) = N \Longleftrightarrow |\mathbf{r}(t)| = \frac{\sqrt{\log N}}{2\pi}, \tag{4.28}$$

$$N^2\max\left\{\frac{1}{(2\pi|\mathbf{r}(t)|)^2},1\right\} = N \Longleftrightarrow |\mathbf{r}(t)| \text{ is around } \sqrt{N}, \tag{4.29}$$

$$N^2\max\left\{\frac{9}{(2\pi|\mathbf{r}(t)|)^4},1\right\} = N \Longleftrightarrow |\mathbf{r}(t)| \text{ is around } N^{1/4}. \tag{4.30}$$

This *intuition* suggests that in the general 3-dimensional torus model of the first or second or third kind, starting from Big Bang, the typical time evolution of the N-element point system reaches LA-equilibrium when the threshold

$$\inf_{\tau\ge t}|\mathbf{r}(\tau)|$$

equals $\frac{\sqrt{\log N}}{2\pi}$ (the case of the first kind), and for the other two kinds the threshold is in the range $\sqrt{N}$ (second kind) and $N^{1/4}$ (third kind).

We emphasize that (4.28)–(4.30) is hardly more than an intuition. Indeed,

$$\overline{\mathbf{D}^2}_{\Omega_i}(\mathcal{Y};\mathbf{r};f;t) = \int_{\Omega_i} \overline{\mathbf{D}^2}(\mathcal{Y};\mathbf{r};\omega;f;t)\,d\omega, \quad i=1,2,3 \tag{4.31}$$

represent only the *expected value* (taken over the corresponding space $\Omega_1, \Omega_2, \Omega_3$ of initial velocities). What we are really interested in is the *typical behaviour*, that is, we want to know what happens for the overwhelming majority of the initial velocities. To answer this question, we also need to know the *standard deviation*. By using the expectation and the standard deviation together, it is easy to describe the typical behaviour. Indeed, it is just a routine application of Chebyshev's inequality. In the next section we study the variance (= the square of the standard deviation); see Theorem 5.1.

5. Estimating a crucial "variance", and more on starting from Big Bang

What is in Section 5? *First we introduce and study an important technical concept: the Variance of the ω-dependent Average Square-Discrepancy. This Variance is a key tool to prove Theorems 3.1, 4.1, and 4.2.*

Then we discuss a Monotonicity Property, which is the first sign of the "entropy increase" in our time-quantitative second law.

Finally, we study Instant Equilibrium, which is the fastest possible approach to equilibrium.

Definition of the Variance of the ω-Dependent Average Square-Discrepancy. For $i = 1, 2, 3$ write (Var stands for Variance)

$$\mathrm{Var}_{\omega\in\Omega_i}\overline{\mathbf{D}^2}(\mathcal{Y},\mathcal{M};\mathbf{r};\omega;f;t)$$

$$= \int_{\Omega_i}\left(\overline{\mathbf{D}^2}(\mathcal{Y},\mathcal{M};\mathbf{r};\omega;f;t) - \overline{\mathbf{D}^2}_{\Omega_i}(\mathcal{Y},\mathcal{M};\mathbf{r};f;t)\right)^2 d\omega. \tag{5.1}$$

In Theorem 5.1 below cases (a), (b) and (c) correspond to the general torus model of the first (= Gaussian), second (= like photons), third kind (= uniform in the unit ball), and the general 3-dimensional isotropic case, respectively.

Theorem 5.1. *(a) Similarly to Theorem 3.1, we study a general N-particle torus model of the first kind in the 3-dimensional unit torus $I^3 = [0,1)^3$. Consider the special case $\mathcal{R} = \mathbf{r}$. By using the notation*

$$\text{Eqlbrm} = \text{Eqlbrm}(\mathcal{M}; f) = \sigma_0^2(f)\left(\sum_{k=1}^{N} \mu_k^2\right)$$

for the equilibrium part and

$$\overline{\mathbf{D}^2}_{1,uneven}(t) = \overline{\mathbf{D}^2}_{1,uneven}(\mathcal{Y}, \mathcal{M}; \mathbf{r}; f; t)$$

for the "unevenness", we have

$$\text{Variance}_{\omega \in \Omega_1} \overline{\mathbf{D}^2}(\mathcal{Y}, \mathcal{M}; \mathbf{r}; \omega; f; t)$$

$$\leq 14\text{Eqlbrm} \cdot \overline{\mathbf{D}^2}_{1,uneven}(t) + 14\text{Eqlbrm}^{3/2}\left(\overline{\mathbf{D}^2}_{1,uneven}(t)\right)^{1/2} + 14\text{Eqlbrm}^2. \tag{5.2}$$

By taking square root in (5.2), we obtain the following upper bound for the standard deviation

$$\text{StandardDeviation}_{\omega \in \Omega_1} \overline{\mathbf{D}^2}(\mathcal{Y}, \mathcal{M}; \mathbf{r}; \omega; f; t)$$

$$= \left(\text{Variance}_{\omega \in \Omega_1} \overline{\mathbf{D}^2}(\mathcal{Y}, \mathcal{M}; \mathbf{r}; \omega; f; t)\right)^{1/2}$$

$$\leq 4\text{Eqlbrm} + 4\sqrt{\text{Eqlbrm}}\sqrt{\overline{\mathbf{D}^2}_{1,uneven}(t)} + 4\text{Eqlbrm}^{3/4}\left(\overline{\mathbf{D}^2}_{1,uneven}(t)\right)^{1/4}. \tag{5.3}$$

These results extend to every dimension $d \geq 1$ without any modification.

(b) Next, similarly to Theorem 4.1 and Theorem 4.2(a), we study a general N-particle torus model of the second kind and third kind in the 3-dimensional unit torus I^3. Consider the special case $\mathcal{R} = \mathbf{r}$. By using the notation

$$\overline{\mathbf{D}^2}_{i,uneven}(t) = \overline{\mathbf{D}^2}_{i,uneven}(\mathcal{Y}, \mathcal{M}; \mathbf{r}; f; t), \ i = 2, 3$$

for the "unevenness", we have the perfect analog of (5.2):

$$\text{Variance}_{\omega \in \Omega_i} \overline{\mathbf{D}^2}(\mathcal{Y}, \mathcal{M}; \mathbf{r}; \omega; f; t)$$

$$\leq 14\text{Eqlbrm} \cdot \overline{\mathbf{D}^2}_{i,uneven}(t) + 14\text{Eqlbrm}^{3/2}\left(\overline{\mathbf{D}^2}_{i,uneven}(t)\right)^{1/2} + 14\text{Eqlbrm}^2. \tag{5.4}$$

Again taking square-root, we have the analog of (5.3):

$$\text{StandardDeviation}_{\omega \in \Omega_i} \overline{\mathbf{D}^2}(\mathcal{Y}, \mathcal{M}; \mathbf{r}; \omega; f; t)$$

$$= \left(\text{Variance}_{\omega\in\Omega_i}\overline{\mathbf{D}^2}(\mathcal{Y},\mathcal{M};\mathbf{r};\omega;f;t)\right)^{1/2}$$

$$\le 4\text{Eqlbrm} + 4\sqrt{\text{Eqlbrm}}\sqrt{\overline{\mathbf{D}^2}_{i,uneven}(t)} + 4\text{Eqlbrm}^{3/4}\left(\overline{\mathbf{D}^2}_{i,uneven}(t)\right)^{1/4},$$

$$i = 2, 3. \tag{5.5}$$

For the second kind these results extend to every dimension $d \ge 2$ *with the modification that the critical function* $\frac{\sin z}{z}$ *is replaced by*

$$(d-2)!!\frac{J_{(d-2)/2}(z)}{z^{(d-2)/2}}\cdot\left\{1 \text{ or } \sqrt{\frac{\pi}{2}}\right\}, \tag{5.6}$$

depending on whether $d \ge 2$ *is even or odd, accordingly.*

For the third kind these results extend to every dimension with the modification that the critical function

$$-3\left(\frac{\cos z}{z^2} - \frac{\sin z}{z^3}\right)$$

is replaced by $\frac{\sin z}{z}$ *for* $d = 1$, *and replaced by*

$$d!!\frac{J_{d/2}(z)}{z^{d/2}}\cdot\left\{1 \text{ or } \sqrt{\frac{\pi}{2}}\right\} \tag{5.7}$$

for the rest of the dimensions, depending on whether $d \ge 2$ *is even or odd, accordingly.*

(c) Finally, similarly to Theorem 4.2(b), consider the general isotropic case in the standard 3-space. It is described by a (basically arbitrary) initial speed density function $g(y)$ *(i.e.,* $\Pr[\rho_j \le u] = \int_0^u g(y)\,dy$*) such that*

$$g(u) \ge 0 \text{ for } 0 \le u < \infty, \quad \int_0^\infty g(u)\,du = 1, \quad \int_0^\infty g^2(u)\,du < \infty,$$

$$\text{and} \quad \int_0^\infty u g(u)\,du < \infty.$$

Then we can formulate the perfect analog of (5.2)–(5.5) with the straightforward modification that the key functions

$$F(w) = e^{-w^2/2}, \quad \frac{\sin w}{w} \quad \text{and} \quad -3\frac{\cos w}{w^2} + 3\frac{\sin w}{w^3}$$

in Theorem 5.1(a) and (b) are replaced by the corresponding function, (see Theorem 4.2(b))

$$\mathbf{F}_g(w) = \mathbf{F}_g(\mathbf{w}) = \int_0^\infty \frac{1}{4\pi}\int_{\mathbf{v}\in\mathbf{S}^2} e^{iu\mathbf{v}\cdot\mathbf{w}} g(u)\,du\,d\text{SA}(\mathbf{v})$$

$$= \int_0^\infty \frac{1}{4\pi} \int_{\mathbf{v}\in\mathbf{S}^2} \cos(u\mathbf{v}\cdot\mathbf{w}) g(u)\, du\, d\mathrm{SA}(\mathbf{v}),$$

where $\mathbf{w}$ *is an arbitrary 3-dimensional vector,* $w = |\mathbf{w}|$ *is the length, and* $d\mathrm{SA}(\mathbf{v})$ *represents the surface area.* $\mathbf{F}_g$ *is called the isotropic Fourier transform of* g*.*

The analog of (5.2)–(5.5) in the general case goes as follows. As usual, we consider the special case $\mathcal{R} = \mathbf{r}$*. By using the notation*

$$\overline{\mathbf{D}^2}_{F_g,uneven}(t) = \overline{\mathbf{D}^2}_{F_g,uneven}(\mathcal{Y},\mathcal{M};\mathbf{r};f;t)$$

$$= \sum_{\substack{\mathbf{n}\in\mathbb{Z}^3:\\ \mathbf{n}\neq\mathbf{0}}} |a_{\mathbf{n}}|^2 \left| \sum_{k=1}^{N} \mu_k e^{2\pi i \mathbf{n}\cdot\mathbf{y}_k} \right|^2 \mathbf{F}_g^2(2\pi|\mathbf{r}(t)|\,|\mathbf{n}|)$$

for the "unevenness part", we have

$$\mathrm{Variance}_{\omega\in\Omega_g} \overline{\mathbf{D}^2}(\mathcal{Y},\mathcal{M};\mathbf{r};\omega;f;t)$$

$$\le 14\mathrm{Eqlbrm}\cdot\overline{\mathbf{D}^2}_{F_g,uneven}(t) + 14\mathrm{Eqlbrm}^{3/2}\left(\overline{\mathbf{D}^2}_{F_g,uneven}(t)\right)^{1/2} + 14\mathrm{Eqlbrm}^2.$$

Taking square-root, we conclude

$$\mathrm{StandardDeviation}_{\omega\in\Omega_g} \overline{\mathbf{D}^2}(\mathcal{Y},\mathcal{M};\mathbf{r};\omega;f;t)$$

$$= \left(\mathrm{Variance}_{\omega\in\Omega_g} \overline{\mathbf{D}^2}(\mathcal{Y},\mathcal{M};\mathbf{r};\omega;f;t)\right)^{1/2}$$

$$\le 4\mathrm{Eqlbrm} + 4\sqrt{\mathrm{Eqlbrm}}\sqrt{\overline{\mathbf{D}^2}_{F_g,uneven}(t)} + 4\mathrm{Eqlbrm}^{3/4}\left(\overline{\mathbf{D}^2}_{F_g,uneven}(t)\right)^{1/4}.$$

For the proof; see Section 9.

By using Theorems 3.1, 4.1, 4.2 and 5.1, it is easy to convert the intuition (4.24)–(4.30) into a precise argument. Again we focus on the 3-dimensional case, and study the special case where $\mathcal{R} = \mathbf{r}$, the particles have unit mass: $\mu_k = 1$ for all $1 \le k \le N$, and the initial point configuration is a Big Bang $\mathbf{y}_1 = \mathbf{y}_2 = \ldots = \mathbf{y}_N$: formally $\mathcal{Y} = \mathcal{Y}_{BB}$. Let $\tau \ge t$ be fixed. By Chebyshev's inequality, for the *overwhelming majority* of the initial velocities $\omega \in \Omega_i$, $i = 1, 2, 3$,

$$\overline{\mathbf{D}^2}(\mathcal{Y}_{BB};\mathbf{r};\omega;f;\tau) = \mathrm{Expectation} + O(\mathrm{StandardDeviation})$$

$$= \overline{\mathbf{D}^2}_{\Omega_i}(\mathcal{Y}_{BB};\mathbf{r};f;\tau) + O(\mathrm{StandardDeviation}_{\omega\in\Omega_i}\overline{\mathbf{D}^2}(\mathcal{Y}_{BB};\mathbf{r};\omega;f;\tau))$$

$$= \overline{\mathbf{D}^2}_{\Omega_i}(\mathcal{Y}_{BB};\mathbf{r};f;\tau) + O(\mathrm{Eqlbrm}) + O(\sqrt{\mathrm{Eqlbrm}}\sqrt{\overline{\mathbf{D}^2}_{i,uneven}(\tau)})$$

$$= O(\mathrm{Eqlbrm}) + O(\overline{\mathbf{D}^2}_{i,uneven}(\tau)), \tag{5.8}$$

where we used (4.8)–(4.11), (4.20)–(4.23), (5.5) and (5.7), and the implicit constant factor in the O-notation is absolute.

In the special case of Bing Bang, by (4.24),

$$\sup_{\tau\ge t}\overline{\mathbf{D}^2}_{1,uneven}(\tau) = \sup_{\tau\ge t}\overline{\mathbf{D}^2}_{1,uneven}(\mathcal{Y}_{BB};\mathbf{r};f;\tau) \approx \sigma_0^2(f)N^2\sup_{\tau\ge t}e^{-(2\pi|\mathbf{r}(\tau)|)^2} \tag{5.9}$$

for the first kind if the test function f is "reasonable", similarly by (4.25),

$$\sup_{\tau\ge t}\overline{\mathbf{D}^2}_{2,uneven}(\tau) = \sup_{\tau\ge t}\overline{\mathbf{D}^2}_{2,uneven}(\mathcal{Y}_{BB};\mathbf{r};f;\tau)$$

$$\approx\sigma_0^2(f)N^2\max\left\{\sup_{\tau\ge t}\frac{\sin^2(2\pi|\mathbf{r}(\tau)|)}{(2\pi|\mathbf{r}(\tau)|)^2},1\right\}\approx\sigma_0^2(f)N^2\max\left\{\sup_{\tau\ge t}\frac{1}{(2\pi|\mathbf{r}(\tau)|)^2},1\right\} \tag{5.10}$$

for the second kind if the test function f is "reasonable", and finally, by (4.26),

$$\sup_{\tau\ge t}\overline{\mathbf{D}^2}_{3,uneven}(\tau) = \sup_{\tau\ge t}\overline{\mathbf{D}^2}_{3,uneven}(\mathcal{Y}_{BB};\mathbf{r};f;\tau)$$

$$\approx\sigma_0^2(f)N^2\max\left\{\sup_{\tau\ge t}\frac{9\cos^2(2\pi|\mathbf{r}(\tau)|)}{(2\pi|\mathbf{r}(\tau)|)^4},1\right\}\approx\sigma_0^2(f)N^2\max\left\{\sup_{\tau\ge t}\frac{9}{(2\pi|\mathbf{r}(\tau)|)^4},1\right\} \tag{5.11}$$

for the third kind if the test function f is "reasonable"; where of course the equilibrium part is

$$\mathrm{Eqlbrm} = \mathrm{Eqlbrm}(f) = \sigma_0^2(f)N.$$

We recall (5.8): for the *overwhelming majority* of the initial velocities $\omega\in\Omega_i$, $i=1,2,3$,

$$\overline{\mathbf{D}^2}(\mathcal{Y}_{BB};\mathbf{r};\omega;f;\tau) = O(\mathrm{Eqlbrm}) + O(\overline{\mathbf{D}^2}_{i,uneven}(\tau)),$$

so the system reaches LA-equilibrium (see Definition 2.4) exactly when the "unevenness" goes down to the size of the equilibrium part. By (5.9)–(5.11) this happens when

$$N^2e^{-(2\pi|\mathbf{r}(t)|)^2} = N \Longleftrightarrow |\mathbf{r}(t)| = \frac{\sqrt{\log N}}{2\pi} \tag{5.12}$$

$$N^2 \max\left\{\frac{1}{(2\pi|\mathbf{r}(t)|)^2}, 1\right\} \approx N \Longleftrightarrow |\mathbf{r}(t)| \text{ is around } \sqrt{N} \tag{5.13}$$

and

$$N^2 \max\left\{\frac{9}{(2\pi|\mathbf{r}(t)|)^4}, 1\right\} \approx N \Longleftrightarrow |\mathbf{r}(t)| \text{ is around } N^{1/4} \tag{5.14}$$

in the general torus model of the first, second and third kind, respectively. Notice that (5.12)–(5.14) is exactly (4.28)–(4.30), justifying our intuition in (4.24)–(4.30). We may refer to (5.12)–(5.14) as the *typical relaxation distance starting from Big Bang.*

It means that, starting from Big Bang, for the overwhelming majority of the initial velocities $\omega \in \Omega_i$, $i = 2, 3$, the general 3-dimensional torus model of the second or third kind exhibits a much slower approach to equilibrium. The threshold ("number of necessary jumps" or "relaxation distance")

$$\inf_{\tau \geq t} |\mathbf{r}(\tau)|$$

switches from the true value $\sqrt{\log N}/2\pi$ in (5.12) (= Gaussian velicity distribution) to the much larger order of $\sqrt{N}$ or $N^{1/4}$, respectively (see (5.13)–(5.14)).

For example, in a realistic gas model with a cubic container of side length 1 meter, N is around 10^{27}, and the average speed of the particles (= point billiards) is roughly around 10^3 meter per second at room temperature (depending on the gas). Then in the case of the Gaussian initial velocity distribution, starting from Big Bang, the typical time evolution of this point billiard system reaches LA-equilibrium in a few milliseconds (even for "ugly" test sets). On the other hand, replacing the Gaussian distribution with the same constant speed 10^3 m/s (i.e., switching from the first kind to the second kind), the typical time evolution of the system reaches LA-equilibrium in about 100 years.

Even if the initial point configuration is much less restrictive than Big Bang; for example, if at the beginning all particles are in the right half of the box, then, as the explicit formula (4.11) shows, the case of same constant speed 10^3 m/s still requires more than 10 years to reach LA-equilibrium. This is based on the elementary fact

$$\min_{0 \leq u \leq \pi} \max\{\sin u, \cos(2u)\} \geq 1/2.$$

The proof goes as follows. The condition "at the start all particles are in the right half of the box" means an initial point configuration $\mathcal{Y} = \{\mathbf{y}_1, \ldots, \mathbf{y}_N\}$

such that the first coordinates satisfy the inequality $0 \le y_{k,1} \le 1/2$, $1 \le k \le N$. We apply (4.11) with the following two test functions

$$f_1(x_1, x_2, x_3) = e^{2\pi i x_1} \quad \text{and} \quad f_2(x_1, x_2, x_3) = e^{4\pi i x_1}.$$

Thus we obtain

$$\overline{\mathbf{D}^2}_{2,uneven}(\mathcal{Y}; \mathbf{r}; f_1; t) = \left|\sum_{k=1}^{N} e^{2\pi i y_{k,1}}\right|^2 \left(\frac{\sin(2\pi|\mathbf{r}(t)|)}{2\pi|\mathbf{r}(t)|}\right)^2$$

$$\approx \left|\sum_{k=1}^{N} e^{2\pi i y_{k,1}}\right|^2 \cdot \frac{1/2}{(2\pi|\mathbf{r}(t)|)^2}, \tag{5.15}$$

and

$$\overline{\mathbf{D}^2}_{2,uneven}(\mathcal{Y}; \mathbf{r}; f_2; t) = \left|\sum_{k=1}^{N} e^{4\pi i y_{k,1}}\right|^2 \left(\frac{\sin(4\pi|\mathbf{r}(t)|)}{4\pi|\mathbf{r}(t)|}\right)^2$$

$$\approx \frac{1}{4}\left|\sum_{k=1}^{N} e^{4\pi i y_{k,1}}\right|^2 \cdot \frac{1/2}{(2\pi|\mathbf{r}(t)|)^2}. \tag{5.15$'$}$$

If there are at least $N/5$ values of $k \in \{1, 2, \ldots, N\}$ such that $1/12 \le y_{k,1} \le 5/12$, then

$$\left|\sum_{k=1}^{N} e^{2\pi i y_{k,1}}\right|^2 \ge \left(\sum_{k=1}^{N} \sin(2\pi i y_{k,1})\right)^2 \ge \left(\frac{N}{5} \cdot \frac{1}{2}\right)^2 = \frac{N^2}{100}. \tag{5.16}$$

On the other hand, if there are at least $4N/5$ values of $k \in \{1, 2, \ldots, N\}$ such that $y_{k,1} \notin [1/12, 5/12]$, then

$$\left|\sum_{k=1}^{N} e^{4\pi i y_{k,1}}\right| \ge \sum_{\substack{1 \le k \le N: \\ y_{k,1} \notin [1/12,5/12]}} \cos(4\pi i y_{k,1}) - \sum_{\substack{1 \le k \le N: \\ y_{k,1} \in [1/12,5/12]}} 1 \ge \frac{4N}{5} \cdot \frac{1}{2} - \frac{N}{5} = \frac{N}{5}.$$

So in this case

$$\frac{1}{4}\left|\sum_{k=1}^{N} e^{4\pi i y_{k,1}}\right|^2 \ge \frac{1}{4} \cdot \frac{N^2}{25} = \frac{N^2}{100}. \tag{5.17}$$

Combining (5.15)–(5.17), we see that the switch from Big Bang to "at the start all particles are in the right half of the box" results a factor of 100 in (at least) one of the two "unevenness" parts

$$\overline{\mathbf{D}^2}_{2,uneven}(\mathcal{Y}; \mathbf{r}; f_1; t) \quad \text{and} \quad \overline{\mathbf{D}^2}_{2,uneven}(\mathcal{Y}; \mathbf{r}; f_2; t).$$

Since both (5.15) and (5.15′) have the second power of $|\mathbf{r}(t)|$ in the denominator, the actual effect of the factor of 100 on the time scale is only the square-root $10 = \sqrt{100}$ (with respect to $f_1(x_1, x_2, x_3) = e^{2\pi i x_1}$ or $f_2(x_1, x_2, x_3) = e^{4\pi i x_1}$, depending on the initial configuration). That is, the actual effect is reduced to a factor of 10. We can summarize this argument in the inequality

$$\max\left\{\left|\sum_{k=1}^{N} e^{2\pi i y_{k,1}}\right|, \frac{1}{2}\left|\sum_{k=1}^{N} e^{4\pi i y_{k,1}}\right|\right\} \geq \frac{N}{10}, \tag{5.18}$$

where (5.18) holds for all $0 \leq y_{k,1} \leq 1/2$, $1 \leq k \leq N$. Since N represents Big Bang, (5.18) explains the switch from (at least) 100 years in Big Bang to (at least) $10 = 100/10$ years if at the start all particles are in the right half, i.e., we divide 100 by the factor 10.

(By using Euler's formula $e^{iz} = \cos z + i \sin z$, we can replace the complex test functions $e^{2\pi i x_1}$ and $e^{4\pi i x_1}$ with the real ones $\cos(2\pi x_1)$, $\sin(2\pi x_1)$, $\cos(4\pi x_1)$, $\sin(4\pi x_1)$.)

What intuitive explanation can one give to the striking difference between a few milliseconds and more than 10 years? Here is one attempt: if all particles start from the same point (Big Bang), and go with the same speed (with uniformly distributed random directions in the 3-space: second kind), the particles continue to stay on a *growing sphere*, radiating outwards. It clearly takes a long time (= many jumps) for the growing sphere to sufficiently "fold and twist" (due to the modulo one effect of the torus) before the points on the sphere themselves become "randomly distributed" in the unit torus (in the sense of "square-root" equilibrium). However, if we switch from Big Bang to the much less restrictive condition that at the beginning all points are in the (say) right half of the unit torus, then the "growing sphere" argument collapses, and it is not clear at all why the system still needs more than 10 years to reach "square-root" equilibrium in the torus. We have to admit, we do not have any convincing intuitive explanation for this surprising phenomenon.

Note that there is a generalization of (5.18) as follows. Big Bang represents the most extreme nonequilibrium initial point configuration, and the case of "all particles start from the right half of the box" is another extreme nonequilibrium initial point configuration. The message of (5.18) is that the quantitative difference between the time scales to reach ("square-root") equilibrium in these two extreme cases of initial configurations — with same constant speed — is (at most) a factor of 10, independently of the number of particles N. What happens in a less extreme nonequilibrium initial point configuration? The case of "all particles start from the right half of the box" is equivalent to the condition that at the start the left half of the box is empty. A general case of extreme nonequilibrium initial point configuration is where at the start some axis-parallel box $B \subset I^3 = [-\frac{1}{2}, \frac{1}{2})^3$ is empty — let $\beta = \text{vol}(B) > 0$ denote the volume of the empty box. The following lemma can be interpreted as a sort of analog of (5.18) in the general case.

Lemma 5.1. *Let $\mathcal{Y} = \{\mathbf{y}_1, \mathbf{y}_2, \ldots, \mathbf{y}_N\} \subset I^3$ be an N-element point set in the unit cube $I^3 = [-\frac{1}{2}, \frac{1}{2})^3$ such that $\mathcal{Y}$ is disjoint from some axis-parallel box $B \subset I^3$ with volume $\text{vol}(B) = \beta > 0$. Then there exists a nonzero lattice point $\mathbf{n} = (n_1, n_2, n_3) \in \mathbb{Z}^3 \setminus \mathbf{0}$ with $\max\{|n_1|, |n_2|, |n_3|\} \le 14/\beta$ such that*

$$\frac{1}{|\mathbf{n}|}\left|\sum_{k=1}^{N} \mu_k e^{2\pi i \mathbf{n}\cdot\mathbf{y}_k}\right| > 10^{-5}\beta^4 N.$$

We postpone the proof of Lemma 5.1 to Section 8.

We can use Lemma 5.1 the same way as we used (5.18) above. As we said, Big Bang represents the most extreme nonequilibrium initial point configuration in the unit cube I^3, and the case of "at the start there is no particle in some axis-parallel box $B \subset I^3$ of volume $\beta > 0$" is another extreme nonequilibrium initial point configuration. The message of Lemma 5.1 is that the quantitative difference between the time scales to reach ("square-root") equilibrium in these two extreme cases of initial configurations — for the torus model of the second kind — is (at most) a factor of $10^5\beta^{-4}$, independently of the number of particles N. In other words, all of these extreme nonequilibrium initial point configurations are basically the same as Big Bang (apart from some absolute constant factor).

It is easy to see that the switch from the torus model of the second kind to the third kind basically means that we have to take *square-root*, and the switch to the first kind basically means that we have to take the *square-root of the logarithm*.

Next we switch from Big Bang to an *arbitrary* initial point configuration $\mathcal{Y}$ and an *arbitrary* general mass distribution $\mathcal{Y}$ with $\sum_{k=1}^{N} \mu_k = N$. We repeat the standard Chebyshev inequality argument; see (5.8). More precisely, we combine Chebyshev's inequality with (5.3) for the first kind and with (5.5) for the second and third kinds. We obtain that for the overwhelming majority of the initial velocities $\omega \in \Omega_i$, $i = 1, 2, 3$,

$$\overline{\mathbf{D}^2}(\mathcal{Y}, \mathcal{M}; \mathbf{r}; \omega; f; t) = \text{Expectation} + O(\text{StandardDeviation})$$

$$= \overline{\mathbf{D}^2}_{\Omega_i}(\mathcal{Y},\mathcal{M};\mathbf{r};f;t) + O\left(\text{StandardDeviation}_{\omega\in\Omega_i}\overline{\mathbf{D}^2}(\mathcal{Y},\mathcal{M};\mathbf{r};\omega;f;t)\right)$$

$$= \overline{\mathbf{D}^2}_{\Omega_i}(\mathcal{Y},\mathcal{M};\mathbf{r};f;t) + O(\text{Eqlbrm}) + O\left(\sqrt{\text{Eqlbrm}}\sqrt{\overline{\mathbf{D}^2}_{i,uneven}(t)}\right)$$

$$+ O\left(\text{Eqlbrm}^{3/4}\left(\overline{\mathbf{D}^2}_{i,uneven}(t)\right)^{1/4}\right)$$

$$= \overline{\mathbf{D}^2}_{i,uneven}(t) + O(\text{Eqlbrm}) + O\left(\sqrt{\text{Eqlbrm}}\sqrt{\overline{\mathbf{D}^2}_{i,uneven}(t)}\right)$$

$$+ O\left(\text{Eqlbrm}^{3/4}\left(\overline{\mathbf{D}^2}_{i,uneven}(t)\right)^{1/4}\right)$$

$$= \overline{\mathbf{D}^2}_{i,uneven}(t) + O(\text{Eqlbrm})$$

$$+ O\left(\text{Eqlbrm}^{3/4}\left(\overline{\mathbf{D}^2}_{i,uneven}(t)\right)^{1/4}\right), \tag{5.19}$$

where

$$\overline{\mathbf{D}^2}_{i,uneven}(t) = \overline{\mathbf{D}^2}_{i,uneven}(\mathcal{Y},\mathcal{M};\mathbf{r};f;t)$$

and

$$\overline{\mathbf{D}^2}_{1,uneven}(t) = \sum_{\substack{\mathbf{n}\in\mathbb{Z}^3:\\ \mathbf{n}\neq\mathbf{0}}} |a_{\mathbf{n}}|^2 \left|\sum_{k=1}^{N}\mu_k e^{2\pi i\mathbf{n}\cdot\mathbf{y}_k}\right|^2 \exp\left(-(2\pi|\mathbf{r}(t)|\ |\mathbf{n}|)^2\right), \tag{5.20}$$

$$\overline{\mathbf{D}^2}_{2,uneven}(t) = \sum_{\substack{\mathbf{n}\in\mathbb{Z}^3:\\ \mathbf{n}\neq\mathbf{0}}} |a_{\mathbf{n}}|^2 \left|\sum_{k=1}^{N}\mu_k e^{2\pi i\mathbf{n}\cdot\mathbf{y}_k}\right|^2 \left(\frac{\sin(2\pi|\mathbf{r}(t)|\ |\mathbf{n}|)}{2\pi|\mathbf{r}(t)|\ |\mathbf{n}|}\right)^2, \tag{5.20$'$}$$

$$\overline{\mathbf{D}^2}_{3,uneven}(t) = \sum_{\substack{\mathbf{n}\in\mathbb{Z}^3:\\ \mathbf{n}\neq\mathbf{0}}} |a_{\mathbf{n}}|^2 \left|\sum_{k=1}^{N}\mu_k e^{2\pi i\mathbf{n}\cdot\mathbf{y}_k}\right|^2$$

$$\cdot 9\left(\frac{\cos(2\pi|\mathbf{r}(t)|\ |\mathbf{n}|)}{(2\pi|\mathbf{r}(t)|\ |\mathbf{n}|)^2} - \frac{\sin(2\pi|\mathbf{r}(t)|\ |\mathbf{n}|)}{(2\pi|\mathbf{r}(t)|\ |\mathbf{n}|)^3}\right)^2, \tag{5.20$''$}$$

and, as usual,

$$\text{Eqlbrm} = \text{Eqlbrm}(\mathcal{M}; f) = \sigma_0^2(f)\left(\sum_{k=1}^{N} \mu_k^2\right), \tag{5.21}$$

and, finally, the implicit constant factor in the O-notation is absolute.

If the system starts from off-equilibrium, then the "unevenness" $\overline{\mathbf{D}^2}_{i,uneven}(t)$ is much larger than the equilibrium part Eqlbrm $=$ Eqlbrm$(\mathcal{M}; f)$, so by (5.19), for the overwhelming majority of the initial velocities $\omega \in \Omega_i$, $i = 1, 2, 3$

$$\overline{\mathbf{D}^2}(\mathcal{Y}, \mathcal{M}; \mathbf{r}; \omega; f; t) = \overline{\mathbf{D}^2}_{i,uneven}(t) + \text{negligible} \tag{5.22}$$

until the system reaches equilibrium — and then

$$\overline{\mathbf{D}^2}(\mathcal{Y}, \mathcal{M}; \mathbf{r}; \omega; f; t) \approx \text{Eqlbrm}(\mathcal{M}; f).$$

Monotonicity: the first sign of a "second law". The function $h_1(x) = e^{-x^2}$ is strictly monotone deceasing in the interval $0 < x < \infty$. It means that we have *strict monotonicity* for the torus model of the first kind (3-dimensional Gaussian initial velocity distribution) in the following sense. The dominating part in (5.22) (see (5.20))

$$\overline{\mathbf{D}^2}_{1,uneven}(t) = \sum_{\substack{\mathbf{n} \in \mathbb{Z}^3: \\ \mathbf{n} \neq \mathbf{0}}} |a_{\mathbf{n}}|^2 \left|\sum_{k=1}^{N} \mu_k e^{2\pi i \mathbf{n} \cdot \mathbf{y}_k}\right|^2 \exp\left(-(2\pi |\mathbf{r}(t)|\, |\mathbf{n}|)^2\right) \tag{5.23}$$

is a **monotone decreasing function of the drifting distance** $|\mathbf{r}(t)|$. Therefore, if the drifting distance is monotone increasing — which is a perfectly natural assumption — then, for the overwhelming majority of the initial velocities $\omega \in \Omega_1$, the time evolution of $\overline{\mathbf{D}^2}(\mathcal{Y}, \mathcal{M}; \mathbf{r}; \omega; f; t)$ is monotone decreasing — apart from a negligible error — until the system reaches equilibrium (more precisely, LA-equilibrium).

This *monotonicity* of the typical time evolution of $\overline{\mathbf{D}^2}(\mathcal{Y}, \mathcal{M}; \mathbf{r}; \omega; f; t)$ for *every* $f \in L^2$ resembles the *monotonicity* of the thermodynamic entropy (i.e., the Second Law). We will return to this idea in Chapter 3, and develop it into a new concept of entropy, called micro-entropy; see also Section 6.

Note that $\mathbf{D}^2(\mathcal{Y}, \mathcal{M}; \mathbf{r}; \omega; f; t)$ is simpler than $\overline{\mathbf{D}^2}(\mathcal{Y}, \mathcal{M}; \mathbf{r}; \omega; f; t)$ (see (2.28) and (2.31)). It is interesting to point out that for the typical time evolution of the simpler $\mathbf{D}^2(\mathcal{Y}, \mathcal{M}; \mathbf{r}; \omega; f; t)$ we cannot expect a similar monotone decreasing behavior — at least not for every $f \in L^2$. Indeed, consider the following 1-dimensional model: the underlying set is the unit interval $[0, 1)$, the test set S is the left half $S = [0, 1/2)$ (so $f = \chi_S$ is the characteristic function of S), the initial point configuration $\mathcal{Y}$ is a "two-point Big Bang", $N/2$ points in $\frac{1}{2} - \varepsilon$ with some very small $\varepsilon > 0$ and $N/2$ points in $3/4$, and every point has unit mass. At the start the test set S has zero discrepancy (the length of S is $1/2$, and it contains exactly $N/2$ points). For a typical time evolution, close to half of the $N/2$ points starting from $\frac{1}{2} - \varepsilon$ move right, and quickly leave S (since $\varepsilon > 0$ is very small). Then the error of S (= absolute value of the discrepancy) jumps up from zero to close to $N/4$, because the points starting from $3/4$ are still "far" from S. Until the points starting from $3/4$ reach S, the error of S remains close to $N/4$. When the points starting from $3/4$ reach S, then the error goes down substantially. So the typical time evolution of this particular system $\mathbf{D}^2(\mathcal{Y}, \mathcal{M}; \mathbf{r}; \omega; f; t)$ is *not* monotonic: first it makes the big jump up from 0 to $N/4$, and after that it falls drastically. This dramatic up-and-down behavior is in sharp contrast with the monotone decrease for the typical time evolution of the average $\overline{\mathbf{D}^2}(\mathcal{Y}, \mathcal{M}; \mathbf{r}; \omega; f; t)$ for the first kind $\omega \in \Omega_1$. Notice that this argument works for any kind of "reasonable" initial velocity distribution.

This simple example suggests that it is probably more promising to work with the more complicated average concept $\overline{\mathbf{D}^2}(\mathcal{Y}, \mathcal{M}; \mathbf{r}; \omega; f; t)$ in order to define a meaningful "entropy increase", which reflects the Second Law. And indeed, we start with the Second Law (and a corresponding Micro-Entropy) based on $\overline{\mathbf{D}^2}(\mathcal{Y}, \mathcal{M}; \mathbf{r}; \omega; f; t)$, which turns out to be technically simpler. (It is simpler, but, unfortunately, it is still quite long and complicated!)

It does not mean, however, that we cannot prove a Second Law based on the concept of $\mathbf{D}^2(\mathcal{Y}, \mathcal{M}; \mathbf{r}; \omega; f; t)$ instead of $\overline{\mathbf{D}^2}(\mathcal{Y}, \mathcal{M}; \mathbf{r}; \omega; f; t)$. Later — in the second half of Section 24 — we explain the changes required to prove a Second Law (expressed in terms of a corresponding Micro-Entropy); see Theorem 24.1, which is based on $\mathbf{D}^2(\mathcal{Y}, \mathcal{M}; \mathbf{r}; \omega; f; t)$.

Let us return to (5.23). The analog results for the second and third kinds are more complicated. For the second kind the function $h_2(x) = (\frac{\sin x}{x})^2$ is not monotone deceasing in the interval $0 < x < \infty$. Nevertheless, we can say, intuitively speaking, that $h_2(x)$ has a "tendency to decrease". Indeed, the mean value of $\sin^2 x$ is $1/2$ (due to the fact $\int_0^\pi \sin^2 x\, dx = \pi/2$), and the function $\frac{1}{2}x^{-2}$ is monotone deceasing in the interval $0 < x < \infty$. So, for the second kind the dominating part in (5.22) (see (5.20′))

$$\overline{\mathbf{D}^2}_{2,uneven}(t) = \sum_{\substack{\mathbf{n} \in \mathbb{Z}^3: \\ \mathbf{n} \neq \mathbf{0}}} |a_{\mathbf{n}}|^2 \left| \sum_{k=1}^{N} \mu_k e^{2\pi i \mathbf{n} \cdot \mathbf{y}_k} \right|^2 \left(\frac{\sin(2\pi |\mathbf{r}(t)|\, |\mathbf{n}|)}{2\pi |\mathbf{r}(t)|\, |\mathbf{n}|} \right)^2, \tag{5.23′}$$

is a **basically monotone decreasing function of the drifting distance** $|\mathbf{r}(t)|$.

For the third kind we replace $h_2(x)$ with $h_3(x) = x^{-4}\cos^2 x$, which is also not monotone deceasing in the interval $1 < x < \infty$. Nevertheless, we can say, intuitively speaking, that $h_3(x)$ has a "tendency to decrease". Indeed, the mean value of $\cos^2 x$ is $1/2$, and the function $\frac{1}{2}x^{-4}$ is monotone deceasing in the interval $1 < x < \infty$. So, for the third kind the dominating part in (5.22) (see (5.20″))

$$\overline{\mathbf{D}^2}_{3,uneven}(t) = \sum_{\substack{\mathbf{n}\in\mathbb{Z}^3:\\ \mathbf{n}\neq\mathbf{0}}} |a_{\mathbf{n}}|^2 \left|\sum_{k=1}^{N} \mu_k e^{2\pi i\mathbf{n}\cdot\mathbf{y}_k}\right|^2$$

$$\cdot 9\left(\frac{\cos(2\pi|\mathbf{r}(t)|\,|\mathbf{n}|)}{(2\pi|\mathbf{r}(t)|\,|\mathbf{n}|)^2} - \frac{\sin(2\pi|\mathbf{r}(t)|\,|\mathbf{n}|)}{(2\pi|\mathbf{r}(t)|\,|\mathbf{n}|)^3}\right)^2, \qquad (5.23'')$$

is also a **basically monotone decreasing function of the drifting distance** $|\mathbf{r}(t)|$.

Next question: What happens if we change the dimension; for example, $d = 3$ is replaced by $d = 2$ or $d = 4$ or any other value? For the first kind the dimension is basically irrelevant, but for the second (and third) kind the dimension makes a big difference. According to Theorem 4.1 and Theorem 5.1(b), in the general d-dimensional torus model of the second kind the crucial change is that the key function $\frac{\sin z}{z}$ in (4.7) is replaced by

$$(d-2)!!\frac{J_{(d-2)/2}(z)}{z^{(d-2)/2}} \cdot \left\{1 \text{ or } \sqrt{\frac{\pi}{2}}\right\}, \qquad (5.24)$$

depending on whether $d \geq 2$ is even or odd, accordingly.

If $\nu \geq 0$ is a half-integer, then $J_\nu(z)$ has a finite form, but for $\nu \geq 0$ integer the Bessel function $J_\nu(z)$ does not have a finite expression. What we can still use is an asymptotic expression — called Hankel's formula — that works for all $\nu \geq 0$. The first order approximation is

$$J_\nu(z) = \sqrt{\frac{2}{\pi z}}\cos\left(z - \frac{(2\nu+1)\pi}{4}\right) + O(z^{-3/2}) \text{ for } z \geq 1, \qquad (5.25)$$

and for illustration we also mention the second order approximation

$$J_\nu(z) = \sqrt{\frac{2}{\pi z}}\left(\cos\left(z - \frac{(2\nu+1)\pi}{4}\right) - \left(\nu^2 - \frac{1}{4}\right)\frac{\sin\left(z - \frac{(2\nu+1)\pi}{4}\right)}{z}\right)$$

$$+O(z^{-5/2}) \text{ for } z \geq 1 \qquad (5.26)$$

and $J_\nu(z) = O(1)$ for $0 \le z < 1$; here the implicit constants in the O-notations depend only on the value of $\nu \ge 0$.

Applying (5.25) in (5.24), we can reformulate the crucial change as follows: changing from the 3-dimensional torus to the d-dimensional torus means that $\frac{\sin z}{z}$ is replaced by

$$(d-2)!!\frac{\cos\left(z - \frac{(d-1)\pi}{4}\right)}{z^{(d-1)/2}} \cdot \left\{\sqrt{\frac{2}{\pi}} \text{ or } 1\right\} + O(z^{-(d+1)/2}) \tag{5.27}$$

(depending on whether $d \ge 2$ is even or odd) for $z \ge 1$.

For example, by choosing $d = 2$ in (5.27), we have that $\frac{\sin z}{z}$ is replaced by

$$\sqrt{\frac{2}{\pi}}\frac{\cos\left(z - \frac{\pi}{4}\right)}{z^{1/2}} + O(z^{-3/2}), \tag{5.28}$$

and choosing $d = 4$ in (5.27), we have that $\frac{\sin z}{z}$ is replaced by

$$2\sqrt{\frac{2}{\pi}}\frac{\cos\left(z - \frac{3\pi}{4}\right)}{z^{3/2}} + O(z^{-5/2}) \tag{5.29}$$

(for $z \ge 1$ in both cases).

It means that, the general 2-dimensional torus model of the second kind exhibits a much, much slower approach to equilibrium than the first kind. Indeed, starting from a Big Bang, or close to Big Bang where (say) all points start from the right half of the unit square, for the overwhelming majority of the initial velocities $\omega \in \Omega_i$, $i = 1, 2$, the number of necessary jumps (= relaxation drifting distance) to reach equilibrium switches from the order $\sqrt{\log N}$ in the Gaussian case to the much larger N for the second kind. For example, if N is in the range of the Avogadro number, the "gas container" is a square of side length 1 meter, and the speed is 10^3 m/s, then $N10^{-3}$ seconds is much more than the estimated age of the Universe. On the other hand, in the analog Gaussian case it takes only a few milliseconds for the system to reach equilibrium. The sharp contrast between the two cases is truly shocking.

In the general 4-dimensional torus model of the second kind, starting from a Big Bang, or close to Big Bang like all points start from the right half of the unit cube $[0, 1]^4$, for the overwhelming majority of the initial velocities $\omega \in \Omega_2$, the number of necessary jumps (= relaxation drifting distance) to reach equilibrium is around $N^{1/3}$.

In the general d-dimensional torus model of the second kind, the number of jumps to reach equilibrium is around $N^{1/(d-1)}$ (assuming that N is much larger than the dimension d). In contrast, for the first kind the number of jumps remains the same $\sqrt{\log N}$.

Finally, we switch form the second kind to the third kind. According to Theorem 4.2 and Theorem 5.1(b), in the general d-dimensional torus model of the third kind the crucial change is that the key function

$$-3\left(\frac{\cos z}{z^2} - \frac{\sin z}{z^3}\right) \tag{5.30}$$

is replaced by

$$d!!\frac{J_{d/2}(z)}{z^{d/2}} \cdot \left\{1 \text{ or } \sqrt{\frac{\pi}{2}}\right\} \tag{5.31}$$

depending on whether $d \geq 2$ is even or odd, accordingly. (In the 1-dimensional case (5.30) is replaced by $\frac{\sin z}{z}$ for $d = 1$. This part is trivial.)

Applying (5.25) in (5.31), and repeating the usual argument, we conclude that, in the general d-dimensional torus model of the third kind, the number of necessary jumps (= relaxation drifting distance) to reach equilibrium is around $N^{1/(d+1)}$ (again we assume that N is much larger than the dimension d). Note that the threshold $N^{1/(d+1)}$ remains true even for $d = 1$.

We summarize these results in the following corollary.

Corollary 5.1. *Corollaries 3.1, 4.1 and 4.2 are best possible in the following sense. Consider a general N-particle torus model in the 3-dimensional unit torus I^3 with initial point configuration*

$$\mathcal{Y} = \{\mathbf{y}_1, \mathbf{y}_2, \ldots, \mathbf{y}_N\} \subset I^3$$

equipped with a parametrized space curve $\mathbf{r}(t)$ such that $\mathbf{r}(0) = \mathbf{0}$; moreover, every particle has mass one.

Assume that the initial point configuration is off-equilibrium in the precise sense that $\mathcal{Y}$ is disjoint from some axis-parallel box $B \subset I^3$ of volume $\beta > 0$. Then for the overwhelming majority of the initial velocities $\omega \in \Omega_i$, $i = 1, 2, 3$ the time evolution of the system $(\mathcal{Y}; \mathbf{r}; \omega)$ does not reach LA-equilibrium until the threshold ("number of jumps" or "relaxation drifting distance") $\inf_{\tau \geq t} |\mathbf{r}(\tau)|$ becomes as large as

$$\inf_{\tau \geq t} |\mathbf{r}(\tau)| \geq \frac{\sqrt{\log N}}{2\pi} \quad \text{or} \quad \geq C(\beta) \cdot \sqrt{N} \quad \text{or} \quad \geq C(\beta) \cdot N^{1/4} \tag{5.32}$$

depending on whether it is the torus model of the first or second or third kind, in this order. The absolute constant factor $C(\beta) > 0$ in (5.32) depends only on the value of the volume $\beta > 0$.

Switching from dimension $d = 3$ to an arbitrary $d \geq 2$, $\sqrt{N}$ and $N^{1/4}$ in (5.32) generalize to $N^{1/(d-1)}$ and $N^{1/(d+1)}$, respectively; $\sqrt{\log N}/2\pi$, on the other hand, remains unchanged. □

We prove Theorem 3.1 in Section 7, we prove Theorems 4.1 and 4.2 and Lemmas 5.1 and 5.2 in Section 8, we prove Theorem 5.1 in Section 9, and finally, we prove Lemma 3.2 in Section 10.

What is the fastest approach to equilibrium? Instant equilibrium via "super-duper fast" speed distributions. The Gaussian initial velocity distribution ("first kind") represents a superexponentially fast approach to equilibrium. This leads to the natural question: Is it true that the Gaussian case represents the fastest approach to equilibrium? Well, the answer is no: the extremely fast approach of the Gaussian case is *not* the fastest. The isotropic velocity distribution with speed (5.45) — see below — is *faster*. In fact, there is an infinite class of speed distributions with the same property.

To explain this surprising fact, we need some simple results from Fourier analysis. First we introduce the so-called "roof function"

$$f_2(y) = \begin{cases} \frac{1}{2} - \frac{|y|}{4}, & if |y| \leq 2; \\ 0, & if |y| > 2. \end{cases} \tag{5.33}$$

The roof function $f_2(y)$ is familiar from probability theory: it is the density function of the convolution (denoted as usual by $*$) of the uniform distribution in $[-1, 1]$ with itself; formally,

$$f_2(y) = f_1 * f_1(y) = \int_{-\infty}^{\infty} f_1(y - z) f_1(z)\, dz,$$

where

$$f_1(y) = \begin{cases} \frac{1}{2}, & if |y| \leq 1; \\ 0, & if |y| > 1, \end{cases} \tag{5.34}$$

is the probability density function of the uniform distribution in $[-1, 1]$.

It is a standard exercise in probability theory to compute the higher convolution powers of the uniform distribution in $[-1, 1]$. For example, a routine calculation gives

$$f_3(y) = f_1 * f_1 * f_1(y) = \begin{cases} (3 - y^2)/8, & if |y| \leq 1; \\ (3 - |y|)^2/16, & if 1 \leq |y| \leq 3; \\ 0, & if |y| > 3, \end{cases} \tag{5.35}$$

and

$$f_4(y) = f_1 * f_1 * f_1 * f_1(y) = \begin{cases} (3|y|^3 - 12y^2 + 32)/96, & \textit{if } |y| \le 2; \\ (-|y|^3 + 12y^2 - 48|y| + 64)/96, & \textit{if } 2 \le |y| \le 4; \\ 0, & \textit{if } |y| > 4. \end{cases} \tag{5.36}$$

In general, let $\kappa \ge 2$ be an arbitrary integer, and consider the κ-th convolution power of f_1:

$$f_\kappa(y) = f_1 * f_1 * \cdots * f_1(y) \quad (\kappa \text{ times convolution}). \tag{5.37}$$

It is not very hard to prove by induction on κ the following generalization of (5.33) and (5.35)–(5.36) (see e.g. Rényi's book [Re62]):

$$f_\kappa(y) = \frac{1}{2^\kappa(\kappa-1)!} \sum_{j=0}^{\lfloor(\kappa+|y|)/2\rfloor} (-1)^j \binom{\kappa}{j} (\kappa+|y|-2j)^{\kappa-1} \text{ if } |y| \le \kappa, \tag{5.38}$$

and 0 if $|y| > \kappa$. Note that $f_\kappa(y)$ is $(\kappa-2)$-times differentiable, and consists of a few generalized parabola arcs of degree $\kappa - 2$ (due to the jumps of the lower integral part function $\lfloor(\kappa+y)/2\rfloor$ as y runs in $0 \le y \le \kappa$) that smoothly fit together at the endpoints if $\kappa \ge 3$.

We need the Fourier transform of f_1:

$$\widehat{f_1}(u) = \frac{1}{\sqrt{2\pi}} \int_{-1}^{1} \frac{1}{2} e^{-iyu}\, dy = \frac{1}{\sqrt{2\pi}} \frac{\sin u}{u}. \tag{5.39}$$

We use the well known fact that the Fourier transform of a convolution is the product of the Fourier transforms; formally,

$$\widehat{f * h} = \widehat{f}\, \widehat{h}. \tag{5.40}$$

By (5.39) and (5.40) we have

$$\widehat{f_\kappa}(u) = \left(\frac{1}{\sqrt{2\pi}} \frac{\sin u}{u} \right)^\kappa. \tag{5.41}$$

Write

$$g_\kappa(u) = (1 - \cos u) \left(\frac{1}{\sqrt{2\pi}} \frac{\sin u}{u} \right)^\kappa$$

$$= \left(1 - \frac{1}{2}e^{iu} - \frac{1}{2}e^{-iu}\right) \left(\frac{1}{\sqrt{2\pi}} \frac{\sin u}{u} \right)^\kappa. \tag{5.42}$$

Then by (5.41) and (5.42),

$$\widehat{g_\kappa}(y) = f_\kappa(y) - \frac{1}{2} f_\kappa(y+1) - \frac{1}{2} f_\kappa(y-1), \tag{5.43}$$

and so $\widehat{g_\kappa}(y) = 0$ for all $|y| \ge \kappa + 1$.

If $\kappa \ge 2$ is an even integer, then $g_\kappa(u) \ge 0$, so $c_\kappa g_\kappa(u)$ defines a probability density function in $0 \le u < \infty$ with the constant factor c_κ satisfying

$$\frac{1}{c_\kappa} = \int_0^\infty g_\kappa(u)\,du = \frac{1}{2}\int_{-\infty}^\infty g_\kappa(u)\,du = \sqrt{\pi/2}\cdot\frac{1}{\sqrt{2\pi}}\int_{-\infty}^\infty g_\kappa(u)\,du$$

$$= \sqrt{\pi/2}\cdot\widehat{g_\kappa}(0) = \sqrt{\pi/2}\left(f_\kappa(0) - \frac{1}{2}f_\kappa(1) - \frac{1}{2}f_\kappa(-1)\right). \tag{5.44}$$

Let $\kappa = 4$; then by (5.36) we have

$$\frac{1}{c_4} = \sqrt{\pi/2}\left(f_4(0) - \frac{1}{2}f_4(1) - \frac{1}{2}f_4(-1)\right) = \frac{32-23}{96}\sqrt{\pi/2} = \frac{3}{32}\sqrt{\pi/2}.$$

Combining these facts we obtain that

$$g(u) = c_4 g_4(u) = \frac{32}{3}\sqrt{\frac{2}{\pi}}g_4(u)$$

$$= \frac{32}{3}\sqrt{\frac{2}{\pi}}(1-\cos u)\left(\frac{1}{\sqrt{2\pi}}\frac{\sin u}{u}\right)^4 = \frac{8\sqrt{2}}{3\pi^2\sqrt{\pi}}(1-\cos u)\left(\frac{\sin u}{u}\right)^4 \tag{5.45}$$

defines a probability density function in $0 \le u < \infty$ such that $\widehat{g}(y) = 0$ for all $|y| \ge 5$.

It easily follows from (5.45) that

$$\int_0^\infty u g(u)\,du < \infty \quad \text{and} \quad \int_0^\infty u^2 g(u)\,du < \infty, \tag{5.46}$$

that is, the first two moments of the corresponding random variable are both finite. (The reason why we did not choose $\kappa = 2$ is that the corresponding expected value $\int_0^\infty u c_2 g_2(u)\,du$ is infinite.)

The corresponding initial velocity space is

$$\omega = (\rho_1, \vartheta_1, \rho_2, \vartheta_2, \ldots, \rho_N, \vartheta_N) \in \Omega_g = ([0,\infty) \times SO(3))^N \tag{5.47}$$

equipped with the following product measure: Haar measure for $SO(3)$ and the probability density function $g = c_4 g_4$ (defined in (5.45)) for $[0,\infty)$).

Then we have

$$\mathbf{D}^2_{\Omega_g}(\mathcal{Y}, \mathcal{M}; \mathcal{R}; f; t) = \int_{\Omega_g} \mathbf{D}^2(\mathcal{Y}, \mathcal{M}; \mathcal{R}; \omega; f; t)\,d\omega$$

$$= \sum_{k=1}^N \mu_k^2 \sum_{\substack{\mathbf{n}\in\mathbb{Z}^3:\\ \mathbf{n}\ne\mathbf{0}}} |a_\mathbf{n}|^2 + \sum_{\substack{\mathbf{n}\in\mathbb{Z}^3:\\ \mathbf{n}\ne\mathbf{0}}} |a_\mathbf{n}|^2 \sum_{1\le k,\ell\le N: k\ne\ell} \mu_k\mu_\ell e^{2\pi i\mathbf{n}\cdot(\mathbf{y}_k - \mathbf{y}_\ell)}$$

$$\cdot \mathbf{F}_g(2\pi|\mathbf{r}_k(t)|\;|\mathbf{n}|)\mathbf{F}_g(2\pi|\mathbf{r}_\ell(t)|\;|\mathbf{n}|)$$

$$+ \sum_{\substack{\mathbf{n}_j \in \mathbb{Z}^3 : j=1,2 \\ \mathbf{0} \neq \mathbf{n}_1 \neq \mathbf{n}_2 \neq \mathbf{0}}} a_{\mathbf{n}_1}$$

$$\cdot \overline{a_{\mathbf{n}_2}} \sum_{1 \le k, \ell \le N : k \neq \ell} \mu_k \mu_\ell e^{2\pi i (\mathbf{n}_1 \cdot \mathbf{y}_k - \mathbf{n}_2 \cdot \mathbf{y}_\ell)} \mathbf{F}_g(2\pi |\mathbf{r}_k(t)|\, |\mathbf{n}_1|) \mathbf{F}_g(2\pi |\mathbf{r}_\ell(t)|\, |\mathbf{n}_2|)$$

$$+ \sum_{\substack{\mathbf{n}_j \in \mathbb{Z}^3 : j=1,2 \\ \mathbf{0} \neq \mathbf{n}_1 \neq \mathbf{n}_2 \neq \mathbf{0}}} a_{\mathbf{n}_1} \overline{a_{\mathbf{n}_2}} \sum_{1 \le k \le N} \mu_k^2 e^{2\pi i (\mathbf{n}_1 - \mathbf{n}_2) \cdot \mathbf{y}_k} \mathbf{F}_g(2\pi |\mathbf{r}_k(t)|\, |\mathbf{n}_1 - \mathbf{n}_2|), \quad (5.48)$$

where (see Theorem 4.2(b))

$$\mathbf{F}_g(w) = \frac{\sqrt{\pi}}{2\sqrt{2}w} \left(g(0)\sqrt{2\pi} - \int_{|x| \ge w} \widehat{g}(x)\, dx \right) = -\frac{\sqrt{\pi}}{2\sqrt{2}w} \int_{|x| \ge w} \widehat{g}(x)\, dx, \quad (5.49)$$

since $g(0) = 0$ (see (5.45)).

Note that (5.48) is basically an analog of formula (3.2) in Theorem 3.1. More precisely, (5.48) is the perfect analog of (7.9), which is an intermediate step in the proof of (3.2). (Of course *analog* means that Ω_1 is replaced by Ω_g and $F(w) = e^{-w^2/2}$ is replaced by $F(w) = \mathbf{F}_g(w)$ in (5.49).)

Since $\widehat{g}(x) = \widehat{g_4}(x) = 0$ for all $|x| \ge 5$, under the condition

$$\mathcal{R}(t) = \min_{1 \le k \le N} |\mathbf{r}_k(t)| \ge \frac{5}{2\pi}, \quad (5.50)$$

we have $\mathbf{F}_g(2\pi |\mathbf{r}_k(t)|\, |\mathbf{n}|) = 0$ for all k and $\mathbf{0} \neq \mathbf{n}$, and $\mathbf{F}_g(2\pi |\mathbf{r}_k(t)|\, |\mathbf{n}_1 - \mathbf{n}_2|) = 0$ for all k and $\mathbf{0} \neq \mathbf{n}_1 \neq \mathbf{n}_2 \neq \mathbf{0}$, which imply that the equation (5.48)–(5.49) simplifies to

$$\mathbf{D}^2_{\Omega_g}(\mathcal{Y}, \mathcal{M}; \mathcal{R}; f; t) = \int_{\Omega_g} \mathbf{D}^2(\mathcal{Y}, \mathcal{M}; \mathcal{R}; \omega; f; t)\, d\omega$$

$$= \sum_{k=1}^{N} \mu_k^2 \sum_{\substack{\mathbf{n} \in \mathbb{Z}^3 : \\ \mathbf{n} \neq \mathbf{0}}} |a_{\mathbf{n}}|^2 = \sigma_0^2(f) \sum_{k=1}^{N} \mu_k^2 = \text{Eqlbrm}(\mathcal{M}; f). \quad (5.51)$$

For comparison, we recall (3.15): if

$$\mathcal{R}(t) \ge \frac{\sqrt{\log N}}{2\pi}, \quad (5.52)$$

and every mass is one, then

$$\int_{\Omega_1} \mathbf{D}^2(\mathcal{Y}; \mathcal{R}; \omega; f; t)\, d\omega \le 2N\sigma_0^2(f). \quad (5.53)$$

We can say that (5.50)–(5.51) is an "upgrading" of (5.52)–(5.53) in the sense that the threshold $\sqrt{\log N}$ (in (5.52)) is replaced by an absolute constant (in (5.50)) independent of the number N of the particles. The existence of an initial velocity distribution exhibiting such an unexpected/strange property is very surprising.

Note that Corollary 3.2 was obtained by applying Markov inequality in (3.15). Similarly, applying Markov inequality in (5.51), we obtain the following shocking "instant equilibrium" analog of Corollary 3.2.

Corollary 5.2. *Consider a general N-particle torus model $(\mathcal{Y};\mathcal{M};\mathcal{R};\Omega_g)$ in the 3-dimensional unit torus $I^3 = [0,1)^3$ with super-duper fast speed distribution defined by (5.45) and (5.47). Let t be a time satisfying*

$$\inf_{\tau\ge t}\mathcal{R}(\tau) \ge \frac{5}{2\pi}.$$

Then, for every $0<\varepsilon<1$, the system $(\mathcal{Y};\mathcal{M};\mathcal{R};\Omega_g)$ is in L-equilibrium with parameters ε and $C=1/\varepsilon$ at time t. □

By choosing $\kappa = 4$, i.e., the probability density function $g(u) = c_4 g_4(u)$ in (5.45), the first two moments of the corresponding random variable are finite (see (5.46)), but the third moment is infinite. By choosing a larger even number κ (i.e., $g(u) = c_\kappa g_\kappa(u)$) we can guarantee that many more higher moments are finite. In fact, the first $\kappa - 2$ moments of the corresponding random variable are finite.

Next we show another, more general new method to find super-duper fast speed distributions. The old method breaks down in even dimensions $d \ge 2$, but the new method works in every dimension $d \ge 2$. Let $\mathbf{h}(\mathbf{x})$, $\mathbf{x}\in\mathbf{R}^d$ be a rotation-invariant real valued function in d variables with compact support, i.e., $\mathbf{h}(\mathbf{x}) = \mathbf{h}(x,0,\ldots,0)$ for all $|\mathbf{x}| = x$, and there is a constant C such that $\mathbf{h}(\mathbf{x}) = 0$ for all $|\mathbf{x}| > C$. The convolution $\mathbf{h}_2 = \mathbf{h} * \mathbf{h}$ defines another rotation-invariant function in d variables with compact support. In fact, $\mathbf{h}_2(\mathbf{x}) = 0$ for all $|\mathbf{x}| > 2C$, and we refer to this as the *key property* of $\mathbf{h}_2$.

Using the well known fact that the Fourier transform of a convolution is the product of the Fourier transforms, we have

$$\widehat{\mathbf{h}_2} = \widehat{\mathbf{h}*\mathbf{h}} = \left(\widehat{\mathbf{h}}\right)^2 \ge 0.$$

Since $\widehat{\mathbf{h}_2}$ is a non-negative rotation-invariant function in d variables, the equation

$$g(u) = u^{d-1}\widehat{\mathbf{h}_2}(\mathbf{u}) \ \text{ with } \ u = |\mathbf{u}| \tag{5.54}$$

defines a non-negative function in one variable $u \geq 0$. By assuming some smoothness of the initial function $\mathbf{h}(\mathbf{x})$, $\mathbf{x} \in \mathsf{R}^d$, we can easily guarantee that $g(u)$ has a finite integral

$$\int_0^\infty g(u)\, du < \infty.$$

What is more, multiplying the initial function $\mathbf{h}(\mathbf{x})$, $\mathbf{x} \in \mathsf{R}^d$ with some constant, we can also guarantee that $g(u)$, $u \geq 0$ is a probability density function

$$\int_0^\infty g(u)\, du = 1.$$

By raising the level of smoothness of $\mathbf{h}(\mathbf{x})$, we can even guarantee that the corresponding random variable has finite expectation, finite variance, finite third moment, finite fourth moment, and so on.

We claim that the isotropic Fourier transform $\mathbf{F}_g$ of this g has a compact support; in fact,

$$\mathbf{F}_g(w) = 0 \quad \text{for all} \quad w > 2C. \tag{5.55}$$

The proof of Corollary 5.2 explains why the compact support property (5.55) is exactly what we need to guarantee that $g(u)$ is a super-duper fast speed density function, exhibiting *instant equilibrium* independently of the initial configuration.

Fact (5.55) immediately follows from the last formula in Theorem 4.2:

$$\mathbf{F}_g(\mathbf{w}) = \int_{\mathsf{R}^d} e^{i\mathbf{x}\cdot\mathbf{w}} \frac{g(|\mathbf{x}|)}{|\mathbf{x}|^{d-1}}\, d\mathbf{x} = (2\pi)^{d/2}\widehat{\mathfrak{g}_d}(-\mathbf{w}), \tag{5.56}$$

where

$$\mathfrak{g}_d(\mathbf{x}) = \frac{g(|\mathbf{x}|)}{|\mathbf{x}|^{d-1}}$$

and $\widehat{\mathfrak{g}_d}(\mathbf{w})$ is its Fourier transform. Indeed, in this special case (5.54) yields

$$\mathfrak{g}_d(\mathbf{x}) = \frac{g(|\mathbf{x}|)}{|\mathbf{x}|^{d-1}} = \widehat{\mathbf{h}_2}(\mathbf{x}),$$

and so

$$\widehat{\mathfrak{g}_d}(\mathbf{w}) = \mathbf{h}_2(\mathbf{w}) \tag{5.57}$$

(where we used the trivial fact that rotation-invariant functions are symmetric).

Using (5.57) in (5.56), we have

$$\begin{aligned} \mathbf{F}_g(w) &= \mathbf{F}_g(\mathbf{w}) = (2\pi)^{d/2}\widehat{\mathfrak{g}_d}(-\mathbf{w}) \\ &= (2\pi)^{d/2}\mathbf{h}_2(-\mathbf{w}) = 0 \quad \text{for all} \quad w = |\mathbf{w}| > 2C, \end{aligned} \tag{5.58}$$

where the last line follows from the *key property* of $\mathbf{h}_2$. (5.58) proves (5.55).

Finally note that $C > 0$ can be arbitrary, but if C is very small or very large, then the mean speed — i.e., the expected value of the probability density function g — turns out to be, respectively, very large or very small (they have a roughly reciprocal relation, due to the "uncertainty principle" of the Fourier transform). So, we have to apply some simple rescaling operation to force the mean speed to be around one (like we have in the 3 basic cases).

In the next section we give a somewhat informal preview of our Second Law.

6. Micro-Entropy and a time-quantitative Second Law: A preview

What is in Section 6? *We recall the one-sentence summary of the book from the preface one more time:*
(1) first we show that in our implicit interaction models the typical time evolution of the large closed system approaches a well-defined Micro-Equilibrium; (2) then we define a new entropy concept on a microscopic level that measures the distance from this microscopic equilibrium (in fact, our Micro-Entropy = absolute constant minus this "distance"); and finally, (3) we prove that this Micro-Entropy satisfies a time-quantitative Second Law, i.e., the typical time evolution of the system exhibits Micro-Entropy increase in an explicit time-dependent form such that reaching maximum Micro-Entropy coincides with the state of Micro-Equilibrium.

Here we elaborate on this.

Sections 3–5 clarify step (1) (for the missing proofs; see Chapter 2). Next comes step (2): Micro-Entropy. It means that, given an arbitrary configuration of the particles and a velocity distribution (say, Gaussian), we want to define a single number — called the corresponding Micro-Entropy of the system in this instant — with the property that the typical time evolution of the system exhibits an increase of this Micro-Entropy.

How can a single number — Micro-Entropy — describe the "disorder" of the whole large multi-particle system in an instant?

Unfortunately, the only way we can carry out this plan rigorously is rather complicated. It is outlined by the following long diagram (which is recalled from the beginning of Section 2). The diagram lists the key steps of how starting from Micro-Equilibrium we eventually reach Micro-Entropy and our time-quantitative Second Law:

$$\text{Square Root (Snapshot) Equilibrium}$$

$$\longrightarrow \text{Relative Disparity (a Boltzmann entropy like quantity)}$$

$$\longrightarrow \text{Disparity Function} \longrightarrow \text{Unevenness Function}$$

$$\longrightarrow \text{Relaxation Time and Relaxation Drifting Distance}$$

$$\longrightarrow \text{Imbalance} \longrightarrow \text{MicroEntropy}$$

$$\longrightarrow \text{MicroEntropy increase described by the Time Evolution Function}$$

$$= \text{Time} - \text{quantitative Second Law},$$

where all of these concepts are defined by explicit computable functions.

We elaborate on this diagram. We already introduced L-equilibrium = Square-Root (Snapshot) Equilibrium in Definitions 2.3–4 — it is our rigorous definition of Micro-Equilibrium. Later in the section we formulate a somewhat vague statement, called Metatheorem, that includes many of the precise definitions of the remaining key concepts. After the Metatheorem, we formulate our Time-quantitative Second Law, which describes the typical time evolution of the Micro-Entropy, including of course the entropy increase (which is the crucial property).

We work out the details of the clarification/proof of the Metatheorem, and the proof of our Second Law, later in Chapters 2–4.

An important intermediate step in the definition of our Micro-Entropy is to introduce the Unevenness Function. The definition of the Unevenness Function requires some preparation. We want to find a definition that describes the time evolution of the maximum "disparity" or "unevenness" (microscopic and macroscopic) of the system at time $t > 0$, starting from $(\mathcal{Y}, \mathcal{M}; \mathbf{r}; \omega)$ that represents $t = 0$; here $\omega \in \Omega_i$ for some $i \in \{1, 2, 3\}$.

It is instructive to start with a wrong choice. It is very tempting to introduce the following ω-dependent function that we call the Individual Disparity Function

$$\text{IndDispFun}(\omega; t) = \text{IndDispFun}(\mathcal{Y}, \mathcal{M}; \mathbf{r}; \omega; t)$$

$$= \max \left\{ \sup_{f \in L^2:\ \sigma_0^2(f) > 0} \frac{1}{\sigma_0^2(f)} \overline{\mathbf{D}^2}(\mathcal{Y}, \mathcal{M}; \mathbf{r}; \omega; f; t), \sum_{k=1}^{N} \mu_k^2 \right\}. \tag{6.1}$$

Note that $\overline{\mathbf{D}^2}(\mathcal{Y},\mathcal{M};\mathbf{r};\omega;f;\tau)$ is the Average Square-Discrepancy defined in (2.31).

Unfortunately, the Individual Disparity Function turns out to be *useless*, because $\text{IndDispFun}(\omega;t) = N^2$ for every ω and t.

We "save" (6.1) in the following way. We recall (5.19): if at time $t > 0$ the system is still far from equilibrium, then for the overwhelming majority of the initial velocities $\omega \in \Omega_i$, $i = 1,2,3$, the ω-dependent $\overline{\mathbf{D}^2}(\mathcal{Y},\mathcal{M};\mathbf{r};\omega;f;t)$ is almost equal to the ω-independent $\overline{\mathbf{D}^2}_{\Omega_i}(\mathcal{Y},\mathcal{M};\mathbf{r};f;t)$, i.e.,

$$\overline{\mathbf{D}^2}(\mathcal{Y},\mathcal{M};\mathbf{r};\omega;f;t) = \overline{\mathbf{D}^2}_{\Omega_i}(\mathcal{Y},\mathcal{M};\mathbf{r};f;t) + \text{negligible},$$

and it holds until the system reaches equilibrium.

This motivates the introduction of the following ω-independent *Disparity Function*

$$\text{DispFun}_i(t) = \text{DispFun}_i(\mathcal{Y},\mathcal{M};\mathbf{r};t)$$

$$= \max\left\{\sup_{f\in L^2:\ \sigma_0^2(f)>0} \frac{1}{\sigma_0^2(f)}\overline{\mathbf{D}^2}_{\Omega_i}(\mathcal{Y},\mathcal{M};\mathbf{r};f;t), \sum_{k=1}^{N}\mu_k^2\right\}, \tag{6.2}$$

see (2.32), (4.4) and (4.17).

In order to obtain a function that is monotone decreasing in t, we introduce the *Sup-Disparity Function*

$$\text{SupDispFun}_i(t) = \text{SupDispFun}_i(\mathcal{Y},\mathcal{M};\mathbf{r};t)$$

$$= \sup_{\tau\geq t}\text{DispFun}_i(\mathcal{Y},\mathcal{M};\mathbf{r};\tau)$$

$$= \max\left\{\sup_{f\in L^2:\ \sigma_0^2(f)>0}\ \sup_{\tau\geq t} \frac{1}{\sigma_0^2(f)}\overline{\mathbf{D}^2}_{\Omega_i}(\mathcal{Y},\mathcal{M};\mathbf{r};f;\tau), \sum_{k=1}^{N}\mu_k^2\right\}. \tag{6.3}$$

Note that

$$0 \leq \overline{\mathbf{D}^2}_{\Omega_1}(\mathcal{Y},\mathcal{M};\mathbf{r};f;t) - \overline{\mathbf{D}^2}_{1,uneven}(\mathcal{Y},\mathcal{M};\mathbf{r};f;t)$$

$$= \text{Eqlbrm}(\mathcal{M};f) + \overline{\mathbf{D}^2}_{1,negl}(\mathcal{Y},\mathcal{M};\mathbf{r};f;t)$$

$$\leq \text{Eqlbrm}(\mathcal{M};f) = \sigma_0^2(f)\sum_{k=1}^{N}\mu_k^2, \tag{6.4}$$

see (3.5)–(3.8), (4.10) and (4.22).

In view of (6.4), it makes only a negligible difference to replace $\overline{\mathbf{D}^2}_{\Omega_1}(\mathcal{Y},\mathcal{M};\mathbf{r};f;t)$ in (6.3) with $\overline{\mathbf{D}^2}_{1,uneven}(\mathcal{Y},\mathcal{M};\mathbf{r};f;t)$ ($\sum_{k=1}^{N}\mu_k^2$ is the "white noise of equilibrium"). Thus we obtain the *Unevenness Function*

$$\text{UnevenFun}_i(t) = \text{UnevenFun}_i(\mathcal{Y},\mathcal{M};\mathbf{r};t)$$

$$= \max\left\{ \sup_{f\in L^2:\ \sigma_0^2(f)>0}\ \sup_{\tau\geq t} \frac{1}{\sigma_0^2(f)}\overline{\mathbf{D}^2}_{i,uneven}(\mathcal{Y},\mathcal{M};\mathbf{r};f;\tau), \sum_{k=1}^{N}\mu_k^2 \right\}, \quad (6.5)$$

see (3.9), (4.11) and (4.23). Notice that in (6.1), (6.2), (6.3) and (6.5) we take the maximum over all possible measurable test sets; what is more, over all $f \in L^2$ (this is a key idea that was already discussed in detail at the beginning of Section 2, see Theorem A there). The case of (6.1) is the trivial/useless way, and (6.2), (6.3) and (6.5) are the subtle/useful ways to do that.

Next we formulate a somewhat vague statement, called Metatheorem, that introduces our Micro-Entropy. It is the key step to our Time-quantitative Second Law. The clarification and the proof of the Metatheorem comes later in Chapters 3–4.

Metatheorem. *Consider a general 3-dimensional N-particle torus model with initial point configuration $(\mathcal{Y},\mathcal{M})$ where each particle has the same space curve*

$$\mathbf{r} = \mathbf{r}(t) = \mathbf{r}_1(t) = \mathbf{r}_2(t) = \ldots = \mathbf{r}_N(t).$$

Then the Unevenness Function, defined in (6.5), has the alternative form expressed in terms of the initial configuration

$$\text{UnevenFun}_1(\mathcal{Y},\mathcal{M};\mathbf{r};t)$$

$$= \max\left\{ \max_{\mathbf{n}\in\mathbb{Z}^3\setminus\mathbf{0}} \left|\sum_{k=1}^{N}\mu_k e^{2\pi i\mathbf{n}\cdot\mathbf{y}_k}\right|^2 \cdot \sup_{\tau\geq t}\exp\left(-(2\pi|\mathbf{r}(\tau)|\ |\mathbf{n}|)^2\right), \sum_{k=1}^{N}\mu_k^2 \right\} \quad (6.6)$$

for the first kind (Gaussian), and

$$\text{UnevenFun}_2(\mathcal{Y},\mathcal{M};\mathbf{r};t)$$

$$= \max\left\{ \max_{\mathbf{n}\in\mathbb{Z}^3\setminus\mathbf{0}} \left|\sum_{k=1}^{N}\mu_k e^{2\pi i\mathbf{n}\cdot\mathbf{y}_k}\right|^2 \sup_{\tau\geq t}\frac{\sin^2(2\pi|\mathbf{r}(\tau)|\ |\mathbf{n}|)}{(2\pi|\mathbf{r}(\tau)|\ |\mathbf{n}|)^2}, \sum_{k=1}^{N}\mu_k^2 \right\} \quad (6.7)$$

for the second kind.

Finally,

$$\text{UnevenFun}_3(\mathcal{Y},\mathcal{M};\mathbf{r};t) = \max\left\{ \max_{\mathbf{n}\in\mathbb{Z}^3\setminus\mathbf{0}} \left|\sum_{k=1}^{N} \mu_k e^{2\pi i\mathbf{n}\cdot\mathbf{y}_k}\right|^2 \right.$$

$$\left. \cdot 9\sup_{\tau\geq t}\left(\frac{\cos(2\pi|\mathbf{r}(\tau)|\,|\mathbf{n}|)}{(2\pi|\mathbf{r}(\tau)|\,|\mathbf{n}|)^2} - \frac{\sin(2\pi|\mathbf{r}(\tau)|\,|\mathbf{n}|)}{(2\pi|\mathbf{r}(\tau)|\,|\mathbf{n}|)^3}\right)^2, \sum_{k=1}^{N}\mu_k^2 \right\} \tag{6.8}$$

for the third kind.

It follows from (6.6)–(6.8) that the function $\text{UnevenFun}_i(t) = \text{UnevenFun}_i(\mathcal{Y},\mathcal{M};\mathbf{r};t)$ *is monotone decreasing in terms of the drifting distance* $|\mathbf{r}(t)|$. *Moreover, the function* $\text{UnevenFun}_i(t)$ *is* N^2 *at the start* $t = 0$, *and the system reaches LA-equilibrium when* $\text{UnevenFun}_i(t)$ *gets close to its minimum* $\sum_{k=1}^{N}\mu_k^2$ *in the sense that*

$$\text{UnevenFun}_i(\tau) = O\left(\sum_{k=1}^{N}\mu_k^2\right)$$

(which is $O(N)$ *if every mass* μ_k *is one).*

The torus model of the first kind reaches LA-equilibrium at time t_0 *when (see (6.6))*

$$2\pi\inf_{\tau\geq t_0}|\mathbf{r}(\tau)| = \max_{\mathbf{n}\in\mathbb{Z}^3\setminus\mathbf{0}}\frac{1}{|\mathbf{n}|}\left(\log\left(\frac{\left|\sum_{k=1}^{N}\mu_k e^{2\pi i\mathbf{n}\cdot\mathbf{y}_k}\right|^2}{\sum_{k=1}^{N}\mu_k^2} \text{ or } 1\right)_{\max}\right)^{1/2}, \tag{6.9}$$

where "$(x \text{ or } 1)_{\max}$*" means the larger number between* x *and 1, that is,* $(x \text{ or } 1)_{\max} = \max\{x,1\}$ *(we often use this unconventional notation in the rest to make the formulas more transparent). We refer to the time* t_0 *and the* R^3*-distance* $\inf_{\tau\geq t_0}|\mathbf{r}(\tau)|$ *defined by the equation (6.9) as the Relaxation Time and the Relaxation Drifting Distance, respectively, of the first kind system* $(\mathcal{Y},\mathcal{M};\mathbf{r};\Omega_1)$.

In view of (6.9) the expression

$$\frac{1}{2\pi}\max_{\mathbf{n}\in\mathbb{Z}^3\setminus\mathbf{0}}\frac{1}{|\mathbf{n}|}\left(\log\left(\frac{\left|\sum_{k=1}^{N}\mu_k e^{2\pi i\mathbf{n}\cdot\mathbf{y}_k}\right|^2}{\sum_{k=1}^{N}\mu_k^2} \text{ or } 1\right)_{\max}\right)^{1/2} \tag{6.9'}$$

represents the R^3*-distance for a typical particle in a typical time evolution of the first kind system* $(\mathcal{Y},\mathcal{M};\mathbf{r};\Omega_1)$ *between the start and the LA-equilibrium.*

In the interesting special case $|\mathbf{r}(t)| = t$ (which includes, among others, the Bernoulli billiard/torus model) we can replace "distance" with "time".

In (6.9) we basically used the function $\frac{1}{2\pi}\sqrt{\log x}$. Next we define the Imbalance, for which we apply the inverse function $e^{(2\pi x)^2}$ in (6.9): for the first kind we have

$$\text{Imbalance}_1(\mathcal{Y},\mathcal{M}) = \max_{\mathbf{n}\in\mathbb{Z}^3\setminus\mathbf{0}} \exp\left(\frac{1}{|\mathbf{n}|^2}\log\left(\frac{\left|\sum_{k=1}^N \mu_k e^{2\pi i\mathbf{n}\cdot\mathbf{y}_k}\right|^2}{\sum_{k=1}^N \mu_k^2} \text{ or } 1\right)_{\max}\right)$$

$$= \max_{\mathbf{n}\in\mathbb{Z}^3\setminus\mathbf{0}} \left(\frac{\left|\sum_{k=1}^N \mu_k e^{2\pi i\mathbf{n}\cdot\mathbf{y}_k}\right|^2}{\sum_{k=1}^N \mu_k^2} \text{ or } 1\right)_{\max}^{1/|\mathbf{n}|^2}. \tag{6.10}$$

Since $e^{(2\pi x)^2}$ is a superexponential function, $\text{Imbalance}_1(\mathcal{Y},\mathcal{M})$ *represents a superexponential function of the Relaxation Drifting Distance for the typical time evolution of $(\mathcal{Y},\mathcal{M};\mathbf{r};\Omega_1)$. In the interesting special case $|\mathbf{r}(t)| = t$ (which includes, among others, the Bernoulli billiard/torus model) we have*

$$\text{Imbalance} = \exp\left(2\pi\cdot\text{RelaxationTime}\right)^2.$$

Note that e^{x^2} is a monotone increasing function, and for $x > 0$, $e^{x^2} = O(1)$ iff $x = O(1)$. Also, the Imbalance is basically the negative Micro-Entropy (see (6.17) below), which completes this long chain of new concepts.

Next we switch to the second kind. Using the elementary fact

$$f_2(x) = \sup_{y\geq x}\left(\frac{\sin y}{y}\right)^2 = \frac{1}{x^2} + O(x^{-3}) \quad \text{for } x > 1,$$

we can rewrite (6.7) in the following simpler form

$$\text{UnevenFun}_2(\mathcal{Y},\mathcal{M};\mathbf{r};t)$$

$$= \max\left\{\max_{\mathbf{n}\in\mathbb{Z}^3\setminus\mathbf{0}}\left|\sum_{k=1}^N \mu_k e^{2\pi i\mathbf{n}\cdot\mathbf{y}_k}\right|^2 \min\left\{\sup_{\tau\geq t}\frac{1}{(2\pi|\mathbf{r}(\tau)|\ |\mathbf{n}|)^2}, 1\right\}, \sum_{k=1}^N \mu_k^2\right\}$$

$$+\text{negligible}. \tag{6.11}$$

It follows from (6.11) that the torus model of the second kind reaches LA-equilibrium at time t_0 when

$$2\pi\inf_{\tau\geq t_0}|\mathbf{r}(\tau)| = \max_{\mathbf{n}\in\mathbb{Z}^3\setminus\mathbf{0}}\frac{1}{|\mathbf{n}|}\left(\frac{\left|\sum_{k=1}^N \mu_k e^{2\pi i\mathbf{n}\cdot\mathbf{y}_k}\right|^2}{\sum_{k=1}^N \mu_k^2} \text{ or } 1\right)_{\max}^{1/2}. \tag{6.12}$$

We refer to the time t_0 and the $\mathbf{R}^3$-distance $\inf_{\tau\geq t_0}|\mathbf{r}(\tau)|$ defined by the equality (6.12) as the Relaxation Time and the Relaxation Drifting Distance, respectively, of the second kind system $(\mathcal{Y}, \mathcal{M}; \mathbf{r}; \Omega_2)$.

In view of (6.12)

$$\frac{1}{2\pi} \max_{\mathbf{n}\in\mathbb{Z}^3\setminus\mathbf{0}} \frac{1}{|\mathbf{n}|} \left(\frac{\left|\sum_{k=1}^{N} \mu_k e^{2\pi i\mathbf{n}\cdot\mathbf{y}_k}\right|^2}{\sum_{k=1}^{N} \mu_k^2} \text{ or } 1 \right)_{\max}^{1/2} \tag{6.12'}$$

represents the $\mathbf{R}^3$-distance for a typical particle in a typical time evolution of the second kind system $(\mathcal{Y}, \mathcal{M}; \mathbf{r}; \Omega_2)$ between the start and the LA-equilibrium. In the interesting special case $|\mathbf{r}(t)| = t$ we can replace "distance" with "time".

In (6.12) we basically used the function $\frac{1}{2\pi}\sqrt{x}$. To define the Imbalance, we apply the inverse function $(2\pi x)^2$ in (6.12): for the second kind we have

$$\text{Imbalance}_2(\mathcal{Y}, \mathcal{M}) = \max_{\mathbf{n}\in\mathbb{Z}^3\setminus\mathbf{0}} \frac{1}{|\mathbf{n}|^2} \left(\frac{\left|\sum_{k=1}^{N} \mu_k e^{2\pi i\mathbf{n}\cdot\mathbf{y}_k}\right|^2}{\sum_{k=1}^{N} \mu_k^2} \text{ or } 1 \right)_{\max}. \tag{6.13}$$

We may say, therefore, that $\text{Imbalance}_2(\mathcal{Y}, \mathcal{M})$ *represents the square of the Relaxation Drifting Distance for the typical time evolution of $(\mathcal{Y}, \mathcal{M}; \mathbf{r}; \Omega_2)$. In particular, for $|\mathbf{r}(t)| = t$ we have*

$$\text{Imbalance} = (2\pi \cdot \text{RelaxationTime})^2.$$

Note that x^2 is a monotone increasing function, and $x^2 = O(1)$ iff $x = O(1)$. Again, the Imbalance is basically the negative Micro-Entropy; see (6.17) below.

Next we switch to the third kind. Using the elementary fact

$$f_3(x) = \sup_{y\geq x} \left(\frac{\cos y}{y^2} - \frac{\sin y}{y^3} \right)^2 = \frac{1}{x^4} + O(x^{-5}) \text{ for } x > 1,$$

we can rewrite (6.8) in the following simpler form

$$\begin{aligned} &\text{UnevenFun}_3(\mathcal{Y}, \mathcal{M}; \mathbf{r}; t) \\ &= \max\left\{ \max_{\mathbf{n}\in\mathbb{Z}^3\setminus\mathbf{0}} \left|\sum_{k=1}^{N} \mu_k e^{2\pi i\mathbf{n}\cdot\mathbf{y}_k}\right|^2 \min\left\{ \sup_{\tau\geq t} \frac{9}{(2\pi|\mathbf{r}(\tau)|\,|\mathbf{n}|)^4}, 1 \right\}, \sum_{k=1}^{N} \mu_k^2 \right\} \\ &\quad + \text{negligible}. \end{aligned} \tag{6.14}$$

The torus model of the third kind reaches LA-equilibrium at time t_0 when (see (6.8))

$$2\pi \inf_{\tau \geq t_0} |\mathbf{r}(\tau)| = \max_{\mathbf{n}\in\mathbb{Z}^3\setminus\mathbf{0}} \frac{1}{|\mathbf{n}|} \left(\frac{9\left|\sum_{k=1}^N \mu_k e^{2\pi i \mathbf{n}\cdot\mathbf{y}_k}\right|^2}{\sum_{k=1}^N \mu_k^2} \text{ or } 1 \right)_{\max}^{1/4}. \tag{6.15}$$

We refer to the time t_0 and the $\mathbf{R}^3$-distance $\inf_{\tau\geq t_0}|\mathbf{r}(\tau)|$ defined by the equality (6.15) as the Relaxation Time and the Relaxation Drifting Distance, respectively, of the third kind system $(\mathcal{Y},\mathcal{M};\mathbf{r};\Omega_3)$.

By (6.15)

$$\frac{1}{2\pi} \max_{\mathbf{n}\in\mathbb{Z}^3\setminus\mathbf{0}} \frac{1}{|\mathbf{n}|} \left(9\frac{\left|\sum_{k=1}^N \mu_k e^{2\pi i \mathbf{n}\cdot\mathbf{y}_k}\right|^2}{\sum_{k=1}^N \mu_k^2} \text{ or } 1 \right)_{\max}^{1/4} \tag{6.15$'$}$$

represents the $\mathbf{R}^3$-distance for a typical particle in a typical time evolution of the third kind system $(\mathcal{Y},\mathcal{M};\mathbf{r};\Omega_3)$ between the start and the LA-equilibrium. In the interesting special case $|\mathbf{r}(t)| = t$ we can replace "distance" with "time".

In (6.15) we basically used the function $\frac{1}{2\pi}(9x)^{1/4}$. To define the Imbalance, we apply the inverse function $(2\pi x)^4/9$ in (6.15): for the third kind we have

$$\text{Imbalance}_3(\mathcal{Y},\mathcal{M}) = \max_{\mathbf{n}\in\mathbb{Z}^3\setminus\mathbf{0}} \frac{1}{|\mathbf{n}|^4} \left(\frac{\left|\sum_{k=1}^N \mu_k e^{2\pi i \mathbf{n}\cdot\mathbf{y}_k}\right|^2}{\sum_{k=1}^N \mu_k^2} \text{ or } 1 \right)_{\max}. \tag{6.16}$$

It means that $\text{Imbalance}_1(\mathcal{Y},\mathcal{M})$ *represents the fourth power of the Relaxation Drifting Distance for the typical time evolution of $(\mathcal{Y},\mathcal{M};\mathbf{r};\Omega_3)$. In particular, for $|\mathbf{r}(t)| = t$ we have*

$$\text{Imbalance} = (2\pi\cdot\text{RelaxationTime})^4.$$

Note that x^4 is a monotone increasing function, and $x^4 = O(1)$ iff $x = O(1)$.

As we already said above, our micro-entropy has a complementary relation with the Imbalance. More precisely, we have

$$\text{MicroEntropy}_i(\mathcal{Y},\mathcal{M}) = \frac{N^2}{\sum_{k=1}^N \mu_k^2} - \text{Imbalance}_i(\mathcal{Y},\mathcal{M}),\ i = 1,2,3. \tag{6.17}$$

Micro-Entropy is a renormalized Relaxation Drifting Distance. It is renormalized in terms of the initial velocity distribution in such a way that Micro-Entropy is nearly independent of the initial velocity distribution.

The typical time evolution of the system exhibits entropy increase; for the details, see our Second Law later in this section.

These results describe the torus model. For the torus-via-unfolding model we have analog results: the only difference is that $e^{2\pi i x}$ *is replaced by* $\cos(\pi x)$*; see Section 20.*

Finally, we have similar formulas in every dimension, and also for every (reasonable) initial velocity distribution beyond the first, second, and third kinds (for the general form of the Imbalance and the Micro-Entropy, see the end of Section 19).

Remarks. The clarification/proof of the Metatheorem is based on the theorems of Sections 3, 4 and 5; for the details, see Chapters 3 and 4.

The Metatheorem represents Case 2: *mass balance* (see at the beginning of Section 2) with the message that "when the micro-entropy reaches its maximum — i.e., in equilibrium — density becomes a well-defined macrovariable". Note that the analog intrinsic micro-entropy in Case 3: *energy balance* (with the message that "when micro-entropy reaches its maximum — i.e., in equilibrium — temperature becomes a well-defined macrovariable") can be defined in a similar way; see Section 24.

Micro-Entropy exhibits "robustness". A remarkable feature of Micro-Entropy is that it is independent of the space curve $\mathbf{r}(t)$. That is, the type of the trajectory (of the particles) does not show up in the formula.

Here is another example of "robustness". The three different kinds have very different Relaxation Drifting Distances (and Relaxation Times); nevertheless, we show that, for a *typical far from equilibrium* N-element point configuration $\mathcal{Y}$ in the unit cube, we have the equality

$$\text{Imbalance}_1(\mathcal{Y}) = \text{Imbalance}_2(\mathcal{Y}) = \text{Imbalance}_3(\mathcal{Y}).$$

(It is not really important in the proof, but for notational simplicity we prefer to assume that every mass is one, so we omit $\mathcal{M}$.) This equality can be extended to basically any initial velocity distribution far beyond the three kinds discussed in the Metatheorem (see Sections 18–19). This proves that our intrinsic micro-entropy concept is basically "universal".

The three Imbalances $\text{Imbalance}_1(\mathcal{Y}, \mathcal{M})$, $i = 1, 2, 3$ (see (6.10), (6.13) and (6.16)) are quite different, but the contribution of the "first six terms" in the maximum, corresponding to $|\mathbf{n}| = 1$, is the same expression that we call Reduced Micro-Entropy:

$$\text{ReducedMicroEntropy}(\mathcal{Y},\mathcal{M}) = \max_{\mathbf{n}\in\mathbb{Z}^3:|\mathbf{n}|=1}\left(\frac{\left|\sum_{k=1}^{N}\mu_k e^{2\pi i\mathbf{n}\cdot\mathbf{y}_k}\right|^2}{\sum_{k=1}^{N}\mu_k^2} \text{ or } 1\right)_{\max}.$$

The name *Reduced Micro-Entropy* indicates that we ignore the contribution of the lattice points $\mathbf{n}\in\mathbb{Z}^3$ with $|\mathbf{n}|>1$.

This observation motivates the idea to compare the following four sets. Given $\lambda>1$, let

$$\mathcal{A}_i(N;\lambda) = \{\mathcal{Y}\subset[0,1]^3 :\ |\mathcal{Y}| = N \text{ and } \text{Imbalance}_1(\mathcal{Y})\geq\lambda\},\ i=1,2,3$$

and

$$\mathcal{B}^*(N;\lambda) = \left\{\mathcal{Y}\subset[0,1]^3 :\ |\mathcal{Y}| = N \text{ and } \max_{\mathbf{n}\in\mathbb{Z}^3:|\mathbf{n}|=1}\left|\sum_{k=1}^{N}e^{2\pi i\mathbf{n}\cdot\mathbf{y}_k}\right|^2\geq\lambda N\right\}.$$

We show that, if λ is *large* then the four sets $\mathcal{A}_i(N;\lambda)$, $i=1,2,3$ and $\mathcal{B}^*(N;\lambda)$ are *almost the same* (of course we always assume that N is large).

To prove it, consider first the second kind, i.e., $i=2$. For every lattice point $\mathbf{n}\in\mathbb{Z}^3$ with $|\mathbf{n}|\geq1$, write

$$\mathcal{B}(N;\lambda;\mathbf{n}) = \left\{\mathcal{Y}\subset[0,1]^3 :\ |\mathcal{Y}| = N \text{ and } \left|\sum_{k=1}^{N}e^{2\pi i\mathbf{n}\cdot\mathbf{y}_k}\right|^2\geq\lambda|\mathbf{n}|^2N\right\}.$$

Notice that

$$\mathcal{B}^*(N;\lambda)\subset\mathcal{A}_2(N;\lambda) = \bigcup_{\mathbf{n}\in\mathbb{Z}^3:|\mathbf{n}|\geq1}\mathcal{B}(N;\lambda;\mathbf{n}).$$

It follows from the large deviation theorem in probability theory (see e.g. [Fel71]) that for $|\mathbf{n}_1|\geq|\mathbf{n}_2|\geq1$,

$$\frac{\text{measure}(\mathcal{B}(N;\lambda;\mathbf{n}_1))}{\text{measure}(\mathcal{B}(N;\lambda;\mathbf{n}_2))} = (1+o(1))\frac{\int_{\lambda|\mathbf{n}_1|^2}^{\infty}e^{-u^2}\,du}{\int_{\lambda|\mathbf{n}_2|^2}^{\infty}e^{-u^2}\,du}$$

$$\leq(1+o(1))\frac{e^{-\lambda^2|\mathbf{n}_1|^4/2}}{e^{-\lambda^2|\mathbf{n}_2|^4/2}} = (1+o(1))e^{-\lambda^2(|\mathbf{n}_1|^4-|\mathbf{n}_2|^4)/2}.$$

Note that

$$\sum_{\mathbf{n}\in\mathbb{Z}^3:|\mathbf{n}|\geq\sqrt{2}}e^{-\lambda^2(|\mathbf{n}|^4-1)/2}\leq\sum_{m\geq2}8m(m-1)e^{-\lambda^2(m^2-1)/2}$$

$$\leq \sum_{m\geq 2} 8m(m-1)e^{-\lambda^2(m-1)} = 16e^{-\lambda^2}\left(1-e^{-\lambda^2}\right)^{-3},$$

where $m = |\mathbf{n}|^2 = n_1^2 + n_2^2 + n_3^2 \geq 2$, and we used the well known fact

$$\sum_{m=2}^{\infty} m(m-1)x^{m-2} = \frac{2}{(1-x)^3} \quad \text{for} \quad |x| < 1.$$

It follows that, for large λ,

$$\frac{\text{measure}\left(\mathcal{A}_2(N;\lambda) \setminus \mathcal{B}(N;\lambda)\right)}{\text{measure}\left(\mathcal{B}(N;\lambda)\right)} \quad \text{is extremely small.}$$

Since the set-difference $\mathcal{A}_2(N;\lambda) \setminus \mathcal{B}^*(N;\lambda)$ is negligible compared to $\mathcal{B}^*(N;\lambda)$, we conclude that, for large λ, $\mathcal{A}_2(N;\lambda)$ and $\mathcal{B}^*(N;\lambda)$ are *almost the same* (of course we always assume that N is large). The same argument works for $\mathcal{A}_i(N;\lambda)$, $i = 1, 3$.

This argument justifies the sentence "Micro-Entropy is a renormalized Relaxation Drifting Distance — it is renormalized in terms of the initial velocity distribution in such a way that Micro-Entropy is nearly independent of the initial velocity distribution" at the end of the Metatheorem. Indeed, Micro-Entropy is "almost the same" as the Reduced Micro-Entropy, and the latter is a configuration function entirely independent of the initial velocity distribution. This is what we mean by saying that Micro-Entropy exhibits "robustness".

Micro-Entropy, Reduced Micro-Entropy and identity (2.2). The key idea in our definitions of *equilibrium* and *micro-entropy* was the bold approach to *take maximum over all possible test functions* $f \in L^2$ ("ugly test functions"). The arguments above imply that for the overwhelming majority of far from equilibrium (initial) configurations $\mathcal{Y}$, it suffices to define *equilibrium* and *micro-entropy* in the much simpler way of *taking maximum over the 6 smooth test functions* $f_{\mathbf{n}}(\mathbf{x}) = e^{2\pi i \mathbf{n}\cdot\mathbf{x}}$, $\mathbf{n} = (\pm 1, 0, 0)$ *or* $(0, \pm 1, 0)$ *or* $(0, 0, \pm 1)$ ("the 6 nicest test functions"), i.e., to take the *Reduced Micro-Entropy.*

The key part of the Reduced Micro-Entropy

$$\text{ReducedMicroEntropy}(\mathcal{Y},\mathcal{M}) = \max_{\mathbf{n}\in\mathbb{Z}^3:|\mathbf{n}|=1} \left(\frac{\left|\sum_{k=1}^N \mu_k e^{2\pi i \mathbf{n}\cdot\mathbf{y}_k}\right|^2}{\sum_{k=1}^N \mu_k^2} \text{ or } 1 \right)_{\max}$$

is the term

$$\left|\sum_{k=1}^{N} \mu_k e^{2\pi i \mathbf{n}\cdot\mathbf{y}_k}\right|^2,$$

which is the square-modulus of a complex exponential sum (for the shortest nonzero lattice points $\mathbf{n} \in \mathbb{Z}^3$). In the special case of equal mass $1 = \mu_1 = \ldots = \mu_N$ this term simplifies to

$$\left| \sum_{k=1}^{N} e^{2\pi i \mathbf{n} \cdot \mathbf{y}_k} \right|^2 .$$

It is interesting to point out that these terms (for all nonzero lattice points $\mathbf{n} \in \mathbb{Z}^3$) already show up in identity (2.2), which is basically the first formula in the book. The surprising intuitive message of the Reduced Micro-Entropy is that "it suffices to focus on the shortest nonzero lattice points $\mathbf{n} \in \mathbb{Z}^3$ in the simple formula (2.2)".

We may summarize it as follows. The precise definition of our Micro-Entropy is quite subtle. But, as it turns out, Micro-Entropy is "almost the same" as the shockingly simple Reduced Micro-Entropy (explaining why we call the latter "entropy"). And Reduced Micro-Entropy is "basically" the beginning of formula (2.2). This is how formula (2.2) plays a crucial role in the book.

Next we formulate a time-quantitative Second Law for the first kind. We use the notation

$$\mathcal{Y}_{\mathbf{r};\omega;t} = \{\mathbf{z}_k = \mathbf{y}_k + \vartheta_k(\rho_k \mathbf{r}(t)) \text{ modulo one}: \ 1 \le k \le N\}, \tag{6.18}$$

where

$$\omega = (\rho_1, \vartheta_1, \rho_2, \vartheta_2, \ldots, \rho_N, \vartheta_N) \in \Omega_1 = ([0,\infty) \times SO(3))^N .$$

By (6.18) we have to study the typical time evolution of the Imbalance defined in (6.10)

$$\text{Imba}_1(\mathcal{Y}_{\mathbf{r};\omega;t}, \mathcal{M}) = \text{Imbalance}_1(\mathcal{Y}_{\mathbf{r};\omega;t}, \mathcal{M})$$

$$= \max_{\mathbf{n} \in \mathbb{Z}^3 \setminus \mathbf{0}} \left(\frac{\left| \sum_{\mathbf{z}_k \in \mathcal{Y}_{\mathbf{r};\omega;t}} \mu_k e^{2\pi i \mathbf{n} \cdot \mathbf{z}_k} \right|^2}{\sum_{k=1}^{N} \mu_k^2} \ \text{or} \ 1 \right)_{\max}^{1/|\mathbf{n}|^2} . \tag{6.19}$$

Actually we study the following question: given a fixed time $t > 0$, what can we say about $\text{Imba}_1(\mathcal{Y}_{\mathbf{r};\omega;t}, \mathcal{M})$ for the majority of $\omega \in \Omega_1$? In particular, we investigate the relation between $\text{Imba}_1(\mathcal{Y}_{\mathbf{r};\omega;t}, \mathcal{M})$ for the majority of $\omega \in \Omega_1$ and the following ω-independent function (t is the time variable)

$$\text{TiEvFun}_1(t) = \text{TiEvFun}_1(\mathcal{Y}, \mathcal{M}; \mathbf{r}; t)$$

$$= \max_{\mathbf{n}\in\mathbb{Z}^3\setminus\mathbf{0}} \left(\frac{\left|\sum_{k=1}^N \mu_k e^{2\pi i\mathbf{n}\cdot\mathbf{y}_k}\right|^2 \exp\left(-(2\pi|\mathbf{r}(t)|\,|\mathbf{n}|)^2\right)}{\sum_{k=1}^N \mu_k^2} \text{ or } 1\right)_{\max}^{1/|\mathbf{n}|^2}$$

$$= \left(e^{-(2\pi|\mathbf{r}(t)|)^2} \max_{\mathbf{n}\in\mathbb{Z}^3\setminus\mathbf{0}} \left(\frac{\left|\sum_{k=1}^N \mu_k e^{2\pi i\mathbf{n}\cdot\mathbf{y}_k}\right|^2}{\sum_{k=1}^N \mu_k^2}\right)^{1/|\mathbf{n}|^2} \text{ or } 1\right)_{\max}. \quad (6.20)$$

We call (6.20) the *Time Evolution Function of the first kind.* Notice that (6.20) is an ω-**independent monotone decreasing** function of the drifting distance $|\mathbf{r}(t)|$. This fact is used to establish **entropy increase** in the following result (which is actually Theorem 13.1; see Section 13).

Time-quantitative Second Law for the first kind. *Consider a 3-dimensional torus model $(\mathcal{Y}, \mathcal{M}; \mathbf{r}; \Omega_1)$ of the first kind. Given $0 < \varepsilon < 10^{-2}$ and a time $t > 0$ with*

$$\text{TiEvFun}_1(\mathcal{Y}, \mathcal{M}; \mathbf{r}; t) \geq 4/\varepsilon,$$

there exists a measurable subset $\Omega_1^* = \Omega_1^*(\varepsilon; t) \subset \Omega_1$ of the initial velocity space of the first kind such that

$$\text{ProdMeas}_1(\Omega_1^*) > 1 - 10^2\varepsilon$$

and

$$\left|\text{MicroEntropy}_1(\mathcal{Y}_{\mathbf{r};\omega;t}, \mathcal{M}) - \left(\frac{N^2}{\sum_{k=1}^N \mu_k^2} - \text{TiEvFun}_1(\mathcal{Y}, \mathcal{M}; \mathbf{r}; t)\right)\right|$$

$$\leq \frac{12}{\sqrt{\varepsilon}\sqrt{\text{TiEvFun}_1(t)}} \cdot \text{TiEvFun}_1(t) \quad (6.21)$$

holds for all $\omega \in \Omega_1^*$. Here $\text{TiEvFun}_1(t) = \text{TiEvFun}_1(\mathcal{Y}, \mathcal{M}; \mathbf{r}; t)$, see (6.20).

Probably the reader is wondering: Why do we call it a "Second Law"? Well, the name is justified by the key fact that $\text{TiEvFun}_1(t)$ is a decreasing function of the drifting distance $|\mathbf{r}(t)|$. So, if the drifting distance $|\mathbf{r}(t)|$ has a tendency to increase (which is a perfectly natural condition), then by (6.21) Micro-Entropy also has a tendency to increase. We will elaborate on this issue later in Section 15.

The case of the second kind is more complicated due to a technical problem: the series

$$\sum_{\mathbf{n}\in\mathbb{Z}^3\setminus\mathbf{0}} \frac{1}{|\mathbf{n}|^2}$$

is divergent. The corresponding sum for the third kind

$$\sum_{\mathbf{n}\in\mathbb{Z}^3\setminus\mathbf{0}} \frac{1}{|\mathbf{n}|^4}$$

is convergent, which makes it easier to prove an analog Second Law for the third kind. (We also have a weaker Second Law for the second kind; see Theorem 14.1.)

Again we use the notation

$$\mathcal{Y}_{\mathbf{r};\omega;t} = \{\mathbf{z}_k = \mathbf{y}_k + \vartheta_k(\rho_k\mathbf{r}(t)) \text{ modulo one}: \ 1 \le k \le N\},$$

where

$$\omega = (\rho_1, \vartheta_1, \rho_2, \vartheta_2, \ldots, \rho_N, \vartheta_N) \in \Omega_3 = \left([0,1) \times SO(3)\right)^N.$$

We investigate the relation between $\mathrm{Imba}_3(\mathcal{Y}_{\mathbf{r};\omega;t}, \mathcal{M})$ (see (6.16)) for the majority of $\omega \in \Omega_3$ and the following ω-independent function (t is the time variable)

$$\mathrm{TiEvFun}_3(t) = \mathrm{TiEvFun}_3(\mathcal{Y}, \mathcal{M}; \mathbf{r}; t) = \max_{\mathbf{n}\in\mathbb{Z}^3\setminus\mathbf{0}} \frac{1}{|\mathbf{n}|^4}$$

$$\left(\frac{\left|\sum_{k=1}^N \mu_k e^{2\pi i \mathbf{n}\cdot\mathbf{y}_k}\right|^2 9\left(\frac{\cos(2\pi|\mathbf{r}(t)|\,|\mathbf{n}|)}{(2\pi|\mathbf{r}(t)|\,|\mathbf{n}|)^2} - \frac{\sin(2\pi|\mathbf{r}(t)|\,|\mathbf{n}|)}{(2\pi|\mathbf{r}(t)|\,|\mathbf{n}|)^3}\right)^2}{\sum_{k=1}^N \mu_k^2} \text{ or } 1 \right)_{\max}$$

$$= \left(\frac{9}{\sum_{k=1}^N \mu_k^2} \max_{\mathbf{n}\in\mathbb{Z}^3\setminus\mathbf{0}} \left|\sum_{k=1}^N \mu_k e^{2\pi i \mathbf{n}\cdot\mathbf{y}_k}\right|^2 \right.$$

$$\left. \left(\frac{\cos(2\pi|\mathbf{r}(t)|\,|\mathbf{n}|)}{4\pi^2|\mathbf{r}(t)|^2|\mathbf{n}|^4} - \frac{\sin(2\pi|\mathbf{r}(t)|\,|\mathbf{n}|)}{8\pi^3|\mathbf{r}(t)|^3|\mathbf{n}|^5} \right)^2 \text{ or } 1 \right)_{\max}, \qquad (6.22)$$

where, as usual, "$(x \text{ or } 1)_{\max}$" means the larger number between x and 1.

We call (6.22) the *Time Evolution Function of the third kind.* Our Second Law for the third kind goes as follows; it is actually Theorem 13.2. For a far-reaching generalization of Theorem 13.2; see Theorem 19.1.

Time-quantitative Second Law for the third kind. *Consider a 3-dimensional torus model* $(\mathcal{Y}, \mathcal{M}; \mathbf{r}; \Omega_3)$ *of the third kind. Given a time* $t > 0$ *and* $0 < \varepsilon < 10^{-2}$, *there exists a measurable subset* $\Omega_3^* = \Omega_3^*(\varepsilon; t) \subset \Omega_3$ *of the initial velocity space of the third kind such that*

$$\mathrm{ProdMeas}_3(\Omega_3^*) > 1 - 10^2\varepsilon$$

and

$$\left|\text{MicroEntropy}_3(\mathcal{Y}_{\mathbf{r};\omega;t},\mathcal{M})-\left(\frac{N^2}{\sum_{k=1}^N\mu_k^2}-\text{TiEvFun}_3(\mathcal{Y},\mathcal{M};\mathbf{r};t)\right)\right|$$

$$\leq\frac{4}{\sqrt{\varepsilon}\sqrt{\text{TiEvFun}_3(t)}}\left(1+\frac{2}{\sqrt{\text{TiEvFun}_3(t)}}\right)\cdot\text{TiEvFun}_3(t)$$

holds for all $\omega\in\Omega_3^*$. *Here* $\text{TiEvFun}_3(t)=\text{TiEvFun}_3(\mathcal{Y},\mathcal{M};\mathbf{r};t)$, *see (6.22).*

We show two illustrations of the Second Law (the proofs are postponed to Sections 15 and 16). In the illustrations we assume, for simplicity, that every mass is one (so we omit $\mathcal{M}$). Then MicroEntropy + Imbalance $=N$, which implies that it suffices to study the typical time evolution of the Imbalance.

The two illustrations below demonstrate that the Second Law describes the typical time evolution of an individual system strikingly well. More precisely, **we obtain a strikingly good approximation of the ω-dependent Imbalance by the explicit ω-independent Time Evolution Function as long as the system is far from equilibrium.**

We begin with the first kind.

Illustration of the Second Law for the first kind. *Let* $\mathcal{Y}=\{\mathbf{y}_1,\ldots,\mathbf{y}_N\}$ *be an arbitrary* N*-element point configuration in the unit cube* $[0,1]^3$ *such that* $|\mathcal{Y}|=N=10^{1000}$ *and*

$$\text{TiEvFun}_1(0)=\text{TiEvFun}_1(\mathcal{Y};\mathbf{r};0)$$

$$=\max_{\mathbf{n}\in\mathbb{Z}^3\setminus\mathbf{0}}\left(\frac{\left|\sum_{k=1}^N e^{2\pi i\mathbf{n}\cdot\mathbf{y}_k}\right|^2}{N}\right)^{1/|\mathbf{n}|^2}>10^{500}.$$

For simplicity, consider the interesting special case $|\mathbf{r}(t)|=t$ *(which includes, among others, the Bernoulli billiard/torus model). Then there is a subset* $\Omega^*\subset\Omega_1$ *of the initial velocity space of the first kind* $\Omega^*\subset\Omega_1$ *with*

$$\text{ProdMeas}_1(\Omega^*)\geq 1-10^{-98},$$

representing an overwhelming majority, such that for every $\omega\in\Omega^*$ *the time evolution of the* ω*-dependent* $\text{Imbalance}_1(\mathcal{Y}_{\mathbf{r};\omega;t})$ *is approximated by the explicit* ω*-independent Time Evolution Function*

$$\text{TiEvFun}_1(t)=\left(e^{-(2\pi t)^2}\text{TiEvFun}_1(0)\ \text{ or }\ 1\right)_{\max}$$

with extreme accuracy. Indeed, we have the time-quantitative result

$$\text{Imbalance}_1(\mathcal{Y}_{\mathbf{r};\omega;t}) = \left(1 \pm 10^{-98}\right) \text{TiEvFun}_1(t)$$

that holds for every time t *in the constant size time interval*

$$0 < t < T = \frac{\sqrt{\log\left(\text{TiEvFun}_1(0) 10^{-400}\right)}}{2\pi}$$

with the possible exception of a set $t \in [0, T]$ *of total length less than* 10^{-98} *(note that* T *is between 2 and 7 depending on the initial configuration* $\mathcal{Y}$*).*

Finally, note that the time interval $[0, T]$ *corresponds to the decrease of the Time Evolution Function of the first kind from* $\text{TiEvFun}_1(0)$, *which is anywhere between* 10^{1000} *and* 10^{500}, *to* 10^{400}. □

Here is a very interesting question.

Open Problem. What happens if we replace the point billiards with more realistic "hard balls" (hard-core billiards)? That is, the billiards are tiny balls, and particle-particle collisions are more frequent than particle-wall collisions. We still assume Gaussian initial velocity distribution. Is it true that, by switching from point billiards to hard-core billiards the typical time evolution of an individual system remains basically the same? Is it true that the ω-independent Time Evolution Function above applies for the hard-core system as well, describing the typical time evolution of an individual hard-core system (as long as the system is far from equilibrium)?

This seems to be a very difficult problem, despite the fact that there is a large literature of hard-core billiards; see e.g. the book of N. Chernov and R. Makarian [Ch-M2006].

Next we switch from the first kind to the third kind. The illustration for the third kind is more special in the sense that the initial point configuration $\mathcal{Y}$ is "pretty close to Big Bang": $\mathcal{Y}$ is in a (say) one-sixth part of the unit cube $[0, 1]^3$.

Illustration of the Second Law for the third kind. *Assume that the initial point configuration* $\mathcal{Y}$ *is in a (say) one-sixth part of the unit cube* $[0, 1]^3$*: let*

$$\mathcal{Y} \subset [0, 1/6] \times [0, 1] \times [0, 1].$$

Let $|\mathcal{Y}| = N = 10^{1000}$, *and again consider the interesting special case* $|\mathbf{r}(t)| = t$. *Then there is a subset* $\Omega^{***} \subset \Omega_3$ *of the initial velocity space of the third kind* $\Omega^{***} \subset \Omega_3$ *with*

$$\text{ProdMeas}_3\left(\Omega^{***}\right) \geq 1 - 10^{-98},$$

representing an overwhelming majority, such that for every $\omega \in \Omega^{***}$ *the time evolution of the* ω*-dependent* $\text{Imbalance}_3(\mathcal{Y}_{\mathbf{r};\omega;t})$ *is approximated by the explicit* ω*-independent Time Evolution Function with extreme accuracy. Indeed, we have the time-quantitative result*

$$\text{Imbalance}_3(\mathcal{Y}_{\mathbf{r};\omega;t}) = \left(1 \pm 10^{-48}\right) \text{TiEvFun}_3(t)$$

$$= \left(1 \pm 10^{-48}\right) \frac{9}{N} \max_{1 \le |\mathbf{n}| \le \sqrt{6}} \left| \sum_{k=1}^{N} e^{2\pi i \mathbf{n} \cdot \mathbf{y}_k} \right|^2 \left(\frac{\cos(2\pi t |\mathbf{n}|)}{4\pi^2 t^2 |\mathbf{n}|^4} - \frac{\sin(2\pi t |\mathbf{n}|)}{8\pi^3 t^3 |\mathbf{n}|^5} \right)^2$$

that holds for every time t *in the extremely long time interval* $0 < t < T = \frac{1}{2\pi} 10^{150}$ *with the possible exception of a totally negligible set of violators* $t \in [0, T]$ *of total length less than* 10^{-50}*. Note that the time interval* $[0, T]$ *corresponds to the decrease of the Time Evolution Function of the third kind from the range of* 10^{1000} *at the start to* 10^{400}*.* □

By definition

$$1 \le \text{Imbalance}_i(\mathcal{Y}, \mathcal{M}) \le \frac{N^2}{\sum_{k=1}^{N} \mu_k^2}.$$

It follows that Micro-Entropy is in the interval

$$0 \le \text{MicroEntropy}_i(\mathcal{Y}, \mathcal{M}) \le \frac{N^2}{\sum_{k=1}^{N} \mu_k^2} - 1,$$

and the minimum 0 is attained exactly for the Big Bang, i.e., where all particles are in the same point. Big Bang represents the worst case off-equilibrium configuration (for all initial velocity distributions).

Next we switch to the opposite direction, where the Imbalance is in the absolute constant range (independent of N):

$$\text{Imbalance}_i(\mathcal{Y}, \mathcal{M}) = O(1). \tag{6.23}$$

Under the somewhat loosely stated condition (6.23), we say that $(\mathcal{Y}, \mathcal{M})$ is an Ω_i-**equilibrium configuration**, or equivalently, we say that $(\mathcal{Y}, \mathcal{M}, \Omega_i)$ is in **spatial equilibrium**.

(6.23) is equivalent to

$$\text{MicroEntropy}_i(\mathcal{Y}, \mathcal{M}) = \frac{N^2}{\sum_{k=1}^{N} \mu_k^2} - O(1). \tag{6.24}$$

Note that (6.24) reflects the well known principle in physics that "equilibrium states" are the same as the "maximum entropy states".

Note that (6.23) is equivalent to the statement that the Relaxation Drifting Distance for the typical time evolution of $(\mathcal{Y}, \mathcal{M}; \mathbf{r}; \Omega_i)$ is $O(1)$.

(In the interesting special case $|\mathbf{r}(t)| = t$ requirement (6.23) is equivalent to the same statement about the Relaxation Time.)

If we generalize the concept of Ω_i-**equilibrium configuration** ($i = 1, 2, 3$) to an arbitrary speed density function Ω_g (see Theorem 4.2(b)), then we obtain a very interesting by-product of Corollary 5.2 (see at the end of Section 5). Consider a 3-dimensional N-particle torus model $(\mathcal{Y}; \mathbf{r}; \Omega_g)$ defined by (5.45) and (5.47) in the special case $|\mathbf{r}(t)| = t$. A surprising consequence of Corollary 5.2 is that *every* initial configuration $\mathcal{Y}$, including Big Bang(!), is an Ω_g-equilibrium configuration. Indeed, by Corollary 5.2 the Relaxation Drifting Distance for the typical time evolution of $(\mathcal{Y}; \mathbf{r}; \Omega_g)$ is $O(1)$, independently of the initial point configuration $\mathcal{Y}$.

Let us return now to Ω_i, $i = 1, 2, 3$: what are the Ω_i-*equilibrium configurations*? How do they look like? It is easy to see that, for a given mass distribution $\mathcal{M}$ (i.e., the k-th particle has mass $\mu_k > 0$, $1 \le k \le N$ and $\sum_{k=1}^{N} \mu_k = N$), a *typical* (or generic, or random) N set $\mathcal{Y}$ in the unit cube I^3 has the property that $(\mathcal{Y}, \mathcal{M})$ is an Ω_i-equilibrium configuration. The proof is particularly simple for $i = 1$ and 3. Then it is a routine application of the second moment method. The case $i = 2$ is somewhat harder: it requires a *large deviation theorem*. First we consider the case $i = 3$ (third kind). For notational convenience, write

$$\Phi(\mathcal{Y}, \mathcal{M}; \mathbf{n}) = \sum_{k=1}^{N} \mu_k e^{2\pi i \mathbf{n} \cdot \mathbf{y}_k}.$$

Let $\mathbf{y}_k^*$, $1 \le k \le N$ be independent random variables, uniformly distributed in the unit torus $I^3 = [0, 1)^3$; and let

$$(\mathcal{Y}^*, \mathcal{M}) = (\mathbf{y}_1^*, \mu_1; \mathbf{y}_2^*, \mu_2; \ldots; \mathbf{y}_N^*, \mu_N)$$

denote the vector obtained by combining the random N-set with the given mass distribution $\mathcal{M}$. Then for every $\mathbf{n} \in \mathbb{Z}^d \setminus \mathbf{0}$,

$$|\Phi(\mathcal{Y}^*, \mathcal{M}; \mathbf{n})|^2 = \left(\sum_{k=1}^{N} \mu_k e^{2\pi i \mathbf{n} \cdot \mathbf{y}_k^*} \right) \left(\sum_{r=1}^{N} \mu_r e^{-2\pi i \mathbf{n} \cdot \mathbf{y}_r^*} \right)$$

$$= \sum_{k=1}^{N} \mu_k^2 + 2 \sum_{1 \le k < r \le N} \mu_k \mu_r e^{2\pi i \mathbf{n} \cdot (\mathbf{y}_k^* - \mathbf{y}_r^*)}. \tag{6.25}$$

Computing the expected value of (6.25), we have

$$\mathbf{E} \, |\Phi(\mathcal{Y}^*, \mathcal{M}; \mathbf{n})|^2 = \sum_{k=1}^{N} \mu_k^2. \tag{6.26}$$

Applying Markov's inequality in (6.26), for every $\lambda > 0$ we have

$$\Pr\left[|\Phi(\mathcal{Y}^*, \mathcal{M}; \mathbf{n})|^2 \geq \lambda \sum_{k=1}^{N} \mu_k^2\right] \leq \frac{1}{\lambda}. \tag{6.27}$$

Choosing

$$\lambda = \lambda(\mathbf{n}, \varepsilon) = \frac{100}{\varepsilon}|\mathbf{n}|^4,$$

(6.27) gives

$$\Pr\left[\max_{\mathbf{n}\in\mathbb{Z}^3\setminus\mathbf{0}} \frac{|\Phi(\mathcal{Y}^*, \mathcal{M}; \mathbf{n})|^2}{|\mathbf{n}|^4 \sum_{k=1}^{N} \mu_k^2} \leq \frac{100}{\varepsilon}\right] \geq 1 - \sum_{\mathbf{n}\in\mathbb{Z}^3\setminus\mathbf{0}} \frac{\varepsilon}{100|\mathbf{n}|^4} > 1 - \varepsilon, \tag{6.28}$$

where we used the simple numerical fact (see also (13.18) later)

$$\sum_{\mathbf{n}\in\mathbb{Z}^3\setminus\mathbf{0}} \frac{1}{|\mathbf{n}|^4} < 10^2.$$

Combining (6.16) and (6.28), we obtain that, given a mass distribution $\mathcal{M}$, for $1 - \varepsilon$ part of all N-element subsets $\mathcal{Y} \subset I^3$, $(\mathcal{Y}, \mathcal{M})$ is an Ω_3-equilibrium configuration; indeed,

$$\text{Imbalance}_3(\mathcal{Y}, \mathcal{M}) = O(1) \tag{6.29}$$

holds with the explicit constant $100/\varepsilon$.

Of course, the same argument works for the first kind as well; in fact, then 100 can be replaced by a smaller constant.

On the other hand, this argument breaks down for the second kind, due to the divergence of the corresponding lattice sum

$$\sum_{\mathbf{n}\in\mathbb{Z}^3\setminus\mathbf{0}} \frac{1}{|\mathbf{n}|^2} = \infty.$$

Nevertheless, (6.29) still holds for the second kind: we just replace the second moment argument above with the more sophisticated *large deviation theorem* in probability theory, which yields exponentially small tail estimation (we omit the details).

As we explained above, for a given mass distribution $\mathcal{M}$, a *random* N-set $\mathcal{Y} \subset I^3$ has the property that $(\mathcal{Y}, \mathcal{M})$ is an Ω_i-equilibrium configuration for $i = 1, 2, 3$. Now what about N-sets that are completely "anti-random"?

For simplicity assume that every mass is one; so we just study N-element point sets $\mathcal{Y} = \{\mathbf{y}_1, \ldots, \mathbf{y}_N\} \subset I^3 = [0, 1)^3$. What about very regular N-sets? For example, a cube lattice is certainly very regular. Is it possible that a cube lattice is in equilibrium? Perhaps surprisingly, the answer is yes.

We study the simplest cube lattice in the unit torus $I^3 = [0,1)^3$. Assume for simplicity that N is a perfect cube $N = m^3$, and consider the N-element set

$$\mathcal{Y}_{m,m,m} = \left\{ \mathbf{y}_{j,k,\ell} = \left(\frac{j}{m}, \frac{k}{m}, \frac{\ell}{m} \right) : \; 0 \le j,k,\ell < m \right\}. \tag{6.30}$$

Since we deal with geometric series, it is easy to evaluate the critical sum explicitly:

$$\Phi(\mathcal{Y}_{m,m,m}; \mathbf{n}) = \sum_{0 \le j,k,\ell < m} e^{2\pi i \mathbf{n} \cdot \mathbf{y}_{j,k,\ell}}$$

$$= \left(\sum_{j=0}^{m-1} e^{2\pi i n_1 j/m} \right) \left(\sum_{k=0}^{m-1} e^{2\pi i n_2 k/m} \right) \left(\sum_{\ell=0}^{m-1} e^{2\pi i n_3 \ell/m} \right)$$

$$= \frac{1 - e^{2\pi i n_1 j}}{1 - e^{2\pi i n_1 j/m}} \cdot \frac{1 - e^{2\pi i n_2 k}}{1 - e^{2\pi i n_2 k/m}} \cdot \frac{1 - e^{2\pi i n_3 \ell}}{1 - e^{2\pi i n_3 \ell/m}} = m^3 N$$

if $\mathbf{n} = (mh_1, mh_2, mh_3)$ for some $(h_1, h_2, h_3) \in \mathbb{Z}^3$, and $\Phi(\mathcal{Y}_{m,m,m}; \mathbf{n}) = 0$ otherwise.

Thus we have

$$\text{Imbalance}_3(\mathcal{Y}_{m,m,m}) = \max_{\mathbf{n} \in \mathbb{Z}^3 \setminus \mathbf{0}} \frac{1}{|\mathbf{n}|^4} \left(\frac{\left| \sum_{0 \le j,k,\ell < m} e^{2\pi i \mathbf{n} \cdot \mathbf{y}_{j,k,\ell}} \right|^2}{N} \text{ or } 1 \right)_{\max}$$

$$= \left(\frac{N^2}{N m^4} \text{ or } 1 \right)_{\max} = 1, \tag{6.31}$$

since $m^4 > m^3 = N$.

Similarly,

$$\text{Imbalance}_1(\mathcal{Y}_{m,m,m}) = \max_{\mathbf{n} \in \mathbb{Z}^3 \setminus \mathbf{0}} \left(\frac{\left| \sum_{0 \le j,k,\ell < m} e^{2\pi i \mathbf{n} \cdot \mathbf{y}_{j,k,\ell}} \right|^2}{N} \text{ or } 1 \right)_{\max}^{1/|\mathbf{n}|^2}$$

$$= \left(\frac{N^2}{N} \right)^{1/m^2} = N^{-N^{2/3}} = 1 + O(N^{-2/3} \log N) = 1 + o(1). \tag{6.32}$$

It follows from (6.31) and (6.32) that the N-element cube lattice $\mathcal{Y}_{m,m,m}$ defined by (6.30) is an Ω_i-equilibrium configuration for $i = 1$ and 3. Indeed,

$\text{Imbalance}_i(\mathcal{Y}_{m,m,m}) = 1$ for $i = 3$ and $1 + o(1)$ for $i = 1$, and 1 is the smallest possible value for the Imbalance. So the corresponding Micro-Entropy is as large as possible, namely $N - 1$ for $i = 3$ and $N - 1 - o(1)$ for $i = 1$.

Notice that the argument breaks down for the case $i = 2$ — second kind — because $m^2 < m^3 = N$. Repeating the argument of (6.31) we obtain

$$\text{Imbalance}_2(\mathcal{Y}_{m,m,m}) = \left(\frac{N^2}{Nm^2} \text{ or } 1\right)_{\max} = m = N^{1/3},$$

that is, the Relaxation Drifting Distance for the typical time evolution of $(\mathcal{Y}_{m,m,m}; \mathbf{r}; \Omega_2)$ with $|\mathbf{r}(t)| = t$ is about the square-root of the Imbalance, i.e., $N^{1/6}$. It follows that the N-element cube lattice $\mathcal{Y}_{m,m,m}$ is *not* an Ω_2-equilibrium configuration. We will return to this case in Section 17; it will lead to an interesting "entropy paradox".

Another very regular N-element point set is an arithmetic progression in the 3-space modulo one. Given a triple $\mathbf{z} = (\alpha, \beta, \gamma)$ of positive real numbers, define the N-element point set $\mathcal{Y}(\mathbf{z}) \subset I^3 = [0,1)^3$ as follows:

$$\mathcal{Y}(\mathbf{z}) = \{\mathbf{y}_k = k\mathbf{z} = (k\alpha, k\beta, k\gamma) \text{ modulo one}: \ 0 \le k \le N-1\}$$
$$= \left\{\mathbf{y}_k = (\{k\alpha\}, \{k\beta\}, \{k\gamma\}): \ 0 \le k \le N-1\right\}, \tag{6.33}$$

where $\{u\}$ denotes the fractional part of a real number u.

Again we deal with a geometric series, so it is very easy to evaluate the corresponding critical sum explicitly:

$$\Phi(\mathcal{Y}(\mathbf{z}); \mathbf{n}) = \sum_{k=0}^{N-1} e^{2\pi i \mathbf{n}\cdot k\mathbf{z}} = \frac{1 - e^{2\pi i N \mathbf{n}\cdot\mathbf{z}}}{1 - e^{2\pi i \mathbf{n}\cdot\mathbf{z}}},$$

which implies the upper bound

$$|\Phi(\mathcal{Y}(\mathbf{z}); \mathbf{n})| \le \min\left\{\frac{2}{|1 - e^{2\pi i \mathbf{n}\cdot\mathbf{z}}|}, N\right\}$$
$$\le \min\left\{\frac{1}{\|\mathbf{n}\cdot\mathbf{z}\|}, N\right\} = \min\left\{\frac{1}{\|n_1\alpha + n_2\beta + n_3\gamma\|}, N\right\}, \tag{6.34}$$

where $\|u\|$ denotes the distance of a real number u from the nearest integer.

Now let $\mathbf{z}^*$ be a random variable uniformly distributed in the unit torus $I^3 = [0,1)^3$. If $\mathbf{n} \in \mathbb{Z}^3 \setminus \mathbf{0}$ is a nonzero lattice point, then the distribution of $\|\mathbf{n}\cdot\mathbf{z}^*\|$ is independent of the choice of the lattice point $\mathbf{n}$: it has the same distribution as that of $\|z^*\|$ where z^* is a random variable uniformly distributed in the unit interval $0 \le z^* < 1$. Combining this fact with (6.34), we have

$$\Pr[|\Phi(\mathcal{Y}(\mathbf{z}^*);\mathbf{n})| \ge \lambda] \le \Pr\left[\|z^*\| \le \frac{1}{\lambda}\right] = \frac{2}{\lambda} \quad \text{for} \quad \lambda \le N, \tag{6.35}$$

and

$$\Pr[|\Phi(\mathcal{Y}(\mathbf{z}^*);\mathbf{n})| \ge \lambda] = 0 \quad \text{for} \quad \lambda > N. \tag{6.36}$$

Choosing

$$\lambda = \lambda(\mathbf{n},\varepsilon) = \frac{2000}{\varepsilon}|\mathbf{n}|^3 \log^2(2+|\mathbf{n}|), \tag{6.37}$$

(6.36) gives

$$\Pr\left[|\Phi(\mathcal{Y}(\mathbf{z}^*);\mathbf{n})| \le \lambda(\mathbf{n},\varepsilon) \text{ holds for all } \mathbf{n} \in \mathbb{Z}^3 \setminus \mathbf{0}\right]$$

$$\ge 1 - \sum_{\mathbf{n}\in\mathbb{Z}^3\setminus\mathbf{0}} 2\frac{\varepsilon}{2000|\mathbf{n}|^3\log^2(2+|\mathbf{n}|)} > 1-\varepsilon, \tag{6.38}$$

where we used the numerical fact

$$\sum_{\mathbf{n}\in\mathbb{Z}^3\setminus\mathbf{0}} \frac{1}{|\mathbf{n}|^3\log^2(2+|\mathbf{n}|)} < 10^3. \tag{6.39}$$

(6.39) can be proved by a routine power-of-two decomposition argument using the elementary geometric fact that the number of lattice points in a large ball of radius R is close to the volume $4\pi R^3/3$. The details go as follows:

$$\sum_{\mathbf{n}\in\mathbb{Z}^3\setminus\mathbf{0}} \frac{1}{|\mathbf{n}|^3\log^2(2+|\mathbf{n}|)} \le 6 + \sum_{m=2}^{\infty} \sum_{\substack{\mathbf{n}\in\mathbb{Z}^3:\\ 2^{m-1}\le|\mathbf{n}|^2<2^m}} \frac{1}{|\mathbf{n}|^3\log^2(2+|\mathbf{n}|)}$$

$$\le 6 + \sum_{m=2}^{\infty} \frac{4\pi(2^m)^{3/2}}{(2^{m-1})^{3/2}((m-1)\log 2/2)^2} \le 6 + \frac{2^{3.5}\pi}{(\log 2)^2}\cdot 4\sum_{m=2}^{\infty}(m-1)^{-2}$$

$$= 6 + \frac{2^{5.5}\pi}{(\log 2)^2}\cdot\frac{\pi^2}{6} < 1000.$$

Combining (6.16) with (6.38), we obtain that for $1-\varepsilon$ part of $\mathbf{z} \in I^3 = [0,1)^3$ the corresponding N-element sets $\mathcal{Y}(\mathbf{z}) \subset I^3$ defined by (6.33) satisfy

$$\text{Imbalance}_3(\mathcal{Y}(\mathbf{z})) = \max_{\mathbf{n}\in\mathbb{Z}^3\setminus\mathbf{0}} \frac{1}{|\mathbf{n}|^4}\left(\frac{\left|\sum_{k=0}^{N-1}\mu_k e^{2\pi i\mathbf{n}\cdot k\mathbf{z}}\right|^2}{N} \text{ or } 1\right)_{\max}$$

$$\le \max_{\mathbf{n}\in\mathbb{Z}^3\setminus\mathbf{0}} \frac{1}{|\mathbf{n}|^4}\left(\frac{\min\{\lambda^2(\mathbf{n},\varepsilon),N^2\}}{N} \text{ or } 1\right)_{\max}$$

$$= \max_{\mathbf{n}\in\mathbb{Z}^3\setminus\mathbf{0}} \frac{1}{|\mathbf{n}|^4}\left(\frac{\min\{4\varepsilon^{-2}10^6|\mathbf{n}|^6\log^4(2+|\mathbf{n}|),N^2\}}{N} \text{ or } 1\right)_{\max}. \quad (6.40)$$

We distinguish two cases.

Case 1: $\frac{2000}{\varepsilon}|\mathbf{n}|^3\log^2(2+|\mathbf{n}|)\le N$.

Then

$$\frac{1}{|\mathbf{n}|^4}\left(\frac{\min\{4\varepsilon^{-2}10^6|\mathbf{n}|^6\log^4(2+|\mathbf{n}|),N^2\}}{N} \text{ or } 1\right)_{\max}$$

$$\le \left(\frac{1}{|\mathbf{n}|^4}\frac{4\varepsilon^{-2}10^6|\mathbf{n}|^6\log^4(2+|\mathbf{n}|)}{N} \text{ or } 1\right)_{\max}$$

$$= \left(\frac{4\varepsilon^{-2}10^6|\mathbf{n}|^2\log^4(2+|\mathbf{n}|)}{N} \text{ or } 1\right)_{\max} = O_\varepsilon(1), \quad (6.41)$$

where in the last step we used the hypothesis of Case 1.

Next comes

Case 2: $\frac{2000}{\varepsilon}|\mathbf{n}|^3\log^2(2+|\mathbf{n}|)> N$.

Then

$$\frac{1}{|\mathbf{n}|^4}\left(\frac{\min\{4\varepsilon^{-2}10^6|\mathbf{n}|^6\log^4(2+|\mathbf{n}|),N^2\}}{N} \text{ or } 1\right)_{\max}$$

$$\le \frac{1}{|\mathbf{n}|^4}\left(\frac{N^2}{N} \text{ or } 1\right)_{\max} \le \left(\frac{N}{|\mathbf{n}|^4} \text{ or } 1\right)_{\max} = O_\varepsilon(1), \quad (6.42)$$

where in the last step we used the hypothesis of Case 2.

Combining (6.40)–(6.42), we obtain that

$$\text{Imbalance}_3(\mathcal{Y}(\mathbf{z})) = O_\varepsilon(1) \quad (6.43)$$

holds for $1-\varepsilon$ part of $\mathbf{z}\in I^3=[0,1)^3$, so the corresponding N-element sets $\mathcal{Y}(\mathbf{z})\subset I^3$ (see (6.33)) are all Ω_3-equilibrium configurations.

Next we repeat the same argument for the first kind. Combining (6.10) with (6.38), we obtain that for $1-\varepsilon$ part of $\mathbf{z}\in I^3=[0,1)^3$ the corresponding N-element sets $\mathcal{Y}(\mathbf{z})\subset I^3$ defined by (6.33) satisfy

$$\text{Imbalance}_1(\mathcal{Y}(\mathbf{z})) = \max_{\mathbf{n}\in\mathbb{Z}^3\setminus\mathbf{0}}\left(\frac{\left|\sum_{k=0}^{N-1}\mu_k e^{2\pi i\mathbf{n}\cdot k\mathbf{z}}\right|^2}{N} \text{ or } 1\right)_{\max}^{1/|\mathbf{n}|^2}$$

$$\leq \max_{\mathbf{n}\in\mathbb{Z}^3\setminus\mathbf{0}} \left(\frac{\min\left\{\lambda^2(\mathbf{n},\varepsilon), N^2\right\}}{N} \text{ or } 1\right)_{\max}^{1/|\mathbf{n}|^2}$$

$$= \max_{\mathbf{n}\in\mathbb{Z}^3\setminus\mathbf{0}} \left(\frac{\min\left\{4\varepsilon^{-2}10^6|\mathbf{n}|^6\log^4(2+|\mathbf{n}|), N^2\right\}}{N} \text{ or } 1\right)_{\max}^{1/|\mathbf{n}|^2}. \tag{6.44}$$

We distinguish two cases.

Case 1: $\frac{2000}{\varepsilon}|\mathbf{n}|^3\log^2(2+|\mathbf{n}|) \leq N$.

Then

$$\left(\frac{\min\left\{4\varepsilon^{-2}10^6|\mathbf{n}|^6\log^4(2+|\mathbf{n}|), N^2\right\}}{N} \text{ or } 1\right)_{\max}^{1/|\mathbf{n}|^2}$$

$$\leq \left(\frac{4\varepsilon^{-2}10^6|\mathbf{n}|^6\log^4(2+|\mathbf{n}|)}{N} \text{ or } 1\right)_{\max}^{1/|\mathbf{n}|^2} = 1 + o_\varepsilon(1), \tag{6.45}$$

where in the last step we used the hypothesis of Case 1.

Next comes

Case 2: $\frac{2000}{\varepsilon}|\mathbf{n}|^3\log^2(2+|\mathbf{n}|) > N$.

Then

$$\left(\frac{\min\left\{4\varepsilon^{-2}10^6|\mathbf{n}|^6\log^4(2+|\mathbf{n}|), N^2\right\}}{N} \text{ or } 1\right)_{\max}^{1/|\mathbf{n}|^2}$$

$$\leq \left(\frac{N^2}{N} \text{ or } 1\right)_{\max}^{1/|\mathbf{n}|^2} = N^{1/|\mathbf{n}|^2} = 1 + o_\varepsilon(1), \tag{6.46}$$

where in the last step we used the hypothesis of Case 2.

Combining (6.44)–(6.46), we obtain that

$$\text{Imbalance}_1(\mathcal{Y}(\mathbf{z})) = 1 + o_\varepsilon(1) \tag{6.47}$$

holds for $1-\varepsilon$ part of $\mathbf{z} \in I^3 = [0,1)^3$, so the corresponding N-element sets $\mathcal{Y}(\mathbf{z}) \subset I^3$ (see (6.33)) are all Ω_1-equilibrium configurations.

Again the argument breaks down for the second kind, due to the usual divergence problem.

Chapters 2–4 are devoted to the proofs of the Metatheorem and the Second Law(s), including several highly recommended illustrations. If the reader is not interested in the technicalities of the proofs, and skipped them, he/she can still read these illustrations. Chapter 5 is about the long-term stability of (square-root) equilibrium.

Chapter 2

Starting the Proofs: Applying Fourier Analysis

7. Proof of Theorem 3.1

What is in Chapter 2? *In Sections 3–5 we formulated several results without proof. Here in Chapter 2 we include the detailed proofs. Our main tool is Fourier analysis.*

We start the proof of Theorem 3.1 with the Fourier expansion of the (complex valued) test function $f \in L^2(I^3)$, $I^3 = [0,1)^3$:

$$f(\mathbf{z}) = \sum_{\mathbf{n}\in\mathbb{Z}^3} a_{\mathbf{n}} e^{2\pi \mathrm{i}\mathbf{n}\cdot\mathbf{z}} \quad \text{with} \quad a_{\mathbf{n}} = a_{\mathbf{n}}(f) = \int_{I^3} f(\mathbf{z}) e^{-2\pi \mathrm{i}\mathbf{n}\cdot\mathbf{z}}\, d\mathbf{z}, \tag{7.1}$$

where of course $\mathbf{n}\cdot\mathbf{z} = n_1 z_1 + n_2 z_2 + n_3 z_3$ denotes the standard inner product and $\mathrm{i} = \sqrt{-1}$. Clearly $a_{\mathbf{0}} = \int_{I^3} f\, d\mathrm{V}$.

If f is replaced with a translation $f_{\mathbf{w}}$ (see (2.30)) then the corresponding Fourier coefficient changes as follows:

$$a_{\mathbf{n}}(f_{\mathbf{w}}) = \int_{I^3} f(\mathbf{z}-\mathbf{w}) e^{-2\pi \mathrm{i}\mathbf{n}\cdot\mathbf{z}}\, d\mathbf{z} = \int_{I^3} f(\mathbf{z}) e^{-2\pi \mathrm{i}\mathbf{n}\cdot(\mathbf{z}+\mathbf{w})}\, d\mathbf{z} = e^{-2\pi \mathrm{i}\mathbf{n}\cdot\mathbf{w}} a_{\mathbf{n}}(f). \tag{7.2}$$

By (7.1) and (7.2) we have

$$N(\mathcal{Y},\mathcal{M};\mathcal{R};\omega;f_{\mathbf{w}};t) - N\int_{I^3} f\, d\mathrm{V} = \sum_{k=1}^{N} \mu_k \left(f_{\mathbf{w}}(\mathbf{x}_k(t)) - \int_{I^3} f\, d\mathrm{V} \right)$$

$$= \sum_{\substack{\mathbf{n}\in\mathbb{Z}^3:\\ \mathbf{n}\neq\mathbf{0}}} e^{-2\pi \mathrm{i}\mathbf{n}\cdot\mathbf{w}} a_{\mathbf{n}} \sum_{k=1}^{N} \mu_k e^{2\pi \mathrm{i}\mathbf{n}\cdot\mathbf{x}_k(t)}$$

$$= \sum_{\substack{\mathbf{n}\in\mathbb{Z}^3:\\ \mathbf{n}\neq\mathbf{0}}} e^{-2\pi \mathrm{i}\mathbf{n}\cdot\mathbf{w}} a_{\mathbf{n}} \sum_{k=1}^{N} \mu_k e^{2\pi \mathrm{i}\mathbf{n}\cdot(\mathbf{y}_k+\vartheta_k(\rho_k\mathbf{r}_k(t)))}. \tag{7.3}$$

By (7.3) we have

$$\mathbf{D}^2(\mathcal{Y},\mathcal{M};\mathcal{R};\omega;f_{\mathbf{w}};t)=\left|N(\mathcal{Y},\mathcal{M};\mathcal{R};\omega;f_{\mathbf{w}};t)-N\int_{I^3}f\,d\mathrm{V}\right|^2$$

$$=\sum_{k=1}^{N}\mu_k^2\sum_{\substack{\mathbf{n}\in\mathbb{Z}^3:\\ \mathbf{n}\neq\mathbf{0}}}|a_{\mathbf{n}}|^2$$

$$+\sum_{\substack{\mathbf{n}\in\mathbb{Z}^3:\\ \mathbf{n}\neq\mathbf{0}}}|a_{\mathbf{n}}|^2\sum_{1\le k,\ell\le N:k\neq\ell}\mu_k\mu_\ell e^{2\pi\mathrm{i}\mathbf{n}\cdot(\mathbf{y}_k-\mathbf{y}_\ell)}$$

$$\cdot e^{2\pi\mathrm{i}\mathbf{n}\cdot(\vartheta_k(\rho_k\mathbf{r}_k(t))-\vartheta_\ell(\rho_\ell\mathbf{r}_\ell(t)))}+\sum_{\substack{\mathbf{n}_j\in\mathbb{Z}^3:j=1,2\\ \mathbf{0}\neq\mathbf{n}_1\neq\mathbf{n}_2\neq\mathbf{0}}}e^{-2\pi\mathrm{i}(\mathbf{n}_1-\mathbf{n}_2)\cdot\mathbf{w}}$$

$$\cdot a_{\mathbf{n}_1}\overline{a_{\mathbf{n}_2}}\sum_{1\le k,\ell\le N}\mu_k\mu_\ell e^{2\pi\mathrm{i}(\mathbf{n}_1\cdot\mathbf{y}_k-\mathbf{n}_2\cdot\mathbf{y}_\ell)}e^{2\pi\mathrm{i}(\vartheta_k(\rho_k\mathbf{r}_k(t))\cdot\mathbf{n}_1-\vartheta_\ell(\rho_\ell\mathbf{r}_\ell(t))\cdot\mathbf{n}_2)},\tag{7.4}$$

where, as usual, $\overline{z}=x-\mathrm{i}y$ denotes the complex conjugate of a complex number $z=x+\mathrm{i}y$. Since

$$\int_{I^3}e^{-2\pi\mathrm{i}\mathbf{n}\cdot\mathbf{w}}\,d\mathbf{w}=0\tag{7.5}$$

for all $\mathbf{n}\in\mathbb{Z}^3$ except the trivial case $\mathbf{n}=\mathbf{0}$, by (7.4) and (7.5) we have

$$\overline{\mathbf{D}^2}(\mathcal{Y},\mathcal{M};\mathcal{R};\omega;f;t)=\int_{I^3}\mathbf{D}^2(\mathcal{Y},\mathcal{M};\mathcal{R};\omega;f_{\mathbf{w}};t)\,d\mathbf{w}$$

$$=\int_{I^3}\left|N(\mathcal{Y},\mathcal{M};\mathcal{R};\omega;f_{\mathbf{w}};t)-N\int_{I^3}f\,d\mathrm{V}\right|^2d\mathbf{w}=\sum_{k=1}^{N}\mu_k^2\sum_{\substack{\mathbf{n}\in\mathbb{Z}^3:\\ \mathbf{n}\neq\mathbf{0}}}|a_{\mathbf{n}}|^2$$

$$+\sum_{\substack{\mathbf{n}\in\mathbb{Z}^3:\\ \mathbf{n}\neq\mathbf{0}}}|a_{\mathbf{n}}|^2\sum_{1\le k,\ell\le N:k\neq\ell}\mu_k\mu_\ell e^{2\pi\mathrm{i}\mathbf{n}\cdot(\mathbf{y}_k-\mathbf{y}_\ell)}e^{2\pi\mathrm{i}\mathbf{n}\cdot(\vartheta_k(\rho_k\mathbf{r}_k(t))-\vartheta_\ell(\rho_\ell\mathbf{r}_\ell(t)))}.\tag{7.6}$$

We can rewrite (7.6) in the form

$$\int_{I^3}\left|N(\mathcal{Y},\mathcal{M};\mathcal{R};\omega;f_{\mathbf{w}};t)-N\int_{I^3}f\,d\mathrm{V}\right|^2d\mathbf{w}$$

$$= \sum_{\substack{\mathbf{n}\in\mathbb{Z}^3:\\ \mathbf{n}\neq\mathbf{0}}} |a_{\mathbf{n}}|^2 \left| \sum_{1\le k\le N} \mu_k e^{2\pi i\mathbf{n}\cdot(\mathbf{y}_k+\vartheta_k(\rho_k\mathbf{r}_k(t)))} \right|^2. \qquad (7.6')$$

Choosing $t = 0$ in identity (7.6′) we obtain (2.2) (it is the 3-dimensional case, but of course (7.6′) holds in every dimension).

It is interesting to point out that identity (7.6′) is in fact a (sophisticated) application of Parseval's formula (note that Parseval's formula holds in the whole L^2-space). Indeed, the first line of (7.6′) is the square integral of a *convolution* (= "convolution of f with a signed measure"), and the square "integral" (= infinite sum over the lattice points) of the *Fourier transform of the convolution* (= "product of the Fourier transforms") is the second line.

Next we integrate (7.6) over the big "initial velocity space" $\Omega_1 = ([0,\infty)\times SO(3))^N$ (see (2.13)):

$$\overline{\mathbf{D}^2}_{\Omega_1}(\mathcal{Y},\mathcal{M};\mathcal{R};f;t) = \int_{\Omega_1} \overline{\mathbf{D}^2}(\mathcal{Y},\mathcal{M};\mathcal{R};\omega;f;t)\,d\omega$$

$$= \sqrt{\frac{2}{\pi}}\int_{\rho_1=0}^{\infty}\int_{\vartheta_1\in SO(3)} \cdots \sqrt{\frac{2}{\pi}}\int_{\rho_N=0}^{\infty}\int_{\vartheta_N\in SO(3)} \mathbf{D}^2(\mathcal{Y},\mathcal{M};\mathcal{R};\omega;f;t)\rho_1^2$$

$$\cdot e^{-\rho_1^2/2}\,d\rho_1\,d\vartheta_1\cdots\rho_N^2 e^{-\rho_N^2/2}\,d\rho_N\,d\vartheta_N$$

$$= \sum_{k=1}^{N}\mu_k^2 \sum_{\substack{\mathbf{n}\in\mathbb{Z}^3:\\ \mathbf{n}\neq\mathbf{0}}} |a_{\mathbf{n}}|^2 + \sum_{\substack{\mathbf{n}\in\mathbb{Z}^3:\\ \mathbf{n}\neq\mathbf{0}}} |a_{\mathbf{n}}|^2 \sum_{1\le k,\ell\le N: k\neq\ell} \mu_k\mu_\ell e^{2\pi i\mathbf{n}\cdot(\mathbf{y}_k-\mathbf{y}_\ell)}$$

$$\cdot\sqrt{\frac{2}{\pi}}\int_{\rho_1=0}^{\infty}\int_{\vartheta_1\in SO(3)} \cdots \sqrt{\frac{2}{\pi}}\int_{\rho_N=0}^{\infty}\int_{\vartheta_N\in SO(3)} e^{2\pi i\mathbf{n}\cdot(\vartheta_k(\rho_k\mathbf{r}_k(t))-\vartheta_\ell(\rho_\ell\mathbf{r}_\ell(t)))}\rho_1^2$$

$$\cdot e^{-\rho_1^2/2}\,d\rho_1\,d\vartheta_1\cdots\rho_N^2 e^{-\rho_N^2/2}\,d\rho_N\,d\vartheta_N$$

$$= \sum_{k=1}^{N}\mu_k^2 \sum_{\substack{\mathbf{n}\in\mathbb{Z}^3:\\ \mathbf{n}\neq\mathbf{0}}} |a_{\mathbf{n}}|^2 + \sum_{\substack{\mathbf{n}\in\mathbb{Z}^3:\\ \mathbf{n}\neq\mathbf{0}}} |a_{\mathbf{n}}|^2 \sum_{1\le k,\ell\le N: k\neq\ell} \mu_k\mu_\ell e^{2\pi i\mathbf{n}\cdot(\mathbf{y}_k-\mathbf{y}_\ell)}$$

$$\cdot\sqrt{\frac{2}{\pi}}\int_0^{\infty}\int_{\vartheta\in SO(3)} e^{2\pi i\vartheta(\rho\mathbf{r}_k(t))\cdot\mathbf{n}}\rho^2 e^{-\rho^2/2}\,d\rho\,d\vartheta$$

$$\cdot \sqrt{\frac{2}{\pi}} \int_0^\infty \int_{\vartheta \in SO(3)} e^{-2\pi i \vartheta(\rho \mathbf{r}_\ell(t))\cdot \mathbf{n}} \rho^2 e^{-\rho^2/2}\, d\rho\, d\vartheta, \tag{7.7}$$

where $d\vartheta$ means the normed Haar measure on $SO(3)$: $\int_{SO(3)} 1\, d\vartheta = 1$, i.e., it is a probability measure. We need

Lemma 7.1. *For every 3-dimensional vector* $\mathbf{w}$ *and unit vector* $\mathbf{e} \in \mathbf{S}^2$, *we have*

$$\sqrt{\frac{2}{\pi}} \int_0^\infty \int_{\vartheta \in SO(3)} e^{-2\pi i \vartheta(\rho \mathbf{e})\cdot \mathbf{w}} \rho^2 e^{-\rho^2/2}\, d\rho\, d\vartheta = e^{-2\pi^2 |\mathbf{w}|^2}.$$

We prove Lemma 7.1 later in the section.

Applying Lemma 7.1 with $\mathbf{w} = -|\mathbf{r}_k(t)|\mathbf{n}$ in (7.7), we have

$$\overline{\mathbf{D}^2}_{\Omega_1}(\mathcal{Y}, \mathcal{M}; \mathcal{R}; f; t) = \int_{\Omega_1} \overline{\mathbf{D}^2}(\mathcal{Y}, \mathcal{M}; \mathcal{R}; \omega; f; t)\, d\omega$$

$$= \sum_{k=1}^N \mu_k^2 \sum_{\substack{\mathbf{n}\in\mathbb{Z}^3:\\ \mathbf{n}\neq\mathbf{0}}} |a_{\mathbf{n}}|^2 + \sum_{\substack{\mathbf{n}\in\mathbb{Z}^3:\\ \mathbf{n}\neq\mathbf{0}}} |a_{\mathbf{n}}|^2 \sum_{1\le k,\ell\le N: k\neq\ell} \mu_k\mu_\ell e^{2\pi i \mathbf{n}\cdot(\mathbf{y}_k - \mathbf{y}_\ell)}$$

$$\cdot \exp\left(-2(\pi|\mathbf{r}_k(t)|\,|\mathbf{n}|)^2\right) \exp\left(-2(\pi|\mathbf{r}_\ell(t)|\,|\mathbf{n}|)^2\right)$$

$$= \sum_{k=1}^N \mu_k^2 \sum_{\substack{\mathbf{n}\in\mathbb{Z}^3:\\ \mathbf{n}\neq\mathbf{0}}} |a_{\mathbf{n}}|^2 + \sum_{\substack{\mathbf{n}\in\mathbb{Z}^3:\\ \mathbf{n}\neq\mathbf{0}}} |a_{\mathbf{n}}|^2 \sum_{1\le k,\ell\le N: k\neq\ell} \mu_k\mu_\ell e^{2\pi i \mathbf{n}\cdot(\mathbf{y}_k - \mathbf{y}_\ell)}$$

$$\cdot \mathbf{F}(2\pi|\mathbf{r}_k(t)|\,|\mathbf{n}|)\mathbf{F}(2\pi|\mathbf{r}_\ell(t)|\,|\mathbf{n}|), \tag{7.8}$$

where $\exp(y) = e^y$ and $\mathbf{F}(x) = e^{-x^2/2}$. We use $\mathbf{F}(x)$ instead of $e^{-x^2/2}$ to emphasize the fact that the proof in the special case of the first kind has a straightforward generalization for the general isotropic case (see Theorems 4.1 and 4.2(a) and (b)): the only change is to replace $\mathbf{F}(x) = e^{-x^2/2}$ with the corresponding isotropic Fourier transform $\mathbf{F} = \mathbf{F}_g$, and of course we also have to replace Ω_1 with Ω_g and replace Lemma 7.1 with the corresponding result (like Lemma 8.1, Lemma 8.2, Lemma 8.3, etc.).

Next we integrate (7.4) in the special case $\mathbf{w} = \mathbf{0}$ over the big "initial velocity space" $\Omega_1 = ([0, \infty) \times SO(3))^N$, and apply Lemma 7.1; thus we obtain the following analog of (7.7)–(7.8):

$$\mathbf{D}^2_{\Omega_1}(\mathcal{Y}, \mathcal{M}; \mathcal{R}; f; t) = \int_{\Omega_1} \mathbf{D}^2(\mathcal{Y}, \mathcal{M}; \mathcal{R}; \omega; f; t)\, d\omega$$

$$= \sum_{k=1}^{N} \mu_k^2 \sum_{\substack{\mathbf{n}\in\mathbb{Z}^3:\\ \mathbf{n}\neq\mathbf{0}}} |a_{\mathbf{n}}|^2 + \sum_{\substack{\mathbf{n}\in\mathbb{Z}^3:\\ \mathbf{n}\neq\mathbf{0}}} |a_{\mathbf{n}}|^2 \sum_{1\le k,\ell\le N: k\neq\ell} \mu_k\mu_\ell e^{2\pi i\mathbf{n}\cdot(\mathbf{y}_k-\mathbf{y}_\ell)}$$

$$\cdot \mathbf{F}(2\pi|\mathbf{r}_k(t)|\,|\mathbf{n}|)\mathbf{F}(2\pi|\mathbf{r}_\ell(t)|\,|\mathbf{n}|)$$

$$+ \sum_{\substack{\mathbf{n}_j\in\mathbb{Z}^3: j=1,2\\ \mathbf{0}\neq\mathbf{n}_1\neq\mathbf{n}_2\neq\mathbf{0}}} a_{\mathbf{n}_1}$$

$$\cdot \overline{a_{\mathbf{n}_2}} \sum_{1\le k,\ell\le N: k\neq\ell} \mu_k\mu_\ell e^{2\pi i(\mathbf{n}_1\cdot\mathbf{y}_k-\mathbf{n}_2\cdot\mathbf{y}_\ell)}\mathbf{F}(2\pi|\mathbf{r}_k(t)|\,|\mathbf{n}_1|)\mathbf{F}(2\pi|\mathbf{r}_\ell(t)|\,|\mathbf{n}_2|)$$

$$+ \sum_{\substack{\mathbf{n}_j\in\mathbb{Z}^3: j=1,2\\ \mathbf{0}\neq\mathbf{n}_1\neq\mathbf{n}_2\neq\mathbf{0}}} a_{\mathbf{n}_1}\overline{a_{\mathbf{n}_2}} \sum_{1\le k\le N} \mu_k^2 e^{2\pi i(\mathbf{n}_1-\mathbf{n}_2)\cdot\mathbf{y}_k}\mathbf{F}(2\pi|\mathbf{r}_k(t)|\,|\mathbf{n}_1-\mathbf{n}_2|). \quad (7.9)$$

Technical Remark. Since $f \in L^2$, the infinite sums in (7.8)–(7.9) with $\mathbf{F}(x) = e^{-x^2/2}$ are absolutely convergent (under the natural assumption $|\mathbf{r}_k(t)| > 0$ for all $1 \le k \le N$), and this justifies in retrospect our informal manipulations with the convergent (almost everywhere) but not absolutely convergent series in (7.1), (7.3) and (7.4). (Note that (7.6) and (7.7) are fine, since Parseval's formula holds in L^2.) The rest of the proof is going to be "safe", because from now on we work exclusively with absolutely convergent series.

We can rewrite (7.8) in the alternative form

$$\overline{\mathbf{D}^2}_{\Omega_1}(\mathcal{Y},\mathcal{M};\mathcal{R};f;t)$$

$$= \sum_{k=1}^{N} \mu_k^2 \sum_{\substack{\mathbf{n}\in\mathbb{Z}^3:\\ \mathbf{n}\neq\mathbf{0}}} |a_{\mathbf{n}}|^2 \left(1-\mathbf{F}^2(2\pi|\mathbf{r}_k(t)|\,|\mathbf{n}|)\right)$$

$$+ \sum_{\substack{\mathbf{n}\in\mathbb{Z}^3:\\ \mathbf{n}\neq\mathbf{0}}} |a_{\mathbf{n}}|^2 \left|\sum_{k=1}^{N} \mu_k e^{2\pi i\mathbf{n}\cdot\mathbf{y}_k}\cdot\mathbf{F}(2\pi|\mathbf{r}_k(t)|\,|\mathbf{n}|)\right|^2. \quad (7.10)$$

By Parseval's formula,

$$\sum_{\substack{\mathbf{n}\in\mathbb{Z}^3:\\ \mathbf{n}\neq\mathbf{0}}} |a_{\mathbf{n}}|^2 = \int_{I^3} |f|^2\,d\mathrm{V} - \left|\int_{I^3} f\,d\mathrm{V}\right|^2 = \sigma_0^2(f). \quad (7.11)$$

Combining (7.10) and (7.11), we have

$$\left|\overline{\mathbf{D}^2}_{\Omega_1}(\mathcal{Y},\mathcal{M};\mathcal{R};f;t)-\sigma_0^2(f)\sum_{k=1}^{N}\mu_k^2\right|$$

$$=\left|\overline{\mathbf{D}^2}_{\Omega_1}(\mathcal{Y},\mathcal{M};\mathcal{R};f;t)-\sum_{k=1}^{N}\mu_k^2\sum_{\substack{\mathbf{n}\in\mathbb{Z}^3:\\ \mathbf{n}\neq\mathbf{0}}}|a_{\mathbf{n}}|^2\right|$$

$$=\left|-\sum_{\substack{\mathbf{n}\in\mathbb{Z}^3:\\ \mathbf{n}\neq\mathbf{0}}}|a_{\mathbf{n}}|^2\sum_{k=1}^{N}\mu_k^2\mathbf{F}^2(2\pi|\mathbf{r}_k(t)|\,|\mathbf{n}|)\right.$$

$$\left.+\sum_{\substack{\mathbf{n}\in\mathbb{Z}^3:\\ \mathbf{n}\neq\mathbf{0}}}|a_{\mathbf{n}}|^2\left|\sum_{k=1}^{N}\mu_k e^{2\pi i\mathbf{n}\cdot\mathbf{y}_k}\cdot\mathbf{F}(2\pi|\mathbf{r}_k(t)|\,|\mathbf{n}|)\right|^2\right|$$

$$\leq\max\left\{\sum_{\substack{\mathbf{n}\in\mathbb{Z}^3:\\ \mathbf{n}\neq\mathbf{0}}}|a_{\mathbf{n}}|^2\sum_{k=1}^{N}\mu_k^2\mathbf{F}^2(2\pi|\mathbf{r}_k(t)|\,|\mathbf{n}|),\right.$$

$$\left.\sum_{\substack{\mathbf{n}\in\mathbb{Z}^3:\\ \mathbf{n}\neq\mathbf{0}}}|a_{\mathbf{n}}|^2\left|\sum_{k=1}^{N}\mu_k e^{2\pi i\mathbf{n}\cdot\mathbf{y}_k}\cdot\mathbf{F}(2\pi|\mathbf{r}_k(t)|\,|\mathbf{n}|)\right|^2\right\}$$

$$\leq\max\left\{\left(\sum_{\substack{\mathbf{n}\in\mathbb{Z}^3:\\ \mathbf{n}\neq\mathbf{0}}}|a_{\mathbf{n}}|^2\right)\left(\sum_{k=1}^{N}\mu_k^2\mathcal{F}^2(2\pi|\mathbf{r}_k(t)|)\right),\right.$$

$$\left.\left(\sum_{\substack{\mathbf{n}\in\mathbb{Z}^3:\\ \mathbf{n}\neq\mathbf{0}}}|a_{\mathbf{n}}|^2\right)\left(\sum_{k=1}^{N}\mu_k\mathcal{F}(2\pi|\mathbf{r}_k(t)|)\right)^2\right\}$$

$$\le \mathcal{F}^2(2\pi\mathcal{R}(t)) \max\left\{ \sigma_0^2(f)\sum_{k=1}^{N}\mu_k^2, \sigma_0^2(f)\left(\sum_{k=1}^{N}\mu_k\right)^2 \right\}$$

$$= \mathcal{F}^2(2\pi\mathcal{R}(t))\sigma_0^2(f)\left(\sum_{k=1}^{N}\mu_k\right)^2 = \mathcal{F}^2(2\pi\mathcal{R}(t))\sigma_0^2(f)N^2, \qquad (7.12)$$

where in the last steps we used the definition of the drifting distance (see (2.20))

$$|\mathbf{r}_k(t)||\mathbf{n}| \ge |\mathbf{r}_k(t)| \ge \mathcal{R}(t),$$

and

$$\mathcal{F}(x) = \sup_{y\ge x}|\mathbf{F}(y)|.$$

If $\mathbf{F}(y)$ is monotone decreasing in $0 \le y < \infty$ then of course $\mathcal{F} = \mathbf{F}$. For example, this is the case for the first kind $\mathbf{F}(y) = e^{-y^2/2}$ (but not for the second and third kinds!) Notice that (7.12) proves (3.1) in Theorem 3.1, proves (4.5) in Theorem 4.1, proves (4.17) in Theorem 4.2(a), and proves (4.17') in Theorem 4.2(b). In the cases of the second and third kinds we replace $\mathbf{F}(y) = e^{-y^2/2}$ with

$$\mathbf{F}(y) = \frac{\sin y}{y} \text{ and } -3\frac{\cos y}{y^2} + 3\frac{\sin y}{y^3}$$

(for the latter, see Lemma 8.1), and of course we replace Ω_1 with Ω_2 and Ω_3, respectively.

Next we prove the second statement in Theorem 3.1. We estimate the end of (7.9). We begin with the subsum

$$\left| \sum_{\substack{\mathbf{n}_j\in\mathbb{Z}^3:j=1,2\\ \mathbf{0}\ne\mathbf{n}_1\ne\mathbf{n}_2\ne\mathbf{0}}} a_{\mathbf{n}_1} \cdot \overline{a_{\mathbf{n}_2}} \sum_{1\le k,\ell\le N:k\ne\ell} \mu_k\mu_\ell e^{2\pi i(\mathbf{n}_1\cdot\mathbf{y}_k - \mathbf{n}_2\cdot\mathbf{y}_\ell)}\mathbf{F}(2\pi|\mathbf{r}_k(t)|\ |\mathbf{n}_1|)\mathbf{F}(2\pi|\mathbf{r}_\ell(t)|\ |\mathbf{n}_2|) \right|$$

$$\le \left| \sum_{\substack{\mathbf{n}_j\in\mathbb{Z}^3:j=1,2\\ \mathbf{0}\ne\mathbf{n}_1\ne\mathbf{n}_2\ne\mathbf{0}}} \frac{1}{2}\left(|a_{\mathbf{n}_1}|^2 + |a_{\mathbf{n}_2}|^2\right)\right.$$

$$\left. \cdot \sum_{1\le k,\ell\le N:k\ne\ell} \mu_k\mu_\ell e^{2\pi i(\mathbf{n}_1\cdot\mathbf{y}_k-\mathbf{n}_2\cdot\mathbf{y}_\ell)}\mathbf{F}(2\pi|\mathbf{r}_k(t)|\,|\mathbf{n}_1|)\mathbf{F}(2\pi|\mathbf{r}_\ell(t)|\,|\mathbf{n}_2|)\right|$$

$$\le \sum_{\substack{\mathbf{n}_1\in\mathbb{Z}^3:\\ \mathbf{n}_1\ne\mathbf{0}}} |a_{\mathbf{n}_1}|^2\,\mathbf{F}(2\pi|\mathbf{r}_k(t)|\,|\mathbf{n}_1|)$$

$$\cdot \sum_{\substack{\mathbf{n}_2\in\mathbb{Z}^3:\\ \mathbf{n}_1\ne\mathbf{n}_2\ne\mathbf{0}}} \left(\sum_{1\le k\le N}\mu_k\right)^2 \mathcal{F}(2\pi\mathcal{R}(t)\,|\mathbf{n}_2|)$$

$$\le \sigma_0^2(f)\mathcal{F}(2\pi\mathcal{R}(t))N^2 \sum_{\substack{\mathbf{n}\in\mathbb{Z}^3:\\ \mathbf{n}\ne\mathbf{0}}} \mathcal{F}(2\pi\mathcal{R}(t)\,|\mathbf{n}|). \tag{7.13}$$

Similarly,

$$\left| \sum_{\substack{\mathbf{n}_j\in\mathbb{Z}^3:j=1,2\\ \mathbf{0}\ne\mathbf{n}_1\ne\mathbf{n}_2\ne\mathbf{0}}} a_{\mathbf{n}_1}\overline{a_{\mathbf{n}_2}} \sum_{1\le k\le N} \mu_k^2 e^{2\pi i(\mathbf{n}_1-\mathbf{n}_2)\cdot\mathbf{y}_k}\mathbf{F}(2\pi|\mathbf{r}_k(t)|\,|\mathbf{n}_1-\mathbf{n}_2|)\right|$$

$$\le \sum_{\substack{\mathbf{n}_j\in\mathbb{Z}^3:j=1,2\\ \mathbf{0}\ne\mathbf{n}_1\ne\mathbf{n}_2\ne\mathbf{0}}} \frac{1}{2}\left(|a_{\mathbf{n}_1}|^2+|a_{\mathbf{n}_2}|^2\right)\left(\sum_{1\le k\le N}\mu_k^2\right)\mathbf{F}(2\pi|\mathbf{r}_k(t)|\,|\mathbf{n}_1-\mathbf{n}_2|)$$

$$\le \sum_{\substack{\mathbf{n}_1\in\mathbb{Z}^3:\\ \mathbf{n}_1\ne\mathbf{0}}} |a_{\mathbf{n}_1}|^2 \left(\sum_{1\le k\le N}\mu_k^2\right) \sum_{\substack{\mathbf{n}_2\in\mathbb{Z}^3:\\ \mathbf{n}_1\ne\mathbf{n}_2\ne\mathbf{0}}} \mathcal{F}(2\pi\mathcal{R}(t)|\mathbf{n}_1-\mathbf{n}_2|)$$

$$\le \sigma_0^2(f)\left(\sum_{1\le k\le N}\mu_k^2\right) \sum_{\substack{\mathbf{n}\in\mathbb{Z}^3:\\ \mathbf{n}\ne\mathbf{0}}} \mathcal{F}(2\pi\mathcal{R}(t)\,|\mathbf{n}|). \tag{7.14}$$

In the special case $\mathbf{F}(x)=\mathcal{F}(x)=e^{-x^2/2}$ we clearly have

$$\sum_{\substack{\mathbf{n}\in\mathbb{Z}^3:\\ \mathbf{n}\ne\mathbf{0}}} \exp\left(-2(\pi\mathcal{R}(t)|\mathbf{n}|)^2\right) = \left(1+2\sum_{n=1}^{\infty}\exp\left(-2(\pi\mathcal{R}(t)n)^2\right)\right)^3 - 1$$

$$\leq \left(1+2\sum_{n=1}^{\infty}\exp\left(-2n(\pi\mathcal{R}(t))^2\right)\right)^3 - 1 = \left(1+\frac{2\exp\left(-2(\pi\mathcal{R}(t))^2\right)}{1-\exp\left(-2(\pi\mathcal{R}(t))^2\right)}\right)^3 - 1$$

$$\leq \frac{2\exp\left(-2(\pi\mathcal{R}(t))^2\right)}{1-\exp\left(-2(\pi\mathcal{R}(t))^2\right)}\left(1+\frac{2\exp\left(-2(\pi\mathcal{R}(t))^2\right)}{1-\exp\left(-2(\pi\mathcal{R}(t))^2\right)}\right)^2. \tag{7.15}$$

Combining (7.12)–(7.15) in the special case $\mathbf{F}(x) = \mathcal{F}(x) = e^{-x^2/2}$, we have

$$\left|\mathbf{D}^2_{\Omega_1}(\mathcal{Y},\mathcal{M};\mathcal{R};f;t) - \sigma_0^2(f)\sum_{k=1}^{N}\mu_k^2\right| \leq e^{-(2\pi\mathcal{R}(t))^2}\sigma_0^2(f)N^2$$

$$+\sigma_0^2(f)N^2\frac{2\exp\left(-4(\pi\mathcal{R}(t))^2\right)}{1-\exp\left(-2(\pi\mathcal{R}(t))^2\right)}\left(1+\frac{2\exp\left(-2(\pi\mathcal{R}(t))^2\right)}{1-\exp\left(-2(\pi\mathcal{R}(t))^2\right)}\right)^2$$

$$+\sigma_0^2(f)\left(\sum_{1\leq k\leq N}\mu_k^2\right)\frac{2\exp\left(-2(\pi\mathcal{R}(t))^2\right)}{1-\exp\left(-2(\pi\mathcal{R}(t))^2\right)}\left(1+\frac{2\exp\left(-2(\pi\mathcal{R}(t))^2\right)}{1-\exp\left(-2(\pi\mathcal{R}(t))^2\right)}\right)^2, \tag{7.16}$$

which proves the second statement (3.2) in Theorem 3.1.

Consider now the special case where the (parametrized) curves are the same $\mathbf{r}_1(t) = \mathbf{r}_2(t) = \ldots = \mathbf{r}_N(t) = \mathbf{r}(t)$. Then using the notation

$$\Phi(\mathcal{Y},\mathcal{M};\mathbf{n}) = \sum_{k=1}^{N}\mu_k e^{2\pi i\mathbf{n}\cdot\mathbf{y}_k}, \tag{7.17}$$

and also $\mathcal{R} = \mathbf{r}$ (because $\mathcal{R}(t) = |\mathbf{r}(t)|$), we can rewrite (7.10) as follows:

$$\overline{\mathbf{D}^2}_{\Omega_1}(\mathcal{Y},\mathcal{M};\mathcal{R}=\mathbf{r};f;t) = \int_{\Omega_1}\overline{\mathbf{D}^2}(\mathcal{Y},\mathcal{M};\mathcal{R}=\mathbf{r};\omega;f;t)\,d\omega$$

$$= \sum_{k=1}^{N}\mu_k^2\sum_{\substack{\mathbf{n}\in\mathbb{Z}^3:\\ \mathbf{n}\neq\mathbf{0}}}|a_\mathbf{n}|^2\left(1-\mathbf{F}^2(2\pi|\mathbf{r}(t)|\,|\mathbf{n}|)\right)$$

$$+\sum_{\substack{\mathbf{n}\in\mathbb{Z}^3:\\ \mathbf{n}\neq\mathbf{0}}}|a_\mathbf{n}|^2\,|\Phi(\mathcal{Y},\mathcal{M};\mathbf{n})|^2\,\mathbf{F}^2(2\pi|\mathbf{r}(t)|\,|\mathbf{n}|). \tag{7.18}$$

Combining (7.17) and (7.18), we obtain (3.4) in Theorem 3.1, obtain (4.7) in Theorem 4.1, obtain (4.19) in Theorem 4.2(a), and obtain (4.19′) in

Theorem 4.2(b). (Of course, in the cases of the second and third kinds we replace $\mathbf{F}(y) = e^{-y^2/2}$ with

$$\mathbf{F}(y) = \frac{\sin y}{y} \text{ and } -3\frac{\cos y}{y^2} + 3\frac{\sin y}{y^3},$$

and replace Ω_1 with Ω_2 and Ω_3, respectively.)

By choosing $t = 0$ in (7.18), we obtain the following simple but useful result:

$$\overline{\mathbf{D}^2}_{\Omega_1}(\mathcal{Y}, \mathcal{M}; \mathcal{R} = \mathbf{r}; f; 0) = \sum_{\substack{\mathbf{n}\in\mathbb{Z}^3:\\ \mathbf{n}\neq\mathbf{0}}} |a_{\mathbf{n}}|^2 \, |\Phi(\mathcal{Y}, \mathcal{M}; \mathbf{n})|^2 . \tag{7.19}$$

Note that (7.19) also follows from (7.6′) with $t = 0$.

By (7.18), Parseval's formula (7.11), and (7.19), we obtain

$$\left| \overline{\mathbf{D}^2}_{\Omega_1}(\mathcal{Y}, \mathcal{M}; \mathcal{R} = \mathbf{r}; f; t) - \sigma_0^2(f) \sum_{k=1}^{N} \mu_k^2 \right|$$
$$\leq \mathcal{F}^2(2\pi|\mathbf{r}(t)|) \max\left\{ \overline{\mathbf{D}^2}_{\Omega_1}(\mathcal{Y}, \mathcal{M}; \mathcal{R} = \mathbf{r}; f; 0), \sigma_0^2(f) \sum_{k=1}^{N} \mu_k^2 \right\} \tag{7.20}$$

with $\mathcal{F}(x) = \sup_{y\geq x} |\mathbf{F}(y)|$, which proves (3.3) in Theorem 3.1, proves (4.6) in Theorem 4.1, proves (4.18) in Theorem 4.2(a), and proves (4.18′) in Theorem 4.2(b). (Again for the second and third kinds we replace $\mathbf{F}(y) = e^{-y^2/2}$ with

$$\mathbf{F}(y) = \frac{\sin y}{y} \text{ and } -3\frac{\cos y}{y^2} + 3\frac{\sin y}{y^3},$$

and replace Ω_1 with Ω_2 and Ω_3, respectively.)

To complete the proof of Theorem 3.1 in the 3-dimensional case, it remains to prove Lemma 7.1. Since the proof of the case of arbitrary dimension $d \geq 1$ goes exactly the same way as that of the 3-dimensional case, to prove Theorem 3.1 in the general case, it suffices to prove the d-dimensional generalization of Lemma 7.1.

Lemma 7.1′. *Let $d \geq 2$ be an integer. For every d-dimensional vector $\mathbf{w} \in \mathbf{R}^d$ and unit vector $\mathbf{e} \in \mathbf{S}^{d-1}$, we have*

$$\sqrt{\frac{2}{\pi}} \int_0^\infty \int_{\vartheta\in SO(d)} e^{-2\pi i\vartheta(\rho\mathbf{e})\cdot\mathbf{w}} \frac{\rho^{d-1} e^{-\rho^2/2}}{(d-2)!!} \, d\rho \, d\vartheta = e^{-2\pi^2|\mathbf{w}|^2}$$

for every odd dimension $d \geq 3$,

$$\int_0^\infty \int_{\vartheta\in SO(d)} e^{-2\pi i\vartheta(\rho\mathbf{e})\cdot\mathbf{w}} \frac{\rho^{d-1} e^{-\rho^2/2}}{2^{\frac{d}{2}-1}(\frac{d}{2}-1)!} \, d\rho \, d\vartheta = e^{-2\pi^2|\mathbf{w}|^2}$$

for every even $d \geq 2$.

Remarks. Here we use the well known notation $k!! = k(k-2)(k-4)\cdots 2$ or 1, where the last factor in the product is 2 or 1 depending on the parity of the integer $k \geq 1$.

Note that the special case $d = 1$ of Lemma 7.1′ is reduced to the well known fact that the Fourier transform of $e^{-x^2/2}$ is itself.

Proof of Lemma 7.1′. We start with the special case $d = 3$, i.e., the original Lemma 7.1. By using the symmetry of the sphere, we clearly have $(\mathbf{e} \in \mathbf{S}^2)$

$$\sqrt{\frac{2}{\pi}} \int_0^\infty \int_{\vartheta \in SO(3)} e^{-2\pi \mathrm{i}\vartheta(\rho\mathbf{e})\cdot\mathbf{w}} \rho^2 e^{-\rho^2/2}\, d\rho\, d\vartheta$$

$$= \sqrt{\frac{2}{\pi}} \int_0^\infty \frac{1}{4\pi} \int_{\mathbf{u}\in\mathbf{S}^2} e^{-2\pi \mathrm{i}\rho\mathbf{u}\cdot\mathbf{w}} \rho^2 e^{-\rho^2/2}\, d\rho\, d\mathrm{SA}(\mathbf{u}).$$

Next we replace the spherically symmetric definition of the Gaussian velocity distribution $\mathbf{v} = \rho\mathbf{u}$ (where $\mathbf{u} \in \mathbf{S}^2$ is uniform on the unit sphere, ρ has distribution (1.9′), and $\rho, \mathbf{u}$ are independent random variables) with the equivalent second definition $\mathbf{v} = (v_1, v_2, v_3)$, where v_1, v_2, v_3 are three independent random variables having common standard normal distribution with expectation 0 and variance 1. This way we can apply (3.36), the second statement of Lemma 3.1. By using (3.36) with the choice $\mathbf{w} = -2\pi\mathbf{w}$, we obtain

$$\sqrt{\frac{2}{\pi}} \int_0^\infty \frac{1}{4\pi} \int_{\mathbf{u}\in\mathbf{S}^2} e^{-2\pi \mathrm{i}\rho\mathbf{u}\cdot\mathbf{w}} \rho^2 e^{-\rho^2/2}\, d\rho\, d\mathrm{SA}(\mathbf{u}) = e^{-2\pi^2|\mathbf{w}|^2},$$

which completes the proof of Lemma 7.1 (i.e., the special case $d = 3$ in Lemma 7.1′).

To prove Lemma 7.1′ in general, we simply repeat the previous argument replacing (3.36) with (3.40).

Note that we could also use the last formula in Theorem 4.2

$$\mathbf{F}_g(\mathbf{w}) = \frac{1}{\mathrm{SuArea}(\mathbf{S}^{d-1})} \int_{\mathbf{v}\in\mathbf{S}^{d-1}} e^{\mathrm{i}u\mathbf{v}\cdot\mathbf{w}} g(u)\, du\, d\mathrm{SA}(\mathbf{v})$$

$$= \int_{\mathbf{R}^d} e^{\mathrm{i}\mathbf{x}\cdot\mathbf{w}} \frac{g(|\mathbf{x}|)}{|\mathbf{x}|^{d-1}}\, d\mathbf{x} = (2\pi)^{d/2} \widehat{\mathbf{g}_d}(-\mathbf{w}),$$

where

$$\mathbf{g}_d(\mathbf{x}) = \frac{g(|\mathbf{x}|)}{|\mathbf{x}|^{d-1}}$$

and $\widehat{\mathbf{g}_d}(\mathbf{w})$ denotes its Fourier transform. In this special case we have (see (3.32))

$$\mathbf{g}_d(\mathbf{x}) = \frac{g(|\mathbf{x}|)}{|\mathbf{x}|^{d-1}}(2\pi)^{-d/2}e^{-|\mathbf{x}|^2},$$

and so

$$(2\pi)^{d/2}\widehat{\mathbf{g}_d}(-\mathbf{w}) = e^{-|\mathbf{x}|^2}$$

(since the multivariable Gauss distribution is a fixed point of the Fourier transform). Finally, applying the substitution $\mathbf{x} = 2\pi\mathbf{w}$, Lemma 7.1′ follows. □

8. Proving Theorems 4.1–2 and Lemma 5.1

To complete the proof of Theorem 4.2 in the 3-dimensional case, we need the following simple result.

Lemma 8.1. *Let* $\mathbf{w}$ *be an arbitrary 3-dimensional vector. We have*

$$\frac{1}{(4\pi/3)}\int_{|\mathbf{u}|\le 1} e^{i\mathbf{w}\cdot\mathbf{u}}\,d\mathbf{u} = -3\frac{\cos|\mathbf{w}|}{|\mathbf{w}|^2} + 3\frac{\sin|\mathbf{w}|}{|\mathbf{w}|^3}.$$

Remark. Lemma 8.1 is about to compute the Fourier transform of the 3-dimensional unit ball (i.e., the Fourier transform of the characteristic function). Dimension 3 is odd, so we get a finite trigonometric expression. In fact, it is closely related to the Bessel function

$$J_{\frac{3}{2}}(z) = \sqrt{\frac{2}{\pi z}}\left(-\cos z + \frac{\sin z}{z}\right).$$

For even dimension d we are not so lucky: then we get the Bessel function $J_{d/2}(z)$, which does not have a finite form (but it has an infinite power series representation; see the formula in the Remarks after Theorem 4.1).

Note that the 1-dimensional analog of Lemma 8.1 is

$$\frac{1}{2}\int_{u=-1}^{1} e^{iwu}\,du = \frac{\sin w}{w},$$

which is closely related to the Bessel function

$$J_{\frac{1}{2}}(z) = \sqrt{\frac{2}{\pi z}}\sin z.$$

As a trivial by-product of Lemma 8.1, we obtain the inequality

$$\left|-3\frac{\cos x}{x^2} + 3\frac{\sin x}{x^3}\right| \le 1.$$

Proof. We recall that $4\pi/3$ is the volume of the unit 3-ball B^3 and $4\pi r^2$ is the surface area of the sphere of radius r. By using the first part (3.35) of Lemma 3.1, we have

$$\frac{1}{(4\pi/3)}\int_{|\mathbf{u}|\le 1} e^{i\mathbf{w}\cdot\mathbf{u}}\,d\mathbf{u} = \frac{3}{4\pi}\int_{r=0}^{1}\left(\int_{|\mathbf{u}|=r} e^{i\mathbf{w}\cdot\mathbf{u}}\,d\mathrm{SA}(\mathbf{u})\right)dr$$

$$= \frac{3}{4\pi}\int_{r=0}^{1} r^2\left(\int_{\mathbf{v}\in\mathbf{S}^2} e^{i\mathbf{w}\cdot r\mathbf{v}}\,d\mathrm{SA}(\mathbf{v})\right)dr = 3\int_{r=0}^{1} r^2\left(\frac{\sin(r|\mathbf{w}|)}{r|\mathbf{w}|}\right)dr$$

$$= \frac{3}{|\mathbf{w}|}\int_{r=0}^{1} r\sin(r|\mathbf{w}|)\,dr,$$

where $d\mathrm{SA}(\mathbf{v})$ indicates the integration on a sphere with respect to the surface area. Using integration by parts in the last line, we have

$$\frac{1}{(4\pi/3)}\int_{|\mathbf{u}|\le 1} e^{i\mathbf{w}\cdot\mathbf{u}}\,d\mathbf{u} = -3\frac{\cos|\mathbf{w}|}{|\mathbf{w}|^2} + 3\frac{\sin|\mathbf{w}|}{|\mathbf{w}|^3},$$

completing the proof of Lemma 8.1. □

We have the perfect analog of the trivial equality (7.19):

$$\overline{\mathbf{D}^2}_{\Omega_i}(\mathcal{Y},\mathcal{M};\mathcal{R}=\mathbf{r};f;0) = \sum_{\substack{\mathbf{n}\in\mathbb{Z}^3:\\ \mathbf{n}\neq\mathbf{0}}} |a_{\mathbf{n}}|^2\,|\Phi(\mathcal{Y},\mathcal{M};\mathbf{n})|^2 \tag{8.1}$$

also holds for $i=2,3$.

In the 3-dimensional case the switch from Theorem 3.1 to Theorem 4.1, or to Theorem 4.2, means that we replace the smooth (probability) density function $\sqrt{2/\pi}\rho^2 e^{-\rho^2/2}$ in Theorem 3.1 with, respectively, the Dirac delta function concentrated on $\rho=1$, or $3\rho^2$ as $0\le\rho\le 1$, or a general probability density function $g(\rho)$, $0\le\rho<\infty$. As we already explained in Section 7, to prove Theorems 4.1, or Theorem 4.2, in the 3-dimensional case, we just repeat the proof of Theorem 3.1 with a minor change: to prove Theorem 4.2(a) we replace Lemma 7.1 with Lemma 8.1, to prove Theorem 4.1 we replace Lemma 7.1 with (3.35): for every 3-dimensional vector $\mathbf{w}$

$$\frac{1}{4\pi}\int_{\mathbf{v}\in\mathbf{S}^2} e^{i\mathbf{w}\cdot\mathbf{v}}\,d\mathrm{SA}(\mathbf{v}) = \frac{\sin w}{w}, \tag{8.2}$$

where $w=|\mathbf{w}|$ is the usual length (and $d\mathrm{SA}(\mathbf{v})$ represents the surface area), and to prove Theorem 4.2(b) we work with the isotropic Fourier transform $\mathbf{F}_g$ of g. In other words, we replace

$$\mathbf{F}(w) = e^{-w^2/2} \text{ with } \mathbf{F}(w) = \frac{\sin w}{w} \text{ and } \mathbf{F}(w) = -3\frac{\cos w}{w^2} + 3\frac{\sin w}{w^3}$$

and $\mathbf{F} = \mathbf{F}_g$ in general, and replace Ω_1 with Ω_2, Ω_3 and Ω_g according to Theorem 4.1, Theorem 4.2(a) and Theorem 4.2(b) (in the 3-dimensional case).

The proof of the case of arbitrary dimension $d \geq 2$ goes along the same lines. The only difference is that we need the d-dimensional version of Lemma 8.1 and the d-dimensional version of (8.2). Both involve the Bessel functions (of the first kind) introduced in the Remarks after Theorem 4.1. First we formulate the general version of (8.2).

Lemma 8.2. *Let $d \geq 2$ be an integer. For every d-dimensional vector $\mathbf{w}$, we obtain with $w = |\mathbf{w}|$,*

$$\frac{1}{\text{SuArea}(\mathbf{S}^{d-1})} \int_{\mathbf{u}\in\mathbf{S}^{d-1}} e^{\mathbf{i}\mathbf{w}\cdot\mathbf{u}}\, d\text{SA}(\mathbf{u}) = (d-2)!!\frac{J_{(d-2)/2}(w)}{w^{(d-2)/2}} \cdot \left\{1 \text{ or } \sqrt{\frac{\pi}{2}}\right\},$$

depending on whether $d \geq 2$ is even or odd, accordingly.

Before proving Lemma 8.2, we formulate a generalization of Lemma 8.1 for arbitrary dimension $d \geq 2$. As usual, let $\mathbf{B}^d$ denote the d-dimensional unit ball centered at the origin.

Lemma 8.3. *Let $d \geq 2$ be an integer. For every d-dimensional vector $\mathbf{w}$, with $w = |\mathbf{w}|$ we have*

$$\frac{1}{\text{vol}(\mathbf{B}^d)} \int_{\mathbf{u}\in\mathbf{B}^d} e^{\mathbf{i}\mathbf{w}\cdot\mathbf{u}}\, d\mathbf{u} = \frac{d!!}{w^{d/2}} J_{d/2}(w) \cdot \left\{1 \text{ or } \sqrt{\frac{\pi}{2}}\right\}$$

depending on whether d is even or odd, accordingly.

It is easier to prove Lemma 8.3 first, and to derive Lemma 8.2 from Lemma 8.3 second.

Proof of Lemma 8.3. Let $\mathbf{B}^d(r)$ denote the d-dimensional ball of radius r centered at the origin (in Lemma 8.3 we just need the special case $r = 1$, but in the proof of Lemma 8.2 we apply the general case of arbitrary radius).

Write

$$f(r, w) = \int_{\mathbf{u}\in\mathbf{B}^d(r)} e^{\mathbf{i}\mathbf{w}\cdot\mathbf{u}}\, d\mathbf{u}. \tag{8.3}$$

To evaluate integral (8.3), we use the well-known fact that

$$V_k = \frac{\pi^{k/2}}{\text{Gamma}(\frac{k}{2}+1)}$$

denotes the volume of the k-dimensional unit ball (where in this book Gamma(z) stands for Euler's Gamma function), and since the ball $\mathbf{B}^d(r)$ is rotationally symmetric ("isotropic"), we can compute (8.3) at the special point $\mathbf{w} = (w, 0, \ldots, 0)$, $w > 0$:

$$f(r,w) = \int_{|\mathbf{x}|\le r} e^{\mathrm{i}x_1 w}\, dx_1\, dx_2 \ldots dx_d$$

$$= \int_{-r}^{r} V_{d-1}(r^2 - x_1^2)^{(d-1)/2} e^{\mathrm{i}x_1 w}\, dx_1$$

$$= V_{d-1} r^d \int_{-1}^{1} (1 - y_1^2)^{(d-1)/2} e^{\mathrm{i}y_1 r w}\, dy_1. \tag{8.4}$$

The good news is that the last integral (8.4) can be expressed in terms of the classical Bessel function $J_\nu(z)$. Indeed, we make use of the Poisson representation formula (see e.g. formula (8) on page 48 in Watson [Wa58])

$$J_\nu(z) = \frac{(z/2)^\nu}{\mathrm{Gamma}(\nu + \frac{1}{2})\sqrt{\pi}} \int_{-1}^{1} e^{\mathrm{i}zx}(1 - x^2)^{\nu - \frac{1}{2}}\, dx, \tag{8.5}$$

which holds for all $\nu > -1/2$ and complex z.

Combining (8.4) and (8.5) with $\nu = d/2$ and $z = rw = r|\mathbf{w}|$, we obtain

$$f(r,w) = V_{d-1} r^d \frac{\mathrm{Gamma}(\frac{d+1}{2})\sqrt{\pi}}{(rw/2)^{d/2}} J_{d/2}(rw)$$

$$= \frac{\pi^{(d-1)/2}}{\mathrm{Gamma}(\frac{d+1}{2})} r^d \frac{\mathrm{Gamma}(\frac{d+1}{2})\sqrt{\pi}}{(rw/2)^{d/2}} J_{d/2}(rw) = \left(\frac{2\pi r}{w}\right)^{d/2} J_{d/2}(rw). \tag{8.6}$$

We are now ready to complete the proof of Lemma 8.3: by using (8.6) with $r = 1$ we have

$$\frac{1}{\mathrm{vol}(\mathbf{B}^d)} \int_{\mathbf{u}\in\mathbf{B}^d} e^{\mathrm{i}\mathbf{w}\cdot\mathbf{u}}\, d\mathbf{u} = \frac{1}{V_d} f(1,w) = \frac{\mathrm{Gamma}(\frac{d}{2}+1)}{\pi^{d/2}} \left(\frac{2\pi}{w}\right)^{d/2} J_{d/2}(w)$$

$$= \frac{2^{d/2}\mathrm{Gamma}(\frac{d}{2}+1)}{w^{d/2}} J_{d/2}(w) = \frac{d!!}{w^{d/2}} J_{d/2}(w) \cdot \left\{1 \text{ or } \sqrt{\frac{\pi}{2}}\right\}$$

depending on whether d is even or odd, accordingly. Thus Lemma 8.3 follows. □

Proof of Lemma 8.2. We recall the well known fact that the surface area of an ordinary sphere of radius R is $4\pi R^2$, and it is the derivative of the volume $4\pi R^3/3$ of the 3-ball of radius R (a simple geometric proof comes from the decomposition of the ball into concentric spheres). The same relation holds in every dimension, so we have

$$\text{SuArea}(\mathbf{S}^{d-1}) = dV_d = d\frac{\pi^{d/2}}{\text{Gamma}(\frac{d}{2}+1)}. \tag{8.7}$$

The same argument proves the following identity

$$h(w) = \int_{\mathbf{u}\in\mathbf{S}^{d-1}} e^{\mathbf{i}\mathbf{w}\cdot\mathbf{u}}\, d\text{SA}(\mathbf{u}) = \frac{\partial f(r,w)}{\partial r}(1,w), \tag{8.8}$$

where the right-hand side of (8.8) is the partial derivative of $f(r,w)$ with respect to r at $(1,w)$.

We make use of a derivative identity for the Bessel functions

$$\frac{d}{dz}\left(z^{\nu} J\nu(z)\right) = z^{\nu} J_{\nu-1}(z). \tag{8.9}$$

The proof of (8.9) is just a straightforward calculation by using the familiar power series representation of the Bessel function:

$$\frac{d}{dz}\left(z^{\nu} J\nu(z)\right) = \frac{d}{dz}\left(z^{\nu}\sum_{m=0}^{\infty}\frac{(-1)^m}{m!\text{Gamma}(m+\nu+1)}\left(\frac{z}{2}\right)^{2m+\nu}\right)$$

$$= \sum_{m=0}^{\infty}(2m+2\nu)\frac{(-1)^m}{m!\text{Gamma}(m+\nu+1)2^{2m+\nu}}z^{2m+2\nu-1}$$

$$= z^{\nu}\sum_{m=0}^{\infty}\frac{(-1)^m}{m!\text{Gamma}(m+\nu)2^{2m+\nu-1}}z^{2m+\nu-1} = z^{\nu}J_{\nu-1}(z),$$

proving (8.9)

Combining (8.6) and (8.8), we obtain that $h(w)$ is the derivative of

$$r^{d/2}\left(\frac{2\pi r}{w}\right)^{d/2} J_{d/2}(rw)$$

with respect to r at $r=1$ (w is fixed), that is, the derivative of

$$\left(\frac{2\pi}{w^2}\right)^{d/2}(rw)^{d/2}J_{d/2}(rw)$$

with respect to r at $r=1$ (w is fixed). Then by using (8.9) we have

$$h(w) = \left(\frac{2\pi}{w^2}\right)^{d/2} w^{d/2}J_{(d/2)-1}(w)\cdot w = (2\pi)^{d/2}w^{1-(d/2)}J_{(d/2)-1}(w). \tag{8.10}$$

Finally, combining (8.7), (8.8) and (8.10), we obtain

$$\frac{1}{\mathrm{SuArea}(\mathbf{S}^{d-1})}\int_{\mathbf{u}\in\mathbf{S}^{d-1}} e^{i\mathbf{w}\cdot\mathbf{u}}\,d\mathrm{SA}(\mathbf{u}) = \frac{1}{\mathrm{SuArea}(\mathbf{S}^{d-1})}h(w)$$

$$= \frac{\mathrm{Gamma}(\frac{d}{2}+1)}{d\pi^{d/2}}(2\pi)^{d/2}w^{1-(d/2)}J_{(d/2)-1}(w)$$

$$= (d-2)!!\frac{J_{(d-2)/2}(w)}{w^{(d-2)/2}}\cdot\left\{1 \text{ or } \sqrt{\frac{\pi}{2}}\right\},$$

depending on whether $d \geq 2$ is even or odd, accordingly. Thus Lemma 8.2 follows. □

Now we are ready to complete the proof of Theorem 4.2(b). By using the first part of Lemma 3.1

$$\frac{1}{4\pi}\int_{\mathbf{v}\in\mathbf{S}^2} e^{iu\mathbf{v}\cdot\mathbf{w}}\,d\mathrm{SA}(\mathbf{v}) = \frac{\sin(uw)}{uw} \quad \text{with } w = |\mathbf{w}|$$

(where $d\mathrm{SA}(\mathbf{v})$ represent the surface area), we obtain

$$\mathbf{F}_g(\mathbf{w}) = \int_0^\infty g(u)\frac{\sin(uw)}{uw}\,du. \tag{8.11}$$

Notice that (8.11) immediately implies the facts

$$\mathbf{F}_g(0) = \mathbf{F}_g(\mathbf{0}) = 1 \quad \text{and} \quad |\mathbf{F}_g(w)| = |\mathbf{F}_g(\mathbf{w})| < 1 \quad \text{for} \quad w = |\mathbf{w}| > 0.$$

Since $g(-x) = g(x)$, we can rewrite (8.11):

$$\mathbf{F}_g(\mathbf{w}) = \frac{1}{2}\int_{-\infty}^\infty g(u)\frac{\sin(uw)}{uw}\,du. \tag{8.12}$$

We recall some well known facts from the theory of Fourier transform:

$$\widehat{f}(t) = \frac{1}{\sqrt{2\pi}}\int_{-\infty}^\infty e^{-ixt}f(x)\,dx \tag{8.13}$$

denotes the Fourier transform of f, and we have the similar inversion formula

$$f(x) = \frac{1}{\sqrt{2\pi}}\int_{-\infty}^\infty e^{ixt}\widehat{f}(t)\,dt. \tag{8.14}$$

It follows from (8.13)–(8.14) that the Fourier inverse of $f(x)$ is $\widehat{f}(-t)$.

We also need the fact

$$\int_{-\infty}^\infty f_1(x)\overline{f_2(x)}\,dx = \int_{-\infty}^\infty \widehat{f_1}(t)\overline{\widehat{f_2}(t)}\,dt, \tag{8.15}$$

which is a corollary of the Parseval–Plancherel formula. Note that (8.13)–(8.15) hold under mild conditions; for example, they hold if the functions are in the space $L^2(\mathbf{R})$.

If $\chi_w(x)$ denotes the 0-1 valued characteristic function of the interval $[-w,w]$, then (8.13) gives via simple calculations

$$\widehat{\chi_w}(u) = \sqrt{\frac{2}{\pi}}\frac{\sin(uw)}{u}. \tag{8.16}$$

Since $\chi_w(x)$ is an even function, the Fourier transform of

$$f(u) = \sqrt{\frac{2}{\pi}}\frac{\sin(uw)}{u}$$

is $\chi_w(x)$. Combining this fact with (8.15), we have

$$\int_{-\infty}^{\infty} g(u)\frac{\sin(uw)}{uw}\,du = \int_{-\infty}^{\infty} \widehat{g}(x)\sqrt{\frac{\pi}{2}}\frac{\chi_w(x)}{w}\,dx = \frac{\sqrt{\pi}}{\sqrt{2}w}\int_{-w}^{w}\widehat{g}(x)\,dx. \tag{8.17}$$

Note that g is not necessarily in the $L^2(\mathbf{R})$-space; nevertheless, (8.15) applies in (8.17), since $\int_{-w}^{w}\widehat{g}(x)\,dx$ is a "safe" finite integral.

By (8.12) and (8.17),

$$\mathbf{F}_g(\mathbf{w}) = \frac{\sqrt{\pi}}{2\sqrt{2}w}\int_{-w}^{w}\widehat{g}(x)\,dx,$$

which finishes the proof of Theorem 4.2(b). Thus the proofs of Theorems 4.1 and 4.2 are complete. □

Proof of Lemma 5.1. We need the discrete version of the Erdős–Turán–Koksma inequality in Uniform Distribution (see e.g. [Dr-Ti97]).

Discrete Erdős–Turán–Koksma Inequality: Let $\mathcal{Y} = \{\mathbf{y}_1, \mathbf{y}_2, \ldots, \mathbf{y}_N\} \subset I^d$ be an arbitrary N-element point set in the unit cube $I^d = [-\frac{1}{2}, \frac{1}{2})^d$ and $M \geq 1$ be an arbitrary integer; then (B denotes axis-parallel boxes)

$$\max_{B=I_1\times\cdots\times I_d\subset I^d}\left|\frac{1}{N}\sum_{\substack{1\le k\le N:\\ \mathbf{y}_k\in B}} 1 - \mathrm{vol}(B)\right|$$

$$\leq \left(\frac{3}{2}\right)^d\left(\frac{2}{M+1} + \sum_{1\le\|\mathbf{n}\|_\infty\le M}\frac{1}{\Lambda(\mathbf{n})}\left|\frac{1}{N}\sum_{k=1}^{N} e^{2\pi i\mathbf{n}\cdot\mathbf{y}_k}\right|\right), \tag{8.18}$$

where

$$\|\mathbf{n}\|_\infty = \max_{1\le j\le d}|n_j|$$

is the maximum norm of the vector $\mathbf{n} = (n_1, \ldots, n_d) \in \mathbb{Z}^d$, and

$$\Lambda(\mathbf{n}) = \prod_{1 \le j \le d} (1 + |n_j|) .$$

We apply (8.18) in the special case $d = 3$: in view of the hypothesis of Lemma 5.1, we have

$$0 < \beta \le \left(\frac{3}{2}\right)^3 \left(\frac{2}{M+1} + \sum_{1 \le \|\mathbf{n}\|_\infty \le M} \frac{1}{\Lambda(\mathbf{n})} \left| \frac{1}{N} \sum_{k=1}^{N} e^{2\pi i \mathbf{n} \cdot \mathbf{y}_k} \right| \right) . \tag{8.19}$$

Since trivially

$$\Lambda(\mathbf{n}) = \prod_{1 \le j \le 3} (1 + |n_j|) \ge |\mathbf{n}| = \sqrt{n_1^2 + n_2^2 + n_3^2},$$

(8.19) implies

$$0 < \beta \le \left(\frac{3}{2}\right)^3 \left(\frac{2}{M+1} + \sum_{1 \le \|\mathbf{n}\|_\infty \le M} \frac{1}{|\mathbf{n}|} \left| \frac{1}{N} \sum_{k=1}^{N} e^{2\pi i \mathbf{n} \cdot \mathbf{y}_k} \right| \right) . \tag{8.20}$$

By choosing $M + 1 = 14/\beta$ in (8.20), we have

$$0 < \frac{\beta}{2} \le \left(\frac{3}{2}\right)^3 \sum_{1 \le \|\mathbf{n}\|_\infty \le M} \frac{1}{|\mathbf{n}|} \left| \frac{1}{N} \sum_{k=1}^{N} e^{2\pi i \mathbf{n} \cdot \mathbf{y}_k} \right| . \tag{8.21}$$

Since there are less than $(2M+1)^3 < (28/\beta)^3$ integer lattice points $\mathbf{n} \in \mathbb{Z}^3 \setminus \mathbf{0}$ such that $1 \le \|\mathbf{n}\|_\infty \le M$, (8.21) implies

$$\max_{1 \le \|\mathbf{n}\|_\infty \le M} \frac{1}{|\mathbf{n}|} \left| \frac{1}{N} \sum_{k=1}^{N} e^{2\pi i \mathbf{n} \cdot \mathbf{y}_k} \right| \ge \frac{\beta}{2(28/\beta)^3} = \frac{\beta^4}{2 \cdot 28^3} > \beta^4 10^{-5},$$

completing the proof of Lemma 5.1. □

9. Proof of Theorem 5.1

We use the notation of Sections 3–5 with arbitrary dimension d.

To settle the different kinds and the different dimensions at the same time, we simplify the notation by using Ω for the different the initial velocity spaces, and, similarly, using the generic notation $\mathbf{F}$ for the corresponding isotropic Fourier transform. That is, we integrate (9.2) below over the spaces

$$\Omega = \Omega_1(\dim = d),\ \Omega_2(\dim = d) \text{ and } \Omega_3(\dim = d), \quad \text{and}$$

$$\Omega = \Omega_g = ([0,\infty) \times SO(3))^N$$

equipped with the product measure (Haar measure for $SO(3)$ and an arbitrary probability density function g for $[0,\infty)$) in the general 3-dimensional isotropic case; see Theorem 4.2(b). Accordingly we write $\mathbf{F}(x) = \mathbf{F}_1(x) = e^{-x^2/2}$ for the first kind in every dimension, write $\mathbf{F}(x) = \mathbf{F}_{2,3}(x) = \frac{\sin x}{x}$ for the second kind in dimension $d = 3$ and write

$$\mathbf{F}(x) = \mathbf{F}_{2,d}(x) = (d-2)!!\frac{J_{(d-2)/2}(z)}{x^{(d-2)/2}} \cdot \left\{1 \text{ or } \sqrt{\frac{\pi}{2}}\right\}$$

for the rest of the dimensions depending on whether $d \geq 2$ is even or odd, write

$$\mathbf{F}(x) = \mathbf{F}_{3,3}(x) = -3\frac{\cos x}{x^2} + 3\frac{\sin x}{x^3}$$

for the third kind in dimension $d = 3$, write

$$\mathbf{F}(x) = \mathbf{F}_{3,d}(x) = d!!\frac{J_{d/2}(x)}{x^{d/2}} \cdot \left\{1 \text{ or } \sqrt{\frac{\pi}{2}}\right\}$$

for the rest of the dimensions $d \geq 2$ depending on whether $d \geq 2$ is even or odd, and for $d = 1$ write $\mathbf{F}(x) = \mathbf{F}_{3,1}(x) = \frac{\sin x}{x}$. Finally, in the general 3-dimensional isotropic case, write

$$\mathbf{F}(x) = \mathbf{F}_g(x) = \mathbf{F}_g(\mathbf{x}) = \int_0^\infty \frac{1}{4\pi}\int_{\mathbf{v}\in\mathbf{S}^2} e^{iu\mathbf{v}\cdot\mathbf{x}} g(u)\, du\, d\text{SA}(\mathbf{v})$$

$$= \int_0^\infty \frac{1}{4\pi}\int_{\mathbf{v}\in\mathbf{S}^2} \cos(u\mathbf{v}\cdot\mathbf{x}) g(u)\, du\, d\text{SA}(\mathbf{v})$$

where $\mathbf{x}$ is an arbitrary 3-dimensional vector (and $d\text{SA}(\mathbf{v})$ represents the surface area). (Note that the integral $\mathbf{F}_g(\mathbf{x})$ defines a real-valued function that depends only on the length $x = |\mathbf{x}|$ of the input vector.)

In the arguments below we constantly use the facts that $\mathbf{F}$ is real-valued and $|\mathbf{F}| \leq 1$.

We evaluate the integral

$$\text{Variance}_{\omega\in\Omega}\overline{\mathbf{D}^2}(\mathcal{Y},\mathcal{M};\mathbf{r};\omega;f;t)$$

$$= \int_\Omega \left(\overline{\mathbf{D}^2}(\mathcal{Y},\mathcal{M};\mathbf{r};\omega;f;t) - \overline{\mathbf{D}^2}_\Omega(\mathcal{Y},\mathcal{M};\mathbf{r};f;t)\right)^2 d\omega$$

$$= \int_\Omega \left(\overline{\mathbf{D}^2}(\mathcal{Y},\mathcal{M};\mathbf{r};\omega;f;t)\right)^2 d\omega - \left(\overline{\mathbf{D}^2}_\Omega(\mathcal{Y},\mathcal{M};\mathbf{r};f;t)\right)^2. \tag{9.1}$$

By (7.6),

$$\left(\overline{\mathbf{D}^2}(\mathcal{Y},\mathcal{M};\mathbf{r};\omega;f;t)\right)^2 = \left(\sum_{j=1}^{N}\mu_j^2 \sum_{\substack{\mathbf{n}\in\mathbb{Z}^d:\\ \mathbf{n}\neq\mathbf{0}}} |a_{\mathbf{n}}|^2\right.$$

$$\left. + \sum_{\substack{\mathbf{n}\in\mathbb{Z}^d:\\ \mathbf{n}\neq\mathbf{0}}} |a_{\mathbf{n}}|^2 \sum_{1\le k,\ell\le N:k\neq\ell} \mu_k\mu_\ell e^{2\pi i\mathbf{n}\cdot(\mathbf{y}_k-\mathbf{y}_\ell)} e^{2\pi i\mathbf{n}\cdot(\vartheta_k(\rho_k\mathbf{r}(t))-\vartheta_\ell(\rho_\ell\mathbf{r}(t)))}\right)^2$$

$$= \left(\sum_{j=1}^{N}\mu_j^2\right)^2 \left(\sum_{\substack{\mathbf{n}\in\mathbb{Z}^d:\\ \mathbf{n}\neq\mathbf{0}}} |a_{\mathbf{n}}|^2\right)^2 + 2\sum_{j=1}^{N}\mu_j^2 \sum_{\substack{\mathbf{n}\in\mathbb{Z}^d:\\ \mathbf{n}\neq\mathbf{0}}} |a_{\mathbf{n}}|^2 \sum_{\substack{\mathbf{m}\in\mathbb{Z}^d:\\ \mathbf{m}\neq\mathbf{0}}} |a_{\mathbf{m}}|^2$$

$$\cdot \sum_{1\le k,\ell\le N:k\neq\ell} e^{2\pi i\mathbf{m}\cdot(\mathbf{y}_k-\mathbf{y}_\ell)}\mu_k\mu_\ell e^{2\pi i\mathbf{m}\cdot(\vartheta_k(\rho_k\mathbf{r}(t))-\vartheta_\ell(\rho_\ell\mathbf{r}(t)))}$$

$$+ \sum_{\substack{\mathbf{n}\in\mathbb{Z}^d:\\ \mathbf{n}\neq\mathbf{0}}} |a_{\mathbf{n}}|^2 \sum_{\substack{\mathbf{m}\in\mathbb{Z}^d:\\ \mathbf{m}\neq\mathbf{0}}} |a_{\mathbf{m}}|^2 \sum_{1\le k_1,\ell_1\le N:k_1\neq\ell_1} \sum_{1\le k_2,\ell_2\le N:k_2\neq\ell_2} \mu_{k_1}\mu_{\ell_1}$$

$$\cdot \mu_{k_2}\mu_{\ell_2} e^{2\pi i\left(\mathbf{n}\cdot(\mathbf{y}_{k_1}-\mathbf{y}_{\ell_1})-\mathbf{m}\cdot(\mathbf{y}_{k_2}-\mathbf{y}_{\ell_2})\right)}$$

$$\cdot e^{2\pi i\left(\mathbf{n}\cdot\left(\vartheta_{k_1}(\rho_{k_1}\mathbf{r}(t))-\vartheta_{\ell_1}(\rho_{\ell_1}\mathbf{r}(t))\right)-\mathbf{m}\cdot\left(\vartheta_{k_2}(\rho_{k_2}\mathbf{r}(t))-\vartheta_{\ell_2}(\rho_{\ell_2}\mathbf{r}(t))\right)\right)}. \tag{9.2}$$

Integrating (9.2) over Ω we have

$$\int_\Omega \left(\overline{\mathbf{D}^2}(\mathcal{Y},\mathcal{M};\mathbf{r};\omega;f;t)\right)^2 d\omega = \mathcal{S}_1+\mathcal{S}_2+\mathcal{S}_3+\widetilde{\mathcal{S}}_3+2\mathcal{S}_4+2\widetilde{\mathcal{S}}_4$$

$$+\mathcal{S}_5+\mathcal{S}_6+\widetilde{\mathcal{S}}_{6,1}+\widetilde{\mathcal{S}}_{6,2}++2\mathcal{S}_7+2\widetilde{\mathcal{S}}_{7,1}+2\widetilde{\mathcal{S}}_{7,2}+\mathcal{S}_8, \tag{9.3}$$

where

$$\mathcal{S}_1 = \left(\sum_{j=1}^{N}\mu_j^2\right)^2 \left(\sum_{\substack{\mathbf{n}\in\mathbb{Z}^d:\\ \mathbf{n}\neq\mathbf{0}}} |a_{\mathbf{n}}|^2\right)^2, \tag{9.4}$$

$$\mathcal{S}_2 = 2\sum_{j=1}^{N}\mu_j^2 \sum_{\substack{\mathbf{n}\in\mathbb{Z}^d:\\ \mathbf{n}\neq\mathbf{0}}} |a_{\mathbf{n}}|^2 \sum_{\substack{\mathbf{m}\in\mathbb{Z}^d:\\ \mathbf{m}\neq\mathbf{0}}} |a_{\mathbf{m}}|^2\, \mathbf{F}^2(2\pi|\mathbf{r}(t)|\ |\mathbf{m}|)$$

$$\cdot \sum_{1\le k,\ell\le N:\ k\ne\ell} \mu_k\mu_\ell e^{2\pi i \mathbf{m}\cdot(\mathbf{y}_k-\mathbf{y}_\ell)}, \tag{9.5}$$

$$\mathcal{S}_3 = \left(\left(\sum_{j=1}^N \mu_j^2\right)^2 - \sum_{j=1}^N \mu_j^4\right) \sum_{\substack{\mathbf{n}\in\mathbb{Z}^d:\\ \mathbf{n}\ne\mathbf{0}}} |a_{\mathbf{n}}|^4, \tag{9.6}$$

which corresponds to the case $\mathbf{n}=\mathbf{m}$, $(k_1,\ell_1)=(k_2,\ell_2)$ in (9.2),

$$\widetilde{\mathcal{S}}_3 = \sum_{\substack{\mathbf{n}\in\mathbb{Z}^d:\\ \mathbf{n}\ne\mathbf{0}}} |a_{\mathbf{n}}|^4 \sum_{1\le k_1,\ell_1\le N:k_1\ne\ell_1} \mu_{k_1}^2\mu_{\ell_1}^2$$

$$\cdot e^{2\pi i 2\mathbf{n}\cdot(\mathbf{y}_{k_1}-\mathbf{y}_{\ell_1})} \cdot \mathbf{F}^2(2\pi|\mathbf{r}(t)|\ |2\mathbf{n}|), \tag{9.6$'$}$$

which corresponds to the case $\mathbf{n}=\mathbf{m}$, $(k_1,\ell_1)=(\ell_2,k_2)$ in (9.2),

$$\mathcal{S}_4 = \sum_{\substack{\mathbf{n}\in\mathbb{Z}^d:\\ \mathbf{n}\ne\mathbf{0}}} |a_{\mathbf{n}}|^4\, \mathbf{F}^2(2\pi|\mathbf{r}(t)|\ |\mathbf{n}|)$$

$$\cdot \sum_{1\le \ell_1,\ell_2\le N:\ell_1\ne\ell_2} \mu_{\ell_1}\mu_{\ell_2} e^{2\pi i \mathbf{n}\cdot(\mathbf{y}_{\ell_1}-\mathbf{y}_{\ell_2})} \sum_{1\le k_1\le N:k_1\ne\ell_1,\ell_2} \mu_{k_1}^2, \tag{9.7}$$

which corresponds to the case $\mathbf{n}=\mathbf{m}$, $k_1=k_2$, $\ell_1\ne\ell_2$ in (9.2) (notice that $\mathcal{S}_4$ has coefficient 2 in (9.3) due to the symmetric case $\mathbf{n}=\mathbf{m}$, $k_1\ne k_2$, $\ell_1=\ell_2$ that has exactly the same contribution),

$$\widetilde{\mathcal{S}}_4 = \sum_{\substack{\mathbf{n}\in\mathbb{Z}^d:\\ \mathbf{n}\ne\mathbf{0}}} |a_{\mathbf{n}}|^4\, \mathbf{F}^2(2\pi|\mathbf{r}(t)|\ |\mathbf{n}|)\mathbf{F}(2\pi|\mathbf{r}(t)|\ |2\mathbf{n}|)$$

$$\cdot \sum_{1\le k_1,\ell_1\le N:k_1\ne\ell_1}\ \sum_{1\le k_2\le N:k_2\ne k_1,\ell_1} \mu_{k_1}^2\mu_{\ell_1}\mu_{k_2} e^{2\pi i \mathbf{n}\cdot(2\mathbf{y}_{k_1}-\mathbf{y}_{\ell_1}-\mathbf{y}_{k_2})}, \tag{9.7$'$}$$

which corresponds to the case $\mathbf{n}=\mathbf{m}$, $k_1=\ell_2$, $\ell_1\ne k_2$ in (9.2) (notice that $\widetilde{\mathcal{S}}_4$ has coefficient 2 in (9.3) due to the symmetric case $\mathbf{n}=\mathbf{m}$, $k_1\ne\ell_2$, $\ell_1=k_2$ that has exactly the same contribution),

$$\mathcal{S}_5 = \sum_{\substack{\mathbf{n}\in\mathbb{Z}^d:\\ \mathbf{n}\ne\mathbf{0}}} |a_{\mathbf{n}}|^4\, \mathbf{F}^4(2\pi|\mathbf{r}(t)|\ |\mathbf{n}|) \sum_{\substack{1\le k_1,\ell_1,k_2,\ell_2\le N:\\ \text{four different indices}}} \mu_{k_1}\mu_{\ell_1}$$

$$\cdot \mu_{k_2}\mu_{\ell_2} e^{2\pi i \mathbf{n}\cdot(\mathbf{y}_{k_1}-\mathbf{y}_{\ell_1}-\mathbf{y}_{k_2}+\mathbf{y}_{\ell_2})}, \tag{9.8}$$

which corresponds to the case $\mathbf{n} = \mathbf{m}$, k_1, ℓ_1, k_2, ℓ_2 are all different,

$$\mathcal{S}_6 = \sum_{\substack{\mathbf{n}\in\mathbb{Z}^d:\\ \mathbf{n}\neq\mathbf{0}}} |a_\mathbf{n}|^2 \sum_{\substack{\mathbf{m}\in\mathbb{Z}^d:\\ \mathbf{m}\neq\mathbf{0},\mathbf{n}}} |a_\mathbf{m}|^2 \mathbf{F}^2(2\pi|\mathbf{r}(t)|\ |\mathbf{n}-\mathbf{m}|)$$

$$\cdot \sum_{1\le k,\ell\le N: k\neq\ell} \mu_k^2 \mu_\ell^2 e^{2\pi i(\mathbf{n}-\mathbf{m})\cdot(\mathbf{y}_k-\mathbf{y}_\ell)}, \tag{9.9}$$

which corresponds to the case $\mathbf{n} \neq \mathbf{m}$, $(k_1, \ell_1) = (k_2, \ell_2)$ in (9.2),

$$\widetilde{\mathcal{S}}_{6,1} = \sum_{\substack{\mathbf{n}\in\mathbb{Z}^d:\\ \mathbf{n}\neq\mathbf{0}}} |a_\mathbf{n}|^2 |a_{-\mathbf{n}}|^2 \sum_{1\le k_1,\ell_1\le N: k_1\neq\ell_1} \mu_{k_1}^2 \mu_{\ell_1}^2$$

$$= \left(\left(\sum_{j=1}^N \mu_j^2 \right)^2 - \sum_{j=1}^N \mu_j^4 \right) \sum_{\substack{\mathbf{n}\in\mathbb{Z}^d:\\ \mathbf{n}\neq\mathbf{0}}} |a_\mathbf{n}|^2 |a_{-\mathbf{n}}|^2, \tag{9.9$'$}$$

which corresponds to the case $\mathbf{n} = -\mathbf{m}$, $(k_1, \ell_1) = (\ell_2, k_2)$ in (9.2),

$$\widetilde{\mathcal{S}}_{6,2} = \sum_{\substack{\mathbf{n}\in\mathbb{Z}^d:\\ \mathbf{n}\neq\mathbf{0}}} |a_\mathbf{n}|^2 \sum_{\substack{\mathbf{m}\in\mathbb{Z}^d:\\ \mathbf{m}\neq\mathbf{0},\pm\mathbf{n}}} |a_\mathbf{m}|^2 \mathbf{F}^2(2\pi|\mathbf{r}(t)|\ |\mathbf{n}+\mathbf{m}|)$$

$$\cdot \sum_{1\le k_1,\ell_1\le N: k_1\neq\ell_1} \mu_{k_1}^2 \mu_{\ell_1}^2 e^{2\pi i(\mathbf{n}+\mathbf{m})\cdot(\mathbf{y}_{k_1}-\mathbf{y}_{\ell_1})}, \tag{9.9$''$}$$

which corresponds to the case $\mathbf{n} \neq \pm\mathbf{m}$, $(k_1, \ell_1) = (\ell_2, k_2)$ in (9.2),

$$\mathcal{S}_7 = \sum_{\substack{\mathbf{n}\in\mathbb{Z}^d:\\ \mathbf{n}\neq\mathbf{0}}} |a_\mathbf{n}|^2 \sum_{\substack{\mathbf{m}\in\mathbb{Z}^d:\\ \mathbf{m}\neq\mathbf{0},\mathbf{n}}} |a_\mathbf{m}|^2 \mathbf{F}(2\pi|\mathbf{r}(t)|\ |\mathbf{n}-\mathbf{m}|)\mathbf{F}(2\pi|\mathbf{r}(t)|\ |\mathbf{n}|)\mathbf{F}(2\pi|\mathbf{r}(t)|\ |\mathbf{m}|)$$

$$\cdot \sum_{1\le \ell_1,\ell_2\le N: \ell_1\neq\ell_2} \mu_{\ell_1}\mu_{\ell_2} \sum_{1\le k_1\le N: k_1\neq\ell_1,\ell_2} \mu_{k_1}^2 e^{2\pi i\left((\mathbf{n}-\mathbf{m})\cdot\mathbf{y}_{k_1} - \mathbf{n}\cdot\mathbf{y}_{\ell_1} + \mathbf{m}\cdot\mathbf{y}_{\ell_2}\right)}, \tag{9.10}$$

which corresponds to the case $\mathbf{n} \neq \mathbf{m}$, $k_1 = k_2$, $\ell_1 \neq \ell_2$ in (9.2), (notice that $\mathcal{S}_7$ has coefficient 2 in (9.3) due to the symmetric case $\mathbf{n} \neq \mathbf{m}$, $k_1 \neq k_2$, $\ell_1 = \ell_2$ that has exactly the same contribution),

$$\widetilde{\mathcal{S}}_{7,1} = \sum_{\substack{\mathbf{n}\in\mathbb{Z}^d:\\ \mathbf{n}\neq\mathbf{0}}} |a_\mathbf{n}|^2 |a_{-\mathbf{n}}|^2 \mathbf{F}^2(2\pi|\mathbf{r}(t)|\ |\mathbf{n}|)$$

$$\cdot \sum_{1\le k_1,\ell_1\le N: k_1\neq\ell_1} \sum_{1\le k_2\le N: k_2\neq k_1,\ell_1} \mu_{k_1}^2 \mu_{\ell_1}\mu_{k_2} e^{2\pi i\mathbf{n}\cdot(\mathbf{y}_{k_2}-\mathbf{y}_{\ell_1})}, \tag{9.10$'$}$$

which corresponds to the case $\mathbf{n} = -\mathbf{m}$, $k_1 = \ell_2$, $\ell_1 \neq k_2$ in (9.2), (notice that $\widetilde{\mathcal{S}}_{7,1}$ has coefficient 2 in (9.3) due to the symmetric case $\mathbf{n} = -\mathbf{m}$, $k_1 \neq \ell_2$, $\ell_1 = k_2$ that has exactly the same contribution),

$$\widetilde{\mathcal{S}}_{7,2} = \sum_{\substack{\mathbf{n}\in\mathbb{Z}^d:\\ \mathbf{n}\neq\mathbf{0}}} |a_{\mathbf{n}}|^2 \sum_{\substack{\mathbf{m}\in\mathbb{Z}^d:\\ \mathbf{m}\neq\mathbf{0},\pm\mathbf{n}}} |a_{\mathbf{m}}|^2\, \mathbf{F}(2\pi|\mathbf{r}(t)|\,|\mathbf{n}+\mathbf{m}|)\mathbf{F}(2\pi|\mathbf{r}(t)|\,|\mathbf{n}|)\mathbf{F}(2\pi|\mathbf{r}(t)|\,|\mathbf{m}|)$$

$$\cdot \sum_{1\leq k_1,\ell_1\leq N: k_1\neq \ell_1}\ \sum_{1\leq k_2\leq N: k_2\neq k_1,\ell_1} \mu_{k_1}^2 \mu_{\ell_1}\mu_{k_2} e^{2\pi i\left((\mathbf{n}+\mathbf{m})\cdot\mathbf{y}_{k_1} - \mathbf{n}\cdot\mathbf{y}_{\ell_1} - \mathbf{m}\cdot\mathbf{y}_{k_2}\right)}, \tag{9.10''}$$

which corresponds to the case $\mathbf{n} \neq \pm\mathbf{m}$, $k_1 = \ell_2$, $\ell_1 \neq k_2$ in (9.2), (notice that $\widetilde{\mathcal{S}}_{7,2}$ has coefficient 2 in (9.3) due to the symmetric case $\mathbf{n} \neq \pm\mathbf{m}$, $k_1 \neq \ell_2$, $\ell_1 = k_2$ that has exactly the same contribution), and finally,

$$\mathcal{S}_8 = \sum_{\substack{\mathbf{n}\in\mathbb{Z}^d:\\ \mathbf{n}\neq\mathbf{0}}} |a_{\mathbf{n}}|^2 \sum_{\substack{\mathbf{m}\in\mathbb{Z}^d:\\ \mathbf{m}\neq\mathbf{0},\mathbf{n}}} |a_{\mathbf{m}}|^2\, \mathbf{F}^2(2\pi|\mathbf{r}(t)|\,|\mathbf{n}|)\mathbf{F}^2(2\pi|\mathbf{r}(t)|\,|\mathbf{m}|)$$

$$\cdot \sum_{\substack{1\leq k_1,\ell_1,k_2,\ell_2\leq N:\\ \text{four different indices}}} \mu_{k_1}\mu_{\ell_1}\mu_{k_2}\mu_{\ell_2} e^{2\pi i\left(\mathbf{n}\cdot(\mathbf{y}_{k_1}-\mathbf{y}_{\ell_1}) - \mathbf{m}\cdot(\mathbf{y}_{k_2}-\mathbf{y}_{\ell_2})\right)}, \tag{9.11}$$

which corresponds to the case $\mathbf{n} \neq \mathbf{m}$, k_1, ℓ_1, k_2, ℓ_2 are all different.

On the other hand, we have

$$\left(\overline{\mathbf{D}^2}_\Omega(\mathcal{Y},\mathcal{M};\mathbf{r};f;t)\right)^2 = \left(\sum_{j=1}^N \mu_j^2 \sum_{\substack{\mathbf{n}\in\mathbb{Z}^d:\\ \mathbf{n}\neq\mathbf{0}}} |a_{\mathbf{n}}|^2\right.$$

$$\left. + \sum_{\substack{\mathbf{n}\in\mathbb{Z}^d:\\ \mathbf{n}\neq\mathbf{0}}} |a_{\mathbf{n}}|^2\, \mathbf{F}^2(\pi|\mathbf{r}(t)|\,|\mathbf{n}|) \sum_{1\leq k,\ell\leq N: k\neq\ell} \mu_k\mu_\ell e^{2\pi i\mathbf{n}\cdot(\mathbf{y}_k-\mathbf{y}_\ell)}\right)^2$$

$$= \left(\sum_{j=1}^N \mu_j^2\right)^2 \left(\sum_{\substack{\mathbf{n}\in\mathbb{Z}^d:\\ \mathbf{n}\neq\mathbf{0}}} |a_{\mathbf{n}}|^2\right)^2 + 2\sum_{j=1}^N \mu_j^2 \sum_{\substack{\mathbf{n}\in\mathbb{Z}^d:\\ \mathbf{n}\neq\mathbf{0}}} |a_{\mathbf{n}}|^2 \sum_{\substack{\mathbf{m}\in\mathbb{Z}^d:\\ \mathbf{m}\neq\mathbf{0}}} |a_{\mathbf{m}}|^2$$

$$\cdot \sum_{1\leq k,\ell\leq N: k\neq\ell} e^{2\pi i\mathbf{m}\cdot(\mathbf{y}_k-\mathbf{y}_\ell)} \mu_k\mu_\ell \mathbf{F}^2(\pi|\mathbf{r}(t)|\,|\mathbf{m}|)$$

$$+ \sum_{\substack{\mathbf{n}\in\mathbb{Z}^d:\\ \mathbf{n}\neq\mathbf{0}}} |a_{\mathbf{n}}|^2 \sum_{\substack{\mathbf{m}\in\mathbb{Z}^d:\\ \mathbf{m}\neq\mathbf{0}}} |a_{\mathbf{m}}|^2 \sum_{1\le k_1,\ell_1\le N: k_1\neq\ell_1} \sum_{1\le k_2,\ell_2\le N: k_2\neq\ell_2} \mu_{k_1}\mu_{\ell_1}$$

$$\cdot \mu_{k_2}\mu_{\ell_2} e^{2\pi i\left(\mathbf{n}\cdot(\mathbf{y}_{k_1}-\mathbf{y}_{\ell_1})-\mathbf{m}\cdot(\mathbf{y}_{k_2}-\mathbf{y}_{\ell_2})\right)}\mathbf{F}^2(\pi|\mathbf{r}(t)|\ |\mathbf{n}|)\mathbf{F}^2(\pi|\mathbf{r}(t)|\ |\mathbf{m}|)$$

$$= \mathcal{S}_1 + \mathcal{S}_2 + \mathcal{S}_3^* + \widetilde{\mathcal{S}}_3^* + 2\mathcal{S}_4^* + 2\widetilde{\mathcal{S}}_4^*$$

$$+ \mathcal{S}_5 + \mathcal{S}_6^* + \widetilde{\mathcal{S}}_{6,1}^* + \widetilde{\mathcal{S}}_{6,2}^* + 2\mathcal{S}_7^* + 2\widetilde{\mathcal{S}}_{7,1}^* + 2\widetilde{\mathcal{S}}_{7,2}^* + \mathcal{S}_8, \tag{9.12}$$

where

$$\mathcal{S}_3^* = \left(\left(\sum_{j=1}^N \mu_j^2\right)^2 - \sum_{j=1}^N \mu_j^4\right) \sum_{\substack{\mathbf{n}\in\mathbb{Z}^d:\\ \mathbf{n}\neq\mathbf{0}}} |a_{\mathbf{n}}|^4\, \mathbf{F}^4(2\pi|\mathbf{r}(t)|\ |\mathbf{n}|) \ge 0, \tag{9.13}$$

which corresponds to the case $\mathbf{n} = \mathbf{m}$, $(k_1, \ell_1) = (k_2, \ell_2)$ in (9.12),

$$\widetilde{\mathcal{S}}_3^* = \sum_{\substack{\mathbf{n}\in\mathbb{Z}^d:\\ \mathbf{n}\neq\mathbf{0}}} |a_{\mathbf{n}}|^4\, \mathbf{F}^4(2\pi|\mathbf{r}(t)|\ |2\mathbf{n}|)$$

$$\cdot \sum_{1\le k_1,\ell_1\le N: k_1\neq\ell_1} \mu_{k_1}^2\mu_{\ell_1}^2 e^{2\pi i 2\mathbf{n}\cdot(\mathbf{y}_{k_1}-\mathbf{y}_{\ell_1})}, \tag{9.13'}$$

which corresponds to the case $\mathbf{n} = \mathbf{m}$, $(k_1, \ell_1) = (\ell_2, k_2)$ in (9.12),

$$\mathcal{S}_4^* = \sum_{\substack{\mathbf{n}\in\mathbb{Z}^d:\\ \mathbf{n}\neq\mathbf{0}}} |a_{\mathbf{n}}|^4\, \mathbf{F}^4(2\pi|\mathbf{r}(t)|\ |\mathbf{n}|)$$

$$\cdot \sum_{1\le \ell_1,\ell_2\le N: \ell_1\neq\ell_2} \mu_{\ell_1}\mu_{\ell_2} e^{2\pi i\mathbf{n}\cdot(\mathbf{y}_{\ell_1}-\mathbf{y}_{\ell_2})} \sum_{1\le k\le N: k\neq\ell_1,\ell_2} \mu_k^2, \tag{9.14}$$

which corresponds to the case $\mathbf{n} = \mathbf{m}$, $k_1 = k_2$, $\ell_1 \neq \ell_2$ in (9.12) (notice that $\mathcal{S}_4^*$ has coefficient 2 in (9.12) due to the symmetric case $\mathbf{n} = \mathbf{m}$, $k_1 \neq k_2$, $\ell_1 = \ell_2$ that has exactly the same contribution),

$$\widetilde{\mathcal{S}}_4^* = \sum_{\substack{\mathbf{n}\in\mathbb{Z}^d:\\ \mathbf{n}\neq\mathbf{0}}} |a_{\mathbf{n}}|^4\, \mathbf{F}^4(2\pi|\mathbf{r}(t)|\ |\mathbf{n}|)$$

$$\cdot \sum_{1\le k_1,\ell_1\le N: k_1\neq\ell_1} \sum_{1\le k_2\le N: k_2\neq k_1,\ell_1} \mu_{k_1}^2\mu_{\ell_1}\mu_{k_2} e^{2\pi i\mathbf{n}\cdot(2\mathbf{y}_{k_1}-\mathbf{y}_{\ell_1}-\mathbf{y}_{k_2})}, \tag{9.14'}$$

which corresponds to the case $\mathbf{n} = \mathbf{m}$, $k_1 = \ell_2$, $\ell_1 \neq k_2$ in (9.12) (notice that $\widetilde{\mathcal{S}}_4^*$ has coefficient 2 in (9.12) due to the symmetric case $\mathbf{n} = \mathbf{m}$, $k_1 \neq \ell_2$, $\ell_1 = k_2$ that has exactly the same contribution),

$$\mathcal{S}_6^* = \sum_{\substack{\mathbf{n}\in\mathbb{Z}^d:\\ \mathbf{n}\neq\mathbf{0}}} |a_{\mathbf{n}}|^2 \sum_{\substack{\mathbf{m}\in\mathbb{Z}^d:\\ \mathbf{m}\neq\mathbf{0},\mathbf{n}}} |a_{\mathbf{m}}|^2 \mathbf{F}^2(2\pi|\mathbf{r}(t)|\ |\mathbf{n}|)\mathbf{F}^2(2\pi|\mathbf{r}(t)|\ |\mathbf{m}|)$$

$$\cdot \sum_{1\le k,\ell\le N:k\neq\ell} \mu_k^2\mu_\ell^2 e^{2\pi i(\mathbf{n}-\mathbf{m})\cdot(\mathbf{y}_k-\mathbf{y}_\ell)}, \tag{9.15}$$

which corresponds to the case $\mathbf{n} \neq \mathbf{m}$, $(k_1, \ell_1) = (k_2, \ell_2)$ in (9.12),

$$\widetilde{\mathcal{S}}_{6,1}^* = \sum_{\substack{\mathbf{n}\in\mathbb{Z}^d:\\ \mathbf{n}\neq\mathbf{0}}} |a_{\mathbf{n}}|^2 |a_{-\mathbf{n}}|^2 \sum_{1\le k_1,\ell_1\le N:k_1\neq\ell_1} \mu_{k_1}^2\mu_{\ell_1}^2$$

$$= \left(\left(\sum_{j=1}^N \mu_j^2\right)^2 - \sum_{j=1}^N \mu_j^4\right) \sum_{\substack{\mathbf{n}\in\mathbb{Z}^d:\\ \mathbf{n}\neq\mathbf{0}}} |a_{\mathbf{n}}|^2 |a_{-\mathbf{n}}|^2, \tag{9.15$'$}$$

which corresponds to the case $\mathbf{n} = -\mathbf{m}$, $(k_1, \ell_1) = (\ell_2.k_2)$ in (9.12),

$$\widetilde{\mathcal{S}}_{6,2}^* = \sum_{\substack{\mathbf{n}\in\mathbb{Z}^d:\\ \mathbf{n}\neq\mathbf{0}}} |a_{\mathbf{n}}|^2 \sum_{\substack{\mathbf{m}\in\mathbb{Z}^d:\\ \mathbf{m}\neq\mathbf{0},\pm\mathbf{n}}} |a_{\mathbf{m}}|^2 \mathbf{F}^2(2\pi|\mathbf{r}(t)|\ |\mathbf{n}+\mathbf{m}|)$$

$$\cdot \sum_{1\le k_1,\ell_1\le N:k_1\neq\ell_1} \mu_{k_1}^2\mu_{\ell_1}^2 e^{2\pi i(\mathbf{n}+\mathbf{m})\cdot(\mathbf{y}_{k_1}-\mathbf{y}_{\ell_1})}, \tag{9.15$''$}$$

which corresponds to the case $\mathbf{n} \neq \pm\mathbf{m}$, $(k_1, \ell_1) = (\ell_2.k_2)$ in (9.12),

$$\mathcal{S}_7^* = \sum_{\substack{\mathbf{n}\in\mathbb{Z}^d:\\ \mathbf{n}\neq\mathbf{0}}} |a_{\mathbf{n}}|^2 \sum_{\substack{\mathbf{m}\in\mathbb{Z}^d:\\ \mathbf{m}\neq\mathbf{0},\mathbf{n}}} |a_{\mathbf{m}}|^2 \mathbf{F}^2(2\pi|\mathbf{r}(t)|\ |\mathbf{n}|)\mathbf{F}^2(2\pi\ |\mathbf{r}(t)|\ |\mathbf{m}|)$$

$$\cdot \sum_{1\le\ell_1,\ell_2\le N:\ell_1\neq\ell_2} \mu_{\ell_1}\mu_{\ell_2} \sum_{1\le k\le N:k\neq\ell_1,\ell_2} \mu_k^2 e^{2\pi i((\mathbf{n}-\mathbf{m})\cdot\mathbf{y}_k - \mathbf{n}\cdot\mathbf{y}_{\ell_1} + \mathbf{m}\cdot\mathbf{y}_{\ell_2})}. \tag{9.16}$$

which corresponds to the case $\mathbf{n} \neq \mathbf{m}$, $k_1 = k_2$, $\ell_1 \neq \ell_2$ in (9.12), (notice that $\mathcal{S}_7^*$ has coefficient 2 in (9.12) due to the symmetric case $\mathbf{n} \neq \mathbf{m}$, $k_1 \neq k_2$, $\ell_1 = \ell_2$ that has exactly the same contribution),

$$\widetilde{\mathcal{S}}_{7,1}^* = \sum_{\substack{\mathbf{n}\in\mathbb{Z}^d:\\ \mathbf{n}\neq\mathbf{0}}} |a_{\mathbf{n}}|^2 |a_{-\mathbf{n}}|^2 \mathbf{F}^4(2\pi|\mathbf{r}(t)|\ |\mathbf{n}|)$$

$$\cdot \sum_{1\le k_1,\ell_1\le N:k_1\ne\ell_1}\ \sum_{1\le k_2\le N:k_2\ne k_1,\ell_1} \mu_{k_1}^2\mu_{\ell_1}\mu_{k_2}e^{2\pi i\mathbf{n}\cdot(\mathbf{y}_{k_2}-\mathbf{y}_{\ell_1})}, \tag{9.16'}$$

which corresponds to the case $\mathbf{n}=-\mathbf{m}$, $k_1=\ell_2$, $\ell_1\ne k_2$ in (9.12), (notice that $\widetilde{\mathcal{S}}^*_{7,1}$ has coefficient 2 in (9.12) due to the symmetric case $\mathbf{n}=-\mathbf{m}$, $k_1\ne\ell_2$, $\ell_1=k_2$ that has exactly the same contribution), and finally,

$$\widetilde{\mathcal{S}}^*_{7,2}=\sum_{\substack{\mathbf{n}\in\mathbb{Z}^d:\\ \mathbf{n}\ne\mathbf{0}}}|a_{\mathbf{n}}|^2\sum_{\substack{\mathbf{m}\in\mathbb{Z}^d:\\ \mathbf{m}\ne\mathbf{0},\pm\mathbf{n}}}|a_{\mathbf{m}}|^2\,\mathbf{F}^2(2\pi|\mathbf{r}(t)|\ |\mathbf{n}|)\mathbf{F}^2(2\pi|\mathbf{r}(t)|\ |\mathbf{m}|)$$

$$\cdot \sum_{1\le k_1,\ell_1\le N:k_1\ne\ell_1}\ \sum_{1\le k_2\le N:k_2\ne k_1,\ell_1} \mu_{k_1}^2\mu_{\ell_1}\mu_{k_2}e^{2\pi i\left((\mathbf{n}+\mathbf{m})\cdot\mathbf{y}_{k_1}-\mathbf{n}\cdot\mathbf{y}_{\ell_1}-\mathbf{m}\cdot\mathbf{y}_{k_2}\right)}, \tag{9.16''}$$

which corresponds to the case $\mathbf{n}\ne\pm\mathbf{m}$, $k_1=\ell_2$, $\ell_1\ne k_2$ in (9.12), (notice that $\widetilde{\mathcal{S}}^*_{7,2}$ has coefficient 2 in (9.12) due to the symmetric case $\mathbf{n}\ne\pm\mathbf{m}$, $k_1\ne\ell_2$, $\ell_1=k_2$ that has exactly the same contribution).

Combining (9.1) with (9.3)–(9.16″), we have

$$\text{Variance}_{\omega\in\Omega}\overline{\mathbf{D}^2}(\mathcal{Y},\mathcal{M};\mathbf{r};\omega;f;t)$$

$$=\int_\Omega\left(\overline{\mathbf{D}^2}(\mathcal{Y},\mathcal{M};\mathbf{r};\omega;f;t)\right)^2 d\omega-\left(\overline{\mathbf{D}^2}_\Omega(\mathcal{Y},\mathcal{M};\mathbf{r};f;t)\right)^2$$

$$\le|\mathcal{S}_3-\mathcal{S}_3^*|+\left|\widetilde{\mathcal{S}}_3-\widetilde{\mathcal{S}}_3^*\right|+2|\mathcal{S}_4-\mathcal{S}_4^*|+2\left|\widetilde{\mathcal{S}}_4-\widetilde{\mathcal{S}}_4^*\right|+|\mathcal{S}_6-\mathcal{S}_6^*|$$

$$+\left|\widetilde{\mathcal{S}}_{6,1}-\widetilde{\mathcal{S}}^*_{6,1}\right|+\left|\widetilde{\mathcal{S}}_{6,2}-\widetilde{\mathcal{S}}^*_{6,2}\right|+2|\mathcal{S}_7-\mathcal{S}_7^*|+2\left|\widetilde{\mathcal{S}}_{7,1}-\widetilde{\mathcal{S}}^*_{7,1}\right|+2\left|\widetilde{\mathcal{S}}_{7,2}-\widetilde{\mathcal{S}}^*_{7,2}\right|, \tag{9.17}$$

since $\mathcal{S}_1,\mathcal{S}_2,\mathcal{S}_5,\mathcal{S}_8$ cancel out.

Using $\mathcal{S}_3^*\ge 0$ (see (9.13)), by (9.6) we have

$$|\mathcal{S}_3-\mathcal{S}_3^*|\le\left(\left(\sum_{j=1}^N\mu_j^2\right)^2-\sum_{j=1}^N\mu_j^4\right)\sum_{\substack{\mathbf{n}\in\mathbb{Z}^d:\\ \mathbf{n}\ne\mathbf{0}}}|a_{\mathbf{n}}|^4$$

$$\le\left(\sum_{j=1}^N\mu_j^2\right)^2\left(\sum_{\substack{\mathbf{n}\in\mathbb{Z}^d:\\ \mathbf{n}\ne\mathbf{0}}}|a_{\mathbf{n}}|^2\right)^2=\left(\sum_{j=1}^N\mu_j^2\right)^2\sigma_0^4(f), \tag{9.18}$$

where we used Parseval's formula

$$\sigma_0^2(f) = \sum_{\substack{\mathbf{n}\in\mathbb{Z}^d:\\ \mathbf{n}\neq\mathbf{0}}} |a_\mathbf{n}|^2 .$$

Similarly, by (9.6′) and (9.13′),

$$\left|\widetilde{S}_3 - \widetilde{S}_3^*\right| \le \left|\widetilde{S}_3\right| + \left|\widetilde{S}_3^*\right|$$

$$\le \left(\sum_{j=1}^N \mu_j^2\right)^2 \sigma_0^4(f) + \left(\sum_{j=1}^N \mu_j^2\right)^2 \sigma_0^4(f). \tag{9.18′}$$

By (9.7), (9.14), and using the notation (7.17)

$$\Phi(\mathcal{Y},\mathcal{M};\mathbf{n}) = \sum_{k=1}^N \mu_k e^{2\pi i\mathbf{n}\cdot\mathbf{y}_k}$$

combined with the simple fact

$$\sum_{1\le\ell_1,\ell_2\le N:\ell_1\neq\ell_2} \mu_{\ell_1}\mu_{\ell_2} e^{2\pi i\mathbf{n}\cdot(\mathbf{y}_{\ell_1}-\mathbf{y}_{\ell_2})} = |\Phi(\mathcal{Y},\mathcal{M};\mathbf{n})|^2 - \sum_{j=1}^N \mu_j^2,$$

we have,

$$|\mathcal{S}_4 - \mathcal{S}_4^*| = \sum_{\substack{\mathbf{n}\in\mathbb{Z}^d:\\ \mathbf{n}\neq\mathbf{0}}} |a_\mathbf{n}|^4 \left(\mathbf{F}^2(2\pi|\mathbf{r}(t)|\,|\mathbf{n}|) - \mathbf{F}^4(2\pi|\mathbf{r}(t)|\,|\mathbf{n}|)\right)$$

$$\cdot\left|\sum_{1\le\ell_1,\ell_2\le N:\ell_1\neq\ell_2} \mu_{\ell_1}\mu_{\ell_2} e^{2\pi i\mathbf{n}\cdot(\mathbf{y}_{\ell_1}-\mathbf{y}_{\ell_2})} \sum_{1\le k\le N:k\neq\ell_1,\ell_2} \mu_k^2\right|$$

$$\le \sum_{\substack{\mathbf{n}\in\mathbb{Z}^d:\\ \mathbf{n}\neq\mathbf{0}}} |a_\mathbf{n}|^4 \left(\mathbf{F}^2(2\pi|\mathbf{r}(t)|\,|\mathbf{n}|) - \mathbf{F}^4(2\pi|\mathbf{r}(t)|\,|\mathbf{n}|)\right) \cdot |\Phi(\mathcal{Y},\mathcal{M};\mathbf{n})|^2 \sum_{k=1}^N \mu_k^2,$$

which implies

$$|\mathcal{S}_4 - \mathcal{S}_4^*| \le \sum_{\substack{\mathbf{n}\in\mathbb{Z}^d:\\ \mathbf{n}\neq\mathbf{0}}} |a_\mathbf{n}|^4\, \mathbf{F}^2(2\pi|\mathbf{r}(t)|\,|\mathbf{n}|)|\Phi(\mathcal{Y},\mathcal{M};\mathbf{n})|^2 \sum_{k=1}^N \mu_k^2$$

$$\le \sigma_0^2(f) \sum_{\mathbf{n}\in\mathbb{Z}^d\setminus\mathbf{0}} |a_\mathbf{n}|^2\, \mathbf{F}^2(2\pi|\mathbf{r}(t)|\,|\mathbf{n}|)|\Phi(\mathcal{Y},\mathcal{M};\mathbf{n})|^2 \sum_{k=1}^N \mu_k^2. \tag{9.19}$$

By (9.7′)

$$\left|\widetilde{\mathcal{S}}_4\right| = \left| \sum_{\substack{\mathbf{n}\in\mathbb{Z}^d:\\ \mathbf{n}\neq\mathbf{0}}} |a_{\mathbf{n}}|^4 \, \mathbf{F}^2(2\pi|\mathbf{r}(t)|\ |\mathbf{n}|)\mathbf{F}(2\pi|\mathbf{r}(t)|\ |2\mathbf{n}|)\right.$$

$$\left. \cdot \sum_{1\le k_2,\ell_1\le N: k_1\neq \ell_1} \mu_{k_2} e^{-2\pi i\mathbf{n}\cdot\mathbf{y}_{k_2}} \mu_{\ell_1} e^{-2\pi i\mathbf{n}\cdot\mathbf{y}_{\ell_1}} \sum_{1\le k_1\le N: k_1\neq k_2,\ell_1} \mu_{k_1}^2 e^{2\pi i\mathbf{n}\cdot 2\mathbf{y}_{k_1}} \right|$$

$$\le \sum_{\substack{\mathbf{n}\in\mathbb{Z}^d:\\ \mathbf{n}\neq\mathbf{0}}} |a_{\mathbf{n}}|^4 \, \mathbf{F}^2(2\pi|\mathbf{r}(t)|\ |\mathbf{n}|) \left|\Phi^2(-\mathcal{Y},\mathcal{M};\mathbf{n})\right| \left(\sum_{1\le k_1\le N} \mu_{k_1}^2\right)$$

$$\le \sigma_0^2(f) \sum_{\mathbf{n}\in\mathbb{Z}^d\setminus\mathbf{0}} |a_{\mathbf{n}}|^2 \, \mathbf{F}^2(2\pi|\mathbf{r}(t)|\ |\mathbf{n}|)|\Phi(\mathcal{Y},\mathcal{M};\mathbf{n})|^2 \sum_{k=1}^N \mu_k^2.$$

Similarly, by (9.14′)

$$\left|\widetilde{\mathcal{S}}_4^*\right| \le \sigma_0^2(f) \sum_{\mathbf{n}\in\mathbb{Z}^d\setminus\mathbf{0}} |a_{\mathbf{n}}|^2 \, \mathbf{F}^2(2\pi|\mathbf{r}(t)|\ |\mathbf{n}|)|\Phi(\mathcal{Y},\mathcal{M};\mathbf{n})|^2 \sum_{k=1}^N \mu_k^2.$$

Combining the last two inequalities, we have

$$\left|\widetilde{\mathcal{S}}_4 - \widetilde{\mathcal{S}}_4^*\right|$$

$$\le 2\sigma_0^2(f) \sum_{\mathbf{n}\in\mathbb{Z}^d\setminus\mathbf{0}} |a_{\mathbf{n}}|^2 \, \mathbf{F}^2(2\pi|\mathbf{r}(t)|\ |\mathbf{n}|)|\Phi(\mathcal{Y},\mathcal{M};\mathbf{n})|^2 \sum_{k=1}^N \mu_k^2. \qquad (9.19′).$$

By (9.9), (9.15) and Parseval's formula

$$\sigma_0^2(f) = \sum_{\substack{\mathbf{n}\in\mathbb{Z}^d:\\ \mathbf{n}\neq\mathbf{0}}} |a_{\mathbf{n}}|^2, \qquad (9.20)$$

we have the trivial upper bound

$$|\mathcal{S}_6 - \mathcal{S}_6^*| \le \sigma_0^4(f) \left(\sum_{k=1}^N \mu_k^2\right)^2 \sup_{u\ge 2\pi|\mathbf{r}(t)|} \mathbf{F}^2(u). \qquad (9.21)$$

Similarly, by (9.9′), (9.9″), (9.15′), and (9.15″) we have the same trivial upper bound

$$\max\left\{\left|\widetilde{\mathcal{S}}_{6,1}\right|, \left|\widetilde{\mathcal{S}}_{6,2}\right|, \left|\widetilde{\mathcal{S}}^*_{6,1}\right|, \left|\widetilde{\mathcal{S}}^*_{6,2}\right|\right\} \leq \sigma_0^4(f)\left(\sum_{k=1}^{N}\mu_k^2\right)^2 \sup_{u\geq 2\pi|\mathbf{r}(t)|}\mathbf{F}^2(u),$$

which implies

$$\left|\widetilde{\mathcal{S}}_{6,1}-\widetilde{\mathcal{S}}^*_{6,1}\right|+\left|\widetilde{\mathcal{S}}_{6,2}-\widetilde{\mathcal{S}}^*_{6,2}\right|$$

$$\leq 4\sigma_0^4(f)\left(\sum_{k=1}^{N}\mu_k^2\right)^2 \sup_{u\geq 2\pi|\mathbf{r}(t)|}\mathbf{F}^2(u). \tag{9.21′}$$

The terms $\left|\widetilde{\mathcal{S}}_{7,1}\right|$ and $\left|\widetilde{\mathcal{S}}^*_{7,1}\right|$ can be estimated similarly to $\left|\widetilde{\mathcal{S}}_{4}\right|$ and $\left|\widetilde{\mathcal{S}}^*_{4}\right|$, and we obtain the same bound

$$\left|\widetilde{\mathcal{S}}_{7,1}-\widetilde{\mathcal{S}}^*_{7,1}\right| \leq 2\sigma_0^2(f)\sum_{\mathbf{n}\in\mathbb{Z}^d\setminus\mathbf{0}}|a_{\mathbf{n}}|^2\,\mathbf{F}^2(2\pi|\mathbf{r}(t)|\;|\mathbf{n}|)|\Phi(\mathcal{Y},\mathcal{M};\mathbf{n})|^2\sum_{k=1}^{N}\mu_k^2. \tag{9.22}$$

Finally, we estimate $|\mathcal{S}_7-\mathcal{S}^*_7|$ and $\left|\widetilde{\mathcal{S}}_{7,2}-\widetilde{\mathcal{S}}^*_{7,2}\right|$ which represent the difficult part in (9.17). We begin with the estimation of $|\mathcal{S}_7-\mathcal{S}^*_7|$ in the special case $1=\mu_1=\ldots=\mu_N$, for which we have a simpler shortcut argument that goes as follows (to indicate this special case, in the notation we omit $\mathcal{M}$). To estimate $|\mathcal{S}_7-\mathcal{S}^*_7|$, we focus on the last double sum in (9.10) and (9.16): by using the notation

$$\Phi(\mathcal{Y};\mathbf{n})=\sum_{k=1}^{N}e^{2\pi i\mathbf{n}\cdot\mathbf{y}_k}$$

(which is the special case of (7.17) with $\mu_k=1$, $1\leq k\leq N$), a simple inclusion-exclusion argument gives

$$\sum_{1\leq\ell_1,\ell_2\leq N:\ell_1\neq\ell_2}\;\sum_{1\leq k\leq N:k\neq\ell_1,\ell_2}e^{2\pi i\left((\mathbf{n}-\mathbf{m})\cdot\mathbf{y}_k-\mathbf{n}\cdot\mathbf{y}_{\ell_1}+\mathbf{m}\cdot\mathbf{y}_{\ell_2}\right)}$$

$$=\sum_{\substack{1\leq\ell_1,\ell_2,k\leq N:\\ \text{three different indices}}}e^{2\pi i(\mathbf{n}-\mathbf{m})\cdot\mathbf{y}_k}e^{-2\pi i\mathbf{n}\cdot\mathbf{y}_{\ell_1}}e^{2\pi i\mathbf{m}\cdot\mathbf{y}_{\ell_2}}$$

$$=\Phi(\mathcal{Y};\mathbf{m})\Phi(\mathcal{Y};-\mathbf{n})\Phi(\mathcal{Y};\mathbf{n}-\mathbf{m})-|\Phi(\mathcal{Y};\mathbf{m})|^2-|\Phi(\mathcal{Y};\mathbf{n})|^2-|\Phi(\mathcal{Y};\mathbf{n}-\mathbf{m})|^2+N. \tag{9.23}$$

By (9.10), (9.16), (9.20) and (9.23),

$$\left|\mathcal{S}_7 - \mathcal{S}_7^2\right|$$

$$\leq \sum_{\substack{\mathbf{n}\in\mathbb{Z}^d:\\ \mathbf{n}\neq\mathbf{0}}} |a_{\mathbf{n}}|^2 \sum_{\substack{\mathbf{m}\in\mathbb{Z}^d:\\ \mathbf{m}\neq\mathbf{0},\mathbf{n}}} |a_{\mathbf{m}}|^2 \left|\mathbf{F}(2\pi|\mathbf{r}(t)|\ |\mathbf{n}-\mathbf{m}|)\mathbf{F}(2\pi|\mathbf{r}(t)|\ |\mathbf{n}|)\mathbf{F}(2\pi|\mathbf{r}(t)|\ |\mathbf{m}|)\right|$$

$$\cdot \left(|\Phi(\mathcal{Y};\mathbf{m})\Phi(\mathcal{Y};\mathbf{n})\Phi(\mathcal{Y};\mathbf{n}-\mathbf{m})| + N\right)$$

$$\leq \sigma_0^4(f)N \sup_{u\geq 2\pi|\mathbf{r}(t)|} \mathbf{F}^2(u) + 2\sum_{\substack{\mathbf{n}\in\mathbb{Z}^d:\\ \mathbf{n}\neq\mathbf{0}}} |a_{\mathbf{n}}|^2 \sum_{\substack{\mathbf{m}\in\mathbb{Z}^d:\\ \mathbf{m}\neq\mathbf{0},\mathbf{n}}} |a_{\mathbf{m}}|^2$$

$$\cdot N\left|\mathbf{F}(2\pi|\mathbf{r}(t)|\ |\mathbf{m}|)\Phi(\mathcal{Y};\mathbf{m})\right| \cdot \left|\mathbf{F}(2\pi|\mathbf{r}(t)|\ |\mathbf{n}|)\Phi(\mathcal{Y};\mathbf{n})\right|$$

$$\leq \sigma_0^4(f)N \sup_{u\geq 2\pi|\mathbf{r}(t)|} \mathbf{F}^2(u) + 2\sum_{\substack{\mathbf{n}\in\mathbb{Z}^d:\\ \mathbf{n}\neq\mathbf{0}}} |a_{\mathbf{n}}|^2 \sum_{\substack{\mathbf{m}\in\mathbb{Z}^d:\\ \mathbf{m}\neq\mathbf{0},\mathbf{n}}} |a_{\mathbf{m}}|^2$$

$$\cdot N\left(\mathbf{F}^2(2\pi|\mathbf{r}(t)|\ |\mathbf{m}|)|\Phi(\mathcal{Y};\mathbf{m})|^2 + \mathbf{F}^2(2\pi|\mathbf{r}(t)|\ |\mathbf{n}|)|\Phi(\mathcal{Y};\mathbf{n})|^2\right)/2$$

$$\leq \sigma_0^4(f)N \sup_{u\geq 2\pi|\mathbf{r}(t)|} \mathbf{F}^2(u) + 2\sum_{\substack{\mathbf{n}\in\mathbb{Z}^d:\\ \mathbf{n}\neq\mathbf{0}}} |a_{\mathbf{n}}|^2 \sum_{\substack{\mathbf{m}\in\mathbb{Z}^d:\\ \mathbf{m}\neq\mathbf{0},\mathbf{n}}} |a_{\mathbf{m}}|^2$$

$$\cdot N\mathbf{F}^2(2\pi|\mathbf{r}(t)|\ |\mathbf{n}|)|\Phi(\mathcal{Y};\mathbf{n})|^2$$

$$\leq \sigma_0^4(f)N \sup_{u\geq 2\pi|\mathbf{r}(t)|} \mathbf{F}^2(u) + 2\sigma_0^2(f)N \sum_{\substack{\mathbf{n}\in\mathbb{Z}^d:\\ \mathbf{n}\neq\mathbf{0}}} |a_{\mathbf{n}}|^2\, \mathbf{F}^2(2\pi|\mathbf{r}(t)|\ |\mathbf{n}|)|\Phi(\mathcal{Y};\mathbf{n})|^2, \tag{9.24}$$

where in the middle of the argument we applied the inequality $|ab| \leq (|a|^2 + |b|^2)/2$.

Similar argument works for $\left|\widetilde{\mathcal{S}}_{7,2} - \widetilde{\mathcal{S}}_{7,2}^*\right|$, and we obtain the same upper bound

$$\left|\widetilde{\mathcal{S}}_{7,2} - \widetilde{\mathcal{S}}_{7,2}^*\right|$$

$$\leq \sigma_0^4(f)N \sup_{u\geq 2\pi|\mathbf{r}(t)|} \mathbf{F}^2(u) + 2\sigma_0^2(f)N \sum_{\substack{\mathbf{n}\in\mathbb{Z}^d:\\ \mathbf{n}\neq\mathbf{0}}} |a_{\mathbf{n}}|^2\, \mathbf{F}^2(2\pi|\mathbf{r}(t)|\ |\mathbf{n}|)|\Phi(\mathcal{Y};\mathbf{n})|^2. \tag{9.24$'$}$$

Now we are ready to finish the special case $1 = \mu_1 = \ldots = \mu_N$. Since the variance is a square integral (see the first line in (9.1)), we have the trivial lower bound

$$\text{Variance}_{\omega\in\Omega}\overline{\mathbf{D}^2}(\mathcal{Y};\mathbf{r};\omega;f;t) \geq 0.$$

On the other hand, by using (9.18)–(9.24′) in (9.17), we obtain the upper bound

$$0 \leq \text{Variance}_{\omega\in\Omega}\overline{\mathbf{D}^2}(\mathcal{Y};\mathbf{r};\omega;f;t) \leq |\mathcal{S}_3 - \mathcal{S}_3^*| + \left|\widetilde{\mathcal{S}}_3 - \widetilde{\mathcal{S}}_3^*\right|$$

$$+ 2|\mathcal{S}_4 - \mathcal{S}_4^*| + 2\left|\widetilde{\mathcal{S}}_4 - \widetilde{\mathcal{S}}_4^*\right| + |\mathcal{S}_6 - \mathcal{S}_6^*| + \left|\widetilde{\mathcal{S}}_{6,1} - \widetilde{\mathcal{S}}_{6,1}^*\right|$$

$$+ \left|\widetilde{\mathcal{S}}_{6,2} - \widetilde{\mathcal{S}}_{6,2}^*\right| + 2|\mathcal{S}_7 - \mathcal{S}_7^*| + 2\left|\widetilde{\mathcal{S}}_{7,1} - \widetilde{\mathcal{S}}_{7,1}^*\right| + 2\left|\widetilde{\mathcal{S}}_{7,2} - \widetilde{\mathcal{S}}_{7,2}^*\right| \leq 3N^2\sigma_0^4(f)$$

$$+ 6\sigma_0^2(f) \sum_{\mathbf{n}\in\mathbb{Z}^d\setminus\mathbf{0}} |a_{\mathbf{n}}|^2 \mathbf{F}^2(2\pi|\mathbf{r}(t)|\,|\mathbf{n}|)|\Phi(\mathcal{Y},\mathcal{M};\mathbf{n})|^2 N$$

$$+ 5\sigma_0^4(f)N^2 \sup_{u\geq 2\pi|\mathbf{r}(t)|} \mathbf{F}^2(u)$$

$$+ 4\sigma_0^2(f) \sum_{\mathbf{n}\in\mathbb{Z}^d\setminus\mathbf{0}} |a_{\mathbf{n}}|^2 \mathbf{F}^2(2\pi|\mathbf{r}(t)|\,|\mathbf{n}|)|\Phi(\mathcal{Y},\mathcal{M};\mathbf{n})|^2 N$$

$$+ 4\sigma_0^4(f)N \sup_{u\geq 2\pi|\mathbf{r}(t)|} \mathbf{F}^2(u) + 4\sigma_0^2(f)N \sum_{\substack{\mathbf{n}\in\mathbb{Z}^d:\\ \mathbf{n}\neq\mathbf{0}}} |a_{\mathbf{n}}|^2 \mathbf{F}^2(2\pi|\mathbf{r}(t)|\,|\mathbf{n}|)|\Phi(\mathcal{Y};\mathbf{n})|^2$$

$$\leq (8N^2 + 4N)\sigma_0^4(f) + 14\sigma_0^2(f)N \sum_{\mathbf{n}\in\mathbb{Z}^d\setminus\mathbf{0}} |a_{\mathbf{n}}|^2 \mathbf{F}^2(2\pi|\mathbf{r}(t)|\,|\mathbf{n}|)|\Phi(\mathcal{Y};\mathbf{n})|^2. \tag{9.25}$$

By taking square root in (9.25), we obtain the following upper bound for the standard deviation (as ω runs in Ω):

$$\text{StandardDeviation}_{\omega\in\Omega}\overline{\mathbf{D}^2}(\mathcal{Y};\mathbf{r};\omega;f;t) = \left(\text{Variance}_{\omega\in\Omega}\overline{\mathbf{D}^2}(\mathcal{Y};\mathbf{r};\omega;f;t)\right)^{1/2}$$

$$\leq 4\sigma_0^2(f)N + 4\sigma_0(f)\sqrt{N}\left(\sum_{\mathbf{n}\in\mathbb{Z}^d\setminus\mathbf{0}} |a_{\mathbf{n}}|^2 \mathbf{F}^2(2\pi|\mathbf{r}(t)|\,|\mathbf{n}|)|\Phi(\mathcal{Y};\mathbf{n})|^2\right)^{1/2}. \tag{9.26}$$

Next we estimate $|\mathcal{S}_7 - \mathcal{S}_7^*|$ in the general case, where different particles may have different masses (so $\mathcal{M}$ shows up again in the notation). This requires a similar but somewhat more complicated argument. Again we first focus on the last double sum in (9.10) and (9.16): by using the notation (7.17), a simple inclusion-exclusion argument gives

$$\sum_{1\le \ell_1,\ell_2\le N:\ell_1\ne\ell_2} \mu_{\ell_1}\mu_{\ell_2} \sum_{1\le k\le N:k\ne \ell_1,\ell_2} \mu_k^2 e^{2\pi i\left((\mathbf{n}-\mathbf{m})\cdot\mathbf{y}_k-\mathbf{n}\cdot\mathbf{y}_{\ell_1}+\mathbf{m}\cdot\mathbf{y}_{\ell_2}\right)}$$

$$= \sum_{\substack{1\le \ell_1,\ell_2,k\le N:\\ \text{three different indices}}} \mu_k^2 e^{2\pi i(\mathbf{n}-\mathbf{m})\cdot\mathbf{y}_k}\cdot\mu_{\ell_1}e^{-2\pi i\mathbf{n}\cdot\mathbf{y}_{\ell_1}}\cdot\mu_{\ell_2}e^{2\pi i\mathbf{m}\cdot\mathbf{y}_{\ell_2}}$$

$$= \Phi(\mathcal{Y},\mathcal{M};\mathbf{m})\Phi(\mathcal{Y},\mathcal{M};-\mathbf{n})\Psi_2(\mathcal{Y},\mathcal{M};\mathbf{n}-\mathbf{m})-\Psi_3(\mathcal{Y},\mathcal{M};-\mathbf{m})\Phi(\mathcal{Y},\mathcal{M};\mathbf{m})$$

$$-\Psi_3(\mathcal{Y},\mathcal{M};\mathbf{n})\Phi(\mathcal{Y},\mathcal{M};-\mathbf{n})-|\Psi_2(\mathcal{Y},\mathcal{M};\mathbf{n}-\mathbf{m})|^2+\sum_{k=1}^N\mu_k^4, \qquad (9.27)$$

where

$$\Psi_j(\mathcal{Y},\mathcal{M};\mathbf{n})=\sum_{k=1}^N \mu_k^j e^{2\pi i\mathbf{n}\cdot\mathbf{y}_k},\ 1\le j\le 3, \qquad (9.28)$$

and

$$\Phi(\mathcal{Y},\mathcal{M};\mathbf{n})=\Psi_1(\mathcal{Y},\mathcal{M};\mathbf{n}). \qquad (9.29)$$

By (9.10), (9.16), (9.20) and (9.27)–(9.29),

$$\left|\mathcal{S}_7-\mathcal{S}_7^2\right|$$

$$\le \sum_{\substack{\mathbf{n}\in\mathbb{Z}^d:\\ \mathbf{n}\ne\mathbf{0}}} |a_{\mathbf{n}}|^2 \sum_{\substack{\mathbf{m}\in\mathbb{Z}^d:\\ \mathbf{m}\ne\mathbf{0},\mathbf{n}}} |a_{\mathbf{m}}|^2 \left|\mathbf{F}(2\pi|\mathbf{r}(t)|\ |\mathbf{n}-\mathbf{m}|)\mathbf{F}(2\pi|\mathbf{r}(t)|\ |\mathbf{n}|)\mathbf{F}(2\pi|\mathbf{r}(t)|\ |\mathbf{m}|)\right|$$

$$\cdot\Bigg(|\Phi(\mathcal{Y},\mathcal{M};\mathbf{m})\Phi(\mathcal{Y},\mathcal{M};-\mathbf{n})\Psi_2(\mathcal{Y},\mathcal{M};\mathbf{n}-\mathbf{m})|$$

$$+|\Psi_3(\mathcal{Y},\mathcal{M};-\mathbf{m})\Phi(\mathcal{Y},\mathcal{M};\mathbf{m})|+|\Psi_3(\mathcal{Y},\mathcal{M};\mathbf{n})\Phi(\mathcal{Y},\mathcal{M};-\mathbf{n})|+\sum_{k=1}^N\mu_k^4\Bigg)$$

$$\le \sigma_0^4(f)\sup_{u\ge 2\pi|\mathbf{r}(t)|}\mathbf{F}^2(u)\sum_{k=1}^N\mu_k^4+2\sum_{\substack{\mathbf{n}\in\mathbb{Z}^d:\\ \mathbf{n}\ne\mathbf{0}}}|a_{\mathbf{n}}|^2\sum_{\substack{\mathbf{m}\in\mathbb{Z}^d:\\ \mathbf{m}\ne\mathbf{0},\mathbf{n}}}|a_{\mathbf{m}}|^2$$

$$\cdot\left(\left(\sum_{k=1}^{N}\mu_k^2\right)|\mathbf{F}(2\pi|\mathbf{r}(t)|\,|\mathbf{m}|)\Phi(\mathcal{Y},\mathcal{M};\mathbf{m})|\cdot|\mathbf{F}(2\pi|\mathbf{r}(t)|\,|\mathbf{n}|)\Phi(\mathcal{Y},\mathcal{M};\mathbf{n})|\right.$$

$$+\left(\sum_{k=1}^{N}\mu_k^3\right)|\mathbf{F}(2\pi|\mathbf{r}(t)|\,|\mathbf{m}|)\Phi(\mathcal{Y},\mathcal{M};\mathbf{m})|$$

$$\left.+\left(\sum_{k=1}^{N}\mu_k^3\right)|\mathbf{F}(2\pi|\mathbf{r}(t)|\,|\mathbf{n}|)\Phi(\mathcal{Y},\mathcal{M};\mathbf{n})|\right). \tag{9.30}$$

We start the estimation of (9.30) with the part

$$\text{PartOne}(9.30) = 2\sum_{\substack{\mathbf{n}\in\mathbb{Z}^d:\\ \mathbf{n}\neq\mathbf{0}}}|a_{\mathbf{n}}|^2\sum_{\substack{\mathbf{m}\in\mathbb{Z}^d:\\ \mathbf{m}\neq\mathbf{0},\mathbf{n}}}|a_{\mathbf{m}}|^2$$

$$\cdot\left(\left(\sum_{k=1}^{N}\mu_k^2\right)|\mathbf{F}(2\pi|\mathbf{r}(t)|\,|\mathbf{m}|)\Phi(\mathcal{Y},\mathcal{M};\mathbf{m})|\cdot|\mathbf{F}(2\pi|\mathbf{r}(t)|\,|\mathbf{n}|)\Phi(\mathcal{Y},\mathcal{M};\mathbf{n})|,\right. \tag{9.31}$$

and repeat the argument of (9.24) applying the simple inequality $|ab|\leq(|a|^2+|b|^2)/2$:

$$\text{PartOne}(9.30) \leq \sum_{\substack{\mathbf{n}\in\mathbb{Z}^d:\\ \mathbf{n}\neq\mathbf{0}}}|a_{\mathbf{n}}|^2\sum_{\substack{\mathbf{m}\in\mathbb{Z}^d:\\ \mathbf{m}\neq\mathbf{0},\mathbf{n}}}|a_{\mathbf{m}}|^2\left(\sum_{k=1}^{N}\mu_k^2\right)$$

$$\cdot\left(\mathbf{F}^2(2\pi|\mathbf{r}(t)|\,|\mathbf{m}|)|\Phi(\mathcal{Y},\mathcal{M};\mathbf{m})|^2+\mathbf{F}^2(2\pi|\mathbf{r}(t)|\,|\mathbf{n}|)|\Phi(\mathcal{Y},\mathcal{M};\mathbf{n})|^2\right)/2$$

$$\leq \sum_{\substack{\mathbf{n}\in\mathbb{Z}^d:\\ \mathbf{n}\neq\mathbf{0}}}|a_{\mathbf{n}}|^2\sum_{\substack{\mathbf{m}\in\mathbb{Z}^d:\\ \mathbf{m}\neq\mathbf{0},\mathbf{n}}}|a_{\mathbf{m}}|^2$$

$$\cdot\left(\sum_{k=1}^{N}\mu_k^2\right)\mathbf{F}^2(2\pi|\mathbf{r}(t)|\,|\mathbf{n}|)|\Phi(\mathcal{Y},\mathcal{M};\mathbf{n})|^2$$

$$\leq \sigma_0^2(f)\left(\sum_{k=1}^{N}\mu_k^2\right)\sum_{\substack{\mathbf{n}\in\mathbb{Z}^d:\\ \mathbf{n}\neq\mathbf{0}}}|a_{\mathbf{n}}|^2\,\mathbf{F}^2(2\pi|\mathbf{r}(t)|\,|\mathbf{n}|)|\Phi(\mathcal{Y},\mathcal{M};\mathbf{n})|^2. \tag{9.32}$$

Next we estimate the second part of (9.30):

$$\text{PartTwo}(9.30) = \sum_{\substack{\mathbf{n}\in\mathbb{Z}^d:\\ \mathbf{n}\neq\mathbf{0}}}|a_{\mathbf{n}}|^2\sum_{\substack{\mathbf{m}\in\mathbb{Z}^d:\\ \mathbf{m}\neq\mathbf{0},\mathbf{n}}}|a_{\mathbf{m}}|^2$$

$$\cdot\left(\sum_{k=1}^{N}\mu_k^3\right)\left(|\mathbf{F}(2\pi|\mathbf{r}(t)|\ |\mathbf{m}|)\Phi(\mathcal{Y},\mathcal{M};\mathbf{m})|+|\mathbf{F}(2\pi|\mathbf{r}(t)|\ |\mathbf{n}|)\Phi(\mathcal{Y},\mathcal{M};\mathbf{n})|\right). \tag{9.33}$$

By using (9.20) in (9.33), we have

$$\text{PartTwo}(9.30)=2\left(\sum_{k=1}^{N}\mu_k^3\right)\sum_{\substack{\mathbf{n}\in\mathbb{Z}^d:\\ \mathbf{n}\neq\mathbf{0}}}|a_{\mathbf{n}}|^2\,|\mathbf{F}(2\pi|\mathbf{r}(t)|\ |\mathbf{n}|)\Phi(\mathcal{Y},\mathcal{M};\mathbf{n})|\sum_{\substack{\mathbf{m}\in\mathbb{Z}^d:\\ \mathbf{m}\neq\mathbf{0},\mathbf{n}}}|a_{\mathbf{m}}|^2$$

$$\leq 2\sigma_0^2(f)\left(\sum_{k=1}^{N}\mu_k^3\right)\sum_{\substack{\mathbf{n}\in\mathbb{Z}^d:\\ \mathbf{n}\neq\mathbf{0}}}|a_{\mathbf{n}}|^2\,|\mathbf{F}(2\pi|\mathbf{r}(t)|\ |\mathbf{n}|)\Phi(\mathcal{Y},\mathcal{M};\mathbf{n})|. \tag{9.34}$$

Applying the Cauchy–Schwarz inequality for the end-sum in (9.34), we have

$$\sum_{\substack{\mathbf{n}\in\mathbb{Z}^d:\\ \mathbf{n}\neq\mathbf{0}}}|a_{\mathbf{n}}|\,|\mathbf{F}(2\pi|\mathbf{r}(t)|\ |\mathbf{n}|)\Phi(\mathcal{Y},\mathcal{M};\mathbf{n})|\cdot|a_{\mathbf{n}}|$$

$$\leq\left(\sum_{\substack{\mathbf{n}\in\mathbb{Z}^d:\\ \mathbf{n}\neq\mathbf{0}}}|a_{\mathbf{n}}|^2\,\mathbf{F}^2(2\pi|\mathbf{r}(t)|\ |\mathbf{n}|)|\Phi(\mathcal{Y},\mathcal{M};\mathbf{n})|^2\right)^{1/2}\left(\sum_{\substack{\mathbf{n}\in\mathbb{Z}^d:\\ \mathbf{n}\neq\mathbf{0}}}|a_{\mathbf{n}}|^2\right)^{1/2}$$

$$=\left(\sum_{\substack{\mathbf{n}\in\mathbb{Z}^d:\\ \mathbf{n}\neq\mathbf{0}}}|a_{\mathbf{n}}|^2\,\mathbf{F}^2(2\pi|\mathbf{r}(t)|\ |\mathbf{n}|)|\Phi(\mathcal{Y},\mathcal{M};\mathbf{n})|^2\right)^{1/2}\cdot\sigma_0(f). \tag{9.35}$$

Using (9.35) in (9.34), we obtain

$$\text{PartTwo}(9.30)\leq 2\sigma_0^3(f)\left(\sum_{k=1}^{N}\mu_k^3\right)$$

$$\left(\sum_{\substack{\mathbf{n}\in\mathbb{Z}^d:\\ \mathbf{n}\neq\mathbf{0}}}|a_{\mathbf{n}}|^2\,\mathbf{F}^2(2\pi|\mathbf{r}(t)|\ |\mathbf{n}|)|\Phi(\mathcal{Y},\mathcal{M};\mathbf{n})|^2\right)^{1/2}. \tag{9.36}$$

Combining (9.26)–(9.34), we have

$$\left|\mathcal{S}_7-\mathcal{S}_7^2\right|\leq\sigma_0^4(f)\sum_{k=1}^{N}\sup_{u\geq 2\pi|\mathbf{r}(t)|}\mathbf{F}^2(u)\mu_k^4$$

$$+\text{PartOne}(9.30)+\text{PartTwo}(9.30)\leq\sigma_0^4(f)\sup_{u\geq 2\pi|\mathbf{r}(t)|}\mathbf{F}^2(u)\sum_{k=1}^{N}\mu_k^4$$

$$+\sigma_0^2(f)\left(\sum_{k=1}^{N}\mu_k^2\right)\sum_{\substack{\mathbf{n}\in\mathbb{Z}^d:\\ \mathbf{n}\neq\mathbf{0}}}|a_{\mathbf{n}}|^2\,\mathbf{F}^2(2\pi|\mathbf{r}(t)|\;|\mathbf{n}|)|\Phi(\mathcal{Y},\mathcal{M};\mathbf{n})|^2$$

$$+2\sigma_0^3(f)\left(\sum_{k=1}^{N}\mu_k^3\right)\left(\sum_{\substack{\mathbf{n}\in\mathbb{Z}^d:\\ \mathbf{n}\neq\mathbf{0}}}|a_{\mathbf{n}}|^2\,\mathbf{F}^2(2\pi\;|\mathbf{r}(t)|\;|\mathbf{n}|)|\Phi(\mathcal{Y},\mathcal{M};\mathbf{n})|^2\right)^{1/2}. \tag{9.37}$$

Similar argument works for $\left|\widetilde{\mathcal{S}}_{7,2}-\widetilde{\mathcal{S}}_{7,2}^*\right|$, and we obtain the same upper bound

$$\left|\widetilde{\mathcal{S}}_{7,2}-\widetilde{\mathcal{S}}_{7,2}^*\right|\leq\sigma_0^4(f)\sup_{u\geq 2\pi|\mathbf{r}(t)|}\mathbf{F}^2(u)\sum_{k=1}^{N}\mu_k^4$$

$$+\sigma_0^2(f)\left(\sum_{k=1}^{N}\mu_k^2\right)\sum_{\substack{\mathbf{n}\in\mathbb{Z}^d:\\ \mathbf{n}\neq\mathbf{0}}}|a_{\mathbf{n}}|^2\,\mathbf{F}^2(2\pi|\mathbf{r}(t)|\;|\mathbf{n}|)|\Phi(\mathcal{Y},\mathcal{M};\mathbf{n})|^2$$

$$+2\sigma_0^3(f)\left(\sum_{k=1}^{N}\mu_k^3\right)\left(\sum_{\substack{\mathbf{n}\in\mathbb{Z}^d:\\ \mathbf{n}\neq\mathbf{0}}}|a_{\mathbf{n}}|^2\,\mathbf{F}^2(2\pi\;|\mathbf{r}(t)|\;|\mathbf{n}|)|\Phi(\mathcal{Y},\mathcal{M};\mathbf{n})|^2\right)^{1/2}. \tag{9.37'}$$

Now we are ready to finish the general case of $\mathcal{M}$ (general mass distribution with average mass one). The variance is a square integral, so it is clearly positive. On the other hand, by using (9.18)–(9.22) and (9.27)–(9.37′) in (9.17), we obtain the upper bound

$$0\leq\text{Variance}_{\omega\in\Omega}\overline{\mathbf{D}^2}(\mathcal{Y};\mathcal{M};\mathbf{r};\omega;f;t)\leq|\mathcal{S}_3-\mathcal{S}_3^*|+\left|\widetilde{\mathcal{S}}_3-\widetilde{\mathcal{S}}_3^*\right|$$

$$+2\,|\mathcal{S}_4-\mathcal{S}_4^*|+2\left|\widetilde{\mathcal{S}}_4-\widetilde{\mathcal{S}}_4^*\right|+|\mathcal{S}_6-\mathcal{S}_6^*|+\left|\widetilde{\mathcal{S}}_{6,1}-\widetilde{\mathcal{S}}_{6,1}^*\right|+\left|\widetilde{\mathcal{S}}_{6,2}-\widetilde{\mathcal{S}}_{6,2}^*\right|$$

$$+2\,|\mathcal{S}_7-\mathcal{S}_7^*|+2\left|\widetilde{\mathcal{S}}_{7,1}-\widetilde{\mathcal{S}}_{7,1}^*\right|+2\left|\widetilde{\mathcal{S}}_{7,2}-\widetilde{\mathcal{S}}_{7,2}^*\right|\leq 3\left(\sum_{k=1}^{N}\mu_k^2\right)^2\sigma_0^4(f)$$

$$+ 6\sigma_0^2(f) \sum_{\mathbf{n}\in\mathbb{Z}^d\setminus\mathbf{0}} |a_{\mathbf{n}}|^2 \mathbf{F}^2(2\pi|\mathbf{r}(t)|\ |\mathbf{n}|)|\Phi(\mathcal{Y},\mathcal{M};\mathbf{n})|^2 \sum_{k=1}^N \mu_k^2$$

$$+ 5\sigma_0^4(f) \left(\sum_{k=1}^N \mu_k^2\right)^2 \sup_{u\geq 2\pi|\mathbf{r}(t)|} \mathbf{F}^2(u)$$

$$+ 4\sigma_0^2(f) \sum_{\mathbf{n}\in\mathbb{Z}^d\setminus\mathbf{0}} |a_{\mathbf{n}}|^2 \mathbf{F}^2(2\pi|\mathbf{r}(t)|\ |\mathbf{n}|)|\Phi(\mathcal{Y},\mathcal{M};\mathbf{n})|^2 \sum_{k=1}^N \mu_k^2$$

$$+ 4\sigma_0^4(f) \sup_{u\geq 2\pi|\mathbf{r}(t)|} \mathbf{F}^2(u) \sum_{k=1}^N \mu_k^4$$

$$+ 4\sigma_0^2(f) \left(\sum_{k=1}^N \mu_k^2\right) \sum_{\substack{\mathbf{n}\in\mathbb{Z}^d:\\ \mathbf{n}\neq\mathbf{0}}} |a_{\mathbf{n}}|^2 \mathbf{F}^2(2\pi|\mathbf{r}(t)|\ |\mathbf{n}|)|\Phi(\mathcal{Y},\mathcal{M};\mathbf{n})|^2$$

$$+ 8\sigma_0^3(f) \left(\sum_{k=1}^N \mu_k^3\right) \left(\sum_{\substack{\mathbf{n}\in\mathbb{Z}^d:\\ \mathbf{n}\neq\mathbf{0}}} |a_{\mathbf{n}}|^2 \mathbf{F}^2(2\pi\ |\mathbf{r}(t)|\ |\mathbf{n}|)|\Phi(\mathcal{Y},\mathcal{M};\mathbf{n})|^2\right)^{1/2}$$

$$\leq 12 \left(\sum_{k=1}^N \mu_k^2\right)^2 \sigma_0^4(f)$$

$$+ 14\sigma_0^2(f) \sum_{\mathbf{n}\in\mathbb{Z}^d\setminus\mathbf{0}} |a_{\mathbf{n}}|^2 \mathbf{F}^2(2\pi|\mathbf{r}(t)|\ |\mathbf{n}|)|\Phi(\mathcal{Y},\mathcal{M};\mathbf{n})|^2 \sum_{k=1}^N \mu_k^2$$

$$+ 8\sigma_0^3(f) \left(\sum_{k=1}^N \mu_k^3\right) \left(\sum_{\substack{\mathbf{n}\in\mathbb{Z}^d:\\ \mathbf{n}\neq\mathbf{0}}} |a_{\mathbf{n}}|^2 \mathbf{F}^2(2\pi\ |\mathbf{r}(t)|\ |\mathbf{n}|)|\Phi(\mathcal{Y},\mathcal{M};\mathbf{n})|^2\right)^{1/2}, \tag{9.38}$$

where we used the facts $|\mathbf{F}| \leq 1$ and

$$\sum_{k=1}^N \mu_k^4 \leq \left(\sum_{k=1}^N \mu_k^2\right)^2.$$

As a generalization of the concept of the "unevenness part" for the three kinds $\overline{\mathbf{D}^2}_{i,uneven}(t)$, $i = 1, 2, 3$, write

$$\overline{\mathbf{D}^2}_{\mathbf{F},uneven}(t) = \overline{\mathbf{D}^2}_{\mathbf{F},uneven}(\mathcal{Y}, \mathcal{M}; \mathbf{r}; f; t)$$

$$= \sum_{\mathbf{n}\in\mathbb{Z}^d\setminus\mathbf{0}} |a_{\mathbf{n}}|^2 \mathbf{F}^2(2\pi|\mathbf{r}(t)|\ |\mathbf{n}|)|\Phi(\mathcal{Y}, \mathcal{M}; \mathbf{n})|^2$$

$$= \sum_{\substack{\mathbf{n}\in\mathbb{Z}^d: \\ \mathbf{n}\neq\mathbf{0}}} |a_{\mathbf{n}}|^2 \left|\sum_{k=1}^{N} \mu_k e^{2\pi i\mathbf{n}\cdot\mathbf{y}_k}\right|^2 \mathbf{F}^2(2\pi|\mathbf{r}(t)|\ |\mathbf{n}|). \tag{9.39}$$

By using (9.39) we can rewrite (9.38) in the form

$$0 \leq \text{Variance}_{\omega\in\Omega}\overline{\mathbf{D}^2}(\mathcal{Y}, \mathcal{M}; \mathbf{r}; \omega; f; t) \leq 12 \left(\sum_{k=1}^{N} \mu_k^2\right)^2 \sigma_0^4(f)$$

$$+ 14\sigma_0^2(f)\overline{\mathbf{D}^2}_{\mathbf{F},uneven}(t) \sum_{k=1}^{N} \mu_k^2 + 8\sigma_0^3(f) \left(\sum_{k=1}^{N} \mu_k^3\right) \left(\overline{\mathbf{D}^2}_{\mathbf{F},uneven}(t)\right)^{1/2}. \tag{9.40}$$

We need the inequality

$$\sum_{k=1}^{N} \mu_k^3 \leq \left(\sum_{k=1}^{N} \mu_k^2\right)^{3/2} \tag{9.41}$$

(we prove it at the end of the section), and also use the notation

$$\text{Eqlbrm} = \text{Eqlbrm}(\mathcal{M}; f) = \sigma_0^2(f) \left(\sum_{k=1}^{N} \mu_k^2\right). \tag{9.42}$$

Applying (9.41) and (9.42) in (9.40), we have

$$0 \leq \text{Variance}_{\omega\in\Omega}\overline{\mathbf{D}^2}(\mathcal{Y}, \mathcal{M}; \mathbf{r}; \omega; f; t)$$

$$\leq 14\text{Eqlbrm}\cdot\overline{\mathbf{D}^2}_{\mathbf{F},uneven}(t) + 8\text{Eqlbrm}^{3/2} \left(\overline{\mathbf{D}^2}_{\mathbf{F},uneven}(t)\right)^{1/2} + 12\text{Eqlbrm}^2. \tag{9.43}$$

By taking square root in (9.43), we obtain the following upper bound for the standard deviation (as ω runs in Ω):

$$\text{StandardDeviation}_{\omega\in\Omega}\overline{\mathbf{D}^2}(\mathcal{Y}, \mathcal{M}; \mathbf{r}; \omega; f; t)$$

$$= \left(\text{Variance}_{\omega\in\Omega}\overline{\mathbf{D}^2}(\mathcal{Y}, \mathcal{M}; \mathbf{r}; \omega; f; t)\right)^{1/2}$$

$$\leq 4\text{Eqlbrm}+4\sqrt{\text{Eqlbrm}}\sqrt{\overline{\mathbf{D}^2}_{\mathbf{F},uneven}(t)}+4\text{Eqlbrm}^{3/4}\left(\overline{\mathbf{D}^2}_{\mathbf{F},uneven}(t)\right)^{1/4}. \tag{9.44}$$

By using the same notation, we can rewrite (9.26) — representing the special case $1 = \mu_1 = \ldots = \mu_N$ — as follows:

$$\text{StandardDeviation}_{\omega\in\Omega}\overline{\mathbf{D}^2}(\mathcal{Y};\mathbf{r};\omega;f;t) = \left(\text{Variance}_{\omega\in\Omega}\overline{\mathbf{D}^2}(\mathcal{Y};\mathbf{r};\omega;f;t)\right)^{1/2}$$

$$\leq 4\text{Eqlbrm} + 4\sqrt{\text{Eqlbrm}}\sqrt{\overline{\mathbf{D}^2}_{\mathbf{F},uneven}(t)}, \tag{9.45}$$

which is slightly simpler than (9.44).

It remains to prove inequality (9.41). We have

$$\left(\sum_{k=1}^{N}\mu_k^2\right)^3 = \sum_{k=1}^{N}\mu_k^6 + 3\sum_{1\leq k_1<k_2\leq N}\mu_{k_1}^4\mu_{k_2}^2 + 3\sum_{1\leq k_1<k_2\leq N}\mu_{k_1}^2\mu_{k_2}^4 + \text{positive}$$

$$\geq \sum_{k=1}^{N}\mu_k^6 + \sum_{1\leq k_1<k_2\leq N}\left(\mu_{k_1}^4\mu_{k_2}^2 + \mu_{k_1}^2\mu_{k_2}^4\right)$$

$$\geq \sum_{k=1}^{N}\mu_k^6 + 2\sum_{1\leq k_1<k_2\leq N}\mu_{k_1}^3\mu_{k_2}^3 = \left(\sum_{k=1}^{N}\mu_k^3\right)^2,$$

proving (9.41). Note that at the end of the argument we used the simple fact

$$\mu_{k_1}^4\mu_{k_2}^2 + \mu_{k_1}^2\mu_{k_2}^4 \geq 2\mu_{k_1}^3\mu_{k_2}^3,$$

which is a special case of the general inequality between the arithmetic and geometric means.

This completes the proof of Theorem 5.1. □

10. Proving Theorem 3.2 and the Quick-Jump-Up Phenomenon

Proof of Theorem 3.2. We need the following two lemmas.

Lemma 10.1. *If*

$$|\mathbf{r}(t)| \geq \sqrt{\log N}, \tag{10.1}$$

then

$$\text{Variance}_{\omega\in\Omega_1}\overline{\mathbf{D}^2}(\mathcal{Y},\mathcal{M};\mathbf{r};\omega;f;t)$$

$$= \int_{\Omega_1} \left(\overline{\mathbf{D}^2}(\mathcal{Y}, \mathcal{M}; \mathbf{r}; \omega; f; t) - \overline{\mathbf{D}^2}_{\Omega_1}(\mathcal{Y}, \mathcal{M}; \mathbf{r}; f; t) \right)^2 d\omega$$

$$\geq \left(\left(\sum_{j=1}^{N} \mu_j^2 \right)^2 - \sum_{j=1}^{N} \mu_j^4 \right) \sum_{\substack{\mathbf{n} \in \mathbb{Z}^d: \\ \mathbf{n} \neq \mathbf{0}}} |a_{\mathbf{n}}|^4 + O(1). \tag{10.2}$$

Lemma 10.2. *If (10.1) holds then*

$$\text{FourthMoment}_{\omega \in \Omega_1} \overline{\mathbf{D}^2}(\mathcal{Y}, \mathcal{M}; \mathbf{r}; \omega; f; t)$$

$$= \int_{\Omega_1} \left(\overline{\mathbf{D}^2}(\mathcal{Y}, \mathcal{M}; \mathbf{r}; \omega; f; t) - \overline{\mathbf{D}^2}_{\Omega_1}(\mathcal{Y}, \mathcal{M}; \mathbf{r}; f; t) \right)^4 d\omega$$

$$\leq 6 \left(\left(\left(\left(\sum_{j=1}^{N} \mu_j^2 \right)^2 - \sum_{j=1}^{N} \mu_j^4 \right) \sum_{\substack{\mathbf{n} \in \mathbb{Z}^d: \\ \mathbf{n} \neq \mathbf{0}}} |a_{\mathbf{n}}|^4 \right)^2 + \left(\sum_{j=1}^{N} \mu_j^4 \right)^2 \sigma_0^8(f) + O(1). \right. \tag{10.3}$$

First we prove the two lemmas, and then explain how they imply the theorem.

Proof of Lemma 10.1. By (7.6) and (7.8),

$$\overline{\mathbf{D}^2}(\mathcal{Y}, \mathcal{M}; \mathbf{r}; \omega; f; t) - \overline{\mathbf{D}^2}_{\Omega_1}(\mathcal{Y}, \mathcal{M}; \mathbf{r}; f; t)$$

$$= \sum_{\substack{\mathbf{n} \in \mathbb{Z}^d: \\ \mathbf{n} \neq \mathbf{0}}} |a_{\mathbf{n}}|^2 \sum_{1 \leq k, \ell \leq N: k \neq \ell} \mu_k \mu_\ell e^{2\pi i \mathbf{n} \cdot (\mathbf{y}_k - \mathbf{y}_\ell)} e^{2\pi i \mathbf{n} \cdot (\vartheta_k(\rho_k \mathbf{r}(t)) - \vartheta_\ell(\rho_\ell \mathbf{r}(t)))}$$

$$- \sum_{\substack{\mathbf{n} \in \mathbb{Z}^d: \\ \mathbf{n} \neq \mathbf{0}}} |a_{\mathbf{n}}|^2 e^{-4(\pi |\mathbf{r}(t)|\ |\mathbf{n}|)^2} \sum_{1 \leq k, \ell \leq N: k \neq \ell} \mu_k \mu_\ell e^{2\pi i \mathbf{n} \cdot (\mathbf{y}_k - \mathbf{y}_\ell)}. \tag{10.4}$$

We have to raise (10.4) to the second power, multiply it out, and integrate the terms over the initial velocity space $\omega \in \Omega_1 = \Omega_1(\dim = d)$. Under the condition (10.1), most of the terms give negligible contribution to the integral due to the superexponentially small factor

$$e^{-(\pi |\mathbf{r}(t)|\ |\mathbf{n}|)^2}. \tag{10.5}$$

We do not have a superexponentially small factor (10.5) in the following single case, called Case 1. Multiplying out the second power.

Case 1: a pair (k,ℓ) shows up with nonzero lattice points $\mathbf{n}_1,\mathbf{n}_2 \in \mathbb{Z}^d\setminus\mathbf{0}$ such that $\mathbf{n}_1+\mathbf{n}_2=\mathbf{0}$.

Therefore, under the condition (10.1), we have

$$\text{Variance}_{\omega\in\Omega_1}\overline{\mathbf{D}^2}(\mathcal{Y},\mathcal{M};\mathbf{r};\omega;f;t)$$

$$=\int_{\Omega_1}\left(\overline{\mathbf{D}^2}(\mathcal{Y},\mathcal{M};\mathbf{r};\omega;f;t)-\overline{\mathbf{D}^2}_{\Omega_1}(\mathcal{Y},\mathcal{M};\mathbf{r};f;t)\right)^2 d\omega$$

$$=\text{ContributionCase1}+O(1), \tag{10.6}$$

where the negligible $O(1)$ is the total contribution of the terms that have a superexponentially small factor (10.5), and

$$\text{ContributionCase1}=\sum_{\substack{(k,\ell):\\ 1\le k\ne\ell\le N}}\mu_k^2\mu_\ell^2\sum_{\substack{\mathbf{n}_1,\mathbf{n}_2\in\mathbb{Z}^d\setminus\mathbf{0}:\\ \mathbf{n}_1+\mathbf{n}_2=\mathbf{0}}}|a_{\mathbf{n}_1}|^2\,|a_{\mathbf{n}_2}|^2. \tag{10.7}$$

Notice that (10.7) equals (10.2), so (10.6) and (10.7) complete the proof of Lemma 10.1. □

Proof of Lemma 10.2. It is similar to the proof of Lemma 10.1, but the details are more complicated. Here we have to raise (10.4) to the fourth power, multiply it out, and integrate the terms over the initial velocity space $\omega\in\Omega_1$. Under the condition (10.1), most of the terms give negligible contribution due to the superexponentially small factor (10.5). We do not have a superexponentially small factor (10.5) in the following three cases (see Cases 1, 2 and 3 below). Multiplying out the fourth power, we have

Case 1: two pairs (k_1,ℓ_1) and (k_2,ℓ_2) such that, (k_1,ℓ_1) shows up with a nonzero lattice point $\mathbf{n}_1\in\mathbb{Z}^d\setminus\mathbf{0}$ and also with its negative $-\mathbf{n}_1$, similarly, (k_2,ℓ_2) shows up with a nonzero lattice point $\mathbf{n}_2\in\mathbb{Z}^d\setminus\mathbf{0}$ and also with its negative $-\mathbf{n}_2$, and finally, $1\le k_1,\ell_1,k_2,\ell_2\le N$ are four different indices;

Case 2: it is Case 1 with the modification that $1\le k_1,\ell_1,k_2,\ell_2\le N$ represent three different indices;

Case 3: a pair (k,ℓ) shows up with nonzero lattice points $\mathbf{n}_1,\mathbf{n}_2,\mathbf{n}_3,\mathbf{n}_4\in\mathbb{Z}^d\setminus\mathbf{0}$ such that $\mathbf{n}_1+\mathbf{n}_2+\mathbf{n}_3+\mathbf{n}_4=\mathbf{0}$.

Therefore, under the condition (10.1), we have

$$\text{FourthMoment}_{\omega\in\Omega_1}\overline{\mathbf{D}^2}(\mathcal{Y},\mathcal{M};\mathbf{r};\omega;f;t)$$

$$=\int_{\Omega_1}\left(\overline{\mathbf{D}^2}(\mathcal{Y},\mathcal{M};\mathbf{r};\omega;f;t)-\overline{\mathbf{D}^2}_{\Omega_1}(\mathcal{Y},\mathcal{M};\mathbf{r};f;t)\right)^4 d\omega$$

$$=\text{ContributionCase1}+\text{ContributionCase2}+\text{ContributionCase3}+O(1), \tag{10.8}$$

where the negligible $O(1)$ is the total contribution of the terms that have a superexponentially small factor (10.5),

$$\text{ContributionCase1} = 6 \sum_{\substack{(k_1,\ell_1):\\ 1\le k_1\ne\ell_1\le N}} \mu_{k_1}^2\mu_{\ell_1}^2 \sum_{\substack{\mathbf{n}_1\in\mathbb{Z}^d:\\ \mathbf{n}_1\ne\mathbf{0}}} |a_{\mathbf{n}_1}|^4$$

$$\cdot \sum_{\substack{(k_2,\ell_2):\ 1\le k_2\ne\ell_2\le N\\ k_1,\ell_1,k_2,\ell_2 \text{ are four different indices}}} \mu_{k_2}^2\mu_{\ell_2}^2 \sum_{\substack{\mathbf{n}_2\in\mathbb{Z}^d:\\ \mathbf{n}_2\ne\mathbf{0}}} |a_{\mathbf{n}_2}|^4, \tag{10.9}$$

$$\text{ContributionCase2} \le 6 \sum_{\substack{(k_1,\ell_1):\\ 1\le k_1\ne\ell_1\le N}} \mu_{k_1}^2\mu_{\ell_1}^2 \sum_{\substack{\mathbf{n}_1\in\mathbb{Z}^d:\\ \mathbf{n}_1\ne\mathbf{0}}} |a_{\mathbf{n}_1}|^4$$

$$\cdot \sum_{\substack{(k_2,\ell_2):\ 1\le k_2\ne\ell_2\le N\\ k_1,\ell_1,k_2,\ell_2 \text{ are three different indices}}} \mu_{k_2}^2\mu_{\ell_2}^2 \sum_{\substack{\mathbf{n}_2\in\mathbb{Z}^d:\\ \mathbf{n}_2\ne\mathbf{0}}} |a_{\mathbf{n}_2}|^4, \tag{10.10}$$

and finally,

$$\text{ContributionCase3} = \sum_{\substack{(k,\ell):\\ 1\le k\ne\ell\le N}} \mu_k^4\mu_\ell^4 \sum_{\substack{\mathbf{n}_1,\mathbf{n}_2,\mathbf{n}_3,\mathbf{n}_4\in\mathbb{Z}^d\setminus\mathbf{0}:\\ \mathbf{n}_1+\mathbf{n}_2+\mathbf{n}_3+\mathbf{n}_4=\mathbf{0}}} |a_{\mathbf{n}_1}|^2|a_{\mathbf{n}_2}|^2|a_{\mathbf{n}_3}|^2|a_{\mathbf{n}_4}|^2. \tag{10.11}$$

Note that the factor $6 = \binom{4}{2}$ in (10.9) comes from the following six arrangements:

$$\langle\mathbf{n}_1,-\mathbf{n}_1,\mathbf{n}_2,-\mathbf{n}_2\rangle, \langle\mathbf{n}_1,\mathbf{n}_2,-\mathbf{n}_1,-\mathbf{n}_2\rangle, \langle\mathbf{n}_1,\mathbf{n}_2,-\mathbf{n}_2,-\mathbf{n}_1\rangle,$$

and the remaining three arise by switching the roles of $\mathbf{n}_1$ and $\mathbf{n}_2$.

Combining (10.8)–(10.11), we have

$$\text{FourthMoment}_{\omega\in\Omega_1}\overline{\mathbf{D}^2}(\mathcal{Y},\mathcal{M};\mathbf{r};\omega;f;t)$$

$$= \int_{\Omega_1}\left(\overline{\mathbf{D}^2}(\mathcal{Y},\mathcal{M};\mathbf{r};\omega;f;t) - \overline{\mathbf{D}^2}_{\Omega_1}(\mathcal{Y},\mathcal{M};\mathbf{r};f;t)\right)^4 d\omega$$

$$\le 6\left(\left(\left(\sum_{j=1}^N \mu_j^2\right)^2 - \sum_{j=1}^N \mu_j^4\right)\sum_{\substack{\mathbf{n}\in\mathbb{Z}^d:\\ \mathbf{n}\ne\mathbf{0}}}|a_{\mathbf{n}}|^4\right)^2$$

$$+\left(\sum_{j=1}^N \mu_j^4\right)^2\left(\sum_{\substack{\mathbf{n}\in\mathbb{Z}^d:\\ \mathbf{n}\ne\mathbf{0}}}|a_{\mathbf{n}}|^2\right)^4 + O(1). \tag{10.12}$$

Since

$$\sigma_0^2(f) = \sum_{\substack{\mathbf{n}\in\mathbb{Z}^d:\\ \mathbf{n}\neq\mathbf{0}}} |a_{\mathbf{n}}|^2,$$

(10.12) completes the proof of Lemma 10.2. □

To finish the proof of Theorem 3.2, we need a simple average argument. Chebyshev's inequality does not help; instead we use the following equally simple "fourth moment lemma".

Lemma 10.3. *Let $(\Omega, \mathcal{G}, \mu)$ be a probability space. Let $g \in \mathcal{G}$ be a μ-measurable function on Ω such that*

$$\int_\Omega g^2\, d\mu \geq K \quad \text{and} \quad \int_\Omega g^4\, d\mu \leq L^2$$

for some $0 < K \leq L < \infty$. Then

$$\mu\left\{g^2 \geq K/4\right\} \geq \left(\frac{3K}{4L}\right)^2.$$

Proof. Clearly

$$\int_{g^2<K/4} g^2\, d\mu < K/4,$$

so by using the Cauchy–Schwartz inequality, we have

$$\left(1-\frac{1}{4}\right)^2 K^2 \leq \left(\int_{g^2\geq K/4} g^2\cdot 1\, d\mu\right)^2$$

$$\leq \int_\Omega g^4\, d\mu \cdot \mu\left\{g^2 \geq K/4\right\} \leq L^2 \mu\left\{g^2 \geq K/4\right\},$$

which implies

$$\mu\left\{g^2 \geq K/4\right\} \geq \left(1-\frac{1}{4}\right)^2 K^2 L^{-2},$$

completing the proof of Lemma 10.3. □

Combining Lemmas 10.1–3 and using the notation

$$\kappa_0^4(f) = \sum_{\substack{\mathbf{n}\in\mathbb{Z}^d:\\ \mathbf{n}\neq\mathbf{0}}} |a_{\mathbf{n}}|^4,$$

we obtain

$$\text{ProdMeas}_1\Bigg\{\omega\in\Omega_1:\ \left(\overline{\mathbf{D}^2}(\mathcal{Y},\mathcal{M};\mathbf{r};\omega;f;t)-\overline{\mathbf{D}^2}_{\Omega_1}(\mathcal{Y},\mathcal{M};\mathbf{r};f;t)\right)^2$$

$$\geq\frac{1}{4}\left(\left(\sum_{j=1}^N\mu_j^2\right)^2-\sum_{j=1}^N\mu_j^4\right)\kappa_0^4(f)+O(1)\Bigg\}$$

$$\geq\frac{9}{16}\cdot\frac{\left(\left(\left(\sum_{j=1}^N\mu_j^2\right)^2-\sum_{j=1}^N\mu_j^4\right)\kappa_0^4(f)+O(1)\right)^2}{6\left(\left(\left(\sum_{j=1}^N\mu_j^2\right)^2-\sum_{j=1}^N\mu_j^4\right)\kappa_0^4(f)\right)^2+\left(\sum_{j=1}^N\mu_j^4\right)^2\sigma_0^8(f)+O(1)}$$

$$=\frac{3}{32}\cdot\frac{1}{1+\delta_1}+\delta_2, \tag{10.13}$$

where

$$\delta_1=\frac{6\left(\sum_{j=1}^N\mu_j^4\right)^2\sigma_0^8(f)+O(1)}{\left(\left(\left(\sum_{j=1}^N\mu_j^2\right)^2-\sum_{j=1}^N\mu_j^4\right)\kappa_0^4(f)\right)^2} \tag{10.14}$$

and

$$\delta_2=\frac{O(1)}{\left(\left(\left(\sum_{j=1}^N\mu_j^2\right)^2-\sum_{j=1}^N\mu_j^4\right)\kappa_0^4(f)\right)^2+\left(\sum_{j=1}^N\mu_j^4\right)^2\sigma_0^8(f)+O(1)}. \tag{10.15}$$

In the "reasonable" cases both δ_1 and δ_2 are close to zero. For example, if every mass is one: $1=\mu_1=\ldots=\mu_N$, and the test function f has a one-term Fourier series: $f(\mathbf{x})=e^{2\pi i\mathbf{n}\cdot\mathbf{x}}$ for some fixed lattice point $\mathbf{n}\in\mathbb{Z}^d\setminus\mathbf{0}$, then (10.13)–(10.15) give

$$\text{ProdMeas}_1\left\{\omega\in\Omega_1:\ \left|\overline{\mathbf{D}^2}(\mathcal{Y};\mathbf{r};\omega;f;t)-\overline{\mathbf{D}^2}_{\Omega_1}(\mathcal{Y};\mathbf{r};f;t)\right|\geq\frac{N}{2}+O(1)\right\}$$

$$\geq\frac{3}{32}\cdot\frac{1}{1+\delta_1}+\delta_2,$$

where

$$\delta_1=\frac{6N^2+O(1)}{N^2(N-1)^2}=\frac{6+o(1)}{N^2}\quad\text{and}\quad\delta_2=\frac{O(1)}{N^2}.$$

This completes the proof of Theorem 3.2. □

The Quick-Jump-Up phenomenon. It is clear that Big Bang represents the worst case initial point configuration. Indeed, for $\mu_k \geq 0$, $1 \leq k \leq N$ the exponential sum has the trivial upper bound

$$\left|\sum_{k=1}^{N} \mu_k e^{2\pi i \mathbf{n}\cdot\mathbf{y}_k}\right| \leq \sum_{k=1}^{N} \mu_k = N, \tag{10.16}$$

and we have equality if $\mathbf{y}_1 = \ldots = \mathbf{y}_N$. So Big Bang gives the maximum of (10.16).

On the other hand, it is easy to make the exponential sum (10.16) very small. For example, consider the case of a finite geometric series: let every mass μ_k be equal to one, and let $\mathbf{y}_k = k\mathbf{v}$ modulo one, $1 \leq k \leq N$. Then (10.16) has a short form

$$\left|\sum_{k=1}^{N} e^{2\pi i \mathbf{n}\cdot\mathbf{y}_k}\right| = \left|\frac{1-e^{2\pi i N\mathbf{n}\cdot\mathbf{v}}}{1-e^{2\pi i \mathbf{n}\cdot\mathbf{v}}}\right| \leq \frac{2}{\left|1-e^{2\pi i \mathbf{n}\cdot\mathbf{v}}\right|}. \tag{10.17}$$

If the dot product $\mathbf{n}\cdot\mathbf{v}$ is not too close to an integer, then (10.17) represents $O(1)$, which is "small".

Consider now the 3-dimensional torus model of the second kind in the special case $\mathcal{R} = \mathbf{r}$ and $1 = \mathbf{y}_1 = \ldots = \mathbf{y}_N$. We start with the second kind, because it exhibits the slowest approach to (stabil average square-root) equilibrium. This will emphasize the sharp contrast with the Quick-Jump-Up phenomenon.

The equilibrium part is $\text{Eqlbrm}(f) = \sigma_0^2(f)N$. Being close to Big Bang means that the system starts from an extreme off-equilibrium $\mathcal{Y} = \{\mathbf{y}_1, \ldots, \mathbf{y}_N\}$ such that

$$\overline{\mathbf{D}^2}(\mathcal{Y};\mathbf{r};\omega;f;0) = \overline{\mathbf{D}^2}_{\Omega_2}(\mathcal{Y};\mathbf{r};f;0) = \overline{\mathbf{D}^2}_{2,uneven}(\mathcal{Y};\mathbf{r};f;0)$$

$$= \sum_{\mathbf{n}\in\mathbb{Z}^3\setminus\mathbf{0}} |a_{\mathbf{n}}|^2 \left|\sum_{k=1}^{N} e^{2\pi i \mathbf{n}\cdot\mathbf{y}_k}\right|^2 \tag{10.18}$$

is around constant times $\sigma_0^2(f)N^2$ — much larger than the linear equilibrium part $\text{Eqlbrm}(f) = \sigma_0^2(f)N$. Starting from close to Big Bang, it takes a very long time for the typical time evolution $\omega \in \Omega_2$ of the system to reach (stabil) equilibrium: the drifting distance $|\mathbf{r}(t)|$ has to reach the size $\sqrt{N}$.

On the other hand, if (10.18) is smaller than the equilibrium part $\text{Eqlbrm}(f) = \sigma_0^2(f)N$, even as small as $O(1)$, then it always takes a very *short* time for a substantial part of the time evolution $\omega \in \Omega_2$ of the system to reach (stabil) equilibrium: constant lower bound suffices, i.e., we just need $|\mathbf{r}(t)| \geq$constant. In fact, we prove it with $|\mathbf{r}(t)| \geq 2$; see (10.33) below. This is what we call the *Quick-Jump-Up phenomenon.* The name is explained by the sharp contrast with the slow approach starting from constant times $\sigma_0^2(f)N^2$ in (10.18) (close to Big Bang) to the equilibrium $\text{Eqlbrm}(f) = \sigma_0^2(f)N$, which requires $|\mathbf{r}(t)| \geq$const$\sqrt{N}$.

The proof of the Quick-Jump-Up phenomenon is easy: it is very similar to the proof of Theorem 3.2. The first step is to recall (4.8)–(4.11), which describes the time evolution of $\overline{\mathbf{D}^2}_{\Omega_2}(\mathcal{Y}; \mathbf{r}; f; t)$:

$$\overline{\mathbf{D}^2}_{\Omega_2}(\mathcal{Y}; \mathbf{r}; f; t) = \sum_{k=1}^{N} \sum_{\substack{\mathbf{n}\in\mathbb{Z}^3:\\ \mathbf{n}\neq\mathbf{0}}} |a_{\mathbf{n}}|^2 \left(1 - \left(\frac{\sin\left(2\pi|\mathbf{r}(t)|\,|\mathbf{n}|\right)}{2\pi|\mathbf{r}(t)|\,|\mathbf{n}|}\right)^2\right)$$

$$+ \overline{\mathbf{D}^2}_{2,uneven}(t), \tag{10.19}$$

where $a_{\mathbf{n}}$, $\mathbf{n} \in \mathbb{Z}^3$ are the Fourier coefficients of the test function $f \in L^2$, and

$$\overline{\mathbf{D}^2}_{2,uneven}(t) = \overline{\mathbf{D}^2}_{2,uneven}(\mathcal{Y}; \mathbf{r}; f; t)$$

$$= \sum_{\substack{\mathbf{n}\in\mathbb{Z}^3:\\ \mathbf{n}\neq\mathbf{0}}} |a_{\mathbf{n}}|^2 \left|\sum_{k=1}^{N} e^{2\pi i \mathbf{n}\cdot\mathbf{y}_k}\right|^2 \left(\frac{\sin(2\pi|\mathbf{r}(t)|\,|\mathbf{n}|)}{2\pi|\mathbf{r}(t)|\,|\mathbf{n}|}\right)^2. \tag{10.20}$$

What we are really interested in is $\overline{\mathbf{D}^2}(\mathcal{Y}; \mathbf{r}; \omega; f; t)$ for a typical $\omega \in \Omega_2$. We recall that $\overline{\mathbf{D}^2}_{\Omega_2}(\mathcal{Y}; \mathbf{r}; f; t)$ represents the expectation (see (4.4)):

$$\overline{\mathbf{D}^2}_{\Omega_2}(\mathcal{Y}; \mathbf{r}; f; t) = \int_{\Omega_2} \overline{\mathbf{D}^2}(\mathcal{Y}; \mathbf{r}; \omega; f; t)\, d\omega. \tag{10.21}$$

To describe the typical behavior of $\overline{\mathbf{D}^2}(\mathcal{Y}; \mathbf{r}; \omega; f; t)$, $\omega \in \Omega_2$ it is not enough to know the expectation: we also need to know the variance. We use Theorem 5.1(b); in fact, to get a better constant factor, we use the following intermediate step in the proof of Theorem 5.1: by (9.25)

$$\text{Variance}_{\omega\in\Omega}\overline{\mathbf{D}^2}(\mathcal{Y}; \mathbf{r}; \omega; f; t) \leq 3N^2\sigma_0^4(f)$$

$$+ 6\sigma_0^2(f)\overline{\mathbf{D}^2}_{\mathbf{F},uneven}(t)N + 5\sigma_0^4(f)N^2 \sup_{u\geq 2\pi|\mathbf{r}(t)|} \mathbf{F}^2(u)$$

$$+ 4\sigma_0^2(f)\overline{\mathbf{D}^2}_{\mathbf{F},uneven}(t)N + 4\sigma_0^4(f)N \sup_{u\geq 2\pi|\mathbf{r}(t)|} \mathbf{F}^2(u) + 4\sigma_0^2(f)\overline{\mathbf{D}^2}_{\mathbf{F},uneven}(t)N, \tag{10.22}$$

where

$$\overline{\mathbf{D}^2}_{\mathbf{F},uneven}(t) = \overline{\mathbf{D}^2}_{\mathbf{F},uneven}(\mathcal{Y};\mathbf{r};f;t)$$

$$= \sum_{\substack{\mathbf{n}\in\mathbb{Z}^d:\\ \mathbf{n}\neq\mathbf{0}}} |a_{\mathbf{n}}|^2 \left| \sum_{k=1}^{N} \mu_k e^{2\pi i\mathbf{n}\cdot\mathbf{y}_k} \right|^2 \mathbf{F}^2(2\pi|\mathbf{r}(t)|\,|\mathbf{n}|).$$

We apply (10.22) in the special case of the second kind $\mathbf{F}(u) = \frac{\sin u}{u}$, and use the fact $\text{Eqlbrm}(f) = \sigma_0^2(f)N$; thus we have

$$\text{Variance}_{\omega\in\Omega_2}\overline{\mathbf{D}^2}(\mathcal{Y};\mathbf{r};\omega;f;t)$$

$$= \int_{\Omega_2} \left(\overline{\mathbf{D}^2}(\mathcal{Y};\mathbf{r};\omega;f;t)\right)^2 d\omega - \left(\overline{\mathbf{D}^2}_{\Omega_2}(\mathcal{Y};\mathbf{r};f;t)\right)^2 \leq 3\text{Eqlbrm}^2(f)$$

$$+ 14\text{Eqlbrm}(f)\overline{\mathbf{D}^2}_{\mathbf{F},uneven}(t) + 5\text{Eqlbrm}^2(f)\frac{1}{(2\pi|\mathbf{r}(t)|)^2}$$

$$+ \frac{4}{N}\text{Eqlbrm}^2(f)\frac{1}{(2\pi|\mathbf{r}(t)|)^2}. \tag{10.23}$$

By (10.19)–(10.21),

$$\int_{\Omega_2} \overline{\mathbf{D}^2}(\mathcal{Y};\mathbf{r};\omega;f;t)\,d\omega = \overline{\mathbf{D}^2}_{\Omega_2}(\mathcal{Y};\mathbf{r};f;t)$$

$$\geq \sum_{k=1}^{N} \sum_{\substack{\mathbf{n}\in\mathbb{Z}^3:\\ \mathbf{n}\neq\mathbf{0}}} |a_{\mathbf{n}}|^2 \left(1 - \frac{1}{(2\pi|\mathbf{r}(t)|\,|\mathbf{n}|)^2}\right)$$

$$\geq \sum_{k=1}^{N} \sum_{\substack{\mathbf{n}\in\mathbb{Z}^3:\\ \mathbf{n}\neq\mathbf{0}}} |a_{\mathbf{n}}|^2 \left(1 - \frac{1}{(2\pi|\mathbf{r}(t)|)^2}\right) \geq \left(1 - \frac{1}{(2\pi|\mathbf{r}(t)|)^2}\right)\text{Eqlbrm}(f). \tag{10.24}$$

Assume that

$$\overline{\mathbf{D}^2}(\mathcal{Y};\mathbf{r};\omega;f;0) = \overline{\mathbf{D}^2}_{2,uneven}(\mathcal{Y};\mathbf{r};f;0) = \sum_{\mathbf{n}\in\mathbb{Z}^3\setminus\mathbf{0}} |a_{\mathbf{n}}|^2 \left| \sum_{k=1}^{N} e^{2\pi i\mathbf{n}\cdot\mathbf{y}_k} \right|^2$$

$$< \text{Eqlbrm}(f) = \sigma_0^2(f)N, \quad \text{and also } \pi|\mathbf{r}(t)| > 1. \tag{10.25}$$

By (10.19), (10.20) and (10.25),

$$\overline{\mathbf{D}^2}_{\Omega_2}(\mathcal{Y};\mathbf{r};f;t) \le \sum_{k=1}^{N} \sum_{\substack{\mathbf{n}\in\mathbb{Z}^3:\\ \mathbf{n}\neq\mathbf{0}}} |a_{\mathbf{n}}|^2 + \frac{\text{Eqlbrm}(f)}{(2\pi|\mathbf{r}(t)|)^2}$$

$$= \text{Eqlbrm}(f)\left(1 + \frac{1}{(2\pi|\mathbf{r}(t)|)^2}\right), \tag{10.26}$$

and

$$\overline{\mathbf{D}^2}_{2,uneven}(t) \le \frac{\overline{\mathbf{D}^2}_{2,uneven}(\mathcal{Y};\mathbf{r};f;0)}{(2\pi|\mathbf{r}(t)|)^2} \le \frac{\text{Eqlbrm}(f)}{(2\pi|\mathbf{r}(t)|)^2}. \tag{10.27}$$

Moreover, by (10.20) and (10.22),

$$\int_{\Omega_2}\left(\overline{\mathbf{D}^2}(\mathcal{Y};\mathbf{r};\omega;f;t)\right)^2 d\omega \le \left(\overline{\mathbf{D}^2}_{\Omega_2}(\mathcal{Y};\mathbf{r};f;t)\right)^2$$

$$+\left(3 + \frac{5}{(2\pi|\mathbf{r}(t)|)^2} + \frac{4}{N(2\pi|\mathbf{r}(t)|)^2}\right)\text{Eqlbrm}^2(f)$$

$$+ 14\text{Eqlbrm}(f)\cdot\overline{\mathbf{D}^2}_{2,uneven}(t). \tag{10.28}$$

By using (10.26)–(10.27) in (10.28), we have

$$\int_{\Omega_2}\left(\overline{\mathbf{D}^2}(\mathcal{Y};\mathbf{r};\omega;f;t)\right)^2 d\omega$$

$$\le \text{Eqlbrm}^2(f)\left(\left(1+\frac{1}{4\pi^2|\mathbf{r}(t)|^2}\right)^2 + \left(3+\frac{5}{(2\pi|\mathbf{r}(t)|)^2}+\frac{4}{N(2\pi|\mathbf{r}(t)|)^2}\right)\right)$$

$$+ 14\text{Eqlbrm}(f)\cdot\overline{\mathbf{D}^2}_{2,uneven}(t)$$

$$\le \text{Eqlbrm}^2(f)\left(\left(1+\frac{1}{4\pi^2|\mathbf{r}(t)|^2}\right)^2 + \left(3+\frac{5}{(2\pi|\mathbf{r}(t)|)^2}+\frac{4}{N(2\pi|\mathbf{r}(t)|)^2}\right)\right)$$

$$+ 14\frac{\text{Eqlbrm}^2(f)}{(2\pi|\mathbf{r}(t)|)^2} \le 5\text{Eqlbrm}^2(f) \tag{10.29}$$

if $|\mathbf{r}(t)| \ge 2$.

On the other hand, if $|\mathbf{r}(t)| \ge 2$ then by (10.23),

$$\int_{\Omega_2}\overline{\mathbf{D}^2}(\mathcal{Y};\mathbf{r};\omega;f;t)\,d\omega \ge \frac{9}{10}\text{Eqlbrm}(f). \tag{10.30}$$

Let

$$\Omega_2^* = \left\{ \omega \in \Omega_2 : \ \overline{\mathbf{D}^2}(\mathcal{Y}; \mathbf{r}; \omega; f; t)\, d\omega \geq \frac{1}{10}\text{Eqlbrm}(f) \right\}. \tag{10.31}$$

Clearly

$$\int_{\Omega_2 \setminus \Omega_2^*} \overline{\mathbf{D}^2}(\mathcal{Y}; \mathbf{r}; \omega; f; t)\, d\omega \leq \frac{1}{10}\text{Eqlbrm}(f). \tag{10.32}$$

Applying the Cauchy–Schwarz inequality and (10.29)–(10.32), we have

$$\begin{aligned}
\left(\left(\frac{9}{10} - \frac{1}{10} \right) \text{Eqlbrm}(f) \right)^2 &\leq \left(\int_{\Omega_2^*} \overline{\mathbf{D}^2}(\mathcal{Y}; \mathbf{r}; \omega; f; t) \cdot 1\, d\omega \right)^2 \\
&\leq \int_{\Omega_2} \left(\overline{\mathbf{D}^2}(\mathcal{Y}, \mathcal{M}; \mathbf{r}; \omega; f; t) \right)^2 d\omega \cdot \text{ProdMeas}_2(\Omega_2^*) \\
&\leq 5\text{Eqlbrm}^2(f) \cdot \text{ProdMeas}_2(\Omega_2^*),
\end{aligned} \tag{10.33}$$

assuming $|\mathbf{r}(t)| \geq 2$. Comparing the two ends of (10.33), we conclude

$$\text{ProdMeas}_2(\Omega_2^*) \geq \frac{(4/5)^2}{5} > \frac{1}{10}.$$

This means, if at the start

$$\overline{\mathbf{D}^2}(\mathcal{Y}; \mathbf{r}; \omega; f; 0) = \overline{\mathbf{D}^2}_{2,uneven}(\mathcal{Y}; \mathbf{r}; f; 0) < \text{Eqlbrm}(f) = \sigma_0^2(f)N,$$

and $|\mathbf{r}(t)| \geq 2$, then for more than 10 percent of the initial velocities $\omega \in \Omega_2$,

$$\overline{\mathbf{D}^2}(\mathcal{Y}; \mathbf{r}; \omega; f; t) \geq \frac{1}{10}\text{Eqlbrm}(f) = \frac{1}{10}\sigma_0^2(f)N. \tag{10.34}$$

(10.34) proves a time-quantitative form of the Quick-Jump-Up phenomenon.

Note that with more work we could substantially improve 10 percent, but it is not really important for the rest, so we stop here.

Chapter 3

Proving Our Time-Quantitative Second Law

11. First step: Relative disparity — a Boltzmann entropy like quantity

What is in Chapter 3? *Here we work out the details of the results mentioned in Section 6, including the Metatheorem. The proofs are based on the results of Sections 3, 4 and 5. In particular, Theorem 5.1 plays a key role.*

In Chapters 3 and 4 we (mostly) assume the "same curve" condition of Theorem 5.1

$$\mathbf{r}(t) = \mathbf{r}_1(t) = \ldots = \mathbf{r}_N(t). \tag{11.1}$$

We return to the general case of "possibly different types of parametrized curves for different particles" in Chapter 5, where we study *approach to equilibrium and long-term stability of equilibrium*; see Theorem 25.1.

First we define the Relative Disparity in the following special case.

All particles have unit mass. We introduce the concept of the *disparity of a finite point set.* It is closely related to (2.28)–(2.31) in the special case of (11.1) and

$$1 = \mu_1 = \ldots = \mu_N. \tag{11.1'}$$

We recall (2.31) in the special case (11.1)–(11.1′): (2.31) defines the the quadratic average of the discrepancy over all translations of the test function $f \in L^2$ in the torus I^3:

$$\overline{\mathbf{D}^2}(\mathcal{Y}; \mathbf{r}; \omega; f; t) = \int_{I^3} \left| N(\mathcal{Y}; \mathbf{r}; \omega; f_{\mathbf{w}}; t) - N \int_{I^3} f\, dV \right|^2 d\mathbf{w}, \tag{11.2}$$

where $f_{\mathbf{w}}$ is defined as $f_{\mathbf{w}}(\mathbf{x}) = f(\mathbf{x} - \mathbf{w})$ (note that f is periodic in $\mathbf{R}^3$), $\mathcal{Y} = (\mathbf{y}_1, \mathbf{y}_2, \ldots, \mathbf{y}_N) \in (I^3)^N$ describes the "initial point configuration"

(= set of starting points), $\mathbf{r}(t)$ is a continuous rectifiable parametrized space curve with $\mathbf{r}(0) = \mathbf{0}$, describing the (common) type of the trajectories of the particles, and finally, $\omega \in \Omega_i$, $i = 1, 2, 3$ determines the initial-velocities of the system. We recall that Ω_i denotes the initial velocity space

$$\omega = (\rho_1, \vartheta_1, \rho_2, \vartheta_2, \ldots, \rho_N, \vartheta_N) \in \Omega_1 = \left([0, \infty) \times SO(3)\right)^N,$$

with

$$\Pr[\rho_j \le u] = \sqrt{\frac{2}{\pi}} \int_0^u y^2 e^{-y^2/2}\, dy$$

for the first kind,

$$\omega = (\vartheta_1, \vartheta_2, \ldots, \vartheta_N) \in \Omega_2 = \left(SO(3)\right)^N$$

(i.e., $1 = \rho_1 = \rho_2 = \ldots = \rho_N$) for the second kind, and

$$\omega = (\rho_1, \vartheta_1, \rho_2, \vartheta_2, \ldots, \rho_N, \vartheta_N) \in \Omega_3 = \left([0, 1] \times SO(3)\right)^N$$

with

$$\Pr[\rho_k \le u] = 3 \int_0^u y^2\, dy = u^3 \quad \text{for } 0 \le y \le 1, \text{ and 0 otherwise}$$

for the third kind. The triple $(\mathcal{Y}; \mathbf{r}; \omega)$ is the complete *initial condition* of the system, it determines the trajectory of the whole system in the high-dimensional unit torus $[0, 1)^{3N}$ (= configuration space)

$$\Gamma(t) = \Gamma_{\mathcal{Y};\mathbf{r};\omega}(t) = (\mathbf{x}_1(t), \mathbf{x}_2(t), \ldots, \mathbf{x}_N(t)) \quad \text{with}$$

$$\mathbf{x}_k(t) = \mathbf{y}_k + \vartheta_k(\rho_k \mathbf{r}(t)) \text{ modulo one for } (1 \le k \le N) \tag{11.3}$$

for all time t. Note that $\Gamma(t) = \Gamma_{\mathcal{Y};\mathbf{r};\omega}(t)$ is a $3N$-dimensional vector, and we use the notation (see (11.3))

$$\mathcal{Y}_{\mathbf{r};\omega;t} = \{\mathbf{x}_k(t) : 1 \le k \le N\}$$

$$= \{\mathbf{y}_k + \vartheta_k(\rho_k \mathbf{r}(t)) \text{ modulo one} : 1 \le k \le N\} \qquad (11.4))$$

for the corresponding N-element point set in the 3-dimensional unit torus $[0, 1)^3$ at time t.

We recall the notation

$$N(\mathcal{Y}; \mathbf{r}; \omega; f; t) = \sum_{1 \le k \le N} f(\mathbf{x}_k(t)). \tag{11.5}$$

Notice that in the special case $f = \chi_S$ (11.5) gives the point-counting function in S at time t:

$$N(\mathcal{Y}; \mathbf{r}; \omega; S; t) = \sum_{\substack{1 \le k \le N: \\ \mathbf{x}_k(t) \in S}} 1.$$

Let $d \geq 1$ be an arbitrary dimension. Now we are ready to define the *disparity* of an arbitrary N-element point set $\mathcal{Z} = \{\mathbf{z}_1, \mathbf{z}_2, \ldots, \mathbf{z}_N\}$ with respect to a (measurable) test set $S \subset [0,1]^d$ in the d-dimensional unit cube. (In the applications we choose $\mathcal{Z} = \mathcal{Y}_{\mathbf{r};\omega;t}$; see (11.4).) Motivated by (11.2) and (11.5) in the special case $f = \chi_S$, we write

$$\text{AvSqDi}(\mathcal{Z}; \chi_S) = \int_{I^d} \left(N(\mathcal{Z}; S + \mathbf{w}) - N\text{vol}(S)\right)^2 \, d\mathbf{w}, \tag{11.6}$$

where $S + \mathbf{w}$ means the translated copy of the set S modulo one ($\mathbf{w} \in [0,1]^d$ is an arbitrary vector),

$$N(\mathcal{Z}; S') = \sum_{\substack{1 \leq k \leq N: \\ \mathbf{z}_k \in S' \cap \mathcal{Z}}} 1$$

denotes the counting number, and AvSqDi stands for Average Square-Discrepancy (since (11.6) is the square integral of the discrepancy over all translates). Note that we already used notation (11.6) in Section 2; see Definition 2.1.

Here comes a key definition: We call

$$\max\left\{\frac{\text{AvSqDi}(\mathcal{Z}; \chi_S)}{\text{vol}(S)(1 - \text{vol}(S))}, N\right\} \tag{11.7}$$

the *relative disparity* of an arbitrary N-element point set $\mathcal{Z} = \{\mathbf{z}_1, \mathbf{z}_2, \ldots, \mathbf{z}_N\}$ with respect to a (measurable) test set $S \subset [0,1]^d$ in the d-dimensional unit cube. We always assume that $0 < \text{vol}(S) < 1$, and the term *relative* refers to the fact that we work with a fixed test set S.

Note that our Micro-Entropy (see the later sections for a detailed discussion) is completely different from the Boltzmann entropy based on macrostates. (In our large system models, where we introduce and use our intrinsic Micro-Entropy, the Bolzmann entropy and the Gibbs entropy are meaningless.)

Nevertheless, the Relative Disparity — an intermediate concept toward our Micro-Entropy — shows some *formal* similarity to the (negative) Boltzmann entropy, which is worth mentioning. We elaborate on this. The simplest interpretation of the Boltzmann entropy in statistical mechanics is the following. Under the assumption that the microstates are equally probable, the entropy is (some absolute constant times) $\log W$, where W is the number of microstates consistent with the given macrostate. This is the famous Boltzmann(-Planck) formula, engraved on Boltzmann's tombstone. (Boltzmann never published it; the formula is actually due to Planck. Of course Planck was motivated by Boltzmann's work. Despite its fame, the formula is highly intuitive, since there is no general agreement on what "macrostate" and "consistent" mean with mathematical rigour.)

In a continuous case — what we have here — we may attempt to replace the number W of microstates by the corresponding "relative number": the corresponding *probability.* (It is perfectly natural, since the letter W in the Boltzmann–Planck formula in fact represents the German word *Wahrscheinlichkeit*, meaning *probability.* Or we may quote Einstein: "On the basis of the kinetic theory of gases Boltzmann had discovered that, aside from a constant factor, entropy is equivalent to the logarithm of the "probability" of the macrostate under consideration.") Taking a *random* N-set $\mathcal{Z}$ with uniform distribution, the probability that

$$|N(\mathcal{Z}; S + \mathbf{w}) - Np| \approx \lambda\sqrt{Npq}$$

with $p = \text{vol}(S)$ and $q = 1 - p$, is about $e^{-\lambda^2/2}$ if N is large ($e^{-\lambda^2/2}$ comes from the large deviation version of the central limit theorem). So we replace $\log W$ with

$$\log(\text{Probability}) = \log\left(e^{-\lambda^2/2}\right)$$

$$= -\frac{\lambda^2}{2} = -\frac{1}{2}\frac{(N(\mathcal{Z}; S + \mathbf{w}) - Np)^2}{Npq}. \tag{11.8}$$

Notice that this intuition becomes meaningless in the range $0 < \lambda < 1$, so it is natural to define a cut-off: we replace (11.8) with

$$-\frac{1}{2}\max\left\{\frac{(N(\mathcal{Z}; S + \mathbf{w}) - Np)^2}{Npq}, 1\right\}$$

$$= -\frac{1}{2N}\max\left\{\frac{(N(\mathcal{Z}; S + \mathbf{w}) - Np)^2}{pq}, N\right\}. \tag{11.9}$$

The next step is to drop $2N$ in the denominator of the first factor of (11.9). Here 2 is just an irrelevant absolute constant, but dropping N is a major change, so it demands some explanation. The reason why we drop the factor N in (11.9) is to satisfy the second basic requirement of entropy in physics: "entropy is additive in thermodynamical equilibrium". "Thermodynamical equilibrium" means complete randomness, so we consider a *random* N-set $\mathcal{Z}$: its variance with respect to the test set S (or any translate $S + \mathbf{w}$) is Npq. So for a *random* N-set $\mathcal{Z}$,

$$\max\left\{\frac{\mathbf{E}\,(N(\mathcal{Z}; S + \mathbf{w}) - Np)^2}{pq}, N\right\} = N$$

(see (11.9), **E** means expectation), and N, the number of particles, represents an additive function. (Two more reasons to justify the cut-off in (11.9) will be given later.)

As we said above, we drop the factor $2N$ in (11.9), and integrate it over all translations $\mathbf{w} \in I^d = [0,1)^d$ of the unit torus; thus we obtain

$$-\int_{I^d} \max\left\{\frac{(N(\mathcal{Z}; S+\mathbf{w}) - Np)^2}{pq}, N\right\} d\mathbf{w}. \tag{11.10}$$

Notice that (11.10) is much less user-friendly than

$$-\text{RelativeDisparity} = -\max\left\{\int_{I^d} \frac{(N(\mathcal{Z}; S+\mathbf{w}) - Np)^2}{pq} d\mathbf{w}, N\right\}, \tag{11.11}$$

see (11.5)–(11.7). Indeed, it is much harder to evaluate (11.10) with the annoying max inside the integral. This is why we "take out the max", and prefer using (11.11) instead of (11.10); (11.11) is the negative relative disparity.

Note that (11.10) and (11.11) are very close to each other anyway: clearly

$$0 \le \text{Integral}(1) - \text{Integral}(2) \le N, \tag{11.12}$$

where

$$\text{Integral}(1) = \int_{I^d} \max\left\{\frac{(N(\mathcal{Z}; S+\mathbf{w}) - Np)^2}{pq}, N\right\} d\mathbf{w}$$

and

$$\text{Integral}(2) = \max\left\{\int_{I^d} \frac{(N(\mathcal{Z}; S+\mathbf{w}) - Np)^2}{pq} d\mathbf{w}, N\right\}.$$

In our arguments below we consider an additive error N the "standard fluctuation". This is explained by Theorem 3.2, which basically says that an error of size N is inevitable *white noise* even in equilibrium. So (11.12) justifies the use of (11.11) instead of (11.10) as the *relative disparity* of the point set $\mathcal{Z}$ with respect to the given test set $S \subset [0,1]^d$.

The fact that N is the white noise in equilibrium is another good reason to use in (11.9) the cut-off at N.

For brevity we rewrite (11.11) simply as

$$\text{RelDisp}(\mathcal{Z}; \chi_S) = \max\left\{\frac{1}{pq}\int_{I^d} (N(\mathcal{Z}; S+\mathbf{w}) - Np)^2 \, d\mathbf{w}, N\right\}, \tag{11.13}$$

where $p = \text{vol}(S)$ and $q = 1 - p$.

The key question here is whether or not this (negative) Boltzmann entropy like concept exhibits a Second Law. Is it true that the typical time evolution of a large point system exhibits decreasing (relative) disparity? The answer is yes. The simplest case is when *typical* means *majority of* $\omega \in \Omega_1$ (Gaussian initial velocity distribution). Then the decreasing property follows from Theorem 3.1 and Theorem 5.1(a) (we will explain it later).

We can easily generalize/upgrade (11.6)–(11.13) from $f = \chi_S$ to an arbitrary (complex valued) square-integrable test function $f \in L^2$ defined in the unit torus $I^d = [0,1)^d$. Write

$$\text{AvSqDi}(\mathcal{Z}; f) = \int_{I^d} \left| N(\mathcal{Z}; f_{\mathbf{w}}) - N \int_{I^d} f \, d\text{V} \right|^2 d\mathbf{w}, \tag{11.14}$$

where $g = f_{\mathbf{w}}$ means the translated function $g(\mathbf{x}) = f(\mathbf{x} - \mathbf{w})$ modulo one, and

$$N(\mathcal{Z}; g) = \sum_{k=1}^{N} g(\mathbf{z}_k).$$

Notice that using the notation (11.14), we can rewrite (11.2) in the form $(d = 3)$

$$\overline{\mathbf{D}^2}(\mathcal{Y}; \mathbf{r}; \omega; f; t) = \text{AvSqDi}(\mathcal{Y}_{\mathbf{r};\omega;t}; f), \tag{11.15}$$

where

$$\mathcal{Y}_{\mathbf{r};\omega;t} = \{\mathbf{y}_k + \vartheta_k(\rho_k \mathbf{r}(t)) \text{ modulo one} : \ 1 \le k \le N\}. \tag{11.16}$$

We emphasize that (11.15)–(11.16) work in every dimension $d \ge 1$.

As usual, let

$$\sigma_0^2(f) = \int_{I^d} \left| f - \int_{I^d} f \, d\text{V} \right|^2 d\text{V} = \int_{I^d} |f|^2 \, d\text{V} - \left| \int_{I^d} f \, d\text{V} \right|^2 \tag{11.17}$$

denote the "variance" of the test function f; we always assume that $\sigma_0^2(f) > 0$.

As a straightforward generalization of (11.7), we call

$$\max \left\{ \frac{\text{AvSqDi}(\mathcal{Z}; f)}{\sigma_0^2(f)}, N \right\} \tag{11.18}$$

the *relative disparity* of an arbitrary N-element point set $\mathcal{Z} = \{\mathbf{z}_1, \mathbf{z}_2, \ldots, \mathbf{z}_N\}$ with respect to a test function $f \in L^2$ in the d-dimensional unit cube.

We have the analog of (11.12):

$$0 \le \text{Integral}(1) - \text{Integral}(2) \le N, \tag{11.19}$$

where

$$\text{Integral}(1) = \int_{I^d} \max\left\{ \frac{\left|N(\mathcal{Z}; f_{\mathbf{w}}) - N\int_{I^d} f\, dV\right|^2}{\sigma_0^2(f)}, N \right\} d\mathbf{w}$$

and

$$\text{Integral}(2) = \max\left\{ \int_{I^d} \frac{\left|N(\mathcal{Z}; f_{\mathbf{w}}) - N\int_{I^d} f\, dV\right|^2}{\sigma_0^2(f)}\, d\mathbf{w}, N \right\}.$$

Similarly, we have the analog of (11.13):

$$\text{RelDisp}(\mathcal{Z}; f) = \max\left\{ \frac{1}{\sigma_0^2(f)} \int_{I^d} \left|N(\mathcal{Z}; f_{\mathbf{w}}) - N\int_{I^d} f\, dV\right|^2 d\mathbf{w}, N \right\}. \tag{11.20}$$

A third reason why it is natural to define this (negative) Boltzmann entropy like concept in the form

$$\max\left\{ \int_{I^d} \frac{\left|N(\mathcal{Z}; f_{\mathbf{w}}) - N\int_{I^d} f\, dV\right|^2}{\sigma_0^2(f)}\, d\mathbf{w}, N \right\}$$

as a cut-off at N, is that for some point distribution $\mathcal{Z}$ the integral

$$\int_{I^d} \left|N(\mathcal{Z}; f_{\mathbf{w}}) - N\int_{I^d} f\, dV\right|^2 d\mathbf{w}$$

can be much smaller than the linear $\sigma_0^2(f)N$: in some "reasonable" cases it can be polylogarithmic in N. However, the typical time evolution of such an "extremely f-uniform" point set $\mathcal{Z}(t)$, $t > 0$ *very rapidly* moves up the integral

$$\int_{I^d} \left|N(\mathcal{Z}(t); f_{\mathbf{w}}) - N\int_{I^d} f\, dV\right|^2 d\mathbf{w}$$

to around N, representing equilibrium. This follows from the Quick-Jump-Up phenomenon, explained at the end of Section 10.

On the other hand, if for some point distribution $\mathcal{Z}$ the integral

$$\int_{I^d} \left|N(\mathcal{Z}; f_{\mathbf{w}}) - N\int_{I^d} f\, dV\right|^2 d\mathbf{w}$$

is much larger than $\sigma_0^2(f)N$, it takes substantially more time for a typical time evolution to move down the integral

$$\int_{I^d} \left|N(\mathcal{Z}(t); f_{\mathbf{w}}) - N\int_{I^d} f\, dV\right|^2 d\mathbf{w}$$

to the range N, representing equilibrium.

Let us focus on the integral

$$\int_{I^d} \left| N(\mathcal{Z}; f_{\mathbf{w}}) - N \int_{I^d} f \, d\mathrm{V} \right|^2 d\mathbf{w}$$

in (11.20). By (2.2) (a subtle form of Parseval's formula)

$$\int_{I^d} \left| N(\mathcal{Z}; f_{\mathbf{w}}) - N \int_{I^d} f \, d\mathrm{V} \right|^2 d\mathbf{w} = \sum_{\substack{\mathbf{n}\in\mathbb{Z}^d: \\ \mathbf{n}\neq\mathbf{0}}} |a_{\mathbf{n}}|^2 \left|\Phi(\mathcal{Z};\mathbf{n})\right|^2, \tag{11.21}$$

where $a_{\mathbf{n}}$, $\mathbf{n} \in \mathbb{Z}^d$ are the Fourier coefficients of the given test function $f \in L^2$, and

$$\Phi(\mathcal{Z};\mathbf{n}) = \sum_{k=1}^{N} e^{2\pi i \mathbf{n}\cdot\mathbf{z}_k}. \tag{11.22}$$

The simplicity of (11.21)–(11.22) is another good reason why we introduced definition (11.18).

By using (11.22) and Parseval's formula (see (7.11)), we can rewrite (11.20) as follows:

$$\mathrm{RelDisp}(\mathcal{Z}; f) = \max\left\{ \frac{\sum_{\mathbf{n}\in\mathbb{Z}^d\setminus\mathbf{0}} |a_{\mathbf{n}}|^2 \left|\Phi(\mathcal{Z};\mathbf{n})\right|^2}{\sum_{\mathbf{m}\in\mathbb{Z}^d\setminus\mathbf{0}} |a_{\mathbf{m}}|^2}, N \right\}$$

$$= \max\left\{ \frac{1}{\sigma_0^2(f)} \sum_{\mathbf{n}\in\mathbb{Z}^d\setminus\mathbf{0}} |a_{\mathbf{n}}|^2 \left|\Phi(\mathcal{Z};\mathbf{n})\right|^2, N \right\}. \tag{11.23}$$

As a trivial corollary of the first line of (11.23), we obtain the bounds

$$N \le \mathrm{RelDisp}(\mathcal{Z}; f) \le N^2 \tag{11.24}$$

for all N-element sets $\mathcal{Z} \subset I^d$ and for all test functions $f \in L^2(I^d)$.

Another trivial corollary of (11.23) is

$$\mathrm{RelDisp}(\mathcal{Z}_{BB}; f) = N^2 \text{ for every } f \in L^2 \text{ with } \sigma_0^2(f) > 0, \tag{11.25}$$

where $\mathcal{Z}_{BB}$ means a Big Bang: one point with multiplicity N. This shows that the upper bound in (11.24) is sharp.

The following fact is less trivial: for every N-element point set $\mathcal{Z} \subset I^d$ and for every $\varepsilon > 0$ there is a test function $f = f_{\mathcal{Z},\varepsilon} \in L^2(I^d)$ such that

$$\mathrm{RelDisp}(\mathcal{Z}; f) > N^2 - \varepsilon. \tag{11.26}$$

What is more, we can choose a "nice" f: it can be analytic with $|f| \le 1$ (see (11.27)).

The proof of (11.26) uses Dirichlet's famous simultaneous diophantine approximation theorem. The details go as follows. Let $\mathbf{m} \in \mathbb{Z}^d$ an arbitrary nonzero lattice point, and consider the N real numbers

$$\alpha_k = \mathbf{m} \cdot \mathbf{z}_k, \quad 1 \le k \le N.$$

By Dirichlet's theorem, there exists an integer r_0 in the interval $1 \le r \le (4\pi N^2/\varepsilon)^N + 1$ such that

$$\|r_0 \alpha_k\| < \frac{\varepsilon}{4\pi N^2} \quad \text{for all } 1 \le k \le N,$$

where $\|x\|$ denotes the distance of a real number x from the nearest integer. It follows that

$$\left| \sum_{k=1}^{N} e^{2\pi i r_0 \mathbf{m} \cdot \mathbf{z}_k} \right| > N - \frac{\varepsilon}{2N}.$$

By choosing the analytic function

$$f_0(\mathbf{x}) = e^{2\pi i r_0 \mathbf{m} \cdot \mathbf{x}}, \tag{11.27}$$

(11.23) yields

$$\text{RelDisp}(\mathcal{Z}; f_0) = |\Phi(\mathcal{Z}; r_0\mathbf{m})|^2 > \left(N - \frac{\varepsilon}{2N}\right)^2 > N^2 - \varepsilon,$$

which proves (11.26).

Next we show that the lower bound N in (11.24) is sharp in the following sense: given any $N \ge 1$ and any test function $f \in L^2$ with $\sigma_0^2(f) > 0$ there is an N-element point set $\mathcal{Z} \subset I^d$ such that

$$\text{RelDisp}(\mathcal{Z}; f) = N. \tag{11.28}$$

We need (11.23):

$$\text{RelDisp}(\mathcal{Z}; f) = \max\left\{ \frac{\sum_{\mathbf{n} \in \mathbb{Z}^d \setminus \mathbf{0}} |a_{\mathbf{n}}|^2 \, |\Phi(\mathcal{Z}; \mathbf{n})|^2}{\sum_{\mathbf{m} \in \mathbb{Z}^d \setminus \mathbf{0}} |a_{\mathbf{m}}|^2}, N \right\}, \tag{11.29}$$

where $a_{\mathbf{n}}$ are the Fourier coefficients of the given test function f, and

$$\Phi(\mathcal{Z}; \mathbf{n}) = \sum_{k=1}^{N} e^{2\pi i \mathbf{n} \cdot \mathbf{z}_k}.$$

Let $\mathbf{z}_k^*$, $1 \le k \le N$ be independent random variables, uniformly distributed in the unit torus I^d; and let $\mathcal{Z}^*$ denote the random N-set consisting of these randomly chosen points. Then for every $\mathbf{n} \in \mathbb{Z}^d \setminus \mathbf{0}$,

$$|\Phi(\mathcal{Z}^*; \mathbf{n})|^2 = \left(\sum_{k=1}^{N} e^{2\pi i \mathbf{n} \cdot \mathbf{z}_k^*} \right) \left(\sum_{r=1}^{N} e^{-2\pi i \mathbf{n} \cdot \mathbf{z}_r^*} \right)$$

$$= N + 2 \sum_{1 \le k < r \le N} e^{2\pi i \mathbf{n} \cdot (\mathbf{z}_k^* - \mathbf{z}_r^*)}. \tag{11.30}$$

Computing the expected value of (11.30), we have

$$\mathbf{E} \left| \Phi(\mathcal{Z}^*; \mathbf{n}) \right|^2 = N. \tag{11.31}$$

Indeed, since $\mathbf{z}_k^* - \mathbf{z}_r^*$ is uniformly distributed in the unit torus I^d, and

$$\int_{I^d} e^{2\pi i \mathbf{n} \cdot \mathbf{x}} \, d\mathbf{x} = 0,$$

we obtain

$$\mathbf{E} e^{2\pi i \mathbf{n} \cdot (\mathbf{z}_k^* - \mathbf{z}_r^*)} = 0$$

for all $1 \le k < r \le N$, and using it in (11.30), (11.31) follows.

It follows from (11.31) that, for any given set of Fourier coefficients $a_{\mathbf{n}}$ satisfying $\sum |a_{\mathbf{n}}|^2 < \infty$, the expectation

$$\mathbf{E} \frac{\sum_{\mathbf{n} \in \mathbb{Z}^d \setminus \mathbf{0}} |a_{\mathbf{n}}|^2 \left| \Phi(\mathcal{Z}^*; \mathbf{n}) \right|^2}{\sum_{\mathbf{m} \in \mathbb{Z}^d \setminus \mathbf{0}} |a_{\mathbf{m}}|^2} = N.$$

Since the expectation is an average, there must exist a "below average" N-element point set $\widetilde{\mathcal{Z}}$:

$$\frac{\sum_{\mathbf{n} \in \mathbb{Z}^d \setminus \mathbf{0}} |a_{\mathbf{n}}|^2 \left| \Phi(\widetilde{\mathcal{Z}}; \mathbf{n}) \right|^2}{\sum_{\mathbf{m} \in \mathbb{Z}^d \setminus \mathbf{0}} |a_{\mathbf{m}}|^2} \le N,$$

so by (11.29), $\text{RelDisp}(\widetilde{\mathcal{Z}}; f) = N$, proving (11.28).

The same argument — using the complex-valued central limit theorem — shows that, for a *typical* (or generic) N-set $\mathcal{Z}$ in the unit cube I^d, $\text{RelDisp}(\mathcal{Z}; f)$ is in the range of constant times N, always above N.

For comparison, note that the extreme N-set $\mathcal{Z}_{BB}$ (= Big Bang) is at the other end of the spectrum (11.24): $\text{RelDisp}(\mathcal{Z}_{BB}; f) = N^2$. Of course, this is not surprising at all. It is perfectly natural that Big Bang is the worst case initial point configuration.

Different particles may have different masses. As a straightforward generalization, we define the *relative disparity* of an arbitrary N-element point set with given mass distribution

$$(\mathcal{Z}, \mathcal{M}) = (\mathbf{z}_1, \mu_1; \mathbf{z}_2, \mu_2; \ldots; \mathbf{z}_N, \mu_N)$$

(where, as usual, the average mass is 1) with respect to an arbitrary but fixed (complex valued) square-integrable test function $f \in L^2$ defined in the unit torus $I^d = [0,1)^d$. Write

$$\text{AvSqDi}(\mathcal{Z},\mathcal{M};f) = \int_{I^d} \left| N(\mathcal{Z},\mathcal{M};f_{\mathbf{w}}) - N\int_{I^d} f\,d\mathrm{V}\right|^2 d\mathbf{w}, \tag{11.32}$$

where $g = f_{\mathbf{w}}$ means the translated function $g(\mathbf{x}) = f(\mathbf{x}-\mathbf{w})$ modulo one, and

$$N(\mathcal{Z},\mathcal{M};g) = \sum_{k=1}^{N} g(\mathbf{z}_k)\mu_k.$$

We have the analog of (11.15):

$$\overline{\mathbf{D}^2}(\mathcal{Y},\mathcal{M};\mathbf{r};\omega;f;t) = \text{AvSqDi}(\mathcal{Y}_{\mathbf{r};\omega;t},\mathcal{M};f), \tag{11.33}$$

where

$$\mathcal{Y}_{\mathbf{r};\omega;t} = \{\mathbf{y}_k + \vartheta_k(\rho_k\mathbf{r}(t)) \text{ modulo one}: \ 1 \le k \le N\}. \tag{11.34}$$

As usual, let

$$\sigma_0^2(f) = \int_{I^d}\left| f - \int_{I^d} f\,d\mathrm{V}\right|^2 d\mathrm{V} = \int_{I^d} |f|^2\,d\mathrm{V} - \left|\int_{I^d} f\,d\mathrm{V}\right|^2 \tag{11.35}$$

denote the "variance" of the test function f; we always assume that $\sigma_0^2(f) > 0$.

As a perfect generalization of (11.18), we call

$$\max\left\{\frac{\text{AvSqDi}(\mathcal{Z},\mathcal{M};f)}{\sigma_0^2(f)}, \sum_{k=1}^{N}\mu_k^2\right\} \tag{11.36}$$

the *relative disparity* of an arbitrary N-element point set with mass distribution

$$(\mathcal{Z},\mathcal{M}) = (\mathbf{z}_1,\mu_1;\mathbf{z}_2,\mu_2;\ldots;\mathbf{z}_N,\mu_N)$$

with respect to a test function $f \in L^2$ in the d-dimensional unit cube.

We have the analog of (11.19):

$$0 \le \text{Integral}(1) - \text{Integral}(2) \le N, \tag{11.37}$$

where

$$\text{Integral}(1) = \int_{I^d} \max\left\{\frac{\left|N(\mathcal{Z},\mathcal{M};f_{\mathbf{w}}) - N\int_{I^d} f\,d\mathrm{V}\right|^2}{\sigma_0^2(f)}, \sum_{k=1}^{N}\mu_k^2\right\} d\mathbf{w}$$

and

$$\text{Integral}(2) = \max\left\{\int_{I^d} \frac{\left|N(\mathcal{Z},\mathcal{M};f_{\mathbf{w}}) - N\int_{I^d} f\,d\mathrm{V}\right|^2}{\sigma_0^2(f)}\,d\mathbf{w}, \sum_{k=1}^{N}\mu_k^2\right\}.$$

Combining (11.32) and (11.36), we obtain the analog of (11.20):

$$\text{RelDisp}(\mathcal{Z},\mathcal{M};f) = \max\left\{\int_{I^d} \frac{\left|N(\mathcal{Z},\mathcal{M};f_{\mathbf{w}}) - N\int_{I^d} f\,d\mathrm{V}\right|^2}{\sigma_0^2(f)}\,d\mathbf{w}, \sum_{k=1}^{N}\mu_k^2\right\}. \tag{11.38}$$

Similarly, we have the straightforward analog of (11.21)–(11.22) (it follows from (7.6′) with $t = 0$, noting that (7.6′) holds in every dimension)

$$\int_{I^d} \left|N(\mathcal{Z},\mathcal{M};f_{\mathbf{w}}) - N\int_{I^d} f\,d\mathrm{V}\right|^2 d\mathbf{w} = \sum_{\mathbf{n}\in\mathbb{Z}^d\setminus\mathbf{0}} |a_{\mathbf{n}}|^2\,|\Phi(\mathcal{Z},\mathcal{M};\mathbf{n})|^2, \tag{11.39}$$

where $a_{\mathbf{n}}$, $\mathbf{n} \in \mathbb{Z}^d$ are the Fourier coefficients of the given test function $f \in L^2$, and

$$\Phi(\mathcal{Z},\mathcal{M};\mathbf{n}) = \sum_{k=1}^{N}\mu_k e^{2\pi i\mathbf{n}\cdot\mathbf{z}_k}. \tag{11.40}$$

And also

$$\begin{aligned}\text{RelDisp}(\mathcal{Z},\mathcal{M};f) &= \max\left\{\frac{\sum_{\mathbf{n}\in\mathbb{Z}^d\setminus\mathbf{0}} |a_{\mathbf{n}}|^2\,|\Phi(\mathcal{Z},\mathcal{M};\mathbf{n})|^2}{\sum_{\mathbf{n}\in\mathbb{Z}^d\setminus\mathbf{0}} |a_{\mathbf{n}}|^2}, \sum_{k=1}^{N}\mu_k^2\right\}\\ &= \max\left\{\frac{1}{\sigma_0^2(f)}\sum_{\mathbf{n}\in\mathbb{Z}^d\setminus\mathbf{0}} |a_{\mathbf{n}}|^2\,|\Phi(\mathcal{Z},\mathcal{M};\mathbf{n})|^2, \sum_{k=1}^{N}\mu_k^2\right\}.\end{aligned} \tag{11.41}$$

The first line in (11.41) implies

$$\sum_{k=1}^{N}\mu_k^2 \le \text{RelDisp}(\mathcal{Z},\mathcal{M};f) \le \left(\sum_{k=1}^{N}\mu_k\right)^2 = N^2 \tag{11.42}$$

for all N-element sets $\mathcal{Z} \subset I^d$, for all mass distributions with $\sum_{k=1}^{N}\mu_k = N$, and for all test functions $f \in L^2(I^d)$.

Similarly, for every N-element point set with given mass distribution $(\mathcal{Z},\mathcal{M})$, and for every $\varepsilon > 0$ there is a test function $f = f_{\mathcal{Z},\mathcal{M},\varepsilon} \in L^2(I^d)$ such that

$$\text{RelDisp}(\mathcal{Z},\mathcal{M};f) > N^2 - \varepsilon, \tag{11.43}$$

and again we can choose an analytic f with $|f| \leq 1$ (see (11.26) and (11.27)). The proof is the same using Dirichlet's approximation theorem.

Similarly, given any $N \geq 1$, any mass distribution with $\sum_{k=1}^{N} \mu_k = N$, and any test function $f \in L^2$ with $\sigma_0^2(f) > 0$ there is an N-element point set $\mathrm{y}_1, \ldots, \mathbf{z}_N$ such that the vector

$$(\mathcal{Z}, \mathcal{M}) = (\mathbf{z}_1, \mu_1; \mathbf{z}_2, \mu_2; \ldots; \mathbf{z}_N, \mu_N)$$

has the property

$$\text{RelDisp}(\mathcal{Z}, \mathcal{M}; f) = \sum_{k=1}^{N} \mu_k^2. \tag{11.44}$$

To prove (11.44), we use (11.41):

$$\text{RelDisp}(\mathcal{Z}, \mathcal{M}; f) = \max\left\{ \frac{\sum_{\mathbf{n}\in\mathbb{Z}^d\setminus\mathbf{0}} |a_{\mathbf{n}}|^2 |\Phi(\mathcal{Z}, \mathcal{M}; \mathbf{n})|^2}{\sum_{\mathbf{m}\in\mathbb{Z}^d\setminus\mathbf{0}} |a_{\mathbf{m}}|^2}, \sum_{k=1}^{N} \mu_k^2 \right\}, \tag{11.45}$$

where $a_{\mathbf{n}}$ are the Fourier coefficients of the given test function f, and

$$\Phi(\mathcal{Z}, \mathcal{M}; \mathbf{n}) = \sum_{k=1}^{N} \mu_k e^{2\pi i \mathbf{n}\cdot\mathbf{z}_k}.$$

Let $\mathbf{z}_k^*$, $1 \leq k \leq N$ be independent random variables, uniformly distributed in the unit torus I^d; and let

$$(\mathcal{Z}^*, \mathcal{M}) = (\mathbf{z}_1^*, \mu_1; \mathbf{z}_2^*, \mu_2; \ldots; \mathbf{z}_N^*, \mu_N)$$

denote the vector obtained by combining the random N-set with the given mass distribution $\mathcal{M}$. Then for every $\mathbf{n} \in \mathbb{Z}^d \setminus \mathbf{0}$,

$$|\Phi(\mathcal{Z}^*, \mathcal{M}; \mathbf{n})|^2 = \left(\sum_{k=1}^{N} \mu_k e^{2\pi i \mathbf{n}\cdot\mathbf{z}_k^*}\right)\left(\sum_{r=1}^{N} \mu_r e^{-2\pi i \mathbf{n}\cdot\mathbf{z}_r^*}\right)$$

$$= \sum_{k=1}^{N} \mu_k^2 + 2 \sum_{1\leq k<r\leq N} \mu_k \mu_r e^{2\pi i \mathbf{n}\cdot(\mathbf{z}_k^* - \mathbf{z}_r^*)}. \tag{11.46}$$

Computing the expected value of (11.46), we have

$$\mathbf{E}\, |\Phi(\mathcal{Z}^*, \mathcal{M}; \mathbf{n})|^2 = \sum_{k=1}^{N} \mu_k^2. \tag{11.47}$$

It follows from (11.47) that, for any given set of Fourier coefficients $a_{\mathbf{n}}$ satisfying $\sum |a_{\mathbf{n}}|^2 < \infty$, the expectation

$$\mathbf{E}\frac{\sum_{\mathbf{n}\in\mathbb{Z}^d\setminus\mathbf{0}} |a_{\mathbf{n}}|^2 \, |\Phi(\mathcal{Z}^*;\mathbf{n})|^2}{\sum_{\mathbf{m}\in\mathbb{Z}^d\setminus\mathbf{0}} |a_{\mathbf{m}}|^2} = \sum_{k=1}^{N} \mu_k^2.$$

Since the expectation is an average, there must exist a "below average" N-element point set $\widetilde{\mathcal{Z}}$ such that if we associate with the k-th point the given mass μ_k, $1 \le k \le N$, we obtain

$$\frac{\sum_{\mathbf{n}\in\mathbb{Z}^d\setminus\mathbf{0}} |a_{\mathbf{n}}|^2 \left|\Phi(\widetilde{\mathcal{Z}},\mathcal{M};\mathbf{n})\right|^2}{\sum_{\mathbf{m}\in\mathbb{Z}^d\setminus\mathbf{0}} |a_{\mathbf{m}}|^2} \le \sum_{k=1}^{N} \mu_k^2,$$

so by (11.45), $\mathrm{RelDisp}(\widetilde{\mathcal{Z}},\mathcal{M};f) = \sum_{k=1}^N \mu_k^2$, proving (11.44).

The same argument — using the complex-valued central limit theorem — shows that, for a *typical* (or generic) N set $\mathcal{Z}$ in the unit cube I^d, having the given mass distribution $\mathcal{M}$ (i.e., the k-th particle has mass μ_k, $1 \le k \le N$), $\mathrm{RelDisp}(\mathcal{Z},\mathcal{M};f)$ is in the range of constant times $\sum_{k=1}^N \mu_k^2$. (Noting that by definition $\mathrm{RelDisp}(\mathcal{Z},\mathcal{M};f)$ is always above $\sum_{k=1}^N \mu_k^2$.)

12. Does the disparity decrease?

Our goal is to show that, for a typical time evolution, a large system becomes less and less "uneven", exhibiting decreasing disparity. We study the *typical* time evolution of the relative disparity (see (11.32), (11.33) and (11.36))

$$\mathrm{RelDisp}(\mathcal{Y}_{\mathbf{r};\omega;t},\mathcal{M};f) = \max\left\{\frac{1}{\sigma_0^2(f)}\overline{\mathbf{D}^2}(\mathcal{Y},\mathcal{M};\mathbf{r};\omega;f;t), \sum_{k=1}^{N}\mu_k^2\right\}, \quad (12.1)$$

where

$$\mathcal{Y}_{\mathbf{r};\omega;t} = \{\mathbf{y}_k + \vartheta_k(\rho_k\mathbf{r}(t)) \text{ modulo one}: \ 1 \le k \le N\}. \quad (12.2)$$

We need an initial velocity space to work with. In the 3-dimensional case we study the torus model of the first kind (Gaussian distribution)

$$\omega = (\rho_1,\vartheta_1,\rho_2,\vartheta_2,\ldots,\rho_N,\vartheta_N) \in \Omega_1 = ([0,\infty)\times SO(3))^N,$$

the second kind

$$\omega = (\vartheta_1,\vartheta_2,\ldots,\vartheta_N) \in \Omega_2 = (SO(3))^N,$$

and the third kind

$$\omega = (\rho_1,\vartheta_1,\rho_2,\vartheta_2,\ldots,\rho_N,\vartheta_N) \in \Omega_3 = ([0,1]\times SO(3))^N.$$

In the general d-dimensional case the first kind (Gaussian distribution) means

$$\omega = (\rho_1, \vartheta_1, \rho_2, \vartheta_2, \ldots, \rho_N, \vartheta_N) \in \Omega_1(\dim = d) = ([0, \infty) \times SO(d))^N,$$

with

$$\Pr[\rho_j \le u] = \frac{1}{2^{\frac{d}{2}-1}(\frac{d}{2}-1)!} \int_0^u y^{d-1} e^{-y^2/2}\, dy$$

for every even $d \ge 2$,

$$\Pr[\rho_j \le u] = \sqrt{\frac{2}{\pi}} \frac{1}{(d-2)!!} \int_0^u y^{d-1} e^{-y^2/2}\, dy$$

for every odd $d \ge 3$, and

$$\Pr[\rho_j \le u] = \sqrt{\frac{2}{\pi}} \int_0^u e^{-y^2/2}\, dy$$

for $d = 1$; the second kind means

$$\omega = (\vartheta_1, \vartheta_2, \ldots, \vartheta_N) \in \Omega_2(\dim = d) = (SO(d))^N$$

(i.e., $1 = \rho_1 = \rho_2 = \ldots = \rho_N$), and finally the third kind means

$$\omega = (\rho_1, \vartheta_1, \rho_2, \vartheta_2, \ldots, \rho_N, \vartheta_N) \in \Omega_3(\dim = d) = ([0, 1] \times SO(d))^N$$

with

$$\Pr[\rho_k \le u] = d \int_0^u y^{d-1}\, dy = u^d \quad \text{for } 0 \le y \le 1, \text{ and } 1 \text{ for } u \ge 1.$$

Note that

$$\frac{1}{\sigma_0^2(f)} \overline{\mathbf{D}^2}(\mathcal{Y}, \mathcal{M}; \mathbf{r}; \omega; f; 0) = \frac{1}{\sigma_0^2(f)} \overline{\mathbf{D}^2}_{\Omega_i}(\mathcal{Y}, \mathcal{M}; \mathbf{r}; f; 0)$$

$$= \frac{1}{\sigma_0^2(f)} \overline{\mathbf{D}^2}_{i,uneven}(\mathcal{Y}, \mathcal{M}; \mathbf{r}; f; 0) = \frac{\sum_{\mathbf{n} \in \mathbb{Z}^d \setminus \mathbf{0}} |a_{\mathbf{n}}|^2 \left| \sum_{k=1}^N \mu_k e^{2\pi i \mathbf{n} \cdot \mathbf{y}_k} \right|^2}{\sum_{\mathbf{n} \in \mathbb{Z}^d \setminus \mathbf{0}} |a_{\mathbf{n}}|^2} \tag{12.3}$$

for all $i = 1, 2, 3$. (12.3) follows from (7.19) and (8.1), which hold in every dimension.

We assume that the system starts from off-equilibrium. By using $\mathcal{Y} = \{\mathbf{y}_1, \ldots, \mathbf{y}_N\}$ for the initial point configuration, off-equilibrium means that (12.3) is substantially larger than

$$\frac{1}{\sigma_0^2(f)} \text{Eqlbrm}(\mathcal{M}; f) = \sum_{k=1}^N \mu_k^2. \tag{12.4}$$

We study first the case of Gaussian initial velocity distribution, i.e., the case of a typical $\omega \in \Omega_1(\dim = d)$, where $d \geq 1$ is arbitrary. Consider a time $t > 0$ such that

$$\frac{1}{\sigma_0^2(f)}\overline{\mathbf{D}^2}_{1,uneven}(t) = \frac{1}{\sigma_0^2(f)}\overline{\mathbf{D}^2}_{1,uneven}(\mathcal{Y},\mathcal{M};\mathbf{r};f;t) \tag{12.5}$$

is still substantially larger than

$$\frac{1}{\sigma_0^2(f)}\mathrm{Eqlbrm}(\mathcal{M};f) = \sum_{k=1}^{N}\mu_k^2. \tag{12.6}$$

Then by (5.19)–(5.23) and (12.1)–(12.2), for every dimension $d \geq 1$, for the overwhelming majority of the initial velocities $\omega \in \Omega_1(\dim = d)$,

$$\begin{aligned}\mathrm{RelDisp}(\mathcal{Y}_{\mathbf{r};\omega;t},\mathcal{M};f) &= \frac{1}{\sigma_0^2(f)}\overline{\mathbf{D}^2}(\mathcal{Y},\mathcal{M};\mathbf{r};\omega;f;t)\\ &= \frac{1+o(1)}{\sigma_0^2(f)}\overline{\mathbf{D}^2}_{1,uneven}(\mathcal{Y},\mathcal{M};\mathbf{r};f;t)\\ &= \frac{1}{\sigma_0^2(f)}\overline{\mathbf{D}^2}_{1,uneven}(t) + \text{negligible},\end{aligned} \tag{12.7}$$

where

$$\frac{1}{\sigma_0^2(f)}\overline{\mathbf{D}^2}_{1,uneven}(t) = \frac{1}{\sigma_0^2(f)}\overline{\mathbf{D}^2}_{1,uneven}(\mathcal{Y},\mathcal{M};\mathbf{r};f;t)$$

$$= \frac{\sum_{\mathbf{n}\in\mathbb{Z}^d\setminus\mathbf{0}}|a_{\mathbf{n}}|^2\left|\sum_{k=1}^{N}\mu_k e^{2\pi i\mathbf{n}\cdot\mathbf{y}_k}\right|^2\exp\left(-(2\pi|\mathbf{r}(t)|\,|\mathbf{n}|)^2\right)}{\sum_{\mathbf{n}\in\mathbb{Z}^d\setminus\mathbf{0}}|a_{\mathbf{n}}|^2}. \tag{12.8}$$

Notice that the dominating part $\overline{\mathbf{D}^2}_{1,uneven}(t)/\sigma_0^2(f)$ in (12.8) is a *monotone decreasing function of the drifting distance* $|\mathbf{r}(t)|$. Therefore, by (12.7)–(12.8), if the drifting distance is monotone increasing, then, for the overwhelming majority of the initial velocities $\omega \in \Omega_1(\dim = d)$, the time evolution of the relative disparity $\mathrm{RelDisp}(\mathcal{Y}_{\mathbf{r};\omega;t},\mathcal{M};f)$ is monotone decreasing — apart from a negligible error — until it gets close to (12.6), which means the system reaches equilibrium; more precisely, it reaches LA-equilibrium. The same holds in every dimension $d \geq 1$.

Note that the factor $1 + o(1)$ in (12.7) can be made more explicit. By Theorem 5.1(a), we have

$$1 + o(1) = 1 + \left(\frac{\mathrm{Eqlbrm}}{\overline{\mathbf{D}^2}_{1,uneven}(t)}\right)^{1/2} \tag{12.9}$$

where

$$\text{Eqlbrm} = \text{Eqlbrm}(\mathcal{M}; f) = \sigma_0^2(f) \sum_{k=1}^{N} \mu_k^2$$

and

$$\overline{\mathbf{D}^2}_{1,uneven}(t) = \overline{\mathbf{D}^2}_{1,uneven}(\mathcal{Y}, \mathcal{M}; \mathbf{r}; f; t).$$

In view of (12.9), the larger the ratio

$$\frac{\overline{\mathbf{D}^2}_{1,uneven}(t)}{\text{Eqlbrm}}, \tag{12.10}$$

the better the approximation. The ratio (12.10) is large iff the system at time t is far from LA-equilibrium. Therefore, as long as the system is far from LA-equilibrium, the approximation is very sharp.

On the other hand, when the system reaches the state of LA-equilibrium, the approximation becomes completely false. Indeed, then the term $\overline{\mathbf{D}^2}_{1,uneven}(t)$ becomes negligible in the precise sense

$$\overline{\mathbf{D}^2}_{1,uneven}(t) = o\left(\text{Eqlbrm}(\mathcal{M}; f)\right),$$

and

$$\text{the time evolution of } \text{RelDisp}(\mathcal{Y}_{\mathbf{r};\omega;t}, \mathcal{M}; f) + \sum_{k=1}^{N} \mu_k^2$$

$$\text{exhibits fluctuations of size const} \cdot \sum_{k=1}^{N} \mu_k^2 \tag{12.11}$$

for the majority of the initial velocities $\omega \in \Omega_1$. Of course const in (12.11) means *positive* constants (due to the cut-off).

To justify (12.11) note that, (12.11) follows from Theorem 3.2 applied to (12.1); see also the arguments in (3.27)–(3.29).

So far we were studying relative disparity $\text{RelDisp}(\mathcal{Z}, \mathcal{M}; f)$ of a given N-element point set $\mathcal{Z}$ with given weights $\sum_{k=1}^{N} \mu_k = N$ *relative* to a test function $f \in L^2$. The results above make it possible to define an "absolute disparity function" — we actually call it the *unevenness function.* This represents the worst case scenario: we want to take the *maximum* — or rather the *supremum,* to be precise — of the relative disparity over all square-integrable test functions $f \in L^2$ with $\sigma_0^2(f) > 0$.

The large space $f \in L^2$ means that we attempt to verify the "decrease of unevenness" in every level: both the macroscopic and microscopic level.

For example, the left half of the unit cube is a nice test set, it detects the macroscopic unevenness where (say) all particles are in the left half of the cube. Complicated test sets, on the other hand, detect microscopic unevenness. So, taking the maximum of the relative disparity over the whole space $f \in L^2$ represents a search for the maximum "unevenness" in the system on both the microscopic and macroscopic level.

We emphasize that the straightforward way to take the maximum of the relative disparity over the whole space $f \in L^2$ does not work:

$$\sup_{f\in L^2:\ \sigma_0^2(f)>0} \mathrm{RelDisp}(\mathcal{Z},\mathcal{M};f) = N^2 \tag{12.12}$$

for all N-element point sets $\mathcal{Z}$ with given weights $\sum_{k=1}^{N}\mu_k = N$ (see (11.26) and (11.43)). (12.12) shows that we *cannot* define a meaningful absolute disparity *number* for a given pair $(\mathcal{Z},\mathcal{M})$.

There is, however, a subtle way to define a meaningful absolute disparity function of the time $t > 0$ — called unevenness function — for a given triple $(\mathcal{Y},\mathcal{M};\mathbf{r})$ and a given initial velocity space — the latter happens to be Ω_1 here. Since the straightforward approach

$$\sup_{f\in L^2:\ \sigma_0^2(f)>0} \mathrm{RelDisp}(\mathcal{Y}_{\mathbf{r};\omega;t},\mathcal{M};f)$$

gives useless information (by (12.12) it always gives the same constant value N^2), motivated by (12.7)–(12.8), we replace $\mathrm{RelDisp}(\mathcal{Y}_{\mathbf{r};\omega;t},\mathcal{M};f)$ with its "typical value"

$$\overline{\mathbf{D}^2}_{1,uneven}(\mathcal{Y},\mathcal{M};\mathbf{r};f;t).$$

That is, we take the maximum (supremum, to be precise) of

$$\frac{1}{\sigma_0^2(f)}\overline{\mathbf{D}^2}_{1,uneven}(\mathcal{Y},\mathcal{M};\mathbf{r};f;t)$$

over all square-integrable test functions $f \in L^2$ with $\sigma_0^2(f) > 0$ (we already gave a preview of this idea at the beginning of Section 2; see Theorem A there). We also take maximum over all time $\tau \geq t$ (it comes from the definition of LA-equilibrium in Definition 2.4, and also it guarantees a monotone decreasing function even if $|\mathbf{r}(t)|$ is not monotone increasing). The result is *almost* the Unevenness Function of the first kind — a function of the time t. By (12.8) it has the form

$$\mathrm{AlmostCorrectUnevennessFunction}_1(\dim = d;\mathcal{Y},\mathcal{M};\mathbf{r};t)$$

$$= \sup_{\tau\geq t}\ \sup_{f\in L^2:\ \sigma_0^2(f)>0} \frac{1}{\sigma_0^2(f)}\overline{\mathbf{D}^2}_{1,uneven}(\mathcal{Y},\mathcal{M};\mathbf{r};f;t)$$

$$= \sup_{\tau \geq t} \sup_{f \in L^2:\ \sigma_0^2(f)>0} \frac{\sum_{\mathbf{n}\in\mathbb{Z}^d\setminus\mathbf{0}} |a_{\mathbf{n}}|^2 \left|\sum_{k=1}^N \mu_k e^{2\pi i\mathbf{n}\cdot\mathbf{y}_k}\right|^2 \exp\left(-(2\pi|\mathbf{r}(\tau)|\ |\mathbf{n}|)^2\right)}{\sum_{\mathbf{n}\in\mathbb{Z}^d\setminus\mathbf{0}} |a_{\mathbf{n}}|^2}$$

$$= \sup_{\tau\geq t} \sup_{\substack{(a_{\mathbf{m}}:\mathbf{m}\in\mathbb{Z}^d\setminus\mathbf{0}):\\ \sum_{\mathbf{m}\in\mathbb{Z}^d\setminus\mathbf{0}}|a_{\mathbf{m}}|^2>0}} \sum_{\mathbf{n}\in\mathbb{Z}^d\setminus\mathbf{0}} \frac{|a_{\mathbf{n}}|^2}{\sum_{\mathbf{m}\in\mathbb{Z}^d\setminus\mathbf{0}} |a_{\mathbf{m}}|^2}$$

$$\cdot \left|\sum_{k=1}^N \mu_k e^{2\pi i\mathbf{n}\cdot\mathbf{y}_k}\right|^2 \exp\left(-(2\pi|\mathbf{r}(\tau)|\ |\mathbf{n}|)^2\right). \tag{12.13}$$

Clearly

$$\sum_{\mathbf{n}\in\mathbb{Z}^d\setminus\mathbf{0}} \frac{|a_{\mathbf{n}}|^2}{\sum_{\mathbf{m}\in\mathbb{Z}^d\setminus\mathbf{0}} |a_{\mathbf{m}}|^2} = 1,$$

which implies that

$$\sup_{\substack{(a_{\mathbf{m}}:\mathbf{m}\in\mathbb{Z}^d\setminus\mathbf{0}):\\ \sum_{\mathbf{m}\in\mathbb{Z}^d\setminus\mathbf{0}}|a_{\mathbf{m}}|^2>0}} \sum_{\mathbf{n}\in\mathbb{Z}^d\setminus\mathbf{0}} \frac{|a_{\mathbf{n}}|^2}{\sum_{\mathbf{m}\in\mathbb{Z}^d\setminus\mathbf{0}} |a_{\mathbf{m}}|^2} \cdot \left|\sum_{k=1}^N \mu_k e^{2\pi i\mathbf{n}\cdot\mathbf{y}_k}\right|^2$$

$$\exp\left(-(2\pi|\mathbf{r}(\tau)|\ |\mathbf{n}|)^2\right) = \max_{\mathbf{n}\in\mathbb{Z}^d\setminus\mathbf{0}} \left|\sum_{k=1}^N \mu_k e^{2\pi i\mathbf{n}\cdot\mathbf{y}_k}\right|^2 \exp\left(-(2\pi|\mathbf{r}(\tau)|\ |\mathbf{n}|)^2\right). \tag{12.14}$$

Using (12.14) in (12.13), we obtain

$$\text{AlmostCorrectUnevennessFunction}_1(\dim = d; \mathcal{Y}, \mathcal{M}; \mathbf{r}; t)$$

$$= \sup_{\tau\geq t} \max_{\mathbf{n}\in\mathbb{Z}^d\setminus\mathbf{0}} \left|\sum_{k=1}^N \mu_k e^{2\pi i\mathbf{n}\cdot\mathbf{y}_k}\right|^2 \exp\left(-(2\pi|\mathbf{r}(\tau)|\ |\mathbf{n}|)^2\right)$$

$$= \max_{\mathbf{n}\in\mathbb{Z}^d\setminus\mathbf{0}} \left|\sum_{k=1}^N \mu_k e^{2\pi i\mathbf{n}\cdot\mathbf{y}_k}\right|^2 \cdot \sup_{\tau\geq t} \exp\left(-(2\pi|\mathbf{r}(\tau)|\ |\mathbf{n}|)^2\right), \tag{12.15}$$

where $a_{\mathbf{n}}$ are the Fourier coefficients of the test function $f \in L^2$. Notice that in the last step the the Fourier coefficients $a_{\mathbf{n}}$ actually disappear; nevertheless, the underlying idea of (12.15) is still Fourier analysis — namely, the Parseval formula.

If

$$\frac{1}{\sigma_0^2(f)}\overline{\mathbf{D}^2}_{1,uneven}(t) = \frac{1}{\sigma_0^2(f)}\overline{\mathbf{D}^2}_{1,uneven}(\mathcal{Y},\mathcal{M};\mathbf{r};f;t)$$

is *not* substantially larger than

$$\frac{1}{\sigma_0^2(f)}\text{Eqlbrm}(\mathcal{M};f) = \sum_{k=1}^{N}\mu_k^2$$

(see (12.5)–(12.6)) then the system is already in equilibrium, and the unevenness function should stay in the range of $\sum_{k=1}^{N}\mu_k^2$. So to define the **Unevenness Function of the first kind**, we slightly modify (12.15) as follows:

$$\text{UnevenFun}_1(\dim = d;t) = \text{UnevenFun}_1(\dim = d;\mathcal{Y},\mathcal{M};\mathbf{r};t)$$

$$= \text{UnevennessFunction}_1(\dim = d;\mathcal{Y},\mathcal{M};\mathbf{r};t)$$

$$= \max\left\{\max_{\mathbf{n}\in\mathbb{Z}^d\setminus\mathbf{0}}\left|\sum_{k=1}^{N}\mu_k e^{2\pi i\mathbf{n}\cdot\mathbf{y}_k}\right|^2\cdot\sup_{\tau\ge t}\exp\left(-(2\pi|\mathbf{r}(\tau)|\,|\mathbf{n}|)^2\right), \sum_{k=1}^{N}\mu_k^2\right\}. \tag{12.16}$$

It immediately follows from the explicit formula (12.16) that the Unevenness Function $\text{UnevenFun}_1(\dim = d;t)$ of the triple $(\mathcal{Y},\mathcal{M};\mathbf{r})$ is a monote decreasing function of the time t.

Starting from an N-element torus model of the first kind $(\mathcal{Y},\mathcal{M};\mathbf{r})$, the Unevenness Function $\text{UnevenFun}_1(\dim = d;t) = \text{UnevenFun}_1(\dim = d;\mathcal{Y},\mathcal{M};\mathbf{r};t)$ in (12.16) describes the time evolution of the "worst case renormed Big Average Square-Discrepancy" (at least for typical initial velocities $\omega \in \Omega_1$). At the start $t = 0$ we have

$$\text{UnevenFun}_1(\dim = d;0) = \text{UnevenFun}_1(\dim = d;\mathcal{Y},\mathcal{M};\mathbf{r};0) = N^2.$$

This is just a rephrasing of the dead-end approach (12.12); both follow from (11.26) and (11.43) (application of Dirichlet's approximation theorem).

Given an N-element torus model $(\mathcal{Y},\mathcal{M};\mathbf{r})$, for a typical initial velocity $\omega \in \Omega_1$ the time evolution of the system reaches LA-equilibrium exactly when the Unevenness Function $\text{UnevenFun}_1(\dim = d;t)$ in (12.13) goes down from the quadratic order N^2 to $\sum_{k=1}^{N}\mu_k^2$ (which becomes the linear order N if every mass μ_k is one).

Next we switch to the 3-dimensional torus models of the second and third kinds. We start with the second kind. The argument above about the typical behaviour of the unevenness for $\omega \in \Omega_1$ was based on (5.2) in Theorem 5.1(a) (combined with Chebyshev's inequality). Next we switch to $\omega \in \Omega_2$, replace (5.2) with (5.4) in Theorem 5.1(b), and replace

$$\exp\left(-(2\pi|\mathbf{r}(t)|\ |\mathbf{n}|)^2\right)$$

with

$$\left(\frac{\sin\left(2\pi|\mathbf{r}(t)|\ |\mathbf{n}|\right)}{2\pi|\mathbf{r}(t)|\ |\mathbf{n}|}\right)^2.$$

Thus we obtain the **Unevenness Function of the second kind**, a perfect analog of (12.16):

$$\text{UnevenFun}_2(\text{dim} = 3; t) = \text{UnevenFun}_2(\text{dim} = 3; \mathcal{Y}, \mathcal{M}; \mathbf{r}; t)$$

$$= \text{UnevennessFunction}_2(\text{dim} = 3; \mathcal{Y}, \mathcal{M}; \mathbf{r}; t)$$

$$= \max\left\{ \max_{\mathbf{n}\in\mathbb{Z}^3\setminus\mathbf{0}} \left|\sum_{k=1}^{N} \mu_k e^{2\pi i\mathbf{n}\cdot\mathbf{y}_k}\right|^2 \cdot \sup_{\tau\geq t}\left(\frac{\sin\left(2\pi|\mathbf{r}(\tau)|\ |\mathbf{n}|\right)}{2\pi|\mathbf{r}(\tau)|\ |\mathbf{n}|}\right)^2, \sum_{k=1}^{N}\mu_k^2 \right\}. \tag{12.17}$$

For the third kind we switch to $\omega \in \Omega_3$, and replace

$$\exp\left(-(2\pi|\mathbf{r}(t)|\ |\mathbf{n}|)^2\right)$$

in (5.2), Theorem 5.1(a) with

$$9\left(\frac{\cos(2\pi|\mathbf{r}(t)|\ |\mathbf{n}|)}{(2\pi|\mathbf{r}(t)|\ |\mathbf{n}|)^2} - \frac{\sin(2\pi|\mathbf{r}(t)|\ |\mathbf{n}|)}{(2\pi|\mathbf{r}(t)|\ |\mathbf{n}|)^3}\right)^2$$

in (5.4), Theorem 5.1(b). Thus we obtain the **Unevenness Function of the third kind**:

$$\text{UnevenFun}_3(\text{dim} = 3; t) = \text{UnevenFun}_3(\text{dim} = 3; \mathcal{Y}, \mathcal{M}; \mathbf{r}; t)$$

$$= \text{UnevennessFunction}_3(\text{dim} = 3; \mathcal{Y}, \mathcal{M}; \mathbf{r}; t)$$

$$= \max\left\{ \max_{\mathbf{n}\in\mathbb{Z}^3\setminus\mathbf{0}} \left|\sum_{k=1}^{N} \mu_k e^{2\pi i\mathbf{n}\cdot\mathbf{y}_k}\right|^2 \right.$$

$$\left. \cdot \sup_{\tau\geq t} 9\left(\frac{\cos(2\pi|\mathbf{r}(\tau)|\ |\mathbf{n}|)}{(2\pi|\mathbf{r}(\tau)|\ |\mathbf{n}|)^2} - \frac{\sin(2\pi|\mathbf{r}(\tau)|\ |\mathbf{n}|)}{(2\pi|\mathbf{r}(\tau)|\ |\mathbf{n}|)^3}\right)^2, \sum_{k=1}^{N}\mu_k^2 \right\}. \tag{12.18}$$

Using the elementary fact

$$g(x) = \sup_{y \geq x} \left(\frac{\sin y}{y} \right)^2 = \frac{1}{x^2} + O(x^{-3}) \quad \text{for } x > 1,$$

we can rewrite (12.15) in the following simpler form

$$\text{UnevenFun}_2(\dim = 3; \mathcal{Y}, \mathcal{M}; \mathbf{r}; t)$$

$$= \max \left\{ \max_{\mathbf{n} \in \mathbb{Z}^3 \setminus \mathbf{0}} \left| \sum_{k=1}^{N} \mu_k e^{2\pi i \mathbf{n} \cdot \mathbf{y}_k} \right|^2 \cdot \min \left\{ \sup_{\tau \geq t} \frac{1}{(2\pi |\mathbf{r}(\tau)| \, |\mathbf{n}|)^2}, 1 \right\}, \sum_{k=1}^{N} \mu_k^2 \right\}$$

$$+ \text{negligible}. \tag{12.19}$$

Similarly, by using the elementary fact

$$h(x) = \sup_{y \geq x} \left(\frac{\cos y}{y^2} - \frac{\sin y}{y^3} \right)^2 = \frac{1}{x^4} + O(x^{-5}) \quad \text{for } x > 1,$$

we can rewrite (12.16) in the following simpler form

$$\text{UnevenFun}_3(\dim = 3; \mathcal{Y}, \mathcal{M}; \mathbf{r}; t)$$

$$= \max \left\{ \max_{\mathbf{n} \in \mathbb{Z}^3 \setminus \mathbf{0}} \left| \sum_{k=1}^{N} \mu_k e^{2\pi i \mathbf{n} \cdot \mathbf{y}_k} \right|^2 \cdot \min \left\{ \sup_{\tau \geq t} \frac{9}{(2\pi |\mathbf{r}(\tau)| \, |\mathbf{n}|)^4}, 1 \right\}, \sum_{k=1}^{N} \mu_k^2 \right\}$$

$$+ \text{negligible}. \tag{12.20}$$

Next we switch from the (most important) special case dimension $d = 3$ to the general case of dimension $d \geq 2$ with $d \neq 3$. We start with the second kind. In view of (5.27), the change from the 3-dimensional torus model of the second kind to the d-dimensional torus model of the second kind means that for $z \geq 1$ the function $\frac{\sin z}{z}$ is replaced by

$$(d-2)!! \frac{\cos\left(z - \frac{(d-1)\pi}{4}\right)}{z^{(d-1)/2}} \cdot \left\{ \sqrt{\frac{2}{\pi}} \text{ or } 1 \right\} + O\left(\min\left\{z^{-(d+1)/2}, 1\right\}\right)$$

depending on whether $d \geq 2$ is even or odd, accordingly. Thus we obtain the general d-**dimensional Unevenness Function of the second kind** at time t:

$$\text{UnevenFun}_2(\dim = d; \mathcal{Y}, \mathcal{M}; \mathbf{r}; t)$$

$$= \max \left\{ \max_{\mathbf{n} \in \mathbb{Z}^d \setminus \mathbf{0}} \left| \sum_{k=1}^{N} \mu_k e^{2\pi i \mathbf{n} \cdot \mathbf{y}_k} \right|^2 \right.$$

$$\cdot \min\left\{\sup_{\tau\geq t}\frac{((d-2)!!)^2}{(2\pi|\mathbf{r}(\tau)|\ |\mathbf{n}|)^{d-1}}\left\{\sqrt{\frac{2}{\pi}}\text{ or }1\right\},1\right\},\sum_{k=1}^{N}\mu_k^2\right\} \tag{12.21}$$

depending on whether $d \geq 2$ is even or odd, accordingly.

Next consider the third kind in dimension $d \geq 2$ with $d \neq 3$. In view of (5.30)–(5.31) and Lemma 5.2, the change from the 3-dimensional model of the third kind to the d-dimensional model of the third kind means that for $z \geq 1$

$$-3\left(\frac{\cos z}{z^2}-\frac{\sin z}{z^3}\right)$$

is replaced by

$$d!!\frac{\sin\left(z-\frac{(d-1)\pi}{4}\right)}{z^{(d+1)/2}}\cdot\left\{\sqrt{\frac{2}{\pi}}\text{ or }1\right\}+O(z^{-(d+2)/2}),$$

depending on whether $d \geq 2$ is even or odd, accordingly. Thus we obtain the general d-**dimensional Unevenness Function of the third kind** at time t:

$$\text{UnevenFun}_3(\dim = d;\mathcal{Y},\mathcal{M};\mathbf{r};t)$$

$$= \max\left\{\max_{\mathbf{n}\in\mathbb{Z}^d\setminus\mathbf{0}}\left|\sum_{k=1}^{N}\mu_k e^{2\pi i\mathbf{n}\cdot\mathbf{y}_k}\right|^2\right.$$

$$\max\left\{\sup_{\tau\geq t}\frac{(d!!)^2}{(2\pi|\mathbf{r}(\tau)|\ |\mathbf{n}|)^{d+1}}\left\{\sqrt{\frac{2}{\pi}}\text{ or }1\right\},1\right\},\sum_{k=1}^{N}\mu_k^2\right\} \tag{12.22}$$

depending on whether $d \geq 2$ is even or odd, and finally, in both (12.21) and (12.22) we use the well known notation $k!! = k(k-2)(k-4)\cdots$, where the last factor is 1 or 2 depending on the parity of k.

Notice that all of these unevenness functions are monotone decreasing functions of the time t; we have monotonicity in every dimension. As usual, at the start $t = 0$ all unevenness functions are equal to N^2 for every dimension $d \geq 1$. Moreover, for a typical initial velocity $\omega \in \Omega_i(\dim = d)$, $i = 2, 3$ the time evolution of the system $(\mathcal{Y},\mathcal{M};\mathbf{r};\omega)$ reaches LA-equilibrium exactly when the Unevenness Function $\text{UnevenFun}_i(\dim = d;\mathcal{Y},\mathcal{M};\mathbf{r};t)$ goes down from N^2 to the range of $\sum_{k=1}^{N}\mu_k^2$ (which is N if every mass μ_k is one).

13. Proving a Second Law (I): First kind and second kind

In Section 6 we formulated two Second Laws without proof. Here we begin with the proof of the first kind (Gaussian distribution) in the standard 3-dimensional case. We recall the notation introduced in (12.2)

$$\mathcal{Y}_{\mathbf{r};\omega;t} = \{\mathbf{z}_k = \mathbf{y}_k + \vartheta_k(\rho_k \mathbf{r}(t)) \text{ modulo one}: \ 1 \le k \le N\}. \tag{13.1}$$

Here

$$\omega = (\rho_1, \vartheta_1, \rho_2, \vartheta_2, \ldots, \rho_N, \vartheta_N) \in \Omega_1 = ([0,\infty) \times SO(3))^N,$$

and Ω_1 is equipped with the product measure ProdMeas_1, where on $[0,\infty)$ we have the probability measure

$$\Pr[\rho_j \le u] = \sqrt{\frac{2}{\pi}} \int_0^u y^2 e^{-y^2/2}\, dy. \tag{13.2}$$

By (6.17)

$$\text{MicroEntropy}_1(\mathcal{Y}_{\mathbf{r};\omega;t}, \mathcal{M}) = \frac{N^2}{\sum_{k=1}^N \mu_k^2} - \text{Imbalance}_1(\mathcal{Y}_{\mathbf{r};\omega;t}, \mathcal{M}),$$

so it suffices to study the Imbalance defined in (6.10)

$$\text{Imba}_1(\mathcal{Y}_{\mathbf{r};\omega;t}, \mathcal{M}) = \text{Imbalance}_1(\mathcal{Y}_{\mathbf{r};\omega;t}, \mathcal{M})$$

$$= \max_{\mathbf{n}\in\mathbb{Z}^3\setminus\mathbf{0}} \left(\frac{\left|\sum_{\mathbf{z}_k\in\mathcal{Y}_{\mathbf{r};\omega;t}} \mu_k e^{2\pi i \mathbf{n}\cdot\mathbf{z}_k}\right|^2}{\sum_{k=1}^N \mu_k^2} \text{ or } 1 \right)_{\max}^{1/|\mathbf{n}|^2}, \tag{13.3}$$

where, similarly to Section 6, the unconventional notation "$(x \text{ or } 1)_{\max}$" means the larger number between x and 1. Actually we study the following question: given a fixed time $t > 0$, what can we say about $\text{Imba}_1(\mathcal{Y}_{\mathbf{r};\omega;t}, \mathcal{M})$ for the majority of $\omega \in \Omega_1$? In particular, we investigate the relation between $\text{Imba}_1(\mathcal{Y}_{\mathbf{r};\omega;t}, \mathcal{M})$ for the majority of $\omega \in \Omega_1$ and the following ω-independent function (t is the time variable)

$$\text{TiEvFun}_1(t) = \text{TiEvFun}_1(\mathcal{Y}, \mathcal{M}; \mathbf{r}; t)$$

$$= \max_{\mathbf{n}\in\mathbb{Z}^3\setminus\mathbf{0}} \left(\frac{\left|\sum_{k=1}^N \mu_k e^{2\pi i \mathbf{n}\cdot\mathbf{y}_k}\right|^2 \exp\left(-(2\pi|\mathbf{r}(t)|\,|\mathbf{n}|)^2\right)}{\sum_{k=1}^N \mu_k^2} \text{ or } 1 \right)_{\max}^{1/|\mathbf{n}|^2}$$

$$= \left(e^{-(2\pi|\mathbf{r}(t)|)^2} \max_{\mathbf{n}\in\mathbb{Z}^3\setminus\mathbf{0}} \left(\frac{\left|\sum_{k=1}^{N} \mu_k e^{2\pi i \mathbf{n}\cdot\mathbf{y}_k}\right|^2}{\sum_{k=1}^{N} \mu_k^2} \right)^{1/|\mathbf{n}|^2} \text{ or } 1 \right)_{\max} . \quad (13.4)$$

We call (13.4) the *Time Evolution Function of the first kind.*

Theorem 13.1. *Consider a 3-dimensional torus model* $(\mathcal{Y}, \mathcal{M}; \mathbf{r}; \Omega_1)$ *of the first kind. Given* $0 < \varepsilon < 10^{-2}$ *and a time* $t > 0$ *with*

$$\text{TiEvFun}_1(\mathcal{Y}, \mathcal{M}; \mathbf{r}; t) \geq 4/\varepsilon,$$

there exists a measurable subset $\Omega_1^* = \Omega_1^*(\varepsilon; t) \subset \Omega_1$ *of the initial velocity space of the first kind such that*

$$\text{ProdMeas}_1(\Omega_1^*) > 1 - 10^2\varepsilon$$

and

$$\left| \text{MicroEntropy}_1(\mathcal{Y}_{\mathbf{r};\omega;t}, \mathcal{M}) - \left(\frac{N^2}{\sum_{k=1}^{N} \mu_k^2} - \text{TiEvFun}_1(\mathcal{Y}, \mathcal{M}; \mathbf{r}; t) \right) \right|$$

$$\leq \frac{12}{\sqrt{\varepsilon}\sqrt{\text{TiEvFun}_1(t)}} \cdot \text{TiEvFun}_1(t)$$

holds for all $\omega \in \Omega_1^*$. *Here* $\text{TiEvFun}_1(t) = \text{TiEvFun}_1(\mathcal{Y}, \mathcal{M}; \mathbf{r}; t)$ *is defined in (13.4).*

Proof. The basic idea is simple: we apply Theorem 3.1 and Theorem 5.1(a) — the rest is just estimations. Unfortunately, the estimation part is quite long and tricky.

First a notation: for every nonzero lattice point $\mathbf{n} \in \mathbb{Z}^3 \setminus \mathbf{0}$ write

$$f^{(\mathbf{n})}(\mathbf{x}) = e^{2\pi i \mathbf{n}\cdot\mathbf{x}}, \quad \text{where } \mathbf{x} \in \mathbf{R}^3. \quad (13.5)$$

This function has the special properties

$$f^{(\mathbf{n})}(\mathbf{x} - \mathbf{w}) = e^{-2\pi i \mathbf{n}\cdot\mathbf{w}} f^{(\mathbf{n})}(\mathbf{x}) \quad \text{and} \quad \int_{I^3} f^{(\mathbf{n})}(\mathbf{x})\, d\mathbf{x} = 0,$$

which imply

$$\left| \sum_{\mathbf{z}_k \in \mathcal{Y}_{\mathbf{r};\omega;t}} \mu_k e^{2\pi i \mathbf{n}\cdot\mathbf{z}_k} \right|^2 = \mathbf{D}^2(\mathcal{Y}, \mathcal{M}; \mathbf{r}; \omega; f^{(\mathbf{n})}; t)$$

$$= \overline{\mathbf{D}^2}(\mathcal{Y}, \mathcal{M}; \mathcal{R}; \omega; f^{(\mathbf{n})}; t), \quad (13.6)$$

where we used (2.28), (2.31) and (2.25).

We recall (3.5)–(3.9), which represent a corollary of Theorem 3.1: in the special case $f = f^{(\mathbf{n})}$ we have

$$\overline{\mathbf{D}^2}_{\Omega_1}(\mathcal{Y},\mathcal{M};\mathbf{r};f^{(\mathbf{n})};t) = \text{Eqlbrm}(\mathcal{M};f^{(\mathbf{n})})$$

$$+\overline{\mathbf{D}^2}_{1,negl}(\mathcal{Y},\mathcal{M};\mathbf{r};f^{(\mathbf{n})};t) + \overline{\mathbf{D}^2}_{1,uneven}(\mathcal{Y},\mathcal{M};\mathbf{r};f^{(\mathbf{n})};t), \tag{13.7}$$

where

$$0 \le \text{Eqlbrm}(\mathcal{M};f^{(\mathbf{n})}) + \overline{\mathbf{D}^2}_{1,negl}(\mathcal{Y},\mathcal{M};\mathbf{r};f^{(\mathbf{n})};t) \le \text{Eqlbrm}(\mathcal{M};f^{(\mathbf{n})})$$

$$= \sum_{k=1}^{N} \mu_k^2, \tag{13.8}$$

and

$$\overline{\mathbf{D}^2}_{1,uneven}(\mathcal{Y},\mathcal{M};\mathbf{r};f^{(\mathbf{n})};t) = \left|\sum_{k=1}^{N} \mu_k e^{2\pi i \mathbf{n}\cdot\mathbf{y}_k}\right|^2 \exp\left(-(2\pi|\mathbf{r}(t)|\,|\mathbf{n}|)^2\right). \tag{13.9}$$

Next we recall Theorem 5.1(a): in the special case $f = f^{(\mathbf{n})}$ we have (see (5.1) and (5.2))

$$\text{Var}_{\omega\in\Omega_1}\overline{\mathbf{D}^2}(\mathcal{Y},\mathcal{M};\mathbf{r};\omega;f^{(\mathbf{n})};t)$$

$$= \int_{\Omega_1}\left(\overline{\mathbf{D}^2}(\mathcal{Y},\mathcal{M};\mathbf{r};\omega;f^{(\mathbf{n})};t) - \overline{\mathbf{D}^2}_{\Omega_1}(\mathcal{Y},\mathcal{M};\mathbf{r};f^{(\mathbf{n})};t)\right)^2 d\omega$$

$$\le 14\text{Eqlbrm}\cdot\overline{\mathbf{D}^2}_{1,uneven}(t) + 14\text{Eqlbrm}^{3/2}\left(\overline{\mathbf{D}^2}_{1,uneven}(t)\right)^{1/2} + 14\text{Eqlbrm}^2, \tag{13.10}$$

where

$$\text{Eqlbrm} = \text{Eqlbrm}(\mathcal{M};f^{(\mathbf{n})}) = \sum_{k=1}^{N}\mu_k^2 \tag{13.11}$$

and

$$\overline{\mathbf{D}^2}_{1,uneven}(t) = \overline{\mathbf{D}^2}_{1,uneven}(\mathcal{Y},\mathcal{M};\mathbf{r};f^{(\mathbf{n})};t). \tag{13.12}$$

By (13.12), for every $\lambda > 0$ we have

$$\text{ProdMeas}_1\left\{\omega\in\Omega_1 : \left|\overline{\mathbf{D}^2}(\mathcal{Y},\mathcal{M};\mathbf{r};\omega;f^{(\mathbf{n})};t) - \overline{\mathbf{D}^2}_{\Omega_1}(\mathcal{Y},\mathcal{M};\mathbf{r};f^{(\mathbf{n})};t)\right| \ge \lambda\right\}$$

$$\le \frac{14}{\lambda^2}\left(\text{Eqlbrm}\cdot\overline{\mathbf{D}^2}_{1,uneven}(t) + \text{Eqlbrm}^{3/2}\left(\overline{\mathbf{D}^2}_{1,uneven}(t)\right)^{1/2} + \text{Eqlbrm}^2\right). \tag{13.13}$$

Combining (13.6)–(13.9) and (13.13), we obtain

$$\text{ProdMeas}_1\left\{\omega \in \Omega_1 : \left| \left| \sum_{\mathbf{z}_k \in \mathcal{Y}_{\mathbf{r};\omega;t}} \mu_k e^{2\pi i \mathbf{n}\cdot\mathbf{z}_k} \right|^2 - \overline{\mathbf{D}^2}_1(\mathbf{n};t) \right| \le \lambda + \text{Eq} \right\}$$

$$\ge 1 - \frac{14}{\lambda^2}\left(\text{Eq}\cdot\overline{\mathbf{D}^2}_1(\mathbf{n};t) + \text{Eq}^{3/2}\left(\overline{\mathbf{D}^2}_1(\mathbf{n};t)\right)^{1/2} + \text{Eq}^2\right), \qquad (13.14)$$

where $\mathbf{z}_k$ is defined in (13.1),

$$\text{Eq} = \text{Eqlbrm} = \sum_{k=1}^{N} \mu_k^2 \qquad (13.15)$$

and (see (13.9) and (13.12))

$$\overline{\mathbf{D}^2}_1(\mathbf{n};t) = \overline{\mathbf{D}^2}_{1,uneven}(\mathcal{Y},\mathcal{M};\mathbf{r};f^{(\mathbf{n})};t)$$

$$= \left|\sum_{k=1}^{N} \mu_k e^{2\pi i \mathbf{n}\cdot\mathbf{y}_k}\right|^2 \exp\left(-(2\pi|\mathbf{r}(t)|\ |\mathbf{n}|)^2\right). \qquad (13.16)$$

We choose $\lambda_\varepsilon = \lambda_\varepsilon(\mathbf{n}) > 0$ such that

$$\lambda_\varepsilon^2 = \lambda_\varepsilon^2(\mathbf{n}) = \frac{14}{\varepsilon}|\mathbf{n}|^4\left(\text{Eq}\cdot\overline{\mathbf{D}^2}_1(\mathbf{n};t) + \text{Eq}^{3/2}\left(\overline{\mathbf{D}^2}_1(\mathbf{n};t)\right)^{1/2} + \text{Eq}^2\right), \qquad (13.17)$$

that is,

$$\lambda_\varepsilon = \lambda_\varepsilon(\mathbf{n}) = \frac{\sqrt{14}}{\sqrt{\varepsilon}}|\mathbf{n}|^2\left(\text{Eq}\cdot\overline{\mathbf{D}^2}_1(\mathbf{n};t) + \text{Eq}^{3/2}\left(\overline{\mathbf{D}^2}_1(\mathbf{n};t)\right)^{1/2} + \text{Eq}^2\right)^{1/2}. \qquad (13.17')$$

We need the convergence of the infinite sum below, including the following explicit upper bound

$$\sum_{\mathbf{n}\in\mathbb{Z}^3\setminus\mathbf{0}} \frac{1}{|\mathbf{n}|^4} < 10^2. \qquad (13.18)$$

It can be proved by a routine power-of-two decomposition argument using the elementary geometric fact that the number of lattice points in a large ball of radius R is close to the volume $4\pi R^3/3$. The details go as follows:

$$\sum_{\mathbf{n}\in\mathbb{Z}^3\setminus\mathbf{0}} \frac{1}{|\mathbf{n}|^4} = 6 + \sum_{m=2}^{\infty} \sum_{\substack{\mathbf{n}\in\mathbb{Z}^3: \\ 2^{m-1}\le|\mathbf{n}|^2<2^m}} \frac{1}{|\mathbf{n}|^4}$$

$$\leq 6+\sum_{m=2}^{\infty}\frac{4\pi(2^m)^{3/2}}{(2^{m-1})^2}\leq 6+2^4\pi\sum_{m=2}^{\infty}2^{-m/2}=6+2^4\pi\frac{1/2}{1-\frac{1}{\sqrt{2}}}<10^2.$$

Combining (13.14)–(13.18), we obtain that there is a measurable subset $\Omega_1^*\subset\Omega_1$ such that

$$\mathrm{ProdMeas}_1(\Omega_1^*)\geq 1-\varepsilon\sum_{\mathbf{n}\in\mathbb{Z}^3\setminus\mathbf{0}}\frac{1}{|\mathbf{n}|^4}>1-10^2\varepsilon, \tag{13.19}$$

and for every $\omega\in\Omega_1^*$ and every nonzero lattice point $\mathbf{n}\in\mathbb{Z}^3\setminus\mathbf{0}$ we have

$$\left|\left|\sum_{\mathbf{z}_k\in\mathcal{Y}_{\mathbf{r};\omega;t}}\mu_k e^{2\pi i\mathbf{n}\cdot\mathbf{z}_k}\right|^2-\overline{\mathbf{D}^2}_1(\mathbf{n};t)\right|\leq\lambda_\varepsilon(\mathbf{n})+\mathrm{Eq}$$

$$=\frac{\sqrt{14}|\mathbf{n}|^2}{\sqrt{\varepsilon}}\left(\mathrm{Eq}\cdot\overline{\mathbf{D}^2}_1(\mathbf{n};t)+\mathrm{Eq}^{3/2}\left(\overline{\mathbf{D}^2}_1(\mathbf{n};t)\right)^{1/2}+\mathrm{Eq}^2\right)^{1/2}+\mathrm{Eq}. \tag{13.20}$$

By hypothesis

$$\varepsilon<10^{-2}. \tag{13.21}$$

Motivated by the Imbalance (13.3), we divide (13.20) by the equilibrium part $\mathrm{Eq}=\sum_{k=1}^N\mu_k^2$:

$$\left|\frac{\left|\sum_{\mathbf{z}_k\in\mathcal{Y}_{\mathbf{r};\omega;t}}\mu_k e^{2\pi i\mathbf{n}\cdot\mathbf{z}_k}\right|^2}{\sum_{k=1}^N\mu_k^2}-\frac{\overline{\mathbf{D}^2}_1(\mathbf{n};t)}{\sum_{k=1}^N\mu_k^2}\right|$$

$$\leq\frac{4|\mathbf{n}|^2}{\sqrt{\varepsilon}}\left(\frac{\overline{\mathbf{D}^2}_1(\mathbf{n};t)}{\mathrm{Eq}}+\left(\frac{\overline{\mathbf{D}^2}_1(\mathbf{n};t)}{\mathrm{Eq}}\right)^{1/2}+1\right)^{1/2}+1$$

$$\leq\frac{4|\mathbf{n}|^2}{\sqrt{\varepsilon}}\left(\frac{\overline{\mathbf{D}^2}_1(\mathbf{n};t)}{\mathrm{Eq}}+2\left(\frac{\overline{\mathbf{D}^2}_1(\mathbf{n};t)}{\mathrm{Eq}}\right)^{1/2}+1\right)^{1/2}+1$$

$$=\frac{4|\mathbf{n}|^2}{\sqrt{\varepsilon}}\left(\left(\frac{\overline{\mathbf{D}^2}_1(\mathbf{n};t)}{\sum_{k=1}^N\mu_k^2}\right)^{1/2}+1\right)+1\leq\frac{2|\mathbf{n}|^2}{\sqrt{\varepsilon}}\left(\left(\frac{\overline{\mathbf{D}^2}_1(\mathbf{n};t)}{\sum_{k=1}^N\mu_k^2}\right)^{1/2}+\frac{3}{2}\right),$$

which implies

$$\left|\left(\frac{\left|\sum_{\mathbf{z}_k\in\mathcal{Y}_{\mathbf{r};\omega;t}}\mu_k e^{2\pi i\mathbf{n}\cdot\mathbf{z}_k}\right|^2}{\sum_{k=1}^N\mu_k^2}\text{ or }1\right)_{\max}-\left(\frac{\overline{\mathbf{D}^2}_1(\mathbf{n};t)}{\sum_{k=1}^N\mu_k^2}\text{ or }1\right)_{\max}\right|$$

$$\leq \left| \frac{\left| \sum_{\mathbf{z}_k \in \mathcal{Y}_{\mathbf{r};\omega;t}} \mu_k e^{2\pi i \mathbf{n} \cdot \mathbf{z}_k} \right|^2}{\sum_{k=1}^N \mu_k^2} - \frac{\overline{\mathbf{D}^2}_1(\mathbf{n};t)}{\sum_{k=1}^N \mu_k^2} \right|$$

$$\leq \frac{4|\mathbf{n}|^2}{\sqrt{\varepsilon}} \left(\left(\frac{\overline{\mathbf{D}^2}_1(\mathbf{n};t)}{\sum_{k=1}^N \mu_k^2} \right)^{1/2} + \frac{3}{2} \right). \tag{13.22}$$

Let $\mathbf{n}_1 \in \mathbb{Z}^3 \setminus \mathbf{0}$ denote the nonzero lattice point for which $\text{Imba}_1(\mathcal{Y}_{\mathbf{r};\omega;t}, \mathcal{M})$ attains its maximum (see (13.3)), and let $\mathbf{n}_2 \in \mathbb{Z}^3 \setminus \mathbf{0}$ denote the nonzero lattice point for which $\text{TiEvFun}_1(t)$ attains its maximum (see (13.4) and (13.9)).

By (13.22)

$$\left(\frac{\left| \sum_{\mathbf{z}_k \in \mathcal{Y}_{\mathbf{r};\omega;t}} \mu_k e^{2\pi i \mathbf{n}_2 \cdot \mathbf{z}_k} \right|^2}{\sum_{k=1}^N \mu_k^2} \text{ or } 1 \right)_{\max}$$

$$\geq \left(\frac{\overline{\mathbf{D}^2}_1(\mathbf{n}_2;t)}{\sum_{k=1}^N \mu_k^2} \text{ or } 1 \right)_{\max} - \frac{6|\mathbf{n}_2|^2}{\sqrt{\varepsilon}} \left(\frac{\overline{\mathbf{D}^2}_1(\mathbf{n}_2;t)}{\sum_{k=1}^N \mu_k^2} \text{ or } 1 \right)_{\max}^{1/2}$$

$$= \left(\frac{\overline{\mathbf{D}^2}_1(\mathbf{n}_2;t)}{\sum_{k=1}^N \mu_k^2} \text{ or } 1 \right)_{\max} \left(1 - \frac{6|\mathbf{n}_2|^2}{\sqrt{\varepsilon}} \left(\frac{\overline{\mathbf{D}^2}_1(\mathbf{n}_2;t)}{\sum_{k=1}^N \mu_k^2} \text{ or } 1 \right)_{\max}^{-1/2} \right),$$

and so

$$\left(\frac{\left| \sum_{\mathbf{z}_k \in \mathcal{Y}_{\mathbf{r};\omega;t}} \mu_k e^{2\pi i \mathbf{n}_2 \cdot \mathbf{z}_k} \right|^2}{\sum_{k=1}^N \mu_k^2} \text{ or } 1 \right)_{\max}^{1/|\mathbf{n}_2|^2}$$

$$\geq \left(\frac{\overline{\mathbf{D}^2}_1(\mathbf{n}_2;t)}{\sum_{k=1}^N \mu_k^2} \text{ or } 1 \right)_{\max}^{1/|\mathbf{n}_2|^2} \left(1 - \frac{6|\mathbf{n}_2|^2}{\sqrt{\varepsilon}} \left(\frac{\overline{\mathbf{D}^2}_1(\mathbf{n}_2;t)}{\sum_{k=1}^N \mu_k^2} \text{ or } 1 \right)_{\max}^{-1/2} \right)^{1/|\mathbf{n}_2|^2}$$

$$= \text{TiEvFun}_1(t) \left(1 - \frac{6|\mathbf{n}_2|^2}{\sqrt{\varepsilon}} \left(\frac{\overline{\mathbf{D}^2}_1(\mathbf{n}_2;t)}{\sum_{k=1}^N \mu_k^2} \text{ or } 1 \right)_{\max}^{-1/2} \right)^{1/|\mathbf{n}_2|^2}. \tag{13.23}$$

We recall the hypothesis

$$\varepsilon \cdot \text{TiEvFun}_1(t) \geq 4. \tag{13.24}$$

By (13.24)

$$\left(\frac{\overline{\mathbf{D}^2}_1(\mathbf{n}_2;t)}{\sum_{k=1}^N \mu_k^2} \text{ or } 1\right)_{\max} = (\text{TiEvFun}_1(t))^{|\mathbf{n}_2|^2}$$
$$\geq (4/\varepsilon)^{|\mathbf{n}_2|^2} \geq 400^{|\mathbf{n}_2|^2}, \tag{13.25}$$

where in the last step we used (13.21).

By (13.25)

$$\frac{6|\mathbf{n}_2|^2}{\sqrt{\varepsilon}}\left(\frac{\overline{\mathbf{D}^2}_1(\mathbf{n}_2;t)}{\sum_{k=1}^N \mu_k^2} \text{ or } 1\right)_{\max}^{-1/2} \leq \frac{1}{\sqrt{\varepsilon}}6|\mathbf{n}_2|^2\cdot 20^{-|\mathbf{n}_2|^2} \leq \frac{1}{10}\cdot\frac{6}{20} = \frac{3}{100}, \tag{13.26}$$

where in the last step we again used (13.21).

We need the elementary inequality

$$(1-x)^y \geq 1-2xy \quad \text{for all} \quad 0 \leq x \leq 1/2 \quad \text{and} \quad y > 0, \tag{13.27}$$

which has a simple proof as follows:

$$(1-x)^y \geq \left(e^{-2x}\right)^y = e^{-2xy} \geq 1-2xy.$$

By (13.26) and (13.27),

$$\left(1 - \frac{6|\mathbf{n}_2|^2}{\sqrt{\varepsilon}}\left(\frac{\overline{\mathbf{D}^2}_1(\mathbf{n}_2;t)}{\sum_{k=1}^N \mu_k^2} \text{ or } 1\right)_{\max}^{-1/2}\right)^{1/|\mathbf{n}_2|^2}$$
$$\geq 1 - \frac{12}{\sqrt{\varepsilon}}\left(\frac{\overline{\mathbf{D}^2}_1(\mathbf{n}_2;t)}{\sum_{k=1}^N \mu_k^2} \text{ or } 1\right)_{\max}^{-1/2}. \tag{13.28}$$

Applying (13.28) in (13.23), we have

$$\left(\frac{\left|\sum_{\mathbf{z}_k\in\mathcal{Y}_{\mathbf{r};\omega;t}} \mu_k e^{2\pi i \mathbf{n}_2\cdot\mathbf{z}_k}\right|^2}{\sum_{k=1}^N \mu_k^2} \text{ or } 1\right)_{\max}^{1/|\mathbf{n}_2|^2}$$
$$\geq \text{TiEvFun}_1(t)\left(1 - \frac{12}{\sqrt{\varepsilon}}\left(\frac{\overline{\mathbf{D}^2}_1(\mathbf{n}_2;t)}{\sum_{k=1}^N \mu_k^2} \text{ or } 1\right)_{\max}^{-1/2}\right). \tag{13.29}$$

By (13.29) we obtain the following lower bound for $\text{Imba}_1(\mathcal{Y}_{\mathbf{r};\omega;t},\mathcal{M})$:

$$\text{Imba}_1(\mathcal{Y}_{\mathbf{r};\omega t},\mathcal{M}) = \max_{\mathbf{n}\in\mathbb{Z}^3\setminus\mathbf{0}}\left(\frac{\left|\sum_{\mathbf{z}_k\in\mathcal{Y}_{\mathbf{r};\omega;t}} \mu_k e^{2\pi i \mathbf{n}\cdot\mathbf{z}_k}\right|^2}{\sum_{k=1}^N \mu_k^2} \text{ or } 1\right)_{\max}^{1/|\mathbf{n}|^2}$$

$$\geq \text{TiEvFun}_1(t)\left(1-\frac{12}{\sqrt{\varepsilon}}\left(\frac{\overline{\mathbf{D}^2}_1(\mathbf{n}_2;t)}{\sum_{k=1}^N \mu_k^2} \text{ or } 1\right)_{\max}^{-1/2}\right). \tag{13.30}$$

Next we give an upper bound for $\text{Imba}_1(\mathcal{Y}_{\mathbf{r};\omega;t},\mathcal{M})$. We have two cases: either

$$\text{Imba}_1(\mathcal{Y}_{\mathbf{r};\omega;t},\mathcal{M}) = \left(\frac{\left|\sum_{\mathbf{z}_k\in\mathcal{Y}_{\mathbf{r};\omega;t}} \mu_k e^{2\pi i\mathbf{n}_1\cdot\mathbf{z}_k}\right|^2}{\sum_{k=1}^N \mu_k^2} \text{ or } 1\right)_{\max}^{1/|\mathbf{n}_1|^2}$$

$$\leq \left(\frac{\left|\sum_{k=1}^N \mu_k e^{2\pi i\mathbf{n}_1\cdot\mathbf{y}_k}\right|^2 \exp\left(-(2\pi|\mathbf{r}(t)|\,|\mathbf{n}_1|)^2\right)}{\sum_{k=1}^N \mu_k^2} \text{ or } 1\right)_{\max}^{1/|\mathbf{n}_1|^2}, \tag{13.31}$$

or

$$\text{Imba}_1(\mathcal{Y}_{\mathbf{r};\omega;t},\mathcal{M}) = \left(\frac{\left|\sum_{\mathbf{z}_k\in\mathcal{Y}_{\mathbf{r};\omega;t}} \mu_k e^{2\pi i\mathbf{n}_1\cdot\mathbf{z}_k}\right|^2}{\sum_{k=1}^N \mu_k^2} \text{ or } 1\right)_{\max}^{1/|\mathbf{n}_1|^2}$$

$$> \left(\frac{\left|\sum_{k=1}^N \mu_k e^{2\pi i\mathbf{n}_1\cdot\mathbf{y}_k}\right|^2 \exp\left(-(2\pi|\mathbf{r}(t)|\,|\mathbf{n}_1|)^2\right)}{\sum_{k=1}^N \mu_k^2} \text{ or } 1\right)_{\max}^{1/|\mathbf{n}_1|^2}. \tag{13.32}$$

Since (13.31) is already a good upper bound for $\text{Imba}_1(\mathcal{Y}_{\mathbf{r};\omega;t},\mathcal{M})$, in the rest we *assume* (13.32). We return now to (13.22): we divide it by

$$\left(\frac{\overline{\mathbf{D}^2}_1(\mathbf{n};t)}{\sum_{k=1}^N \mu_k^2} \text{ or } 1\right)_{\max}$$

("$(x \text{ or } 1)_{\max}$" means the larger number between x and 1), and obtain

$$\left|\left(\frac{\left|\sum_{\mathbf{z}_k\in\mathcal{Y}_{\mathbf{r};\omega;t}} \mu_k e^{2\pi i\mathbf{n}\cdot\mathbf{z}_k}\right|^2}{\sum_{k=1}^N \mu_k^2} \text{ or } 1\right)_{\max}\left(\frac{\overline{\mathbf{D}^2}_1(\mathbf{n};t)}{\sum_{k=1}^N \mu_k^2} \text{ or } 1\right)_{\max}^{-1} - 1\right|$$

$$= \left(\frac{\overline{\mathbf{D}^2}_1(\mathbf{n};t)}{\sum_{k=1}^N \mu_k^2} \text{ or } 1\right)_{\max}^{-1}$$

$$\cdot\left|\left(\frac{\left|\sum_{\mathbf{z}_k\in\mathcal{Y}_{\mathbf{r};\omega;t}} \mu_k e^{2\pi i\mathbf{n}\cdot\mathbf{z}_k}\right|^2}{\sum_{k=1}^N \mu_k^2} \text{ or } 1\right)_{\max} - \left(\frac{\overline{\mathbf{D}^2}_1(\mathbf{n};t)}{\sum_{k=1}^N \mu_k^2} \text{ or } 1\right)_{\max}\right|$$

$$\le \frac{4|\mathbf{n}|^2}{\sqrt{\varepsilon}}\left(\left(\frac{\overline{\mathbf{D}^2}_1(\mathbf{n};t)}{\sum_{k=1}^N \mu_k^2}\right)^{1/2}+\frac{3}{2}\right)\left(\frac{\overline{\mathbf{D}^2}_1(\mathbf{n};t)}{\sum_{k=1}^N \mu_k^2}\ \text{or}\ 1\right)_{\max}^{-1}$$

$$\le \frac{4|\mathbf{n}|^2}{\sqrt{\varepsilon}}\left(\left(\frac{\overline{\mathbf{D}^2}_1(\mathbf{n};t)}{\sum_{k=1}^N \mu_k^2}\ \text{or}\ 1\right)_{\max}^{-1/2}+2\left(\frac{\overline{\mathbf{D}^2}_1(\mathbf{n};t)}{\sum_{k=1}^N \mu_k^2}\ \text{or}\ 1\right)_{\max}^{-1}\right). \tag{13.33}$$

Let $A \ge B \ge 1$ be positive real numbers and let $h \ge 1$ be a positive integer, then we have the simple inequality

$$A^{1/h} - B^{1/h} \le \frac{1}{h}(A-B)B^{\frac{1}{h}-1} \le \frac{1}{h}(A-B). \tag{13.34}$$

Note that (13.34) easily follows from the Mean Value theorem in calculus applied to the function $f(x) = x^{1/h}$.

We apply (13.34) with $h = |\mathbf{n}_1|^2$, $B = 1$,

$$A = \left(\frac{\left|\sum_{\mathbf{z}_k\in\mathcal{Y}_{\mathbf{r};\omega;t}} \mu_k e^{2\pi i\mathbf{n}_1\cdot\mathbf{z}_k}\right|^2}{\sum_{k=1}^N \mu_k^2}\ \text{or}\ 1\right)_{\max}\left(\frac{\overline{\mathbf{D}^2}_1(\mathbf{n}_1;t)}{\sum_{k=1}^N \mu_k^2}\ \text{or}\ 1\right)_{\max}^{-1} > 1$$

(see (13.32)), and combine it with (13.33):

$$\left(\frac{\left|\sum_{\mathbf{z}_k\in\mathcal{Y}_{\mathbf{r};\omega;t}} \mu_k e^{2\pi i\mathbf{n}_1\cdot\mathbf{z}_k}\right|^2}{\sum_{k=1}^N \mu_k^2}\ \text{or}\ 1\right)_{\max}^{1/|\mathbf{n}_1|^2}\left(\frac{\overline{\mathbf{D}^2}_1(\mathbf{n}_1;t)}{\sum_{k=1}^N \mu_k^2}\ \text{or}\ 1\right)_{\max}^{-1/|\mathbf{n}_1|^2} - 1$$

$$\le \frac{1}{|\mathbf{n}_1|^2}\frac{4|\mathbf{n}_1|^2}{\sqrt{\varepsilon}}\left(\left(\frac{\overline{\mathbf{D}^2}_1(\mathbf{n}_1;t)}{\sum_{k=1}^N \mu_k^2}\ \text{or}\ 1\right)_{\max}^{-1/2}+2\left(\frac{\overline{\mathbf{D}^2}_1(\mathbf{n}_1;t)}{\sum_{k=1}^N \mu_k^2}\ \text{or}\ 1\right)_{\max}^{-1}\right)$$

$$= \frac{4}{\sqrt{\varepsilon}}\left(\left(\frac{\overline{\mathbf{D}^2}_1(\mathbf{n}_1;t)}{\sum_{k=1}^N \mu_k^2}\ \text{or}\ 1\right)_{\max}^{-1/2}+2\left(\frac{\overline{\mathbf{D}^2}_1(\mathbf{n}_1;t)}{\sum_{k=1}^N \mu_k^2}\ \text{or}\ 1\right)_{\max}^{-1}\right). \tag{13.35}$$

By using the notation (13.16), we can rewrite the definition (13.4) of the Time Evolution Function as follows:

$$\text{TiEvFun}_1(t) = \max_{\mathbf{n}\in\mathbb{Z}^3\setminus\mathbf{0}}\left(\frac{\overline{\mathbf{D}^2}_1(\mathbf{n};t)}{\sum_{k=1}^N \mu_k^2}\ \text{or}\ 1\right)_{\max}^{1/|\mathbf{n}|^2}. \tag{13.36}$$

We distinguish two cases. In the *first case* we assume

$$\left(\frac{\overline{\mathbf{D}^2}_1(\mathbf{n};t)}{\sum_{k=1}^N \mu_k^2}\ \text{or}\ 1\right)_{\max} \ge \text{TiEvFun}_1(t).$$

Then by (13.35),

$$\left(\frac{\left|\sum_{\mathbf{z}_k\in\mathcal{Y}_{\mathbf{r};\omega;t}}\mu_k e^{2\pi i\mathbf{n}_1\cdot\mathbf{z}_k}\right|^2}{\sum_{k=1}^N\mu_k^2}\ \text{ or }\ 1\right)_{\max}^{1/|\mathbf{n}_1|^2}\left(\frac{\overline{\mathbf{D}^2}_1(\mathbf{n}_1;t)}{\sum_{k=1}^N\mu_k^2}\ \text{ or }\ 1\right)_{\max}^{-1/|\mathbf{n}_1|^2}-1$$

$$\leq\frac{4}{\sqrt{\varepsilon}}\left((\text{TiEvFun}_1(t))^{-1/2}+2\,(\text{TiEvFun}_1(t))^{-1}\right),$$

which implies that

$$\left|\left(\frac{\left|\sum_{\mathbf{z}_k\in\mathcal{Y}_{\mathbf{r};\omega;t}}\mu_k e^{2\pi i\mathbf{n}_1\cdot\mathbf{z}_k}\right|^2}{\sum_{k=1}^N\mu_k^2}\ \text{ or }\ 1\right)_{\max}^{1/|\mathbf{n}_1|^2}-\left(\frac{\overline{\mathbf{D}^2}_1(\mathbf{n}_1;t)}{\sum_{k=1}^N\mu_k^2}\ \text{ or }\ 1\right)_{\max}^{1/|\mathbf{n}_1|^2}\right|$$

$$\leq\frac{4}{\sqrt{\varepsilon}\sqrt{\text{TiEvFun}_1(t)}}\left(1+\frac{2}{\sqrt{\text{TiEvFun}_1(t)}}\right)\left(\frac{\overline{\mathbf{D}^2}_1(\mathbf{n}_1;t)}{\sum_{k=1}^N\mu_k^2}\ \text{ or }\ 1\right)_{\max}^{1/|\mathbf{n}_1|^2}$$

$$\leq\frac{4}{\sqrt{\varepsilon}\sqrt{\text{TiEvFun}_1(t)}}\left(1+\frac{2}{\sqrt{\text{TiEvFun}_1(t)}}\right)\cdot\text{TiEvFun}_1(t),\tag{13.37}$$

where in the last step we used (13.36).

In the *second case* we assume

$$\left(\frac{\overline{\mathbf{D}^2}_1(\mathbf{n}_1;t)}{\sum_{k=1}^N\mu_k^2}\ \text{ or }\ 1\right)_{\max}<\text{TiEvFun}_1(t).$$

Then by (13.22) and (13.34),

$$\left|\left(\frac{\left|\sum_{\mathbf{z}_k\in\mathcal{Y}_{\mathbf{r};\omega;t}}\mu_k e^{2\pi i\mathbf{n}_1\cdot\mathbf{z}_k}\right|^2}{\sum_{k=1}^N\mu_k^2}\ \text{ or }\ 1\right)_{\max}^{1/|\mathbf{n}_1|^2}-\left(\frac{\overline{\mathbf{D}^2}_1(\mathbf{n}_1;t)}{\sum_{k=1}^N\mu_k^2}\ \text{ or }\ 1\right)_{\max}^{1/|\mathbf{n}_1|^2}\right|$$

$$\leq\frac{1}{|\mathbf{n}_1|^2}\left|\left(\frac{\left|\sum_{\mathbf{z}_k\in\mathcal{Y}_{\mathbf{r};\omega;t}}\mu_k e^{2\pi i\mathbf{n}_1\cdot\mathbf{z}_k}\right|^2}{\sum_{k=1}^N\mu_k^2}\ \text{ or }\ 1\right)_{\max}-\left(\frac{\overline{\mathbf{D}^2}_1(\mathbf{n}_1;t)}{\sum_{k=1}^N\mu_k^2}\ \text{ or }\ 1\right)_{\max}\right|$$

$$\leq\frac{1}{|\mathbf{n}_1|^2}\frac{4|\mathbf{n}_1|^2}{\sqrt{\varepsilon}}\left(\left(\frac{\overline{\mathbf{D}^2}_1(\mathbf{n}_1;t)}{\sum_{k=1}^N\mu_k^2}\right)^{1/2}+\frac{3}{2}\right)$$

$$\le \frac{4}{\sqrt{\varepsilon}}\left((\mathrm{TiEvFun}_1(t))^{1/2}+2\right)$$

$$= \frac{4}{\sqrt{\varepsilon}\sqrt{\mathrm{TiEvFun}_1(t)}}\left(1+\frac{2}{\sqrt{\mathrm{TiEvFun}_1(t)}}\right)\cdot \mathrm{TiEvFun}_1(t). \qquad (13.38)$$

Combining (13.37) and (13.38) we obtain

$$\mathrm{Imba}_1(\mathcal{Y}_{\mathbf{r};\omega;t},\mathcal{M}) = \left(\frac{\left|\sum_{\mathbf{z}_k\in\mathcal{Y}_{\mathbf{r};\omega;t}}\mu_k e^{2\pi i\mathbf{n}_1\cdot\mathbf{z}_k}\right|^2}{\sum_{k=1}^N \mu_k^2} \text{ or } 1\right)_{\max}^{1/|\mathbf{n}_1|^2}$$

$$\le \left(\frac{\overline{\mathbf{D}^2}_1(\mathbf{n}_1;t)}{\sum_{k=1}^N \mu_k^2} \text{ or } 1\right)_{\max}^{1/|\mathbf{n}_1|^2}$$

$$+\frac{4}{\sqrt{\varepsilon}\sqrt{\mathrm{TiEvFun}_1(t)}}\left(1+\frac{2}{\sqrt{\mathrm{TiEvFun}_1(t)}}\right)\cdot \mathrm{TiEvFun}_1(t)$$

$$\le \mathrm{TiEvFun}_1(t)+\frac{4}{\sqrt{\varepsilon}\sqrt{\mathrm{TiEvFun}_1(t)}}\left(1+\frac{2}{\sqrt{\mathrm{TiEvFun}_1(t)}}\right)\cdot \mathrm{TiEvFun}_1(t), \qquad (13.39)$$

assuming (13.32) holds.

If (13.31) holds, then we have

$$\mathrm{Imba}_1(\mathcal{Y}_{\mathbf{r};\omega;t},\mathcal{M}) = \left(\frac{\left|\sum_{\mathbf{z}_k\in\mathcal{Y}_{\mathbf{r};\omega;t}}\mu_k e^{2\pi i\mathbf{n}_1\cdot\mathbf{z}_k}\right|^2}{\sum_{k=1}^N \mu_k^2} \text{ or } 1\right)_{\max}^{1/|\mathbf{n}_1|^2}$$

$$\le \left(\frac{\left|\sum_{k=1}^N\mu_k e^{2\pi i\mathbf{n}_1\cdot\mathbf{y}_k}\right|^2 \exp\left(-(2\pi|\mathbf{r}(t)|\;|\mathbf{n}_1|)^2\right)}{\sum_{k=1}^N \mu_k^2} \text{ or } 1\right)_{\max}^{1/|\mathbf{n}_1|^2}$$

$$\le \mathrm{TiEvFun}_1(t). \qquad (13.40)$$

In the other direction, we recall (13.30):

$$\mathrm{Imba}_1(\mathcal{Y}_{\mathbf{r};\omega;t},\mathcal{M}) \ge \mathrm{TiEvFun}_1(t)\left(1-\frac{12}{\sqrt{\varepsilon}}\left(\frac{\overline{\mathbf{D}^2}_1(\mathbf{n}_2;t)}{\sum_{k=1}^N \mu_k^2} \text{ or } 1\right)_{\max}^{-1/2}\right)$$

$$\geq \text{TiEvFun}_1(t)\left(1-\frac{12}{\sqrt{\varepsilon}\sqrt{\text{EvolFun}_1(t)}}\right). \tag{13.41}$$

Combining (13.39)–(13.41), we obtain

$$|\text{Imba}_1(\mathcal{Y}_{\mathbf{r};\omega;t},\mathcal{M})-\text{TiEvFun}_1(t)|$$

$$\leq \frac{12}{\sqrt{\varepsilon}\sqrt{\text{TiEvFun}_1(t)}}\cdot \text{TiEvFun}_1(t) \tag{13.42}$$

for all $\omega \in \Omega_1^*$ (see (13.19)), where $\Omega_1^* \subset \Omega_1$ is a measurable subset such that

$$\text{ProdMeas}_1(\Omega_1^*) > 1-10^2\varepsilon.$$

By (6.17)

$$\text{MicroEntropy}_1(\mathcal{Y}_{\mathbf{r};\omega;t},\mathcal{M})=\frac{N^2}{\sum_{k=1}^N \mu_k^2}-\text{Imbalance}_1(\mathcal{Y}_{\mathbf{r};\omega;t},\mathcal{M}),$$

and combining it with (13.42), Theorem 13.1 follows. □

Next we discuss the standard 3-dimensional case of the third kind, and prove an analog of Theorem 13.1. The proof turns out to be a simpler version of the proof of Theorem 13.1 above. (Note that the second kind leads to a technical nuisance — a "divergence problem" — explaining why we discuss it later in Section 14.)

For the third kind (12.2) gives

$$\mathcal{Y}_{\mathbf{r};\omega;t}=\{\mathbf{z}_k=\mathbf{y}_k+\vartheta_k(\rho_k\mathbf{r}(t)) \text{ modulo one}:\ 1\leq k\leq N\},$$

where

$$\omega=(\rho_1,\vartheta_1,\rho_2,\vartheta_2,\ldots,\rho_N,\vartheta_N)\in\Omega_3=([0,1]\times SO(3))^N$$

with

$$\Pr[\rho_k\leq u]=3\int_0^u y^2\,dy=u^3 \text{ for } 0\leq y\leq 1, \text{ and } 1 \text{ for } u\geq 1.$$

By (6.17)

$$\text{MicroEntropy}_3(\mathcal{Y}_{\mathbf{r};\omega;t},\mathcal{M})=\frac{N^2}{\sum_{k=1}^N \mu_k^2}-\text{Imbalance}_3(\mathcal{Y}_{\mathbf{r};\omega;t},\mathcal{M}),$$

so it suffices to study the Imbalance defined in (6.16)

$$\text{Imba}_3(\mathcal{Y}_{\mathbf{r};\omega;t},\mathcal{M})=\text{Imbalance}_3(\mathcal{Y}_{\mathbf{r};\omega;t},\mathcal{M})$$

$$= \max_{\mathbf{n}\in\mathbb{Z}^3\setminus\mathbf{0}} \frac{1}{|\mathbf{n}|^4}\left(\frac{\left|\sum_{\mathbf{z}_k\in\mathcal{Y}_{\mathbf{r};\omega;t}}\mu_k e^{2\pi i\mathbf{n}\cdot\mathbf{z}_k}\right|^2}{\sum_{k=1}^N \mu_k^2} \text{ or } 1\right)_{\max}. \tag{13.43}$$

Again we want to find the relation between $\text{Imba}_3(\mathcal{Y}_{\mathbf{r};\omega;t},\mathcal{M})$ for the majority of $\omega\in\Omega_3$ and the following ω-independent function (t is the time variable)

$$\text{TiEvFun}_3(t) = \text{TiEvFun}_3(\mathcal{Y},\mathcal{M};\mathbf{r};t) = \max_{\mathbf{n}\in\mathbb{Z}^3\setminus\mathbf{0}} \frac{1}{|\mathbf{n}|^4}$$

$$\left(\frac{\left|\sum_{k=1}^N \mu_k e^{2\pi i\mathbf{n}\cdot\mathbf{y}_k}\right|^2 9\left(\frac{\cos(2\pi|\mathbf{r}(t)|\,|\mathbf{n}|)}{(2\pi|\mathbf{r}(t)|\,|\mathbf{n}|)^2}-\frac{\sin(2\pi|\mathbf{r}(t)|\,|\mathbf{n}|)}{(2\pi|\mathbf{r}(t)|\,|\mathbf{n}|)^3}\right)^2}{\sum_{k=1}^N \mu_k^2} \text{ or } 1\right)_{\max}$$

$$= \left(\frac{9}{\sum_{k=1}^N \mu_k^2}\max_{\mathbf{n}\in\mathbb{Z}^3\setminus\mathbf{0}}\left|\sum_{k=1}^N \mu_k e^{2\pi i\mathbf{n}\cdot\mathbf{y}_k}\right|^2\right.$$

$$\left.\left(\frac{\cos(2\pi|\mathbf{r}(t)|\,|\mathbf{n}|)}{4\pi^2|\mathbf{r}(t)|^2|\mathbf{n}|^4}-\frac{\sin(2\pi|\mathbf{r}(t)|\,|\mathbf{n}|)}{8\pi^3|\mathbf{r}(t)|^3|\mathbf{n}|^5}\right)^2 \text{ or } 1\right)_{\max} \tag{13.44}$$

that we call the *Time Evolution Function of the third kind.*

By a straightforward adaptation of the proof argument of Theorem 13.1 up to (13.22), we obtain that there is a measurable subset $\Omega_3^*\subset\Omega_3$ such that

$$\text{ProdMeas}_3(\Omega_3^*) > 1-10^2\varepsilon, \tag{13.45}$$

and for every $\omega\in\Omega_3^*$ and every nonzero lattice point $\mathbf{n}\in\mathbb{Z}^3\setminus\mathbf{0}$ we have the analog of (13.22):

$$\left|\frac{\left|\sum_{\mathbf{z}_k\in\mathcal{Y}_{\mathbf{r};\omega;t}}\mu_k e^{2\pi i\mathbf{n}\cdot\mathbf{z}_k}\right|^2}{\sum_{k=1}^N \mu_k^2}-\frac{\overline{\mathbf{D}^2}_3(\mathbf{n};t)}{\sum_{k=1}^N \mu_k^2}\right|$$

$$\leq \frac{4|\mathbf{n}|^2}{\sqrt{\varepsilon}}\left(\left(\frac{\overline{\mathbf{D}^2}_3(\mathbf{n};t)}{\sum_{k=1}^N \mu_k^2}\right)^{1/2}+\frac{3}{2}\right), \tag{13.46}$$

where

$$\overline{\mathbf{D}^2}_3(\mathbf{n};t) = \left|\sum_{k=1}^N \mu_k e^{2\pi i\mathbf{n}\cdot\mathbf{y}_k}\right|^2 9\left(\frac{\cos(2\pi|\mathbf{r}(t)|\,|\mathbf{n}|)}{(2\pi|\mathbf{r}(t)|\,|\mathbf{n}|)^2}-\frac{\sin(2\pi|\mathbf{r}(t)|\,|\mathbf{n}|)}{(2\pi|\mathbf{r}(t)|\,|\mathbf{n}|)^3}\right)^2. \tag{13.47}$$

By using the notation (13.47), we can rewrite the definition (13.44) of the Time Evolution Function as follows:

$$\mathrm{TiEvFun}_3(t) = \max_{\mathbf{n}\in\mathbb{Z}^3\setminus\mathbf{0}} \frac{1}{|\mathbf{n}|^4}\left(\frac{\overline{\mathbf{D}^2}_3(\mathbf{n};t)}{\sum_{k=1}^N \mu_k^2} \text{ or } 1\right)_{\max}. \tag{13.48}$$

Combining (13.46) and (13.48), we have

$$\left|\frac{1}{|\mathbf{n}|^4}\left(\frac{\left|\sum_{\mathbf{z}_k\in\mathcal{Y}_{\mathbf{r};\omega;t}} \mu_k e^{2\pi i\mathbf{n}\cdot\mathbf{z}_k}\right|^2}{\sum_{k=1}^N \mu_k^2} \text{ or } 1\right)_{\max} - \frac{1}{|\mathbf{n}|^4}\left(\frac{\overline{\mathbf{D}^2}_3(\mathbf{n};t)}{\sum_{k=1}^N \mu_k^2} \text{ or } 1\right)_{\max}\right|$$

$$\le \frac{1}{|\mathbf{n}|^4}\frac{4|\mathbf{n}|^2}{\sqrt{\varepsilon}}\left(\left(\frac{\overline{\mathbf{D}^2}_3(\mathbf{n};t)}{\sum_{k=1}^N \mu_k^2}\right)^{1/2} + \frac{3}{2}\right)$$

$$\le \frac{4}{\sqrt{\varepsilon}}\left(\left(\frac{1}{|\mathbf{n}|^4}\left(\frac{\overline{\mathbf{D}^2}_3(\mathbf{n};t)}{\sum_{k=1}^N \mu_k^2} \text{ or } 1\right)_{\max}\right)^{1/2} + 2\right)$$

$$\le \frac{2}{\sqrt{\varepsilon}}\left((\mathrm{TiEvFun}_3(t))^{1/2} + 2\right)$$

$$= \frac{4}{\sqrt{\varepsilon}\sqrt{\mathrm{TiEvFun}_3(t)}}\left(1 + \frac{2}{\sqrt{\mathrm{TiEvFun}_3(t)}}\right)\cdot \mathrm{TiEvFun}_3(t). \tag{13.49}$$

Let $\mathbf{n}_1 \in \mathbb{Z}^3\setminus\mathbf{0}$ denote the nonzero lattice point for which $\mathrm{Imba}_3(\mathcal{Y}_{\mathbf{r};\omega;t}, \mathcal{M})$ attains its maximum, and let $\mathbf{n}_2 \in \mathbb{Z}^3\setminus\mathbf{0}$ denote the nonzero lattice point for which $\mathrm{TiEvFun}_3(t)$ attains its maximum.

By (13.49) and the triangle inequality,

$$\mathrm{Imba}_3(\mathcal{Y}_{\mathbf{r};\omega;t}, \mathcal{M}) = \frac{1}{|\mathbf{n}_1|^4}\left(\frac{\left|\sum_{\mathbf{z}_k\in\mathcal{Y}_{\mathbf{r};\omega;t}} \mu_k e^{2\pi i\mathbf{n}_1\cdot\mathbf{z}_k}\right|^2}{\sum_{k=1}^N \mu_k^2} \text{ or } 1\right)_{\max}$$

$$\le \frac{1}{|\mathbf{n}_1|^4}\left(\frac{\overline{\mathbf{D}^2}_3(\mathbf{n}_1;t)}{\sum_{k=1}^N \mu_k^2} \text{ or } 1\right)_{\max}$$

$$+ \frac{4}{\sqrt{\varepsilon}\sqrt{\mathrm{TiEvFun}_3(t)}}\left(1 + \frac{2}{\sqrt{\mathrm{TiEvFun}_3(t)}}\right)\cdot \mathrm{TiEvFun}_3(t)$$

$$\leq \text{TiEvFun}_3(t) + \frac{4}{\sqrt{\varepsilon}\sqrt{\text{TiEvFun}_3(t)}}\left(1 + \frac{2}{\sqrt{\text{TiEvFun}_3(t)}}\right) \cdot \text{TiEvFun}_3(t), \tag{13.50}$$

and similarly,

$$\text{Imba}_3(\mathcal{Y}_{\mathbf{r};\omega;t}, \mathcal{M}) \geq \frac{1}{|\mathbf{n}_2|^4}\left(\frac{\left|\sum_{\mathbf{z}_k \in \mathcal{Y}_{\mathbf{r};\omega;t}} \mu_k e^{2\pi i \mathbf{n}_2 \cdot \mathbf{z}_k}\right|^2}{\sum_{k=1}^N \mu_k^2} \text{ or } 1\right)_{\max}$$

$$\geq \frac{1}{|\mathbf{n}_2|^4}\left(\frac{\overline{\mathbf{D}^2}_3(\mathbf{n}_2;t)}{\sum_{k=1}^N \mu_k^2} \text{ or } 1\right)_{\max} - \frac{4}{\sqrt{\varepsilon}\sqrt{\text{TiEvFun}_3(t)}}$$

$$\left(1 + \frac{2}{\sqrt{\text{TiEvFun}_3(t)}}\right) \cdot \text{TiEvFun}_3(t)$$

$$= \text{TiEvFun}_3(t) - \frac{4}{\sqrt{\varepsilon}\sqrt{\text{TiEvFun}_3(t)}}\left(1 + \frac{2}{\sqrt{\text{TiEvFun}_3(t)}}\right) \cdot \text{TiEvFun}_3(t). \tag{13.51}$$

Combining (13.50) and (13.51), we have

$$|\text{Imba}_3(\mathcal{Y}_{\mathbf{r};\omega;t}, \mathcal{M}) - \text{TiEvFun}_3(t)|$$

$$\leq \frac{4}{\sqrt{\varepsilon}\sqrt{\text{TiEvFun}_3(t)}}\left(1 + \frac{2}{\sqrt{\text{TiEvFun}_3(t)}}\right) \cdot \text{TiEvFun}_3(t)$$

for all $\omega \in \Omega_3^*$, where $\Omega_3^* \subset \Omega_3$ is a measurable subset such that

$$\text{ProdMeas}_3(\Omega_3^*) > 1 - 10^2\varepsilon.$$

Since

$$\text{MicroEntropy}_3(\mathcal{Y}_{\mathbf{r};\omega;t}, \mathcal{M}) = \frac{N^2}{\sum_{k=1}^N \mu_k^2} - \text{Imbalance}_3(\mathcal{Y}_{\mathbf{r};\omega;t}, \mathcal{M}),$$

we obtain the following analog of Theorem 13.1.

Theorem 13.2. *Consider a 3-dimensional torus model $(\mathcal{Y}, \mathcal{M}; \mathbf{r}; \Omega_3)$ of the third kind. Given a time $t > 0$ and $0 < \varepsilon < 10^{-2}$, there exists a measurable subset $\Omega_3^* = \Omega_3^*(\varepsilon; t) \subset \Omega_3$ of the initial velocity space of the third kind such that*

$$\text{ProdMeas}_3(\Omega_3^*) > 1 - 10^2\varepsilon$$

and

$$\left|\text{MicroEntropy}_3(\mathcal{Y}_{\mathbf{r};\omega;t},\mathcal{M}) - \left(\frac{N^2}{\sum_{k=1}^N \mu_k^2} - \text{TiEvFun}_3(\mathcal{Y},\mathcal{M};\mathbf{r};t)\right)\right|$$

$$\leq \frac{4}{\sqrt{\varepsilon}\sqrt{\text{TiEvFun}_3(t)}}\left(1+\frac{2}{\sqrt{\text{TiEvFun}_3(t)}}\right)\cdot \text{TiEvFun}_3(t)$$

holds for all $\omega \in \Omega_3^*$. *Here* $\text{TiEvFun}_3(t) = \text{TiEvFun}_3(\mathcal{Y},\mathcal{M};\mathbf{r};t)$ *is defined in (13.44).* □

For a far-reaching generalization of Theorem 13.2; see Theorem 19.1.

We will discuss some interesting consequences of Theorems 13.1 and 13.2 in Sections 15–16.

14. Proving a Second Law (II): Second kind

The case of the second kind is more complicated than that of the third kind. It is due to the divergence of the series

$$\sum_{\mathbf{n}\in\mathbb{Z}^3\setminus\mathbf{0}} \frac{1}{|\mathbf{n}|^2},$$

noting that the corresponding sum for the third kind

$$\sum_{\mathbf{n}\in\mathbb{Z}^3\setminus\mathbf{0}} \frac{1}{|\mathbf{n}|^4}$$

is convergent. Nevertheless, we try to follow the proof argument of Theorem 13.2, at least partially.

We start with basically repeating the argument between (13.43)–(13.49). For the second kind (12.2) gives

$$\mathcal{Y}_{\mathbf{r};\omega;t} = \{\mathbf{z}_k = \mathbf{y}_k + \vartheta_k(\mathbf{r}(t)) \text{ modulo one}: \ 1\leq k\leq N\}, \tag{14.1}$$

where

$$\omega = (\vartheta_1,\vartheta_2,\ldots,\vartheta_N)\in\Omega_2 = (SO(3))^N.$$

By (6.17)

$$\text{MicroEntropy}_2(\mathcal{Y}_{\mathbf{r};\omega;t},\mathcal{M}) = \frac{N^2}{\sum_{k=1}^N \mu_k^2} - \text{Imbalance}_2(\mathcal{Y}_{\mathbf{r};\omega;t},\mathcal{M}),$$

so it suffices to study the Imbalance defined in (6.13)

$$\text{Imba}_2(\mathcal{Y}_{\mathbf{r};\omega;t},\mathcal{M}) = \text{Imbalance}_2(\mathcal{Y}_{\mathbf{r};\omega;t},\mathcal{M})$$

$$= \max_{\mathbf{n}\in\mathbb{Z}^3\setminus\mathbf{0}} \frac{1}{|\mathbf{n}|^2}\left(\frac{\left|\sum_{\mathbf{z}_k\in\mathcal{Y}_{\mathbf{r};\omega;t}} \mu_k e^{2\pi i\mathbf{n}\cdot\mathbf{z}_k}\right|^2}{\sum_{k=1}^N \mu_k^2} \text{ or } 1\right)_{\max}. \tag{14.2}$$

Again we want to find the relation between $\mathrm{Imba}_2(\mathcal{Y}_{\mathbf{r};\omega;t},\mathcal{M})$ for the majority of $\omega\in\Omega_2$ and the following ω-independent function (t is the time variable)

$$\mathrm{TiEvFun}_2(t) = \mathrm{TiEvFun}_2(\mathcal{Y},\mathcal{M};\mathbf{r};t)$$

$$= \max_{\mathbf{n}\in\mathbb{Z}^3\setminus\mathbf{0}} \frac{1}{|\mathbf{n}|^2}\left(\frac{1}{\sum_{k=1}^N \mu_k^2}\left|\sum_{k=1}^N \mu_k e^{2\pi i\mathbf{n}\cdot\mathbf{y}_k}\right|^2 \left(\frac{\sin(2\pi|\mathbf{r}(t)|\,|\mathbf{n}|)}{2\pi|\mathbf{r}(t)|\,|\mathbf{n}|}\right)^2 \text{ or } 1\right)_{\max}$$

$$= \left(\frac{1}{\sum_{k=1}^N \mu_k^2}\max_{\mathbf{n}\in\mathbb{Z}^3\setminus\mathbf{0}}\left|\sum_{k=1}^N \mu_k e^{2\pi i\mathbf{n}\cdot\mathbf{y}_k}\right|^2 \left(\frac{\sin(2\pi|\mathbf{r}(t)|\,|\mathbf{n}|)}{2\pi|\mathbf{r}(t)|\,|\mathbf{n}|^2}\right)^2 \text{ or } 1\right)_{\max} \tag{14.3}$$

that we call the *Time Evolution Function of the second kind.*

For the sake of simplicity in this section we assume that every mass is one: $1=\mu_1=\ldots=\mu_N$ (so we omit $\mathcal{M}$ in the notation).

By a straightforward adaptation of the proof argument of Theorem 13.1 up to (13.22), we obtain that there is a measurable subset $\Omega_2^*\subset\Omega_2$ such that

$$\mathrm{ProdMeas}_2(\Omega_2^*) > 1-10^3\varepsilon, \tag{14.4}$$

and for every $\omega\in\Omega_2^*$ and every nonzero lattice point $\mathbf{n}\in\mathbb{Z}^3\setminus\mathbf{0}$ we have the analog of (13.22):

$$\left|\frac{\left|\sum_{\mathbf{z}_k\in\mathcal{Y}_{\mathbf{r};\omega;t}} e^{2\pi i\mathbf{n}\cdot\mathbf{z}_k}\right|^2}{N} - \frac{\overline{\mathbf{D}^2}_2(\mathbf{n};t)}{N}\right|$$

$$\le \frac{4|\mathbf{n}|^{3/2}\log(2+|\mathbf{n}|)}{\sqrt{\varepsilon}}\left(\left(\frac{\overline{\mathbf{D}^2}_2(\mathbf{n};t)}{N}\right)^{1/2}+2\right), \tag{14.5}$$

where

$$\overline{\mathbf{D}^2}_2(\mathbf{n};t) = \left|\sum_{k=1}^N e^{2\pi i\mathbf{n}\cdot\mathbf{y}_k}\right|^2 \left(\frac{\sin(2\pi|\mathbf{r}(t)|\,|\mathbf{n}|)}{2\pi|\mathbf{r}(t)|\,|\mathbf{n}|}\right)^2. \tag{14.6}$$

The only novelty here is that we replace (13.17′) with

$$\lambda_\varepsilon = \lambda_\varepsilon(\mathbf{n}) = \frac{4}{\sqrt{\varepsilon}}|\mathbf{n}|^{3/2}\log(2+|\mathbf{n}|)$$

$$\left(\mathrm{Eq}\cdot\overline{\mathbf{D}^2}_2(\mathbf{n};t) + \mathrm{Eq}^{3/2}\left(\overline{\mathbf{D}^2}_2(\mathbf{n};t)\right)^{1/2} + \mathrm{Eq}^2\right)^{1/2},$$

and replace the numerical series (13.18)

$$\sum_{\mathbf{n}\in\mathbb{Z}^3\setminus\mathbf{0}} \frac{1}{|\mathbf{n}|^4} < 10^2$$

with the series in (6.39)

$$\sum_{\mathbf{n}\in\mathbb{Z}^3\setminus\mathbf{0}} \frac{1}{|\mathbf{n}|^3\log^2(2+|\mathbf{n}|)} < 10^3.$$

By using the notation (14.6), we can rewrite the definition (14.3) of the Time Evolution Function as follows:

$$\mathrm{TiEvFun}_2(t) = \max_{\mathbf{n}\in\mathbb{Z}^3\setminus\mathbf{0}} \frac{1}{|\mathbf{n}|^2}\left(\frac{\overline{\mathbf{D}^2}_2(\mathbf{n};t)}{N} \text{ or } 1\right)_{\max}. \tag{14.7}$$

By (14.5),

$$\left|\frac{1}{|\mathbf{n}|^2}\left(\frac{\left|\sum_{\mathbf{z}_k\in\mathcal{Y}_{\mathbf{r};\omega;t}} e^{2\pi i\mathbf{n}\cdot\mathbf{z}_k}\right|^2}{N} \text{ or } 1\right)_{\max} - \frac{1}{|\mathbf{n}|^2}\left(\frac{\overline{\mathbf{D}^2}_2(\mathbf{n};t)}{N} \text{ or } 1\right)_{\max}\right|$$

$$\leq \frac{1}{|\mathbf{n}|^2}\frac{4|\mathbf{n}|^{3/2}\log(2+|\mathbf{n}|)}{\sqrt{\varepsilon}}\left(\left(\frac{\overline{\mathbf{D}^2}_2(\mathbf{n};t)}{N}\right)^{1/2} + 2\right)$$

$$= \frac{4\log(2+|\mathbf{n}|)}{\sqrt{\varepsilon}|\mathbf{n}|^{1/2}}\left(\left(\frac{\overline{\mathbf{D}^2}_2(\mathbf{n};t)}{N}\right)^{1/2} + 2\right). \tag{14.8}$$

Combining (14.7) and (14.8), we have

$$\left|\frac{1}{|\mathbf{n}|^2}\left(\frac{\left|\sum_{\mathbf{z}_k\in\mathcal{Y}_{\mathbf{r};\omega;t}} e^{2\pi i\mathbf{n}\cdot\mathbf{z}_k}\right|^2}{N} \text{ or } 1\right)_{\max} - \frac{1}{|\mathbf{n}|^2}\left(\frac{\overline{\mathbf{D}^2}_2(\mathbf{n};t)}{N} \text{ or } 1\right)_{\max}\right|$$

$$\leq \frac{4|\mathbf{n}|^{1/2}\log(2+|\mathbf{n}|)}{\sqrt{\varepsilon}}\left(\left(\frac{1}{|\mathbf{n}|^2}\left(\frac{\overline{\mathbf{D}^2}_2(\mathbf{n};t)}{N} \text{ or } 1\right)_{\max}\right)^{1/2} + 2\right)$$

$$\le \frac{4|\mathbf{n}|^{1/2}\log(2+|\mathbf{n}|)}{\sqrt{\varepsilon}}\left((\text{TiEvFun}_2(t))^{1/2}+2\right)$$

$$= \frac{4|\mathbf{n}|^{1/2}\log(2+|\mathbf{n}|)}{\sqrt{\varepsilon}\sqrt{\text{TiEvFun}_2(t)}}\left(1+\frac{2}{\sqrt{\text{TiEvFun}_2(t)}}\right)\cdot \text{TiEvFun}_2(t). \tag{14.9}$$

Comparing (14.9) with (13.38), we see the extra factor of $|\mathbf{n}|^{1/2}\log(2+|\mathbf{n}|)$. Since this factor can be arbitrarily large, (14.9) alone is not enough to obtain a Second Law. We need a second estimation. It comes from (14.8) and the triangle inequality

$$\frac{1}{|\mathbf{n}|^2}\left(\frac{\left|\sum_{\mathbf{z}_k\in\mathcal{Y}_{\mathbf{r};\omega;t}} e^{2\pi i\mathbf{n}\cdot\mathbf{z}_k}\right|^2}{N} \text{ or } 1\right)_{\max}$$

$$\le \frac{1}{|\mathbf{n}|^2}\left(\frac{\overline{\mathbf{D}^2}_2(\mathbf{n};t)}{N} \text{ or } 1\right)_{\max} + \frac{4\log(2+|\mathbf{n}|)}{\sqrt{\varepsilon}|\mathbf{n}|^{1/2}}\left(\left(\frac{\overline{\mathbf{D}^2}_2(\mathbf{n};t)}{N}\right)^{1/2}+2\right), \tag{14.10}$$

and we will combine it with the trivial upper bound

$$\overline{\mathbf{D}^2}_2(\mathbf{n};t) = \left|\sum_{k=1}^{N} e^{2\pi i\mathbf{n}\cdot\mathbf{y}_k}\right|^2\left(\frac{\sin(2\pi|\mathbf{r}(t)|\,|\mathbf{n}|)}{2\pi|\mathbf{r}(t)|\,|\mathbf{n}|}\right)^2$$

$$\le N^2\min\left\{\frac{1}{(2\pi|\mathbf{r}(t)|\,|\mathbf{n}|)^2},1\right\}. \tag{14.11}$$

Combining (14.9)–(14.11), we can derive a Second Law for the second kind *if* the initial point configuration $\mathcal{Y}=\{\mathbf{y}_1,\ldots,\mathbf{y}_N\}$ is "close to Big Bang" in the sense that more than half of the N points are concentrated in a very small ball. More precisely, we say that $\mathcal{Y}$ is (η,δ)-*close to Big Bang* if there is a point $\mathbf{w}\in[0,1)^3$ in the unit torus such that

$$\sum_{\substack{1\le k\le N:\\ |\mathbf{y}_k-\mathbf{w}|\le\eta}} 1 \ge \left(\frac{1}{2}+\delta\right)N. \tag{14.12}$$

Theorem 14.1. *Consider a 3-dimensional torus model $(\mathcal{Y};\mathbf{r};\Omega_2)$ of the second kind. Assume that the initial point configuration $\mathcal{Y}$ is (η,δ)-close to Big Bang with $\delta=1/10$ and*

$$\eta = \frac{1}{20\pi}e^{-\sqrt{\log N}}.$$

Given a time $t > 0$ *and an* $0 < \varepsilon < 10^{-3}$ *with*

$$\varepsilon \cdot \text{TiEvFun}_2(\mathcal{Y}; \mathbf{r}; t) \geq 10^6(1 + \sqrt{\log N})^2 e^{4\sqrt{\log N}}$$

and

$$3600|\mathbf{r}(t)|^2 e^{6\sqrt{\log(2|\mathbf{r}(t)|+1)}} \leq N e^{-4\sqrt{\log N}},$$

there exists a measurable subset $\Omega_2^* = \Omega_2^*(\varepsilon; t) \subset \Omega_2$ *of the initial velocity space of the second kind such that*

$$\text{ProdMeas}_2(\Omega_2^*) > 1 - 10^3\varepsilon$$

and for all $\omega \in \Omega_2^*$

$$|\text{MicroEntropy}_2(\mathcal{Y}_{\mathbf{r};\omega;t}) - (N - \text{TiEvFun}_2(\mathcal{Y}; \mathbf{r}; t))|$$

$$\leq \frac{6\left(3 + 2\sqrt{\log(2|\mathbf{r}(t)|)}\right) e^{\sqrt{\log(2|\mathbf{r}(t)|)}}}{\sqrt{\varepsilon}\sqrt{\text{TiEvFun}_2(t)}} \left(1 + \frac{2}{\sqrt{\text{TiEvFun}_2(t)}}\right) \cdot \text{TiEvFun}_2(t)$$

for $|\mathbf{r}(t)| \geq 5000$, *and*

$$|\text{MicroEntropy}_2(\mathcal{Y}_{\mathbf{r};\omega;t}) - (N - \text{TiEvFun}_2(\mathcal{Y}; \mathbf{r}; t))|$$

$$\leq \frac{200(2|\mathbf{r}(t)| + 1)(5 + \log(2|\mathbf{r}(t)| + 3))}{\sqrt{\varepsilon}\sqrt{\text{TiEvFun}_2(t)}} \left(1 + \frac{2}{\sqrt{\text{TiEvFun}_2(t)}}\right) \cdot \text{TiEvFun}_2(t)$$

for $0 < |\mathbf{r}(t)| < 5000$. *Here* $\text{TiEvFun}_2(t) = \text{TiEvFun}_2(\mathcal{Y}; \mathbf{r}; t)$, *see (14.3).*

Remark. The condition in Theorem 14.1 seems rather complicated, but it greatly simplifies if we make the observation that $e^{\sqrt{\log x}}$ is a "basically negligible" subpolynomial function of x, i.e.,

$$e^{\sqrt{\log x}} = x^{o(1)} \quad \text{as} \quad x \to \infty.$$

The weird function $e^{\sqrt{\log x}}$ comes from a number-theoretic lemma (see Lemma 14.1 below); it represents the best result that we can prove about a difficult diophantine approximation problem. It is well possible that $e^{\sqrt{\log x}}$ can be replaced by something much smaller — perhaps by $\log x$ — but we cannot prove it. Despite its "ugliness", Theorem 14.1 is a useful result that has some very interesting consequences, together with Theorems 13.1 and 13.2, that we will discuss in Sections 15–16.

Proof of Theorem 14.1. We may say, roughly speaking, that if $\mathcal{Y}$ is (η, δ)-*close to Big Bang* and η is very small then the critical sum in (14.6) is large. More precisely, we have

$$\left|\sum_{k=1}^{N} e^{2\pi i \mathbf{n}\cdot\mathbf{y}_k}\right| \geq \left|\sum_{\substack{1\leq k\leq N: \\ |\mathbf{y}_k - \mathbf{w}|\leq\eta}} e^{2\pi i \mathbf{n}\cdot\mathbf{y}_k}\right| - \left|\sum_{\substack{1\leq k\leq N: \\ |\mathbf{y}_k - \mathbf{w}|>\eta}} e^{2\pi i \mathbf{n}\cdot\mathbf{y}_k}\right|$$

$$\geq \left| \sum_{\substack{1\leq k\leq N:\\ |\mathbf{y}_k-\mathbf{w}|\leq\eta}} e^{2\pi i\mathbf{n}\cdot\mathbf{w}} \right| - \left| \sum_{\substack{1\leq k\leq N:\\ |\mathbf{y}_k-\mathbf{w}|\leq\eta}} \left(e^{2\pi i\mathbf{n}\cdot\mathbf{w}} - e^{2\pi i\mathbf{n}\cdot\mathbf{y}_k}\right) \right| - \left| \sum_{\substack{1\leq k\leq N:\\ |\mathbf{y}_k-\mathbf{w}|>\eta}} 1 \right|$$

$$\geq \left(\frac{1}{2}+\delta\right) N - 2\pi|\mathbf{n}|\eta N - \left(\frac{1}{2}-\delta\right) N = 2N\left(\delta - \pi|\mathbf{n}|\eta\right). \tag{14.13}$$

Moreover, we use the simple lower bound

$$|\sin(\pi x)| \geq \|x\|, \tag{14.14}$$

where $\|x\|$ denotes, as usual, the distance of a real number x from the nearest integer.

If $\mathcal{Y}$ is (η,δ)-*close to Big Bang*, then by (14.6) and (14.13)–(14.14) we have

$$\overline{\mathbf{D}^2}_2(\mathbf{n};t) \geq \left(2N\left(\delta - \pi|\mathbf{n}|\eta\right)^+\right)^2 \cdot \frac{1}{\pi^2}\left(\frac{\|(2|\mathbf{r}(t)|\ |\mathbf{n}|)\|}{2|\mathbf{r}(t)|\ |\mathbf{n}|}\right)^2, \tag{14.15}$$

where $(x)^+ = x$ if $x > 0$ and 0 otherwise.

To make use of (14.15), we need a number-theoretic result which says that, given an arbitrarily large real number x $(= 2|\mathbf{r}(t)|)$, there is a relatively small lattice point $\mathbf{n} = (n_1, n_2, n_3) \in \mathbb{Z}^3 \setminus \mathbf{0}$ such that

$$\frac{\|x\sqrt{n_1^2+n_2^2+n_3^2}\|}{x\sqrt{n_1^2+n_2^2+n_3^2}}$$

is not too small.

Lemma 14.1. *Given an arbitrary real number $x \geq 10^4$, there is a lattice point $\mathbf{n} = (n_1, n_2, n_3) \in \mathbb{Z}^3 \setminus \mathbf{0}$ such that*

$$|\mathbf{n}| = \sqrt{n_1^2+n_2^2+n_3^2} \leq e^{\sqrt{\log x}}$$

and

$$\frac{\|x\sqrt{n_1^2+n_2^2+n_3^2}\|}{x\sqrt{n_1^2+n_2^2+n_3^2}} \geq \frac{1}{3x}e^{-2\sqrt{\log x}}.$$

If $0 < x < 10^4$ then there is a lattice point $\mathbf{n} = (n_1, n_2, n_3) \in \mathbb{Z}^3 \setminus \mathbf{0}$ such that

$$|\mathbf{n}| = \sqrt{n_1^2+n_2^2+n_3^2} \leq 3\sqrt{x+1}$$

and

$$\frac{\|x\sqrt{n_1^2+n_2^2+n_3^2}\|}{x\sqrt{n_1^2+n_2^2+n_3^2}} \geq \frac{1}{24}(x+1)^{-3/2}.$$

We postpone the proof of Lemma 14.1 to the end of Section 16.

Let $\mathbf{n}_1 \in \mathbb{Z}^3 \setminus \mathbf{0}$ denote the nonzero lattice point for which $\text{Imba}_2(\mathcal{Y}_{\mathbf{r};\omega;t})$ attains its maximum (see (14.2)), and let $\mathbf{n}_2 \in \mathbb{Z}^3 \setminus \mathbf{0}$ denote the nonzero lattice point for which $\text{TiEvFun}_2(t)$ attains its maximum (see (14.3) and (14.6)).

We recall the hypothesis

$$\delta \geq 10^{-1} \quad \text{and} \quad 3600\pi^2|\mathbf{r}(t)|^2 e^{6\sqrt{\log(2|\mathbf{r}(t)|)}} \leq Ne^{-4\sqrt{\log N}}. \tag{14.16}$$

We distinguish two cases. The first case represents the main difficulty.

Case 1: Let

$$2|\mathbf{r}(t)| \geq 10^4.$$

We are now ready to make use of (14.15). We apply Lemma 14.1 with $x = 2|\mathbf{r}(t)|$: we obtain a lattice point $\mathbf{n}^* \in \mathbb{Z}^3 \setminus \mathbf{0}$ with

$$|\mathbf{n}^*| \leq e^{\sqrt{\log(2|\mathbf{r}(t)|)}} \tag{14.17}$$

such that

$$\frac{\|(2|\mathbf{r}(t)|\ |\mathbf{n}^*|)\|}{2|\mathbf{r}(t)|\ |\mathbf{n}^*|} \geq \frac{1}{6|\mathbf{r}(t)|} e^{-2\sqrt{\log(2|\mathbf{r}(t)|)}}. \tag{14.18}$$

By hypothesis

$$\frac{\delta}{2\eta\pi} \geq e^{\sqrt{\log(2|\mathbf{r}(t)|)}},$$

so we can rewrite (14.17) as

$$|\mathbf{n}^*| \leq e^{\sqrt{\log(2|\mathbf{r}(t)|)}} \leq \frac{\delta}{2\eta\pi}. \tag{14.19}$$

Then by (14.7), (14.15) and (14.18) we have

$$\text{TiEvFun}_2(t) \geq \frac{1}{|\mathbf{n}^*|^2} \left(\frac{\overline{\mathbf{D}^2}_2(\mathbf{n}^*;t)}{N} \ \text{ or } \ 1 \right)_{\max}$$

$$\geq \frac{1}{|\mathbf{n}^*|^2} \frac{\left(2N\left(\delta - \pi|\mathbf{n}^*|\eta\right)^+\right)^2}{N} \frac{1}{\pi^2} \left(\frac{\|(2|\mathbf{r}(t)|\ |\mathbf{n}^*|)\|}{2|\mathbf{r}(t)|\ |\mathbf{n}^*|} \right)^2$$

$$\geq \frac{1}{|\mathbf{n}^*|^2} \frac{\left(2N\left(\delta - \pi|\mathbf{n}^*|\eta\right)^+\right)^2}{N} \frac{1}{\pi^2} \frac{1}{36|\mathbf{r}(t)|^2} e^{-4\sqrt{\log(2|\mathbf{r}(t)|)}}$$

$$\geq \frac{N\,(\delta/2)^2}{9\pi^2|\mathbf{r}(t)|^2}e^{-6\sqrt{\log(2|\mathbf{r}(t)|)}} \geq \frac{N}{3600\pi^2|\mathbf{r}(t)|^2}e^{-6\sqrt{\log(2|\mathbf{r}(t)|)}}, \tag{14.20}$$

where in the last step we used (14.19) and the first half of (14.16) (x^+ denotes the "positive part of x", i.e., $x^+ = x$ if $x > 0$ and 0 otherwise).

By (14.9) and the triangle inequality,

$$\text{Imba}_2(\mathcal{Y}_{\mathbf{r};\omega;t}) = \max_{\mathbf{n}\in\mathbb{Z}^3\setminus\mathbf{0}} \frac{1}{|\mathbf{n}|^2}\left(\frac{\left|\sum_{\mathbf{z}_k\in\mathcal{Y}_{\mathbf{r};\omega;t}} e^{2\pi i\mathbf{n}\cdot\mathbf{z}_k}\right|^2}{N} \text{ or } 1\right)_{\max}$$

$$\geq \frac{1}{|\mathbf{n}_2|^2}\left(\frac{\left|\sum_{\mathbf{z}_k\in\mathcal{Y}_{\mathbf{r};\omega;t}} e^{2\pi i\mathbf{n}_2\cdot\mathbf{z}_k}\right|^2}{N} \text{ or } 1\right)_{\max} \geq \frac{1}{|\mathbf{n}_2|^2}\left(\frac{\overline{\mathbf{D}^2}_2(\mathbf{n}_2;t)}{N} \text{ or } 1\right)_{\max}$$

$$-\frac{4|\mathbf{n}_2|^{1/2}\log(2+|\mathbf{n}_2|)}{\sqrt{\varepsilon}\sqrt{\text{TiEvFun}_2(t)}}\left(1+\frac{2}{\sqrt{\text{TiEvFun}_2(t)}}\right)\cdot\text{TiEvFun}_2(t)$$

$$= \text{TiEvFun}_2(t)\left(1-\frac{4|\mathbf{n}_2|^{1/2}\log(2+|\mathbf{n}_2|)}{\sqrt{\varepsilon}\sqrt{\text{TiEvFun}_2(t)}}\left(1+\frac{2}{\sqrt{\text{TiEvFun}_2(t)}}\right)\right). \tag{14.21}$$

Next we return to (14.10)–(14.11); we have

$$\frac{1}{|\mathbf{n}|^2}\left(\frac{\left|\sum_{\mathbf{z}_k\in\mathcal{Y}_{\mathbf{r};\omega;t}} e^{2\pi i\mathbf{n}\cdot\mathbf{z}_k}\right|^2}{N} \text{ or } 1\right)_{\max}$$

$$\leq \frac{1}{|\mathbf{n}|^2}\left(\frac{N}{(2\pi|\mathbf{r}(t)|\;|\mathbf{n}|)^2} \text{ or } 1\right)_{\max}$$

$$+\frac{4\log(2+|\mathbf{n}|)}{\sqrt{\varepsilon}|\mathbf{n}|^{1/2}}\left(\left(\frac{N}{(2\pi|\mathbf{r}(t)|\;|\mathbf{n}|)^2}\right)^{1/2}+2\right). \tag{14.22}$$

Moreover, by (14.11) we have the trivial upper bound

$$\frac{1}{|\mathbf{n}|^2}\frac{\overline{\mathbf{D}^2}_2(\mathbf{n};t)}{N} \leq \frac{N}{4\pi^2|\mathbf{r}(t)|^2\;|\mathbf{n}|^4}. \tag{14.23}$$

Using the second half of (14.16) in (14.20), we have the lower bound

$$\frac{1}{|\mathbf{n}_2|^2}\left(\frac{\overline{\mathbf{D}^2}_2(\mathbf{n}_2;t)}{N} \text{ or } 1\right)_{\max} = \text{TiEvFun}_2(t)$$

$$\geq \frac{N}{3600\pi^2|\mathbf{r}(t)|^2}e^{-6\sqrt{\log(2|\mathbf{r}(t)|)}} \geq e^{4\sqrt{\log N}}. \tag{14.24}$$

For every lattice point $\mathbf{n} \in \mathbb{Z}^3$ with

$$|\mathbf{n}| > 8e^{2\sqrt{\log(2|\mathbf{r}(t)|)}} \tag{14.25}$$

(14.23) gives the lower bound

$$\frac{1}{|\mathbf{n}|^2}\frac{\overline{\mathbf{D}^2}_2(\mathbf{n};t)}{N} \leq \frac{N}{4\pi^2|\mathbf{r}(t)|^2\ |\mathbf{n}|^4}$$

$$\leq \frac{N}{8^4\cdot 4\pi^2|\mathbf{r}(t)|^2}e^{-8\sqrt{\log(2|\mathbf{r}(t)|)}} < \frac{N}{3600\pi^2|\mathbf{r}(t)|^2}e^{-6\sqrt{\log(2|\mathbf{r}(t)|)}}$$

$$\leq \frac{1}{|\mathbf{n}_2|^2}\left(\frac{\overline{\mathbf{D}^2}_2(\mathbf{n}_2;t)}{N} \text{ or } 1\right)_{\max} = \text{TiEvFun}_2(t), \tag{14.26}$$

where in the last step we used (14.24). It follows from (14.25)–(14.26) that

$$|\mathbf{n}_2| \leq 8e^{2\sqrt{\log(2|\mathbf{r}(t)|)}}. \tag{14.27}$$

By using (14.27) in (14.21), we obtain

$$\frac{1}{|\mathbf{n}_1|^2}\left(\frac{\left|\sum_{\mathbf{z}_k\in\mathcal{Y}_{\mathbf{r};\omega;t}} e^{2\pi i\mathbf{n}_1\cdot\mathbf{z}_k}\right|^2}{N} \text{ or } 1\right)_{\max} = \text{Imba}_2(\mathcal{Y}_{\mathbf{r};\omega;t})$$

$$\geq \text{TiEvFun}_2(t)\left(1 - \frac{4|\mathbf{n}_2|^{1/2}\log(2+|\mathbf{n}_2|)}{\sqrt{\varepsilon}\sqrt{\text{TiEvFun}_2(t)}} - \frac{8|\mathbf{n}_2|^{1/2}\log(2+|\mathbf{n}_2|)}{\sqrt{\varepsilon}\cdot\text{TiEvFun}_2(t)}\right)$$

$$\geq \text{TiEvFun}_2(t)\left(1 - 2\left(\frac{2\sqrt{8}e^{\sqrt{\log(2|\mathbf{r}(t)|)}} + 4\sqrt{8}e^{\sqrt{\log(2|\mathbf{r}(t)|)}}}{\sqrt{\varepsilon}\cdot\text{TiEvFun}_2(t)}\right)\right.$$

$$\left.\log\left(2+8e^{2\sqrt{\log(2|\mathbf{r}(t)|)}}\right)\right)$$

$$\geq \text{TiEvFun}_2(t)\left(1 - 2\left(\frac{2\sqrt{8}e^{\sqrt{\log(2|\mathbf{r}(t)|)}} + 4\sqrt{8}e^{\sqrt{\log(2|\mathbf{r}(t)|)}}}{\sqrt{\varepsilon}\cdot\text{TiEvFun}_2(t)}\right)\right.$$

$$\left.\left(3+2\sqrt{\log N}\right)\right). \tag{14.28}$$

Note that the upper bound

$$\left(\frac{2\sqrt{8}e^{\sqrt{\log(2|\mathbf{r}(t)|)}}+4\sqrt{8}e^{\sqrt{\log(2|\mathbf{r}(t)|)}}}{\sqrt{\varepsilon}\cdot\mathrm{TiEvFun}_2(t)}\right)\left(3+2\sqrt{\log N}\right)\le\frac{1}{4}$$

immediately follows from the hypothesis

$$\varepsilon\cdot\mathrm{TiEvFun}_2(t)\ge 10^6(1+\sqrt{\log N})^2e^{4\sqrt{\log N}}. \tag{14.29}$$

Using this upper bound in (14.28) we have

$$\frac{1}{|\mathbf{n}_1|^2}\left(\frac{\left|\sum_{\mathbf{z}_k\in\mathcal{Y}_{\mathbf{r};\omega;t}}e^{2\pi i\mathbf{n}_1\cdot\mathbf{z}_k}\right|^2}{N}\ \text{or}\ 1\right)_{\max}=\mathrm{Imba}_2(\mathcal{Y}_{\mathbf{r};\omega;t})$$

$$\ge\frac{1}{2}\mathrm{TiEvFun}_2(t)\ge\frac{N}{7200\pi^2|\mathbf{r}(t)|^2}e^{-6\sqrt{\log(2|\mathbf{r}(t)|)}}, \tag{14.30}$$

where in the last step we used (14.24).

On the other hand, for every lattice point $\mathbf{n}\in\mathbb{Z}^3$ with

$$|\mathbf{n}|>8e^{2\sqrt{\log(2|\mathbf{r}(t)|)}} \tag{14.31}$$

(14.22) gives the upper bound

$$\frac{1}{|\mathbf{n}|^2}\left(\frac{\left|\sum_{\mathbf{z}_k\in\mathcal{Y}_{\mathbf{r};\omega;t}}e^{2\pi i\mathbf{n}\cdot\mathbf{z}_k}\right|^2}{N}\ \text{or}\ 1\right)_{\max}$$

$$\le\frac{N}{4\pi^2|\mathbf{r}(t)|^2\ |\mathbf{n}|^4}+\frac{1}{|\mathbf{n}|^2}+\frac{4\log(2+|\mathbf{n}|)}{\sqrt{\varepsilon}|\mathbf{n}|^{1/2}}\left(\left(\frac{N}{(2\pi|\mathbf{r}(t)|\ |\mathbf{n}|)^2}\right)^{1/2}+2\right)$$

$$<\frac{N}{7200\pi^2|\mathbf{r}(t)|^2}e^{-6\sqrt{\log(2|\mathbf{r}(t)|)}}, \tag{14.32}$$

where the last inequality in (14.32) is just a routine calculation (it easily comes from (14.16), (14.24), (14.29) and (14.31) via trivial estimations).

Comparing (14.30) and (14.32) we obtain that

$$|\mathbf{n}_1|\le 8e^{2\sqrt{\log(2|\mathbf{r}(t)|)}}. \tag{14.33}$$

The next step is to repeat the argument of (13.50)–(13.51). By (14.9) and the triangle inequality,

$$\mathrm{Imba}_2(\mathcal{Y}_{\mathbf{r};\omega;t})=\frac{1}{|\mathbf{n}_1|^2}\left(\frac{\left|\sum_{\mathbf{z}_k\in\mathcal{Y}_{\mathbf{r};\omega;t}}e^{2\pi i\mathbf{n}_1\cdot\mathbf{z}_k}\right|^2}{N}\ \text{or}\ 1\right)_{\max}$$

$$\le \frac{1}{|\mathbf{n}_1|^2}\left(\frac{\overline{\mathbf{D}^2}_2(\mathbf{n}_1;t)}{N} \text{ or } 1\right)_{\max}$$

$$+\frac{4|\mathbf{n}_1|^{1/2}\log(2+|\mathbf{n}_1|)}{\sqrt{\varepsilon}\sqrt{\text{TiEvFun}_2(t)}}\left(1+\frac{2}{\sqrt{\text{TiEvFun}_2(t)}}\right)\cdot\text{TiEvFun}_2(t)$$

$$\le\left(1+\frac{4|\mathbf{n}_1|^{1/2}\log(2+|\mathbf{n}_1|)}{\sqrt{\varepsilon}\sqrt{\text{TiEvFun}_2(t)}}+\frac{8|\mathbf{n}_1|^{1/2}\log(2+|\mathbf{n}_1|)}{\sqrt{\varepsilon}\cdot\text{TiEvFun}_2(t)}\right)\cdot\text{TiEvFun}_2(t). \tag{14.34}$$

Similarly,

$$\text{Imba}_2(\mathcal{Y}_{\mathbf{r};\omega;t})\ge\frac{1}{|\mathbf{n}_2|^2}\left(\frac{\left|\sum_{\mathbf{z}_k\in\mathcal{Y}_{\mathbf{r};\omega;t}}e^{2\pi i\mathbf{n}_2\cdot\mathbf{z}_k}\right|^2}{N} \text{ or } 1\right)_{\max}$$

$$\ge \frac{1}{|\mathbf{n}_2|^2}\left(\frac{\overline{\mathbf{D}^2}_2(\mathbf{n}_2;t)}{N} \text{ or } 1\right)_{\max}$$

$$-\left(\frac{4|\mathbf{n}_2|^{1/2}\log(2+|\mathbf{n}_2|)}{\sqrt{\varepsilon}\sqrt{\text{TiEvFun}_2(t)}}+\frac{8|\mathbf{n}_2|^{1/2}\log(2+|\mathbf{n}_2|)}{\sqrt{\varepsilon}\cdot\text{TiEvFun}_2(t)}\right)\cdot\text{TiEvFun}_2(t)$$

$$=\left(1-\frac{4|\mathbf{n}_2|^{1/2}\log(2+|\mathbf{n}_2|)}{\sqrt{\varepsilon}\sqrt{\text{TiEvFun}_2(t)}}-\frac{8|\mathbf{n}_2|^{1/2}\log(2+|\mathbf{n}_2|)}{\sqrt{\varepsilon}\cdot\text{TiEvFun}_2(t)}\right)\cdot\text{TiEvFun}_2(t). \tag{14.35}$$

Combining (14.27) and (14.33)–(14.35), we obtain

$$|\text{Imba}_2(\mathcal{Y}_{\mathbf{r};\omega;t})-\text{TiEvFun}_2(t)|$$

$$\le\frac{4|n_3^{1/2}\log(2+n_3))|)}{\sqrt{\varepsilon}\sqrt{\text{TiEvFun}_2(t)}}\left(1+\frac{2}{\sqrt{\text{TiEvFun}_2(t)}}\right)\cdot\text{TiEvFun}_2(t), \tag{14.36}$$

where

$$n_3=\max\{|\mathbf{n}_1|,|\mathbf{n}_2|\}\le 8e^{2\sqrt{\log(2|\mathbf{r}(t)|)}}. \tag{14.37}$$

By (14.37),

$$n_3^{1/2}\log(2+n_3)\le\sqrt{8}e^{\sqrt{\log(2|\mathbf{r}(t)|)}}\left(3+2\sqrt{\log(2|\mathbf{r}(t)|)}\right),$$

and using it (14.36), we have

$$|\text{Imba}_2(\mathcal{Y}_{\mathbf{r};\omega;t})-\text{TiEvFun}_2(t)|$$

$$\leq \frac{6\left(3+2\sqrt{\log(2|\mathbf{r}(t)|)}\right)e^{\sqrt{\log(2|\mathbf{r}(t)|)}}}{\sqrt{\varepsilon}\sqrt{\text{TiEvFun}_2(t)}}\left(1+\frac{2}{\sqrt{\text{TiEvFun}_2(t)}}\right)\cdot\text{TiEvFun}_2(t), \tag{14.38}$$

which completes Case 1.

Next comes

Case 2: Let

$$0 < 2|\mathbf{r}(t)| < 10^4.$$

We basically repeat the argument of Case 1. In fact, Case 2 is simpler. Again we make use of (14.15). We apply Lemma 14.1 with $x = 2|\mathbf{r}(t)|$: we obtain a lattice point $\mathbf{n}^* \in \mathbb{Z}^3 \setminus \mathbf{0}$ with

$$|\mathbf{n}^*| \leq 3\sqrt{2|\mathbf{r}(t)|+1} \tag{14.39}$$

such that

$$\frac{\|(2|\mathbf{r}(t)|\ |\mathbf{n}^*|)\|}{2|\mathbf{r}(t)|\ |\mathbf{n}^*|} \geq \frac{1}{24}(2|\mathbf{r}(t)|+1)^{-3/2}. \tag{14.40}$$

By hypothesis

$$\frac{\delta}{2\eta\pi} \geq 3\sqrt{2|\mathbf{r}(t)|+1},$$

so we can rewrite (14.39) as

$$|\mathbf{n}^*| \leq 3\sqrt{2|\mathbf{r}(t)|+1} \leq \frac{\delta}{2\eta\pi}. \tag{14.41}$$

Then by (14.7), (14.15) and (14.40) we have

$$\text{TiEvFun}_2(t) \geq \frac{1}{|\mathbf{n}^*|^2}\left(\frac{\overline{\mathbf{D}^2}_2(\mathbf{n}^*;t)}{N} \text{ or } 1\right)_{\max}$$

$$\geq \frac{1}{|\mathbf{n}^*|^2}\frac{\left(2N\left(\delta-\pi|\mathbf{n}^*|\eta\right)^+\right)^2}{N}\frac{1}{\pi^2}\left(\frac{\|(2|\mathbf{r}(t)|\ |\mathbf{n}^*|)\|}{2|\mathbf{r}(t)|\ |\mathbf{n}^*|}\right)^2$$

$$\geq \frac{1}{|\mathbf{n}^*|^2}\frac{\left(2N\left(\delta-\pi|\mathbf{n}^*|\eta\right)^+\right)^2}{N}\frac{1}{576\pi^2(2|\mathbf{r}(t)|+1)^3}$$

$$\geq \frac{N\left(\delta/2\right)^2}{144\pi^2(2|\mathbf{r}(t)|+1)^3} \geq \frac{20^{-2}N}{144\pi^2(2|\mathbf{r}(t)|+1)^3}, \tag{14.42}$$

where in the last step we used (14.41) and the first half of (14.16).

We recall (14.21):

$$\text{Imba}_2(\mathcal{Y}_{\mathbf{r};\omega;t}) = \max_{\mathbf{n}\in\mathbb{Z}^3\setminus\mathbf{0}} \frac{1}{|\mathbf{n}|^2}\left(\frac{\left|\sum_{\mathbf{z}_k\in\mathcal{Y}_{\mathbf{r};\omega;t}} e^{2\pi i\mathbf{n}\cdot\mathbf{z}_k}\right|^2}{N} \text{ or } 1\right)_{\max}$$

$$\geq \text{TiEvFun}_2(t)\left(1 - \frac{4|\mathbf{n}_2|^{1/2}\log(2+|\mathbf{n}_2|)}{\sqrt{\varepsilon}\sqrt{\text{TiEvFun}_2(t)}} - \frac{8|\mathbf{n}_2|^{1/2}\log(2+|\mathbf{n}_2|)}{\sqrt{\varepsilon}\cdot\text{TiEvFun}_2(t)}\right). \tag{14.43}$$

Next we return to (14.10)–(14.11); since $2|\mathbf{r}(t)| \geq 10^4$, we have

$$\frac{1}{|\mathbf{n}|^2}\left(\frac{\left|\sum_{\mathbf{z}_k\in\mathcal{Y}_{\mathbf{r};\omega;t}} e^{2\pi i\mathbf{n}\cdot\mathbf{z}_k}\right|^2}{N} \text{ or } 1\right)_{\max}$$

$$\leq \frac{1}{|\mathbf{n}|^2}\left(\min\left\{\frac{N}{(2\pi|\mathbf{r}(t)|\,|\mathbf{n}|)^2}, N\right\} \text{ or } 1\right)_{\max}$$

$$+\frac{4\log(2+|\mathbf{n}|)}{\sqrt{\varepsilon}|\mathbf{n}|^{1/2}}\left(\left(\min\left\{\frac{N}{(2\pi|\mathbf{r}(t)|\,|\mathbf{n}|)^2}, N\right\}\right)^{1/2} + 2\right). \tag{14.44}$$

Moreover, by (14.11) we have the trivial upper bound

$$\frac{1}{|\mathbf{n}|^2}\frac{\overline{\mathbf{D}^2}_2(\mathbf{n};t)}{N} \leq \min\left\{\frac{N}{4\pi^2|\mathbf{r}(t)|^2\,|\mathbf{n}|^4}, \frac{N}{|\mathbf{n}|^2}\right\}. \tag{14.45}$$

By (14.42) we have the lower bound

$$\frac{1}{|\mathbf{n}_2|^2}\left(\frac{\overline{\mathbf{D}^2}_2(\mathbf{n}_2;t)}{N} \text{ or } 1\right)_{\max} = \text{TiEvFun}_2(t)$$

$$\geq \frac{20^{-2}N}{144\pi^2(2|\mathbf{r}(t)|+1)^3} \geq \sqrt{N}, \tag{14.46}$$

since $2|\mathbf{r}(t)| < 10^4$. For every lattice point $\mathbf{n} \in \mathbb{Z}^3$ with

$$|\mathbf{n}| > 10^3(2|\mathbf{r}(t)|+1)^2 \tag{14.47}$$

(14.45) gives the lower bound

$$\frac{1}{|\mathbf{n}|^2}\frac{\overline{\mathbf{D}^2}_2(\mathbf{n};t)}{N} \leq \min\left\{\frac{N}{4\pi^2|\mathbf{r}(t)|^2\,|\mathbf{n}|^4}, \frac{N}{|\mathbf{n}|^2}\right\}$$

$$\leq \min\left\{\frac{N}{10^{13}\cdot 4\pi^2|\mathbf{r}(t)|^2}(2|\mathbf{r}(t)|+1)^{-8}, 10^{-6}(2|\mathbf{r}(t)|+1)^{-4}N\right\}$$

$$< \frac{20^{-2}N}{144\pi^2(2|\mathbf{r}(t)|+1)^3} \leq \frac{1}{|\mathbf{n}_2|^2}\left(\frac{\overline{\mathbf{D}^2}_2(\mathbf{n}_2;t)}{N} \text{ or } 1\right)_{\max} = \text{TiEvFun}_2(t), \tag{14.48}$$

where in the last step we used (14.46). It follows from (14.47)–(14.48) that

$$|\mathbf{n}_2| \leq 10^3(2|\mathbf{r}(t)|+1)^2. \tag{14.49}$$

By using (14.46) and (14.49) in (14.43), we obtain

$$\frac{1}{|\mathbf{n}_1|^2}\left(\frac{\left|\sum_{\mathbf{z}_k\in\mathcal{Y}_{\mathbf{r};\omega;t}} e^{2\pi i\mathbf{n}_1\cdot\mathbf{z}_k}\right|^2}{N} \text{ or } 1\right)_{\max} = \text{Imba}_2(\mathcal{Y}_{\mathbf{r};\omega;t})$$

$$\geq \text{TiEvFun}_2(t)\left(1 - \frac{4|\mathbf{n}_2|^{1/2}\log(2+|\mathbf{n}_2|)}{\sqrt{\varepsilon}\sqrt{\text{TiEvFun}_2(t)}} - \frac{8|\mathbf{n}_2|^{1/2}\log(2+|\mathbf{n}_2|)}{\sqrt{\varepsilon}\cdot\text{TiEvFun}_2(t)}\right)$$

$$\geq \text{TiEvFun}_2(t)\left(1 - \frac{2\cdot 10^{3/2}(2|\mathbf{r}(t)|+1)}{\sqrt{\varepsilon}N^{1/4}}\log\left(2+10^3(2|\mathbf{r}(t)|+1)^2\right)\left(1+\frac{2}{N^{1/4}}\right)\right)$$

$$\geq \frac{1}{2}\text{TiEvFun}_2(t) \geq \frac{20^{-2}N}{288\pi^2(2|\mathbf{r}(t)|+1)^3}. \tag{14.50}$$

On the other hand, for every lattice point $\mathbf{n}\in\mathbb{Z}^3$ with

$$|\mathbf{n}| > 10^3(2|\mathbf{r}(t)|+1)^2 \tag{14.51}$$

(14.44) gives the upper bound

$$\frac{1}{|\mathbf{n}|^2}\left(\frac{\left|\sum_{\mathbf{z}_k\in\mathcal{Y}_{\mathbf{r};\omega;t}} e^{2\pi i\mathbf{n}\cdot\mathbf{z}_k}\right|^2}{N} \text{ or } 1\right)_{\max}$$

$$\leq \frac{1}{|\mathbf{n}|^2}\left(\min\left\{\frac{N}{(2\pi|\mathbf{r}(t)|\,|\mathbf{n}|)^2}, N\right\} \text{ or } 1\right)_{\max}$$

$$+ \frac{4\log(2+|\mathbf{n}|)}{\sqrt{\varepsilon}|\mathbf{n}|^{1/2}}\left(\left(\min\left\{\frac{N}{(2\pi|\mathbf{r}(t)|\,|\mathbf{n}|)^2}, N\right\}\right)^{1/2} + 2\right)$$

$$\leq \frac{20^{-2}N}{288\pi^2(2|\mathbf{r}(t)|+1)^3}, \tag{14.52}$$

where the last inequality in (14.52) is just a trivial calculation.

Comparing (14.50) and (14.52) we obtain that

$$|\mathbf{n}_1| \le 10^3(2|\mathbf{r}(t)|+1)^2. \tag{14.53}$$

Repeating the argument of (14.34)–(14.35), we obtain the following analog of (14.36):

$$|\text{Imba}_2(\mathcal{Y}_{\mathbf{r};\omega;t}) - \text{TiEvFun}_2(t)|$$

$$\le \frac{4n_3^{1/2}\log(2+n_3)}{\sqrt{\varepsilon}\sqrt{\text{TiEvFun}_2(t)}}\left(1+\frac{2}{\sqrt{\text{TiEvFun}_2(t)}}\right)\cdot \text{TiEvFun}_2(t), \tag{14.54}$$

where

$$n_3 = \max\{|\mathbf{n}_1|,|\mathbf{n}_2|\} \le 10^3(2|\mathbf{r}(t)|+1)^2. \tag{14.55}$$

By (14.55),

$$n_3^{1/2}\log(2+n_3) \le 10^{3/2}(2|\mathbf{r}(t)|+1)\left(10+2\log(2|\mathbf{r}(t)|+3)\right),$$

and using it (14.54), we have

$$|\text{Imba}_2(\mathcal{Y}_{\mathbf{r};\omega;t}) - \text{TiEvFun}_2(t)|$$

$$\le \frac{200(2|\mathbf{r}(t)|+1)\left(5+\log(2|\mathbf{r}(t)|+3)\right)}{\sqrt{\varepsilon}\sqrt{\text{TiEvFun}_2(t)}}\left(1+\frac{2}{\sqrt{\text{TiEvFun}_2(t)}}\right)\cdot \text{TiEvFun}_2(t). \tag{14.56}$$

Finally, note that (14.38) and (14.56) complete the proof of Theorem 14.1. □

15. Illustrations of the Second Law (I): Third kind

The key functions in Theorems 13.1, 13.2 and 14.1 are the ω-independent Time Evolution Functions

$$\text{TiEvFun}_1(\mathcal{Y},\mathcal{M};\mathbf{r};t)$$

$$= \left(e^{-(2\pi|\mathbf{r}(t)|)^2}\max_{\mathbf{n}\in\mathbb{Z}^3\setminus\mathbf{0}}\left(\frac{\left|\sum_{k=1}^N \mu_k e^{2\pi i\mathbf{n}\cdot\mathbf{y}_k}\right|^2}{\sum_{k=1}^N \mu_k^2}\right)^{1/|\mathbf{n}|^2} \text{ or } 1\right)_{\max} \tag{15.1}$$

for the first kind,

$$\text{TiEvFun}_3(\mathcal{Y},\mathcal{M};\mathbf{r};t)$$

$$= \left(\frac{9}{\sum_{k=1}^{N} \mu_k^2} \max_{\mathbf{n} \in \mathbb{Z}^3 \setminus \mathbf{0}} \left| \sum_{k=1}^{N} \mu_k e^{2\pi i \mathbf{n} \cdot \mathbf{y}_k} \right|^2 \right.$$

$$\left. \cdot \left(\frac{\cos(2\pi |\mathbf{r}(t)|\ |\mathbf{n}|)}{4\pi^2 |\mathbf{r}(t)|^2 |\mathbf{n}|^4} - \frac{\sin(2\pi |\mathbf{r}(t)|\ |\mathbf{n}|)}{8\pi^3 |\mathbf{r}(t)|^3 |\mathbf{n}|^5} \right)^2 \text{ or } 1 \right)_{\max} \tag{15.2}$$

for the third kind, and finally,

$$\text{TiEvFun}_2(\mathcal{Y}, \mathcal{M}; \mathbf{r}; t)$$

$$= \left(\frac{1}{\sum_{k=1}^{N} \mu_k^2} \max_{\mathbf{n} \in \mathbb{Z}^3 \setminus \mathbf{0}} \left| \sum_{k=1}^{N} \mu_k e^{2\pi i \mathbf{n} \cdot \mathbf{y}_k} \right|^2 \left(\frac{\sin(2\pi |\mathbf{r}(t)|\ |\mathbf{n}|)}{2\pi |\mathbf{r}(t)|\ |\mathbf{n}|^2} \right)^2 \text{ or } 1 \right)_{\max} \tag{15.3}$$

for the second kind (all in the standard 3-dimensional case).

The Time Evolution Function for the first kind is shockingly simple: (15.1) has the form

$$\text{TiEvFun}_1(\mathcal{Y}, \mathcal{M}; \mathbf{r}; t) = \left(C \cdot e^{-(2\pi |\mathbf{r}(t)|)^2} \text{ or } 1 \right)_{\max},$$

where $C = C(\mathcal{Y}, \mathcal{M})$ is just a positive constant depending only on the initial condition $(\mathcal{Y}, \mathcal{M})$.

Formulas (15.2) and (15.3) are more complicated. They are expressed in the form of a maximum taken over *infinitely* many terms (corresponding to the infinitely many nonzero lattice points $\mathbf{n} \in \mathbb{Z}^3 \setminus \mathbf{0}$).

We show that in some interesting special cases (15.2) can be reduced to a maximum taken over a bounded number of terms, where the bound is independent of N.

Consider first the special case where the initial point configuration is a Big Bang $\mathbf{y}_1 = \mathbf{y}_2 = \ldots = \mathbf{y}_N$, formally $\mathcal{Y} = \mathcal{Y}_{BB}$, and the particles have unit mass $\mu_k = 1$ for all $1 \le k \le N$ (so we omit $\mathcal{M}$).

Note that in this special case (15.1) simplifies to

$$\text{TiEvFun}_1(\mathcal{Y}_{BB}; \mathbf{r}; t) = \left(N e^{-(2\pi |\mathbf{r}(t)|)^2} \text{ or } 1 \right)_{\max}. \tag{15.4}$$

In this special case we have the straightforward reduction of formula (15.2) (i.e., the third kind) to

$$\text{TiEvFun}_3(\mathcal{Y}_{BB}; \mathbf{r}; t)$$

$$= \left(9N \max_{\mathbf{n} \in \mathbb{Z}^3 \setminus \mathbf{0}} \left(\frac{\cos(2\pi |\mathbf{r}(t)|\ |\mathbf{n}|)}{4\pi^2 |\mathbf{r}(t)|^2 |\mathbf{n}|^4} - \frac{\sin(2\pi |\mathbf{r}(t)|\ |\mathbf{n}|)}{8\pi^3 |\mathbf{r}(t)|^3 |\mathbf{n}|^5} \right)^2 \text{ or } 1 \right)_{\max}. \tag{15.5}$$

This raises the question: can we make a further reduction of (15.5), something similar to (15.4)?

The answer is yes. We show that, if $|\mathbf{r}(t)| \geq 10$ then the third kind $\text{TiEvFun}_3(\mathcal{Y}_{BB}; \mathbf{r}; t)$ reduces to a maximum taken over *four* terms:

$$\text{TiEvFun}_3(\mathcal{Y}_{BB}; \mathbf{r}; t)$$

$$= \left(9N \max_{1 \leq m \leq 4} \left(\frac{\cos(2\pi\sqrt{m}|\mathbf{r}(t)|)}{4\pi^2 m^2 |\mathbf{r}(t)|^2} - \frac{\sin(2\pi\sqrt{m}|\mathbf{r}(t)|)}{8\pi^3 m^{5/2} |\mathbf{r}(t)|^3} \right)^2 \text{ or } 1 \right)_{\max}. \tag{15.6}$$

The proof of (15.6) is based on the elementary trigonometric fact

$$\max\left\{ \cos^2 x, \frac{1}{256}\cos^2(2x) \right\} \geq \frac{1}{260} \text{ holds for every } x. \tag{15.7}$$

To verify (15.7), we distinguish two cases. If $\cos^2 x \geq 1/260$ then we are done. If $\cos^2 x < 1/260$, then the identity

$$1 - \cos^2(2x) = (1 - \cos(2x))\,(1 + \cos(2x))$$

$$= 2\,(1 - \cos(2x)) \left(\frac{1 + \cos(2x)}{2} \right) = 2\,(1 - \cos(2x)) \cdot \cos^2 x$$

implies that

$$1 - \cos^2(2x) < 4 \cdot \frac{1}{260}, \quad \text{and so} \quad \frac{1}{256}\cos^2(2x) > \frac{1}{256} \cdot \frac{256}{260} = \frac{1}{260},$$

which proves (15.7).

The reason why (15.7) implies (15.6) is that for $y \geq 10$ we have

$$\left| \left(\frac{\cos(2\pi y)}{(2\pi y)^2} - \frac{\sin(2\pi y)}{(2\pi y)^3} \right)^2 - \frac{\cos^2(2\pi y)}{(2\pi y)^4} \right| \leq \frac{2}{(2\pi y)^5} = \text{negligible},$$

and also

$$\left(\frac{1}{\sqrt{5}} \right)^8 = \frac{1}{625}$$

is much smaller than $1/260$ (see (15.7)), which explains why in (15.6) we can restrict ourselves to the four terms $\max_{1 \leq m \leq 4}$, ignoring the rest $m \geq 5$ of the integers of the form $m = n_1^2 + n_2^2 + n_3^2$, $n_i \in \mathbb{Z}$.

Finally consider the second kind: in this special case we have the trivial reduction of (15.3) to

$$\text{TiEvFun}_2(\mathcal{Y}_{BB}; \mathbf{r}; t) = \left(N \max_{\mathbf{n} \in \mathbb{Z}^3 \setminus \mathbf{0}} \left(\frac{\sin(2\pi|\mathbf{r}(t)|\,|\mathbf{n}|)}{2\pi|\mathbf{r}(t)|\,|\mathbf{n}|^2} \right)^2 \text{ or } 1 \right)_{\max}. \tag{15.8}$$

Again it raises the question: can we make a further reduction of (15.8), something similar to (15.4) or (15.6)? That is, can we replace $\max_{\mathbf{n}\in\mathbb{Z}^3\setminus\mathbf{0}}$ with a maximum that is restricted to a bounded number of lattice points, where the bound is independent of N? Here the answer is *no*. The negative answer is a straightforward consequence of the following number-theoretic fact

$$\inf_{x>\pi} \sup_{m=n_1^2+n_2^2+n_3^2\geq 1} \frac{1}{m^2}\sin^2(\sqrt{m}x) = 0. \tag{15.9}$$

Note that (15.9) is the complete opposite of (15.7). Indeed, we can rewrite (15.7) in the form

$$\inf_{x>0} \max\left\{\cos^2 x, \frac{1}{256}\cos^2(2x)\right\} \geq \frac{1}{260},$$

where the lower bound is a positive absolute constant $1/260$ (instead of the zero in (15.9)).

Notice that (15.9) is a simple corollary of the classical Dirichlet's theorem in simultaneous diophantine approximation. Indeed, let $\varepsilon > 0$ be arbitrarily small and let $M < \infty$ be arbitrarily large. Let m run through the (finite) set of integers of the form $1 \leq m = n_1^2 + n_2^2 + n_3^2 \leq M$, $n_i \in \mathbb{Z}$. By Dirichlet's theorem there exists a real number $x_0 > \pi$ such that $\|\sqrt{m}x_0/\pi\| < \varepsilon$ holds for every integer m of the form $1 \leq m = n_1^2+n_2^2+n_3^2 \leq M$. Thus we have

$$\sup_{m=n_1^2+n_2^2+n_3^2\geq 1} \frac{1}{m^2}\sin^2(\sqrt{m}x_0) \leq \max\left\{\varepsilon^2\pi^2, \frac{1}{M^2}\right\},$$

which implies (15.9).

For simplicity assume now the extra hypothesis that $|\mathbf{r}(t)|$ is a *monotone increasing* function of t. Then the Time Evolution Function of the first kind

$$\text{TiEvFun}_1(\mathcal{Y}_{BB}; \mathbf{r}; t) = \left(Ne^{-(2\pi|\mathbf{r}(t)|)^2} \text{ or } 1\right)_{\max}$$

is a *monotone decreasing* function of the time t (note that this remains true even in the most general case (15.1)). On the other hand, the Time Evolution Function of the third kind

$$\text{TiEvFun}_3(\mathcal{Y}_{BB}; \mathbf{r}; t)$$

$$= \left(9N \max_{1\leq m\leq 4}\left(\frac{\cos(2\pi\sqrt{m}|\mathbf{r}(t)|)}{4\pi^2m^2|\mathbf{r}(t)|^2} - \frac{\sin(2\pi\sqrt{m}|\mathbf{r}(t)|)}{8\pi^3m^{5/2}|\mathbf{r}(t)|^3}\right)^2 \text{ or } 1\right)_{\max}$$

is not monotone decreasing, but it still exhibits a *tendency to decrease with oscillations*. Here the oscillations are due to the function $\cos^2 x$, and the

tendency to decrease more precisely means that
the mean value of

$$\left(9N\max_{1\le m\le 4}\left(\frac{\cos(2\pi\sqrt{m}|\mathbf{r}(t)|)}{4\pi^2m^2|\mathbf{r}(t)|^2}-\frac{\sin(2\pi\sqrt{m}|\mathbf{r}(t)|)}{8\pi^3m^{5/2}|\mathbf{r}(t)|^3}\right)^2\text{ or }1\right)_{\max}$$

equals to

$$\left(9N\max_{1\le m\le 4}\frac{1/2}{16\pi^4m^4|\mathbf{r}(t)|^4}\text{ or }1\right)_{\max} \tag{15.10}$$

if $|\mathbf{r}(t)|$ is large (due to the fact $\int_0^\pi \cos^2 x\,dx=\pi/2$). Now (15.10) is clearly monotone decreasing (assuming $|\mathbf{r}(t)|$ is a monotone increasing function of t).

Similarly, the Time Evolution Function of the second kind

$$\text{TiEvFun}_2(\mathcal{Y}_{BB};\mathbf{r};t)=\left(N\max_{\mathbf{n}\in\mathbb{Z}^3\setminus\mathbf{0}}\left(\frac{\sin(2\pi|\mathbf{r}(t)|\;|\mathbf{n}|)}{2\pi|\mathbf{r}(t)|\;|\mathbf{n}|^2}\right)^2\text{ or }1\right)_{\max}$$

is not monotone decreasing, but it also exhibits a *tendency to decrease with oscillations*, where the oscillations are due to the function $\sin^2 x$. Here the "mean value" is

$$\left(N\max_{\mathbf{n}\in\mathbb{Z}^3\setminus\mathbf{0}}\frac{1/2}{4\pi^2|\mathbf{n}|^4|\mathbf{r}(t)|^2}\text{ or }1\right)_{\max} \tag{15.11}$$

if $|\mathbf{r}(t)|$ is large, and again (15.11) is clearly monotone decreasing (assuming $|\mathbf{r}(t)|$ is a monotone increasing function of t).

Note that the oscillations in the second and third kinds are *substantial*. To explain it, we assume, for simplicity, that $|\mathbf{r}(t)|=t$ (this interesting special case includes, among others, the Bernoulli billiard/torus model). In the case of the third kind, it suffices to study the function

$$f_3(t)=\max_{1\le m\le 4}\frac{\cos^2(2\pi\sqrt{m}t)}{m^4},\quad 1<t<\infty$$

for large t. It is obvious that $f_3(t)=\cos^2(2\pi t)$ as long as $\cos^2(2\pi t)>1/16$. So $f_3(t)$ follows the oscillations of $\cos^2(2\pi t)$ for the majority of t.

In the case of the second kind it suffices to study the function

$$f_2(t)=\max_{\mathbf{n}\in\mathbb{Z}^3\setminus\mathbf{0}}\frac{\sin^2(2\pi|\mathbf{n}|t)}{|\mathbf{n}|^4},\quad 1<t<\infty$$

for large t. It is obvious that $f_2(t)=\sin^2(2\pi t)$ as long as $\sin^2(2\pi t)>1/4$. So $f_2(t)$ follows the oscillations of $\sin^2(2\pi t)$ for the majority of t.

Next we study how well Theorems 13.1, 13.2 and 14.1 describe the typical time evolution of an individual system. Again we study the same special case: the initial point configuration is a Big Bang $\mathbf{y}_1 = \mathbf{y}_2 = \ldots = \mathbf{y}_N$, formally $\mathcal{Y} = \mathcal{Y}_{BB}$, and the particles have unit mass $\mu_k = 1$ for all $1 \le k \le N$ (so we omit $\mathcal{M}$). We also assume that $|\mathbf{r}(t)|$ is a linear function of the time: let $|\mathbf{r}(t)| = t$.

In view of the simple complementary relation between the Imbalance and our entropy

$$\text{MicroEntropy} + \text{Imbalance} = N$$

it suffices to study the time evolution of the Imbalance.

We start with the third kind, and, to make the illustration more striking, we assume that the number of particles N is much larger than the Avogadro number: let $N = 10^{1000}$. We recall Theorem 13.2: given a time $t > 0$ and $0 < \varepsilon < 1$, there exists a measurable subset $\Omega_3^* = \Omega_3^*(\varepsilon; t) \subset \Omega_3$ of the initial velocity space of the third kind such that

$$\text{ProdMeas}_3(\Omega_3^*) > 1 - 10^2\varepsilon \tag{15.12}$$

and

$$|\text{Imba}_3(\mathcal{Y}_{\mathbf{r};\omega;t}) - \text{TiEvFun}_3(t)|$$

$$\le \frac{2}{\sqrt{\varepsilon}\sqrt{\text{TiEvFun}_3(t)}}\left(1 + \frac{2}{\sqrt{\text{TiEvFun}_3(t)}}\right) \cdot \text{EvolFun}_3(t) \tag{15.13}$$

holds for all $\omega \in \Omega_3^*$, where $\mathcal{Y} = \mathcal{Y}_{BB}$ is a Big Bang, and

$$\text{TiEvFun}_3(t) = \text{TiEvFun}_3(\mathcal{Y}_{BB}; \mathbf{r}; t)$$

$$= \left(9N \max_{1\le m\le 4}\left(\frac{\cos(2\pi\sqrt{m}t)}{4\pi^2 m^2 t^2} - \frac{\sin(2\pi\sqrt{m}t)}{8\pi^2 m^{5/2} t^3}\right)^2 \text{ or } 1\right)_{\max} \tag{15.14}$$

(we used $|\mathbf{r}(t)| = t$).

Since we start from Big Bang, we have $\text{TiEvFun}_3(0) = \text{TiEvFun}_3(\mathcal{Y}_{BB}; \mathbf{r}; 0) = N = 10^{1000}$. We study the time evolution of a typical system until $\text{TiEvFun}_3(t)$ goes down to $\text{TiEvFun}_3(t) = (1 + o(1))10^{400}$. Since

$$\sup_{\tau \ge t} \text{TiEvFun}_3(\mathcal{Y}_{BB}; \mathbf{r}; \tau) = \left(1 + O\left(\frac{1}{1+t}\right)\right) \frac{9N}{(2\pi t)^4},$$

we have

$$\left(1 + O\left(\frac{1}{1+t}\right)\right) \frac{9N}{(2\pi t)^4} = 10^{400} \iff t = (1 + o(1))\frac{\sqrt{3}}{2\pi}10^{150},$$

that is, we study the time interval

$$0 < t < T = \frac{\sqrt{3}}{2\pi} 10^{150}. \tag{15.15}$$

By (15.7),

$$\text{TiEvFun}_3(t) \geq \frac{10^{400}}{260} \quad \text{for all } 0 < t < T = \frac{\sqrt{3}}{2\pi} 10^{150}. \tag{15.16}$$

Let

$$\varepsilon = 10^{-300}. \tag{15.17}$$

Applying Theorem 13.2 (see (15.12)–(15.13)), for every t in $0 < t < T$ there exists a measurable subset $\Omega_3^*(\varepsilon; t) \subset \Omega_3$ of the initial velocity space of the third kind such that

$$\text{ProdMeas}_3(\Omega_3^*(\varepsilon; t)) > 1 - 10^2\varepsilon \tag{15.18}$$

and

$$|\text{Imba}_3(\mathcal{Y}_{\mathbf{r};\omega;t}) - \text{TiEvFun}_3(t)|$$

$$\leq \frac{4}{\sqrt{\varepsilon}\sqrt{\text{TiEvFun}_3(t)}} \left(1 + \frac{2}{\sqrt{\text{TiEvFun}_3(t)}}\right) \cdot \text{EvolFun}_3(t) \tag{15.19}$$

holds for every $\omega \in \Omega_3^*(\varepsilon; t)$.

Let $0 < \delta < 1$ be a real number, and consider the following subset of Ω_3:

$$\Omega_3^{**}(\delta) = \{\omega \in \Omega_3 : \ \omega \notin \Omega_3^*(\varepsilon; t) \text{ holds for a set of } t \in [0, T]$$

$$\text{with length } \geq \delta T\}, \tag{15.20}$$

where *length* stands for the 1-dimensional Lebesgue measure.

We claim that

$$\text{ProdMeas}_3(\Omega_3^{**}(\delta)) \leq \frac{10^2\varepsilon}{\delta}. \tag{15.21}$$

The proof of (15.21) is a standard "continuous double counting" argument, i.e., an application of Fubini's theorem for double integrals. Indeed, consider the product space $\Omega_3 \times [0, T]$ with the product measure (product of ProdMeas_3 and the 1-dimensional Lebesgue measure), and introduce the function

$$f(\omega, t) = 1 \text{ if } \omega \notin \Omega_3^*(\varepsilon; t) \text{ and } 0 \text{ otherwise.}$$

We have

$$\text{ProdMeas}_3(\Omega_3^{**}(\delta)) \cdot \delta T \leq \int_{\Omega_3^{**}(\delta)\times[0,T]} f \leq \int_{\Omega_3\times[0,T]} f$$

$$= \int_0^T (1 - \text{ProdMeas}_3(\Omega_3^*(\varepsilon;t)))\, dt \leq 10^2\varepsilon T \text{ by (15.18)},$$

which implies (15.21).

Let

$$\delta = 10^{-200}. \tag{15.22}$$

Combining (15.17) and (15.20)–(15.22), we have

$$\text{ProdMeas}_3 \{\omega \in \Omega_3 : \omega \in \Omega_3^*(\varepsilon;t) \text{ holds for a set of } t \in [0,T]$$
$$\text{with length} \geq T - \delta T\} \geq 1 - \frac{10^2\varepsilon}{\delta}, \tag{15.23}$$

where

$$\delta T = \frac{\sqrt{3}}{2\pi}10^{-200+150} = \frac{\sqrt{3}}{2\pi}10^{-50} \text{ and } \frac{10^2\varepsilon}{\delta} = 10^{2+200-300} = 10^{-98}. \tag{15.24}$$

By (15.16) and (15.17),

$$\frac{2}{\sqrt{\varepsilon}\sqrt{\text{TiEvFun}_3(t)}}\left(1 + \frac{2}{\sqrt{\text{TiEvFun}_3(t)}}\right) \leq \frac{3}{10^{-150}\cdot 10^{200}/\sqrt{260}} < 10^{-48}. \tag{15.25}$$

Combining (15.19) and (15.23)–(15.25), we obtain a subset $\Omega^{***} \subset \Omega_3$ such that

$$\text{ProdMeas}_3\,(\Omega^{***}) \geq 1 - \frac{10^2\varepsilon}{\delta} = 1 - 10^{-98}, \tag{15.26}$$

and for every $\omega \in \Omega^{***}$

$$\text{Imba}_3(\mathcal{Y}_{\mathbf{r};\omega;t}) = \left(1 \pm 10^{-48}\right) \text{TiEvFun}_3(t)$$

$$= \left(1 \pm 10^{-48}\right) \max_{1\leq m\leq 4} \frac{9N}{m^2}\left(\frac{\cos(2\pi\sqrt{m}t)}{(2\pi\sqrt{m}t)^2} - \frac{\sin(2\pi\sqrt{m}t)}{(2\pi\sqrt{m}t)^3}\right)^2 \tag{15.27}$$

holds for every time t in the interval $0 < t < T = \frac{\sqrt{3}}{2\pi}10^{150}$ (see (15.15)) with the possible exception of a set $t \in [0,T]$ of total length less than 10^{-50}. Thus we obtain the following result.

First Illustration of Theorem 13.2. *Starting from a Big Bang $\mathcal{Y} = \mathcal{Y}_{BB}$ with $|\mathcal{Y}| = N = 10^{1000}$ and $|\mathbf{r}(t)| = t$, there is a subset $\Omega^{***} \subset \Omega_3$ of the initial velocity space of the third kind $\Omega^{***} \subset \Omega_3$ with*

$$\text{ProdMeas}_3\left(\Omega^{***}\right) \geq 1 - 10^{-98},$$

representing an overwhelming majority, such that for every $\omega \in \Omega^{***}$ the time evolution of the ω-dependent $\text{Imba}_3(\mathcal{Y}_{\mathbf{r};\omega;t})$ is approximated by the explicit ω-independent Time Evolution Function

$$\text{TiEvFun}_3(t) = \left(9N \max_{1\leq m\leq 4}\left(\frac{\cos(2\pi\sqrt{m}t)}{4\pi^2m^2t^2} - \frac{\sin(2\pi\sqrt{m}t)}{8\pi^3m^{5/2}t^3}\right)^2 \text{ or } 1\right)_{\max}$$

with extreme accuracy. Indeed, we have the time-quantitative result that

$$\text{Imba}_3(\mathcal{Y}_{\mathbf{r};\omega;t}) = \left(1 \pm 10^{-48}\right)\text{TiEvFun}_3(t)$$

holds for every time t in the extremely long time interval $0 < t < T = \frac{\sqrt{3}}{2\pi}10^{150}$ with the possible exception of a totally negligible set of violators $t \in [0,T]$ of total length less than 10^{-50}. Note that

$$0 \leq t \leq T = \frac{\sqrt{3}}{2\pi}10^{150} \iff 10^{1000} \geq \text{TiEvFun}_3(t) = \frac{9N}{(2\pi t)^4} \geq 10^{400},$$

that is, the time interval $[0,T]$ corresponds to the decrease of the Time Evolution Function of the third kind from the maximum $N = 10^{1000}$ to 10^{400}. □

We can say, therefore, that the Second Law for the third kind (Theorem 13.2) describes the typical time evolution of an individual system strikingly well. To be more precise, we obtain a strikingly good approximation of the ω-dependent Imbalance by the explicit ω-independent Time Evolution Function as long as the system is far from equilibrium.

Big Bang $\mathcal{Y} = \mathcal{Y}_{BB}$ is a very special initial configuration. What happens if we relax Big Bang, and request that $\mathcal{Y}$ is in a (say) one-sixth part of the unit cube $[0,1]^3$? Formally, let

$$\mathcal{Y} \subset [0,1/6] \times [0,1] \times [0,1], \tag{15.28}$$

and we keep the assumptions $N = 10^{1000}$ and $|\mathbf{r}(t)| = t$. Can we prove an analog of the First Illustration above under the weaker condition (15.28)? The answer is yes. We simply repeat the argument of the First Illustration with some straightforward modifications.

Now we use the following simple inequality: given $0 \leq y_{k,1} \leq 1/6$, $1 \leq k \leq N$, we have

$$\min\left\{\left|\sum_{k=1}^{N} e^{2\pi i y_{k,1}}\right|, \left|\sum_{k=1}^{N} e^{4\pi i y_{k,1}}\right|\right\} \geq \frac{N}{2}. \tag{15.29}$$

To prove (15.29), we note that

$$\min_{|x|\leq\pi/3} \cos x = 1/2,$$

so we have

$$\left|\sum_{k=1}^{N} e^{4\pi \mathrm{i} y_{k,1}}\right| = \left|e^{-\pi\mathrm{i}/6}\sum_{k=1}^{N} e^{4\pi \mathrm{i} y_{k,1}}\right| = \left|\sum_{k=1}^{N} e^{2\pi\mathrm{i}(2y_{k,1}-1/6)}\right|$$

$$\geq \left|\sum_{k=1}^{N} \cos\big(2\pi(2y_{k,1}-1/6)\big)\right| = \sum_{k=1}^{N} \cos\big(2\pi(2y_{k,1}-1/6)\big) \geq N/2. \quad (15.30)$$

Similarly,

$$\left|\sum_{k=1}^{N} e^{2\pi \mathrm{i} y_{k,1}}\right| = \left|e^{-\pi\mathrm{i}/6}\sum_{k=1}^{N} e^{2\pi \mathrm{i} y_{k,1}}\right| = \left|\sum_{k=1}^{N} e^{2\pi\mathrm{i}(y_{k,1}-1/12)}\right|$$

$$\geq \left|\sum_{k=1}^{N} \cos\big(2\pi(y_{k,1}-1/12)\big)\right| = \sum_{k=1}^{N} \cos\big(2\pi(y_{k,1}-1/12)\big) \geq \sqrt{3}N/2 > N/2, \quad (15.31)$$

where in the last step we used

$$\min_{|x|\leq\pi/6} \cos x = \sqrt{3}/2.$$

Combining (15.30) and (15.31), (15.29) follows.

We show that, under the condition (15.28), if $|\mathbf{r}(t)| = t \geq 10$, then the Time Evolution Function of the third kind $\mathrm{TiEvFun}_3(\mathcal{Y};\mathbf{r};t)$ (see (15.2)) reduces to a maximum taken over the 3-dimensional nonzero lattice points of length $\leq \sqrt{6}$:

$$\mathrm{TiEvFun}_3(\mathcal{Y};\mathbf{r};t)$$

$$= \left(\frac{9}{N}\max_{1\leq|\mathbf{n}|\leq\sqrt{6}}\left|\sum_{k=1}^{N} e^{2\pi\mathrm{i}\mathbf{n}\cdot\mathbf{y}_k}\right|^2 \left(\frac{\cos(2\pi t|\mathbf{n}|)}{4\pi^2t^2|\mathbf{n}|^4} - \frac{\sin(2\pi t|\mathbf{n}|)}{8\pi^3t^3|\mathbf{n}|^5}\right)^2 \text{ or } 1\right)_{\max}. \quad (15.32)$$

The proof of (15.32) is based on the elementary facts (15.7) and (15.29). Indeed, by choosing $\mathbf{n}_j = (j,0,0)$, $j = 1,2$, (15.7) and (15.29) imply

$$\max\left\{\left|\sum_{k=1}^{N} e^{2\pi\mathrm{i}\mathbf{n}_j\cdot\mathbf{y}_k}\right|^2 \frac{\cos^2(2\pi t|\mathbf{n}_j|)}{|\mathbf{n}_j|^8} : j=1,2\right\} \geq \frac{1}{260}(N/2)^2 = \frac{N^2}{1040}. \quad (15.33)$$

On the other hand,

$$\left(\frac{1}{\sqrt{8}}\right)^8 = \frac{1}{4096}$$

which is much smaller than $1/1040$; see (15.33). (We have 8 here, because 7 cannot be written as a sum of three squares.) This explains why in (15.32) we can restrict ourselves to the lattice points $\mathbf{n} \in \mathbb{Z}^3$ with $1 \le |\mathbf{n}| \le \sqrt{6}$, ignoring the rest $|\mathbf{n}| > \sqrt{6}$.

Condition (15.28), and its corollary (15.29), yield $10^{1000} = N \ge \text{TiEvFun}_3(0) = \text{TiEvFun}_3(\mathcal{Y};\mathbf{r};0) \ge N/4$. We study the time evolution of a typical system until $\text{TiEvFun}_3(t)$ goes down to $\text{TiEvFun}_3(t) \approx 10^{400}$. Since

$$\sup_{\tau \ge t} \text{TiEvFun}_3(\mathcal{Y};\mathbf{r};\tau) \approx \frac{N}{(2\pi t)^4},$$

we have

$$\frac{N}{(2\pi t)^4} = 10^{400} \iff t = \frac{1}{2\pi}10^{150},$$

that is, we study the time interval

$$0 < t < T = \frac{1}{2\pi}10^{150}. \tag{15.34}$$

By (15.7) and (15.29),

$$\text{TiEvFun}_3(t) \ge \frac{10^{400}}{1040} \quad \text{for all } 0 < t < T = \frac{1}{2\pi}10^{150}. \tag{15.35}$$

Let

$$\varepsilon = 10^{-300}. \tag{15.36}$$

Applying Theorem 13.2, for every t in $0 < t < T$ there exists a measurable subset $\Omega_3^*(\varepsilon;t) \subset \Omega_3$ of the initial velocity space of the third kind such that

$$\text{ProdMeas}_3(\Omega_3^*(\varepsilon;t)) > 1 - 10^2\varepsilon \tag{15.37}$$

and

$$|\text{Imba}_3(\mathcal{Y}_{\mathbf{r};\omega;t}) - \text{TiEvFun}_3(t)|$$

$$\le \frac{4}{\sqrt{\varepsilon}\sqrt{\text{TiEvFun}_3(t)}}\left(1 + \frac{2}{\sqrt{\text{TiEvFun}_3(t)}}\right) \cdot \text{EvolFun}_3(t) \tag{15.38}$$

holds for every $\omega \in \Omega_3^*(\varepsilon;t)$.

Again we work with the set

$$\Omega_3^{**}(\delta) = \{\omega \in \Omega_3 : \ \omega \notin \Omega_3^*(\varepsilon; t) \text{ holds for a set of } t \in [0,T]$$
$$\text{with length } \geq \delta T\}$$

defined in (15.20) for every $0 < \delta < 1$, and use upper bound (15.21)

$$\text{ProdMeas}_3(\Omega_3^{**}(\delta)) \leq \frac{10^2\varepsilon}{\delta}. \tag{15.39}$$

Again let

$$\delta = 10^{-200}, \tag{15.40}$$

and we have the analog of (15.23):

$$\text{ProdMeas}_3 \{\omega \in \Omega_3 : \omega \in \Omega_3^*(\varepsilon; t) \text{ holds for a set of } t \in [0,T]$$
$$\text{with length} \geq T - \delta T\} \geq 1 - \frac{10^2\varepsilon}{\delta}, \tag{15.41}$$

where

$$\delta T = \frac{1}{2\pi}10^{-200+150} = \frac{1}{2\pi}10^{-50} \text{ and } \frac{10^2\varepsilon}{\delta} = 10^{2+200-300} = 10^{-98}. \tag{15.42}$$

By (15.35) and (15.36),

$$\frac{4}{\sqrt{\varepsilon}\sqrt{\text{TiEvFun}_3(t)}}\left(1 + \frac{2}{\sqrt{\text{TiEvFun}_3(t)}}\right) \leq \frac{3}{10^{-150} \cdot 10^{200}/\sqrt{1040}} < 10^{-48}. \tag{15.43}$$

Combining these facts, we obtain a subset $\Omega^{***} \subset \Omega_3$ such that

$$\text{ProdMeas}_3 \left(\Omega^{***}\right) \geq 1 - \frac{10^2\varepsilon}{\delta} = 1 - 10^{-98}, \tag{15.44}$$

and for every $\omega \in \Omega^{***}$

$$\text{Imba}_3(\mathcal{Y}_{\mathbf{r};\omega;t}) = \left(1 \pm 10^{-48}\right) \text{TiEvFun}_3(t)$$
$$= \left(1 \pm 10^{-48}\right) \frac{9}{N} \max_{1 \leq |\mathbf{n}| \leq \sqrt{6}} \left| \sum_{k=1}^{N} e^{2\pi i \mathbf{n} \cdot \mathbf{y}_k} \right|^2 \left(\frac{\cos(2\pi t|\mathbf{n}|)}{4\pi^2 t^2 |\mathbf{n}|^4} - \frac{\sin(2\pi t|\mathbf{n}|)}{8\pi^3 t^3 |\mathbf{n}|^5} \right)^2 \tag{15.45}$$

holds for every time t in the interval $0 < t < T = \frac{1}{2\pi}10^{150}$ with the possible exception of a set $t \in [0,T]$ of total length less than 10^{-50}. Thus we have the following result.

Second Illustration of Theorem 13.2. *Starting from nearly Big Bang in the precise sense of (15.28) with* $|\mathcal{Y}| = N = 10^{1000}$ *and* $|\mathbf{r}(t)| = t$, *the statement of the First Illustration remains true with (15.45) instead of (15.27).* □

16. Illustrations of the Second Law (II): First kind and second kind

Next we study the first kind; again we consider the interesting special case where every particle has unit mass (so we omit $\mathcal{M}$) and $|\mathbf{r}(t)| = t$. Then the Time Evolution Function

$$\text{TiEvFun}_1(\mathcal{Y};\mathbf{r};t)$$

$$= \left(e^{-(2\pi t)^2} \max_{\mathbf{n}\in\mathbb{Z}^3\setminus\mathbf{0}} \left(\frac{\left|\sum_{k=1}^{N} e^{2\pi i \mathbf{n}\cdot\mathbf{y}_k}\right|^2}{N} \right)^{1/|\mathbf{n}|^2} \text{ or } 1 \right)_{\max} \tag{16.1}$$

is monotone decreasing in t. So for the first kind we do not have the "oscillation problem" of the third kind. This is why for the first kind it is easy to study a much larger class of initial conditions far beyond Big Bang, or nearly Big Bang.

Again we choose $N = 10^{1000}$. Again we basically repeat the argument of the case of the third kind; in fact, the version here is somewhat simpler. We recall Theorem 13.1: given $0 < \varepsilon < 1$ and a time $t > 0$ with

$$\text{TiEvFun}_1(\mathcal{Y};\mathbf{r};t) \geq 4/\varepsilon, \tag{16.2}$$

there exists a measurable subset $\Omega_1^* = \Omega_1^*(\varepsilon;t) \subset \Omega_1$ of the initial velocity space of the first kind such that

$$\text{ProdMeas}_1(\Omega_1^*) > 1 - 10^2\varepsilon \tag{16.3}$$

and

$$|\text{Imba}_1(\mathcal{Y}_{\mathbf{r};\omega;t}) - \text{TiEvFun}_1(t)| \leq \frac{12}{\sqrt{\varepsilon}\sqrt{\text{TiEvFun}_1(t)}} \cdot \text{TiEvFun}_1(t) \tag{16.4}$$

holds for every $\omega \in \Omega_1^*$. Here $\mathcal{Y} = \{\mathbf{y}_1, \ldots, \mathbf{y}_N\}$ is an arbitrary N-element point configuration in the unit cube such that $\text{TiEvFun}_1(0) > 10^{500}$, that is,

$$\text{TiEvFun}_1(0) = \text{TiEvFun}_1(\mathcal{Y};\mathbf{r};0)$$

$$= \left(\max_{\mathbf{n}\in\mathbb{Z}^3\setminus\mathbf{0}} \left(\frac{\left|\sum_{k=1}^{N} e^{2\pi i \mathbf{n}\cdot\mathbf{y}_k}\right|^2}{N} \right)^{1/|\mathbf{n}|^2} \text{ or } 1 \right)_{\max}$$

$$= \max_{\mathbf{n}\in\mathbb{Z}^3\setminus\mathbf{0}} \left(\frac{\left|\sum_{k=1}^{N} e^{2\pi i \mathbf{n}\cdot\mathbf{y}_k}\right|^2}{N} \right)^{1/|\mathbf{n}|^2} > 10^{500}. \tag{16.5}$$

Again we study the time evolution of a typical system until $\text{TiEvFun}_1(t)$ goes down from a value above 10^{500} (see (16.5)) to the value 10^{400}. We have

$$\text{TiEvFun}_1(0)e^{-(2\pi t)^2} = 10^{400} \iff t = \frac{\sqrt{\log\left(\text{TiEvFun}_1(0)10^{-400}\right)}}{2\pi},$$

that is, we study the time interval

$$0 < t < T = \frac{\sqrt{\log\left(\text{TiEvFun}_1(0)10^{-400}\right)}}{2\pi}. \tag{16.6}$$

Note that

$$\frac{\sqrt{\log 10^{500-400}}}{2\pi} \le T \le \frac{\sqrt{\log 10^{1000-400}}}{2\pi} \iff \frac{\sqrt{100\log 10}}{2\pi} \le T \le \frac{\sqrt{600\log 10}}{2\pi},$$

implying that T is between 2 and 7, i.e., T is in the constant range.

We are going to use the trivial fact that $\text{TiEvFun}_1(t)$ is a monotone decreasing function of the time t: formally

$$\text{TiEvFun}_1(t) \ge 10^{400} \quad \text{for all } 0 < t < T. \tag{16.7}$$

Now we choose

$$\varepsilon = 10^{-200}. \tag{16.8}$$

Applying Theorem 13.1 (see (16.3)–(16.4)), for every t in $0 < t < T$ there exists a measurable subset $\Omega_1^*(\varepsilon;t) \subset \Omega_1$ of the initial velocity space of the first kind such that

$$\text{ProdMeas}_1(\Omega_1^*(\varepsilon;t)) > 1 - 10^2\varepsilon \tag{16.9}$$

and

$$|\text{Imba}_1(\mathcal{Y}_{\mathbf{r};\omega;t}) - \text{TiEvFun}_1(t)| \le \frac{12}{\sqrt{\varepsilon}\sqrt{\text{TiEvFun}_1(t)}} \cdot \text{EvolFun}_1(t) \tag{16.10}$$

holds for every $\omega \in \Omega_1^*(\varepsilon;t)$.

We need the analog of (15.18): let $0 < \delta < 1$ be a real number, and consider the following subset of Ω_1:

$$\Omega_1^{**}(\delta) = \{\omega \in \Omega_1 :\ \omega \notin \Omega_1^*(\varepsilon;t) \text{ holds for a set of } t \in [0,T]$$
$$\text{with length } \ge \delta T\} \tag{16.11}$$

(here *length* stands for the 1-dimensional Lebesgue measure).

We have the perfect analog of (15.19):

$$\text{ProdMeas}_1(\Omega_1^{**}(\delta)) \leq \frac{10^2\varepsilon}{\delta}. \tag{16.12}$$

Note that the proof of (16.12) is exactly the same "continuous double counting" argument as that of (15.19).

Let

$$\delta = 10^{-100}. \tag{16.13}$$

Again we have the analog of (15.21):

$$\text{ProdMeas}_1\left\{\omega \in \Omega_1 : \omega \in \Omega_1^*(\varepsilon; t) \text{ holds for a set of } t \in [0,T] \right.$$
$$\left. \text{with length} \geq T - \delta T\right\} \geq 1 - \frac{10^2\varepsilon}{\delta}, \tag{16.14}$$

where

$$\delta T = 10^{-100}T < 10^{-99} \quad \text{and} \quad \frac{10^2\varepsilon}{\delta} = 10^{2+100-200} = 10^{-98}. \tag{16.15}$$

By (16.7) and (16.8),

$$\frac{12}{\sqrt{\varepsilon}\sqrt{\text{TiEvFun}_1(t)}} \leq \frac{12}{10^{-100}\cdot 10^{200}} < 10^{-98}. \tag{16.16}$$

Summarizing, we have a subset $\Omega^{***} \subset \Omega_1$ such that

$$\text{ProdMeas}_1\left(\Omega^{***}\right) \geq 1 - \frac{10^2\varepsilon}{\delta} = 1 - 10^{-98}, \tag{16.17}$$

and for every $\omega \in \Omega^{***}$

$$\text{Imba}_1(\mathcal{Y}_{\mathbf{r};\omega;t}) = \left(1 \pm 10^{-98}\right)\text{TiEvFun}_1(t) = \left(1 \pm 10^{-98}\right)e^{-(2\pi t)^2}\text{TiEvFun}_1(0) \tag{16.18}$$

holds for every time t in the interval

$$0 < t < T = \frac{\sqrt{\log\left(\text{TiEvFun}_1(0)10^{-400}\right)}}{2\pi} \tag{16.19}$$

with the possible exception of a set $t \in [0,T]$ of total length less than 10^{-98} (note that T is between 2 and 7 depending on the initial configuration $\mathcal{Y}$).

Thus we obtain the following result, which is much more general than the results in Section 15.

Illustration of Theorem 13.1. *Let $\mathcal{Y} = \{\mathbf{y}_1, \ldots, \mathbf{y}_N\}$ be an arbitrary N-element point configuration in the unit cube $[0,1]^3$ such that $|\mathcal{Y}| = N = 10^{1000}$ and*

$$\text{TiEvFun}_1(0) = \text{TiEvFun}_1(\mathcal{Y}; \mathbf{r}; 0)$$

$$= \max_{\mathbf{n}\in\mathbb{Z}^3\setminus\mathbf{0}} \left(\frac{\left|\sum_{k=1}^{N} e^{2\pi i \mathbf{n}\cdot\mathbf{y}_k}\right|^2}{N} \right)^{1/|\mathbf{n}|^2} > 10^{500}.$$

For simplicity, let $|\mathbf{r}(t)| = t$. *Then there is a subset* $\Omega^{***} \subset \Omega_1$ *of the initial velocity space of the first kind* $\Omega^{***} \subset \Omega_1$ *with*

$$\text{ProdMeas}_1\left(\Omega^{***}\right) \geq 1 - 10^{-98},$$

representing an overwhelming majority, such that for every $\omega \in \Omega^{***}$ *the time evolution of the* ω*-dependent* $\text{Imba}_1(\mathcal{Y}_{\mathbf{r};\omega;t})$ *is approximated by the explicit* ω*-independent Time Evolution Function*

$$\text{TiEvFun}_1(t) = \left(e^{-(2\pi t)^2}\text{TiEvFun}_1(0) \quad \text{or} \quad 1\right)_{\max}$$

with extreme accuracy. Indeed, we have the time-quantitative result that

$$\text{Imba}_1(\mathcal{Y}_{\mathbf{r};\omega;t}) = \left(1 \pm 10^{-98}\right)\text{TiEvFun}_1(t)$$

holds for every time t in the constant size time interval

$$0 < t < T = \frac{\sqrt{\log\left(\text{TiEvFun}_1(0)10^{-400}\right)}}{2\pi}$$

with the possible exception of a negligible set $t \in [0,T]$ *of total length less than* 10^{-98} *(note that* T *is between 2 and 7 depending on the initial configuration* $\mathcal{Y}$*).*

Finally, note that the time interval $[0,T]$ *corresponds to the decrease of the Time Evolution Function of the first kind from* $\text{TiEvFun}_1(0)$, *which is anywhere between* 10^{1000} *and* 10^{500}, *to* 10^{400}. □

We can say, therefore, that, similarly to the third kind, the Second Law for the first kind (Theorem 13.1) describes the typical time evolution of an individual system strikingly well. Again we obtain a strikingly good approximation of the ω-dependent Imbalance by the explicit ω-independent Time Evolution Function as long as the system is far from equilibrium.

How about the second kind? Do we have a similar result for the second kind? To answer this question, we study the second kind in the simplest special case. We start from a Big Bang, again we assume $|\mathbf{r}(t)| = t$, but this time we choose an even larger N: let $N = 10^{10000}$.

The technical reason behind choosing such a large N is that Theorem 14.1 has the unpleasant factor $e^{\sqrt{\log x}}$, due to Lemma 14.1, and we need $e^{10\sqrt{\log x}}$ to be much smaller than x, which requires a *very* large x.

Again we basically repeat the argument of the case of the third kind in Section 15. We recall Theorem 14.1: given an $0 < \varepsilon < 10^{-3}$ and a time t with

$$\varepsilon \cdot \mathrm{TiEvFun}_2(\mathcal{Y}; \mathbf{r}; t) \geq 10^6 (1 + \sqrt{\log N})^2 e^{4\sqrt{\log N}} \tag{16.20}$$

and

$$3600 t^2 e^{6\sqrt{\log(2t+1)}} \leq N e^{-4\sqrt{\log N}}, \tag{16.21}$$

there exists a measurable subset $\Omega_2^* = \Omega_2^*(\varepsilon; t) \subset \Omega_2$ of the initial velocity space of the second kind such that

$$\mathrm{ProdMeas}_2(\Omega_2^*) > 1 - 10^3 \varepsilon$$

and for all $\omega \in \Omega_2^*$

$$|\mathrm{Imba}_2(\mathcal{Y}_{\mathbf{r};\omega;t}) - \mathrm{TiEvFun}_2(t)|$$

$$\leq \frac{6\left(3 + 2\sqrt{\log(2t)}\right) e^{\sqrt{\log(2t)}}}{\sqrt{\varepsilon}\sqrt{\mathrm{TiEvFun}_2(t)}} \left(1 + \frac{2}{\sqrt{\mathrm{TiEvFun}_2(t)}}\right) \cdot \mathrm{TiEvFun}_2(t) \tag{16.22}$$

for $t \geq 5000$, and

$$|\mathrm{Imba}_2(\mathcal{Y}_{\mathbf{r};\omega;t}) - \mathrm{TiEvFun}_2(t)|$$

$$\leq \frac{200(2t+1)\,(5 + \log(2t+3))}{\sqrt{\varepsilon}\sqrt{\mathrm{TiEvFun}_2(t)}} \left(1 + \frac{2}{\sqrt{\mathrm{TiEvFun}_2(t)}}\right) \cdot \mathrm{TiEvFun}_2(t) \tag{16.23}$$

for $0 < t < 5000$ (we used $|\mathbf{r}(t)| = t$). Here $\mathcal{Y} = \mathcal{Y}_{BB}$ is a Big Bang, and

$$\mathrm{TiEvFun}_2(t) = \mathrm{TiEvFun}_2(\mathcal{Y}_{BB}; \mathbf{r}; t)$$

$$= \left(N \max_{\mathbf{n} \in \mathbb{Z}^3 \setminus \mathbf{0}} \left(\frac{\sin(2\pi t\, |\mathbf{n}|)}{2\pi t\, |\mathbf{n}|^2}\right)^2 \text{ or } 1\right)_{\max}. \tag{16.24}$$

Since we start from Big Bang, we have $\mathrm{TiEvFun}_2(0) = \mathrm{TiEvFun}_2(\mathcal{Y}_{BB}; \mathbf{r}; 0) = N = 10^{10000}$. We study the time evolution of a typical system until $\mathrm{TiEvFun}_2(t)$ goes down to $\mathrm{TiEvFun}_2(t) = 10^{4000}$. Since

$$\sup_{\tau \geq t} \mathrm{TiEvFun}_2(\mathcal{Y}_{BB}; \mathbf{r}; \tau) = \left(1 + O\left(\frac{1}{1+t}\right)\right) \frac{N}{(2\pi t)^2},$$

we have

$$\left(1 + O\left(\frac{1}{1+t}\right)\right) \frac{N}{(2\pi t)^2} = 10^{4000} \iff t = (1 + o(1)) \frac{1}{2\pi} 10^{3000},$$

that is, we study the time interval

$$0 < t < T = \frac{1}{2\pi} 10^{3000}. \tag{16.25}$$

By Lemma 14.1,

$$\text{TiEvFun}_2(t) \geq 10^{4000} e^{-6\sqrt{3000 \log 10}} > 10^{3700} \quad \text{for all } 0 < t < T = \frac{1}{2\pi} 10^{3000}. \tag{16.26}$$

Let

$$\varepsilon = 10^{-3400}. \tag{16.27}$$

Since conditions (16.20) and (16.21) are clearly satisfied, we can apply Theorem 14.1 (see (16.21)–(16.23)). For every t in $5000 \leq t < T$ there exists a measurable subset $\Omega_2^*(\varepsilon; t) \subset \Omega_3$ of the initial velocity space of the second kind such that

$$\text{ProdMeas}_2(\Omega_2^*(\varepsilon; t)) > 1 - 10^3 \varepsilon \tag{16.28}$$

and

$$|\text{Imba}_2(\mathcal{Y}_{\mathbf{r};\omega;t}) - \text{TiEvFun}_2(t)|$$
$$\leq \frac{6\left(3 + 2\sqrt{3000 \log 10}\right) e^{\sqrt{3000 \log 10}}}{\sqrt{\varepsilon}\sqrt{\text{TiEvFun}_2(t)}} \left(1 + \frac{2}{\sqrt{\text{TiEvFun}_2(t)}}\right) \cdot \text{TiEvFun}_2(t) \tag{16.29}$$

holds for every $\omega \in \Omega_2^*(\varepsilon; t)$. (Note that the remaining case $0 < t < 5000$ goes similarly; in fact, it is simpler. We omit the routine details.)

Again we need the analog of (15.18): let $0 < \delta < 1$ be a real number, and consider the following subset of Ω_2:

$$\Omega_2^{**}(\delta) = \{\omega \in \Omega_2 : \ \omega \notin \Omega_2^*(\varepsilon; t) \text{ holds for a set of } t \in [0, T]$$
$$\text{with length } \geq \delta T\} \tag{16.30}$$

(as usual, *length* stands for the 1-dimensional Lebesgue measure).

We have the perfect analog of (15.19):

$$\text{ProdMeas}_2(\Omega_2^{**}(\delta)) \leq \frac{10^3 \varepsilon}{\delta}. \tag{16.31}$$

The proof of (16.31) is exactly the same "continuous double counting" argument as that of (15.19).

Let

$$\delta = 10^{-3200}. \tag{16.32}$$

We have the analog of (15.21):

$$\text{ProdMeas}_2\,\{\omega \in \Omega_2 : \omega \in \Omega_2^*(\varepsilon; t) \text{ holds for a set of } t \in [0, T]$$

$$\text{with length} \geq T - \delta T\} \geq 1 - \frac{10^3\varepsilon}{\delta}, \tag{16.33}$$

where

$$\delta T = \frac{1}{2\pi}10^{-3200+3000} = \frac{1}{2\pi}10^{-200} \text{ and } \frac{10^3\varepsilon}{\delta} = 10^{3+3200-3400} = 10^{-197}. \tag{16.34}$$

By (16.26) and (16.27),

$$\frac{6\left(3 + 2\sqrt{3000\log 10}\right)e^{\sqrt{3000\log 10}}}{\sqrt{\varepsilon}\sqrt{\text{TiEvFun}_2(t)}}\left(1 + \frac{2}{\sqrt{\text{TiEvFun}_2(t)}}\right)$$

$$\leq \frac{10^{50}}{10^{-3400/2}10^{3700/2}} = 10^{-100}. \tag{16.35}$$

Summarizing, we have a subset $\Omega^{***} \subset \Omega_2$ such that

$$\text{ProdMeas}_2\left(\Omega^{***}\right) \geq 1 - \frac{10^3\varepsilon}{\delta} = 1 - 10^{-197}, \tag{16.36}$$

and for every $\omega \in \Omega^{***}$

$$\text{Imba}_2(\mathcal{Y}_{\mathbf{r};\omega;t}) = \left(1 \pm 10^{-100}\right)\text{TiEvFun}_2(t)$$

$$= \left(1 \pm 10^{-100}\right)\max_{\mathbf{n}\in\mathbb{Z}^3\setminus\mathbf{0}}\frac{N}{|\mathbf{n}|^2}\left(\frac{\sin(2\pi t|\mathbf{n}|)}{2\pi t|\mathbf{n}|}\right)^2 \tag{16.37}$$

holds for every time t in the interval

$$0 < t < T = \frac{1}{2\pi}10^{3000} \tag{16.38}$$

(see (16.25)) with the possible exception of a set $t \in [0, T]$ of total length less than 10^{-200}.

The following result is an analog of the First Illustration of Theorem 13.2.

Illustration of Theorem 14.1. *Starting from a Big Bang $\mathcal{Y} = \mathcal{Y}_{BB}$ with $|\mathcal{Y}| = N = 10^{10000}$ and $|\mathbf{r}(t)| = t$, there is a subset $\Omega^{***} \subset \Omega_2$ of the initial velocity space of the second kind $\Omega^{***} \subset \Omega_2$ with*

$$\text{ProdMeas}_2\left(\Omega^{***}\right) \geq 1 - 10^{-197},$$

representing an overwhelming majority, such that for every $\omega \in \Omega^{***}$ the time evolution of the ω-dependent $\mathrm{Imba}_2(\mathcal{Y}_{\mathbf{r};\omega;t})$ *is approximated by the explicit ω-independent Time Evolution Function*

$$\mathrm{TiEvFun}_2(t) = \left(N \max_{\mathbf{n}\in\mathbb{Z}^3\setminus\mathbf{0}} \left(\frac{\sin(2\pi t\ |\mathbf{n}|)}{2\pi t|\mathbf{n}|^2}\right)^2 \text{ or } 1\right)_{\max}$$

with extreme accuracy. Indeed, we have the time-quantitative result that

$$\mathrm{Imba}_2(\mathcal{Y}_{\mathbf{r};\omega;t}) = \left(1 \pm 10^{-100}\right) \mathrm{TiEvFun}_2(t)$$

holds for every time t in the extremely long time interval $0 < t < T = \frac{1}{2\pi}10^{3000}$ with the possible exception of a set $t \in [0,T]$ of total length less than 10^{-200}.

Note that the time interval $[0,T]$ corresponds to the decrease of the Time Evolution Function of the second kind from the maximum $N = 10^{10000}$ to 10^{4000}. □

We can say, therefore, that, similarly to the third kind and the first kind, the Second Law for the second kind (Theorem 14.1) also describes the typical time evolution of an individual system strikingly well. As a rule, we obtain a strikingly good approximation of the ω-dependent Imbalance by the explicit ω-independent Time Evolution Function as long as the system is far from equilibrium.

We conclude Section 16 with a proof of Lemma 14.1.

Proof of Lemma 14.1. The statement of the lemma is trivial for $0 < x \leq 1/2$. Indeed, by choosing $\mathbf{n} = (1,0,0)$, we have $\|x\|/x = 1$.

Next we study the case $1/2 < x < 10^4$. A famous theorem of Gauss states that, a positive integer r can be written as a sum of three squares $r = n_1^2 + n_2^2 + n_3^2$, $n_i \in \mathbb{Z}$, $1 \leq i \leq 3$ if and only if r is not of the form $r = 4^a(8b-1)$ with $a, b \in \mathbb{Z}$. It follows that every positive integer r of the form $r = 4s+1$, $s \in \mathbb{Z}$ is a sum of three squares. Let m be the smallest positive integer with the property $m \geq 2x$ (so $m < 2x+1$); then

$$x\left(\sqrt{4(m+1)+1} - \sqrt{4m+1}\right)$$

$$= \frac{4x}{\sqrt{4(m+1)+1}\sqrt{4m+1}} \leq \frac{4x}{4m+1} < \frac{x}{m} \leq \frac{1}{2}. \tag{16.39}$$

In the other direction we have

$$x\left(\sqrt{4(m+1)+1} - \sqrt{4m+1}\right)$$

$$= \frac{4x}{\sqrt{4(m+1)+1}\sqrt{4m+1}} \geq \frac{4x}{4m+5} > \frac{x}{m+2}$$

$$> \frac{x}{(2x+1)+2} = \frac{x}{2x+3}. \tag{16.40}$$

By using (16.39), (16.40), and the triangle inequality, we have

$$\|x\sqrt{4(m+1)+1}\| + \|x\sqrt{4m+1}\| \geq \|x\left(\sqrt{4(m+1)+1} - \sqrt{4m+1}\right)\|$$

$$= x\left(\sqrt{4(m+1)+1} - \sqrt{4m+1}\right) > \frac{x}{2x+3}. \tag{16.41}$$

By (16.41),

$$\frac{\|x\sqrt{4(m+1)+1}\|}{x\sqrt{4(m+1)+1}} + \frac{\|x\sqrt{4m+1}\|}{x\sqrt{4m+1}} \geq \frac{\|x\sqrt{4(m+1)+1}\| + \|x\sqrt{4m+1}\|}{x\sqrt{4(m+1)+1}}$$

$$> \frac{x/(2x+3)}{x\sqrt{4m+5}} = \frac{1}{(2x+3)\sqrt{4m+5}} > \frac{1}{(2x+3)\sqrt{4(2x+1)+5}} > \frac{1}{2(2x+3)^{3/2}},$$

proving the lemma in the range $0 < x < 10^4$.

Finally, assume that $x > 10^4$. Generalizing the idea above, we study the following "iterated difference sequence"

$$S_{m,1} = \sqrt{4(m+1)+1} - \sqrt{4m+1},$$

$$S_{m,2} = \sqrt{4(m+2)+1} - 2\sqrt{4(m+1)+1} + \sqrt{4m+1},$$

and in general,

$$S_{m,\ell} = \sum_{j=0}^{\ell} (-1)^{\ell-j} \binom{\ell}{j} \sqrt{4(m+j)+1}, \tag{16.42}$$

where the integral parameters m and ℓ will be specified later (in terms of x). We need the following lemma from calculus.

Lemma 16.1. *If $f(x)$ is a sufficiently smooth real function then*

$$\sum_{j=0}^{k} (-1)^{k-j} \binom{k}{j} f(x+j) = f^{(k)}(\xi), \tag{16.43}$$

where $f^{(k)}$ is the k-th derivative of f, and ξ is an appropriate real in the interval $x < \xi < x+k$.

This lemma is a rather well known generalization of the Mean Value Theorem; nevertheless, for the sake of completeness, we outline a *proof.*

The case $k = 1$ trivially follows from the Mean Value Theorem. The general case can be proved by induction on k by using the following tricky auxiliary function (x is fixed and y is the variable):

$$g(y) = (y-x)\left(\sum_{j=0}^{k}(-1)^{k-j}\binom{k}{j}f(x+j)\right)$$

$$-\sum_{i=0}^{k-1}(-1)^{k-1-i}\binom{k-1}{i}\left(f(y+i)-f(x+i)\right).$$

Since $g(x) = 0$ (trivial) and $g(x+1) = 0$ (special property of g), by Rolle's theorem there exists a ξ_0 in $x < \xi_0 < x+1$ such that $g'(\xi_0) = 0$. By computing the derivative $g'(y)$ we have

$$0 = g'(\xi_0) = \sum_{j=0}^{k}(-1)^{k-j}\binom{k}{j}f(x+j) - \sum_{i=0}^{k-1}(-1)^{k-1-i}\binom{k-1}{i}f'(\xi_0+i). \tag{16.44}$$

By induction we can apply (16.43) with $(k-1)$ to (16.44), and obtain

$$\sum_{j=0}^{k}(-1)^{k-j}\binom{k}{j}f(x+j) = \sum_{i=0}^{k-1}(-1)^{k-1-i}\binom{k-1}{i}f'(\xi_0+i) = (f')^{(k-1)}(\xi),$$

where $\xi_0 < \xi < \xi_0 + k - 1$. Since $x < \xi_0 < x+1$, Lemma 16.1 follows. □

Let $f(x) = (4x+1)^{1/2}$, then

$$f'(x) = \frac{1}{2}\cdot 4(4x+1)^{-1/2}, \quad f''(x) = \frac{1}{2}\left(-\frac{1}{2}\right)\cdot 4^2(4x+1)^{-3/2},$$

and in general for an arbitrary $k \geq 2$,

$$f^{(k)}(x) = \frac{1}{2}\left(-\frac{1}{2}\right)\left(-\frac{3}{2}\right)\cdots\left(-\frac{2k-3}{2}\right)\cdot 4^k(4x+1)^{-(2k-1)/2}$$

$$= (-1)^{k-1}2^k(2k-3)!!(4x+1)^{-(2k-1)/2}. \tag{16.45}$$

(Here we used the well known notation $s!! = s(s-2)(s-4)\cdots$, where the last factor is 1 or 2 depending on the parity of s.)

We combine (16.42), Lemma 16.1 and (16.45): given positive integers m and ℓ, there exists a real number $\xi = \xi(m,\ell)$ in the interval $m < \xi < m+\ell$ such that

$$S_{m,\ell} = \sum_{j=0}^{\ell}(-1)^{\ell-j}\binom{\ell}{j}\sqrt{4(m+j)+1} = (-1)^{\ell-1}2^\ell(2\ell-3)!!(4\xi+1)^{-(2\ell-1)/2}. \tag{16.46}$$

By (16.46), for $\ell \geq 1$ we have

$$2^\ell(2\ell-3)!!(4m+1)^{-(2\ell-1)/2} \geq |S_{m,\ell}| \geq 2^\ell(2\ell-3)!!(4m+4\ell+1)^{-(2\ell-1)/2}. \tag{16.47}$$

Note that

$$1 \leq \frac{(4m+4\ell+1)^{(2\ell-1)/2}}{(4m+1)^{(2\ell-1)/2}} < \left(1+\frac{4\ell}{4m+1}\right)^\ell < \sqrt{e} < 2 \quad \text{if} \quad m \geq 2\ell^2. \tag{16.48}$$

We specify $\ell = \ell(x)$ as follows (lower integral part):

$$\ell = \left\lfloor \sqrt{\log x} \right\rfloor. \tag{16.49}$$

Combining (16.47)–(16.49), we obtain that there is an integer m with

$$e^{\sqrt{\log x}} \leq 4m+4\ell+1 \leq e^{2\sqrt{\log x}} \tag{16.50}$$

such that

$$\frac{1}{3} \leq x\,|S_{m,\ell}| \leq \frac{2}{3}. \tag{16.51}$$

By (16.51) and the triangle inequality,

$$\frac{1}{3} \leq \|xS_{m,\ell}\| \leq \sum_{j=0}^{\ell}\binom{\ell}{j}\|x\sqrt{4(m+j)+1}\| \leq 2^\ell \max_{0\leq j\leq \ell}\|x\sqrt{4(m+j)+1}\|,$$

which implies

$$\max_{0\leq j\leq\ell}\|x\sqrt{4(m+j)+1}\| \geq \frac{1}{3}2^{-\ell} \geq \frac{1}{3}2^{-\sqrt{\log x}}, \tag{16.52}$$

where in the last step we used (16.49).

Let $0 \leq j_0 \leq \ell$ be the integer with the maximum property

$$\|x\sqrt{4(m+j_0)+1}\| = \max_{0\leq j\leq\ell}\|x\sqrt{4(m+j)+1}\|. \tag{16.53}$$

By Gauss's theorem mentioned above the integer $4(m+j_0)+1$ can be written as a sum of three squares $4(m+j_0)+1 = n_1^2+n_2^2+n_3^2$. So, choosing the lattice point $\mathbf{n} = (n_1, n_2, n_3)$, Lemma 14.1 follows from (16.49)–(16.50) and (16.52)–(16.53). □

17. Illustrations of the Second Law (III): More on the second kind, and the Paradoxes

Here is a nutshell summary of Sections 15–16. The Second Law (proved in Sections 13–14) describes the typical time evolution of an individual system. As a rule, we obtain a strikingly good approximation of the ω-dependent Imbalance by the explicit ω-independent Time Evolution Function as long as the system is far from equilibrium.

For simplicity, consider the interesting special case where $|\mathbf{r}(t)| = t$, and where every mass is one (so we omit $\mathcal{M}$). Then the ω-independent Time Evolution Functions simplify to

$$\text{TiEvFun}_1(\mathcal{Y};\mathbf{r};t)$$

$$= \left(e^{-(2\pi t)^2} \max_{\mathbf{n}\in\mathbb{Z}^3\setminus\mathbf{0}} \left(\frac{\left|\sum_{k=1}^{N} e^{2\pi i\mathbf{n}\cdot\mathbf{y}_k}\right|^2}{N} \right)^{1/|\mathbf{n}|^2} \text{ or } 1 \right)_{\max} \tag{17.1}$$

for the first kind,

$$\text{TiEvFun}_3(\mathcal{Y};\mathbf{r};t)$$

$$= \left(\frac{9}{N} \max_{\mathbf{n}\in\mathbb{Z}^3\setminus\mathbf{0}} \left| \sum_{k=1}^{N} e^{2\pi i\mathbf{n}\cdot\mathbf{y}_k} \right|^2 \left(\frac{\cos(2\pi t|\mathbf{n}|)}{4\pi^2 t^2|\mathbf{n}|^4} - \frac{\sin(2\pi t|\mathbf{n}|)}{8\pi^3 t^3|\mathbf{n}|^5} \right)^2 \text{ or } 1 \right)_{\max} \tag{17.2}$$

for the third kind, and finally,

$$\text{TiEvFun}_2(\mathcal{Y};\mathbf{r};t)$$

$$= \left(\frac{1}{N} \max_{\mathbf{n}\in\mathbb{Z}^3\setminus\mathbf{0}} \left| \sum_{k=1}^{N} e^{2\pi i\mathbf{n}\cdot\mathbf{y}_k} \right|^2 \left(\frac{\sin(2\pi t|\mathbf{n}|)}{2\pi t|\mathbf{n}|^2} \right)^2 \text{ or } 1 \right)_{\max} \tag{17.3}$$

for the second kind (all in the standard 3-dimensional case).

The Time Evolution Function for the first kind (17.1) has a surprisingly simple form, and it has the crucial property of being monotone decreasing.

On the other hand (17.2) and (17.3) are not monotone decreasing, but, as we explained it at the beginning of Section 15, they still exhibit a *tendency to decrease with oscillations*, where the oscillations are due to the functions $\cos^2 x$ and $\sin^2 x$.

To justify the claim of "tendency to decrease", we replace the oscillating function $\sin^2 x$ with its mean value $1/2$ (due to the fact $\int_0^\pi \sin^2 x\,dx = \pi/2$), and then (17.3) reduces to

$$\text{mean value of } \mathrm{TiEvFun}_2(\mathcal{Y};\mathbf{r};t)$$

$$= \left(\frac{1}{N}\max_{\mathbf{n}\in\mathbb{Z}^3\setminus\mathbf{0}}\left|\sum_{k=1}^{N} e^{2\pi i\mathbf{n}\cdot\mathbf{y}_k}\right|^2 \frac{1/2}{4\pi^2t^2|\mathbf{n}|^4} \text{ or } 1\right)_{\max}, \qquad (17.4)$$

which is a decreasing function in t.

Next we switch to the third kind (17.2). If t is large than the factor

$$\left(\frac{\cos(2\pi t|\mathbf{n}|)}{4\pi^2t^2|\mathbf{n}|^4} - \frac{\sin(2\pi t|\mathbf{n}|)}{8\pi^3t^3|\mathbf{n}|^5}\right)^2$$

in (17.2) can we replaced by the simpler

$$\frac{\cos^2(2\pi t|\mathbf{n}|)}{16\pi^4t^4|\mathbf{n}|^8},$$

and repeating the mean value argument above, we obtain

$$\text{mean value of } \mathrm{TiEvFun}_3(\mathcal{Y};\mathbf{r};t)$$

$$= \left(\frac{9}{N}\max_{\mathbf{n}\in\mathbb{Z}^3\setminus\mathbf{0}}\left|\sum_{k=1}^{N} e^{2\pi i\mathbf{n}\cdot\mathbf{y}_k}\right|^2 \frac{1/2}{16\pi^4t^4|\mathbf{n}|^8} \text{ or } 1\right)_{\max}, \qquad (17.5)$$

which is again a decreasing function in t.

Beside the "tendency to decrease", (17.2) and (17.3) also exhibit oscillations. As we pointed out in Section 15, in the special case of Big Bang $\mathcal{Y} = \mathcal{Y}_{BB}$, for large t, the Time Evolution Function of the third kind

$$\mathrm{TiEvFun}_3(\mathcal{Y}_{BB};\mathbf{r};t) = \left(9N\max_{1\le m\le 4}\left(\frac{\cos(2\pi\sqrt{m}t)}{4\pi^2m^2t^2} - \frac{\sin(2\pi\sqrt{m}t)}{8\pi^3m^{5/2}t^3}\right)^2 \text{ or } 1\right)_{\max}$$

follows the oscillation of the single term corresponding to $m = 1$

$$\left(9N\frac{\cos^2(2\pi t)}{16\pi^4t^4} \text{ or } 1\right)_{\max}$$

as long as $\cos^2(2\pi t) > 1/16$, which represents the majority of t. (This holds until the system reaches equilibrium.)

Similarly, the Time Evolution Function of the second kind

$$\mathrm{TiEvFun}_2(\mathcal{Y}_{BB};\mathbf{r};t) = \left(N\max_{\mathbf{n}\in\mathbb{Z}^3\setminus\mathbf{0}}\frac{\sin^2(2\pi t|\mathbf{n}|)}{4\pi^2t^2|\mathbf{n}|^4} \text{ or } 1\right)_{\max}$$

follows the oscillation of the single term corresponding to $|\mathbf{n}| = 1$

$$\left(N\frac{\sin^2(2\pi t)}{4\pi^2 t^2} \text{ or } 1\right)_{\max}$$

as long as $\sin^2(2\pi t) > 1/4$, which still represents the majority of t (it holds until the system reaches equilibrium). We may say, therefore, that both $\mathrm{TiEvFun}_2(\mathcal{Y}_{BB}; \mathbf{r}; t)$ and $\mathrm{TiEvFun}_3(\mathcal{Y}_{BB}; \mathbf{r}; t)$ exhibit substantial oscillations.

In Section 6 we explained the surprising fact that, roughly speaking, the simplest cube lattice in the unit torus $I^3 = [0,1)^3$ is in *equilibrium*. More precisely, assume that N is a perfect cube $N = m^3$, and consider the N-element set

$$\mathcal{Y}_{m,m,m} = \left\{\mathbf{y}_{j,k,\ell} = \left(\frac{j}{m}, \frac{k}{m}, \frac{\ell}{m}\right) : \ 0 \le j,k,\ell < m\right\}. \tag{17.6}$$

We recall (6.31),

$$\mathrm{Imba}_3(\mathcal{Y}_{m,m,m}) = \max_{\mathbf{n}\in\mathbb{Z}^3\setminus\mathbf{0}} \frac{1}{|\mathbf{n}|^4}\left(\frac{\left|\sum_{0\le j,k,\ell<m} e^{2\pi i \mathbf{n}\cdot\mathbf{y}_{j,k,\ell}}\right|^2}{N} \text{ or } 1\right)_{\max}$$

$$= \left(\frac{N^2}{Nm^4} \text{ or } 1\right)_{\max} = 1, \tag{17.7}$$

since $m^4 > m^3 = N$.

Similarly, by (6.32),

$$\mathrm{Imba}_1(\mathcal{Y}_{m,m,m}) = \max_{\mathbf{n}\in\mathbb{Z}^3\setminus\mathbf{0}} \left(\frac{\left|\sum_{0\le j,k,\ell<m} e^{2\pi i \mathbf{n}\cdot\mathbf{y}_{j,k,\ell}}\right|^2}{N} \text{ or } 1\right)_{\max}^{1/|\mathbf{n}|^2}$$

$$= \left(\frac{N^2}{N}\right)^{1/m^2} = N^{-N^{2/3}} = 1 + O(N^{-2/3}\log N) = 1 + o(1). \tag{17.8}$$

It follows from (17.7) and (17.8) that the N-element cube lattice $\mathcal{Y}_{m,m,m}$ defined by (17.6) is an Ω_i-equilibrium configuration for $i = 1$ and 3.

The argument breaks down for the case $i = 2$ — representing the second kind — because $m^2 < m^3 = N$, and so

$$\mathrm{Imba}_2(\mathcal{Y}_{m,m,m}) = \max_{\mathbf{n}\in\mathbb{Z}^3\setminus\mathbf{0}} \frac{1}{|\mathbf{n}|^2}\left(\frac{\left|\sum_{0\le j,k,\ell<m} e^{2\pi i \mathbf{n}\cdot\mathbf{y}_{j,k,\ell}}\right|^2}{N} \text{ or } 1\right)_{\max}$$

$$= \left(\frac{N^2}{Nm^2} \text{ or } 1 \right)_{\max} = m = N^{1/3}.$$

This raises the question: what can we say about the typical time evolution of $\text{Imba}_2(\mathcal{Y}_{\mathbf{r};\omega;t})$ if we start from the cube lattice $\mathcal{Y} = \mathcal{Y}_{m,m,m}$ defined by (17.6)? Theorem 14.1 was a Second Law for the second kind in the special case where $\mathcal{Y}$ was "close to a Big Bang". Here we prove an analog of Theorem 14.1 for $\mathcal{Y} = \mathcal{Y}_{m,m,m}$. The proof of this special case is simpler than that of Theorem 14.1.

For the sake of simplicity we assume that every mass is one (so we omit $\mathcal{M}$ in the notation).

We closely follow Section 14. We recall (14.1) and (14.2):

$$\mathcal{Y}_{\mathbf{r};\omega;t} = \{\mathbf{z}_k = \mathbf{y}_k + \vartheta_k(\mathbf{r}(t)) \text{ modulo one} : \ 1 \le k \le N\}, \tag{17.9}$$

where

$$\omega = (\vartheta_1, \vartheta_2, \ldots, \vartheta_N) \in \Omega_2 = (SO(3))^N,$$

$\mathcal{Y} = \{\mathbf{y}_1, \ldots, \mathbf{y}_N\} = \mathcal{Y}_{m,m,m}$, and we study the Imbalance

$$\text{Imba}_2(\mathcal{Y}_{\mathbf{r};\omega;t}) = \max_{\mathbf{n} \in \mathbb{Z}^3 \setminus \mathbf{0}} \frac{1}{|\mathbf{n}|^2} \left(\frac{\left| \sum_{\mathbf{z}_k \in \mathcal{Y}_{\mathbf{r};\omega;t}} e^{2\pi i \mathbf{n} \cdot \mathbf{z}_k} \right|^2}{N} \text{ or } 1 \right)_{\max}$$

$$= \left(\max_{\mathbf{n} \in \mathbb{Z}^3 \setminus \mathbf{0}} \frac{1}{|\mathbf{n}|^2} \frac{\left| \sum_{\mathbf{z}_k \in \mathcal{Y}_{\mathbf{r};\omega;t}} e^{2\pi i \mathbf{n} \cdot \mathbf{z}_k} \right|^2}{N} \text{ or } 1 \right)_{\max}. \tag{17.10}$$

Again we want to find the relation between $\text{Imba}_2(\mathcal{Y}_{\mathbf{r};\omega;t})$ for the majority of $\omega \in \Omega_2$ and the ω-independent Time Evolution Function

$$\text{TiEvFun}_2(t) = \text{TiEvFun}_2(\mathcal{Y}; \mathbf{r}; t)$$

$$= \max_{\mathbf{n} \in \mathbb{Z}^3 \setminus \mathbf{0}} \frac{1}{|\mathbf{n}|^2} \left(\frac{1}{N} \left| \sum_{k=1}^{N} e^{2\pi i \mathbf{n} \cdot \mathbf{y}_k} \right|^2 \left(\frac{\sin(2\pi |\mathbf{r}(t)| \, |\mathbf{n}|)}{2\pi |\mathbf{r}(t)| \, |\mathbf{n}|} \right)^2 \text{ or } 1 \right)_{\max}$$

$$= \left(\frac{1}{N} \max_{\mathbf{n} \in \mathbb{Z}^3 \setminus \mathbf{0}} \left| \sum_{k=1}^{N} e^{2\pi i \mathbf{n} \cdot \mathbf{y}_k} \right|^2 \left(\frac{\sin(2\pi |\mathbf{r}(t)| \, |\mathbf{n}|)}{2\pi |\mathbf{r}(t)| \, |\mathbf{n}|^2} \right)^2 \text{ or } 1 \right)_{\max}. \tag{17.11}$$

In the special case $\mathcal{Y} = \mathcal{Y}_{m,m,m}$ the critical exponential sum in (17.11)

$$\left| \sum_{k=1}^{N} e^{2\pi i \mathbf{n} \cdot \mathbf{y}_k} \right|^2$$

can be explicitly evaluated by using geometric series (see the calculation between (6.30) and (6.31)), and we obtain

$$\left|\sum_{k=1}^{N} e^{2\pi i \mathbf{n}\cdot \mathbf{y}_k}\right|^2 = N^2$$

if $\mathbf{n} = (mh_1, mh_2, mh_3)$ for some $(h_1, h_2, h_3) \in \mathbb{Z}^3$, and 0 otherwise.

Using this fact, we can simplify (17.11) as follows:

$$\text{TiEvFun}_2(t) = \text{TiEvFun}_2(\mathcal{Y}; \mathbf{r}; t)$$

$$= \max_{\mathbf{h}\in\mathbb{Z}^3\setminus\mathbf{0}} \frac{1}{m^2|\mathbf{h}|^2} \left(N \left(\frac{\sin(2\pi m|\mathbf{r}(t)|\ |\mathbf{h}|)}{2\pi m|\mathbf{r}(t)|\ |\mathbf{h}|} \right)^2 \text{ or } 1 \right)_{\max}$$

$$= \left(N^{1/3} \max_{\mathbf{h}\in\mathbb{Z}^3\setminus\mathbf{0}} \left(\frac{\sin(2\pi m|\mathbf{r}(t)|\ |\mathbf{h}|)}{2\pi m|\mathbf{r}(t)|\ |\mathbf{h}|^2} \right)^2 \text{ or } 1 \right)_{\max}. \tag{17.12}$$

We have the following analog of (14.4)–(14.6): there is a measurable subset $\Omega_2^* \subset \Omega_2$ such that

$$\text{ProdMeas}_2(\Omega_2^*) > 1 - 2\cdot 10^3\varepsilon, \tag{17.13}$$

and for every $\omega \in \Omega_2^*$

$$\left| \frac{\left|\sum_{\mathbf{z}_k\in\mathcal{Y}_{\mathbf{r};\omega;t}} e^{2\pi m i \mathbf{h}\cdot\mathbf{z}_k}\right|^2}{N} - \frac{\overline{\mathbf{D}^2}_2(m\mathbf{h}; t)}{N} \right|$$

$$\le \frac{4|\mathbf{h}|^{3/2}\log(2+|\mathbf{h}|)}{\sqrt{\varepsilon}} \left(\left(\frac{\overline{\mathbf{D}^2}_2(m\mathbf{h}; t)}{N} \right)^{1/2} + 2 \right) \tag{17.14}$$

holds for every nonzero lattice point $\mathbf{h} \in \mathbb{Z}^3 \setminus \mathbf{0}$, and also

$$\frac{\left|\sum_{\mathbf{z}_k\in\mathcal{Y}_{\mathbf{r};\omega;t}} e^{2\pi i \mathbf{n}\cdot\mathbf{z}_k}\right|^2}{N} \le \frac{4|\mathbf{n}|^{3/2}\log(2+|\mathbf{n}|)}{\sqrt{\varepsilon}}, \tag{17.15}$$

holds for every nonzero lattice point $\mathbf{n} \in \mathbb{Z}^3 \setminus \mathbf{0}$ which does not have the form $\mathbf{n} = m\mathbf{h}$, $\mathbf{h} \in \mathbb{Z}^3 \setminus \mathbf{0}$. Here

$$\overline{\mathbf{D}^2}_2(m\mathbf{h}; t) = N^2 \left(\frac{\sin(2\pi m|\mathbf{r}(t)|\ |\mathbf{h}|)}{2\pi m|\mathbf{r}(t)|\ |\mathbf{h}|} \right)^2. \tag{17.16}$$

Note that the extra factor of 2 in (17.13) comes from the fact that we applied the numerical series

$$\sum_{\mathbf{n}\in\mathbb{Z}^3\setminus\mathbf{0}} \frac{1}{|\mathbf{n}|^3\log^2(2+|\mathbf{n}|)} < 10^3$$

twice: first for $\mathbf{h} \in \mathbb{Z}^3 \setminus \mathbf{0}$, and second time for the lattice points $\mathbf{n} \in \mathbb{Z}^3 \setminus \mathbf{0}$ which do not have the form $\mathbf{n} = m\mathbf{h}$, $\mathbf{h} \in \mathbb{Z}^3 \setminus \mathbf{0}$.

We can rewrite the definition (17.12) of the Time Evolution Function as follows:

$$\text{TiEvFun}_2(t) = \left(\frac{1}{m^2}\max_{\mathbf{h}\in\mathbb{Z}^3\setminus\mathbf{0}}\frac{1}{|\mathbf{h}|^2}\frac{\overline{\mathbf{D}^2}_2(m\mathbf{h};t)}{N} \text{ or } 1\right)_{\max}. \tag{17.17}$$

Let $\mathbf{h}_0$ denote the lattice point $\mathbf{h} \in \mathbb{Z}^3 \setminus \mathbf{0}$ for which the maximum in (17.7) is attained:

$$\text{TiEvFun}_2(t) = \frac{1}{m^2}\frac{1}{|\mathbf{h}_0|^2}\left(\frac{\overline{\mathbf{D}^2}_2(m\mathbf{h}_0;t)}{N} \text{ or } 1\right)_{\max}. \tag{17.18}$$

We can assume that

$$|\mathbf{h}_0| \le \sqrt{m} = N^{1/6}, \tag{17.19}$$

since otherwise (see (17.16))

$$\frac{1}{m^2|\mathbf{h}|^2}\frac{\overline{\mathbf{D}^2}_2(m\mathbf{h};t)}{N} < \frac{N}{m^2\cdot m} = 1.$$

By (17.4),

$$\left|\frac{1}{m^2|\mathbf{h}|^2}\left(\frac{\left|\sum_{\mathbf{z}_k\in\mathcal{Y}_{\mathbf{r};\omega;t}} e^{2\pi m i\mathbf{h}\cdot\mathbf{z}_k}\right|^2}{N} \text{ or } 1\right)_{\max} - \frac{1}{m^2|\mathbf{h}|^2}\left(\frac{\overline{\mathbf{D}^2}_2(m\mathbf{h};t)}{N} \text{ or } 1\right)_{\max}\right|$$

$$\le \frac{1}{m^2|\mathbf{h}|^2}\frac{4|\mathbf{h}|^{3/2}\log(2+|\mathbf{h}|)}{\sqrt{\varepsilon}}\left(\left(\frac{\overline{\mathbf{D}^2}_2(m\mathbf{h};t)}{N}\right)^{1/2} + 2\right)$$

$$\le \frac{4\log(2+|\mathbf{h}|)}{\sqrt{\varepsilon}m^2|\mathbf{h}|^{1/2}}\left(\left(\frac{\overline{\mathbf{D}^2}_2(m\mathbf{h};t)}{N} \text{ or } 1\right)^{1/2} + 2\right)$$

$$\leq \frac{4|\mathbf{h}|^{1/2}\log(2+|\mathbf{h}|)}{m\sqrt{\varepsilon}}\left(\left(\frac{1}{m^2|\mathbf{h}|^2}\left(\frac{\overline{\mathbf{D}^2}_2(m\mathbf{h};t)}{N} \text{ or } 1\right)_{\max}\right)^{1/2}+2\right). \tag{17.20}$$

By choosing $\mathbf{h} = \mathbf{h}_0$ in (17.20) (see (17.18)), and applying the triangle inequality, we have

$$\frac{1}{m^2|\mathbf{h}_0|^2}\left(\frac{\left|\sum_{\mathbf{z}_k\in\mathcal{Y}_{\mathbf{r};\omega;t}} e^{2\pi m i\mathbf{h}_0\cdot\mathbf{z}_k}\right|^2}{N} \text{ or } 1\right)_{\max}$$

$$\geq \frac{1}{m^2|\mathbf{h}_0|^2}\left(\frac{\overline{\mathbf{D}^2}_2(m\mathbf{h}_0;t)}{N} \text{ or } 1\right)_{\max}$$

$$-\frac{4|\mathbf{h}_0|^{1/2}\log(2+|\mathbf{h}|_0)}{m\sqrt{\varepsilon}}\left(\left(\frac{1}{m^2|\mathbf{h}_0|^2}\left(\frac{\overline{\mathbf{D}^2}_2(m\mathbf{h}_0;t)}{N} \text{ or } 1\right)_{\max}\right)^{1/2}+2\right)$$

$$= \mathrm{TiEvFun}_2(t) - \frac{4|\mathbf{h}_0|^{1/2}\log(2+|\mathbf{h}_0|)}{m\sqrt{\varepsilon}}\left((\mathrm{TiEvFun}_2(t))^{1/2}+2\right)$$

$$\geq \mathrm{TiEvFun}_2(t) - \frac{4\log N}{N^{1/4}\sqrt{\varepsilon}}\left((\mathrm{TiEvFun}_2(t))^{1/2}+2\right), \tag{17.21}$$

where in the last step we used (17.19).

Combining (17.10) and (17.21), we obtain

$$\mathrm{Imba}_2(\mathcal{Y}_{\mathbf{r};\omega;t}) \geq \mathrm{TiEvFun}_2(t) - \frac{4\log N}{N^{1/4}\sqrt{\varepsilon}}\left((\mathrm{TiEvFun}_2(t))^{1/2}+2\right). \tag{17.22}$$

Next we give an upper bound for $\mathrm{Imba}_2(\mathcal{Y}_{\mathbf{r};\omega;t})$. We define the lattice point $\mathbf{h}_1 \in \mathbb{Z}^3 \setminus \mathbf{0}$ by the equality

$$\frac{1}{|\mathbf{h}_1|^2}\left(\frac{\left|\sum_{\mathbf{z}_k\in\mathcal{Y}_{\mathbf{r};\omega;t}} e^{2\pi m i\mathbf{h}_1\cdot\mathbf{z}_k}\right|^2}{N} \text{ or } 1\right)_{\max}$$

$$= \max_{\mathbf{h}\in\mathbb{Z}^3\setminus\mathbf{0}} \frac{1}{|\mathbf{h}|^2}\left(\frac{\left|\sum_{\mathbf{z}_k\in\mathcal{Y}_{\mathbf{r};\omega;t}} e^{2\pi m i\mathbf{h}\cdot\mathbf{z}_k}\right|^2}{N} \text{ or } 1\right)_{\max}. \tag{17.23}$$

We can assume that

$$|\mathbf{h}_1| \le \sqrt{m} = N^{1/6}, \tag{17.24}$$

since otherwise we have the trivial upper bound

$$\frac{1}{m^2|\mathbf{h}_1|^2} \frac{\left|\sum_{\mathbf{z}_k \in \mathcal{Y}_{\mathbf{r};\omega;t}} e^{2\pi m i \mathbf{h}_1 \cdot \mathbf{z}_k}\right|^2}{N} < \frac{N}{m^2 \cdot m} = 1.$$

By choosing $\mathbf{h} = \mathbf{h}_1$ in (17.20) (see (17.23)), and applying the triangle inequality, we have

$$\max_{\mathbf{h} \in \mathbb{Z}^3 \setminus \mathbf{0}} \frac{1}{m^2|\mathbf{h}|^2} \left(\frac{\left|\sum_{\mathbf{z}_k \in \mathcal{Y}_{\mathbf{r};\omega;t}} e^{2\pi m i \mathbf{h} \cdot \mathbf{z}_k}\right|^2}{N} \text{ or } 1 \right)_{\max}$$

$$\le \frac{1}{m^2|\mathbf{h}_1|^2} \left(\frac{\overline{\mathbf{D}^2}_2(m\mathbf{h}_1; t)}{N} \text{ or } 1 \right)_{\max}$$

$$+ \frac{4|\mathbf{h}_1|^{1/2} \log(2 + |\mathbf{h}|_1)}{m\sqrt{\varepsilon}} \left(\left(\frac{1}{m^2|\mathbf{h}_1|^2} \left(\frac{\overline{\mathbf{D}^2}_2(m\mathbf{h}_1; t)}{N} \text{ or } 1 \right)_{\max} \right)^{1/2} + 2 \right)$$

$$\le \text{TiEvFun}_2(t) + \frac{4|\mathbf{h}_1|^{1/2} \log(2 + |\mathbf{h}_1|)}{m\sqrt{\varepsilon}} \left((\text{TiEvFun}_2(t))^{1/2} + 2 \right)$$

$$\le \text{TiEvFun}_2(t) + \frac{4 \log N}{N^{1/4}\sqrt{\varepsilon}} \left((\text{TiEvFun}_2(t))^{1/2} + 2 \right), \tag{17.25}$$

where in the last step we used (17.24).

Moreover, by (17.15), for every nonzero lattice point $\mathbf{n} \in \mathbb{Z}^3 \setminus \mathbf{0}$ which does not have the form $\mathbf{n} = m\mathbf{h}$, $\mathbf{h} \in \mathbb{Z}^3 \setminus \mathbf{0}$,

$$\frac{1}{|\mathbf{n}|^2} \frac{\left|\sum_{\mathbf{z}_k \in \mathcal{Y}_{\mathbf{r};\omega;t}} e^{2\pi i \mathbf{n} \cdot \mathbf{z}_k}\right|^2}{N} \le \frac{4 \log(2 + |\mathbf{n}|)}{|\mathbf{n}|^{1/2}\sqrt{\varepsilon}} \le \frac{8}{\sqrt{\varepsilon}}. \tag{17.26}$$

Combining (17.25) and (17.26), we have

$$\text{Imba}_2(\mathcal{Y}_{\mathbf{r};\omega;t}) \le \text{TiEvFun}_2(t) + \frac{4 \log N}{N^{1/4}\sqrt{\varepsilon}} \left((\text{TiEvFun}_2(t))^{1/2} + 2 \right) + \frac{8}{\sqrt{\varepsilon}}. \tag{17.27}$$

Combining (17.22) and (17.27), we obtain the following result, which is much simpler than Theorem 14.1. Since MicroEntropy + Imbalance $= N$, we formulate it in terms of the Imbalance.

Theorem 17.1. *Consider the 3-dimensional torus model* $(\mathcal{Y}_{m,m,m};\mathbf{r};\Omega_2)$ *of the second kind such that* N *is a perfect cube* $N=m^3$ *and*

$$\mathcal{Y}=\mathcal{Y}_{m,m,m}=\left\{\mathbf{y}_{j,k,\ell}=\left(\frac{j}{m},\frac{k}{m},\frac{\ell}{m}\right):\ 0\le j,k,\ell<m\right\}$$

is a cube lattice. Given a time $t>0$ *and an* $0<\varepsilon<\frac{1}{2}10^{-3}$, *there exists a measurable subset* $\Omega_2^*=\Omega_2^*(\varepsilon;t)\subset\Omega_2$ *of the initial velocity space of the second kind such that*

$$\text{ProdMeas}_2(\Omega_2^*)>1-2\cdot 10^3\varepsilon$$

and for all $\omega\in\Omega_2^*$

$$|\text{Imba}_2(\mathcal{Y}_{\mathbf{r};\omega;t})-\text{TiEvFun}_2(\mathcal{Y}_{m,m,m};\mathbf{r};t)|$$

$$\le\left(\frac{4\log N}{N^{1/4}\sqrt{\varepsilon}\sqrt{\text{TiEvFun}_2(t)}}\left(1+\frac{2}{\sqrt{\text{TiEvFun}_2(t)}}\right)+\frac{8}{\sqrt{\varepsilon}\text{TiEvFun}_2(t)}\right)$$

$$\cdot\,\text{TiEvFun}_2(t).$$

Here

$$\text{TiEvFun}_2(t)=\text{TiEvFun}_2(\mathcal{Y}_{m,m,m};\mathbf{r};t)$$

$$=\left(N^{1/3}\max_{\mathbf{h}\in\mathbb{Z}^3\setminus\mathbf{0}}\left(\frac{\sin(2\pi N^{1/3}|\mathbf{r}(t)|\ |\mathbf{h}|)}{2\pi N^{1/3}|\mathbf{r}(t)|\ |\mathbf{h}|^2}\right)^2\ \text{or}\ 1\right)_{\max}.$$

□

To illustrate Theorem 17.1, we consider the special case $|\mathbf{r}(t)|=t$, and choose $N=m^3=10^{3000}$, implying $m=10^{1000}$.

At the start $\mathcal{Y}=\mathcal{Y}_{m,m,m}$ we have $\text{TiEvFun}_3(0)=\text{TiEvFun}_3(\mathcal{Y}_{m,m,m};\mathbf{r};0)=m=10^{1000}$. We study the time evolution of a typical system until $\text{TiEvFun}_2(t)$ goes down to $\text{TiEvFun}_2(t)=(1+o(1))10^{200}$. Since

$$\sup_{\tau\ge t}\text{TiEvFun}_2(\mathcal{Y}_{m,m,m};\mathbf{r};\tau)=\left(1+O\left(\frac{1}{1+t}\right)\right)m\cdot\min\left\{\frac{1}{(2\pi mt)^2},1\right\},$$

we have

$$\left(1+O\left(\frac{1}{1+t}\right)\right)\frac{m}{(2\pi mt)^2}=10^{200}\iff t=(1+o(1))\frac{1}{2\pi}10^{-600},$$

that is, we study the extremely short time interval

$$0<t<T=\frac{1}{2\pi}10^{-600}. \tag{17.28}$$

By Lemma 14.1 with $x = 2mt$,

$$\text{TiEvFun}_2(t) \geq 10^{200} e^{-6\sqrt{100 \log 10}} > 10^{101} \quad \text{for all } 0 < t < T = \frac{1}{2\pi} 10^{-600}. \tag{17.29}$$

Let

$$\varepsilon = 10^{-100}. \tag{17.30}$$

Applying Theorem 17.1, for every t in $0 < t < T$ there exists a measurable subset $\Omega_2^*(\varepsilon; t) \subset \Omega_2$ of the initial velocity space of the second kind such that

$$\text{ProdMeas}_2(\Omega_2^*(\varepsilon; t)) > 1 - 2 \cdot 10^3 \varepsilon \tag{17.31}$$

and

$$|\text{Imba}_2(\mathcal{Y}_{\mathbf{r};\omega;t}) - \text{TiEvFun}_2(\mathcal{Y}; \mathbf{r}; t)|$$
$$\leq \left(\frac{4 \log N}{N^{1/4} \sqrt{\varepsilon} \sqrt{\text{TiEvFun}_2(t)}} \left(1 + \frac{2}{\sqrt{\text{TiEvFun}_2(t)}} \right) + \frac{8}{\sqrt{\varepsilon} \text{TiEvFun}_2(t)} \right)$$
$$\cdot \text{TiEvFun}_2(t) \tag{17.32}$$

holds for every $\omega \in \Omega_2^*(\varepsilon; t)$.

As usual, we use the analog of (15.20)–(15.21): let $0 < \delta < 1$ be a real number, and consider the following subset of Ω_2:

$$\Omega_2^{**}(\delta) = \{\omega \in \Omega_2 : \ \omega \notin \Omega_2^*(\varepsilon; t) \text{ holds for a set of } t \in [0, T]$$
$$\text{with length } \geq \delta T\} \tag{17.33}$$

(*length* stands for the 1-dimensional Lebesgue measure).

Repeating the proof of (15.21), we have

$$\text{ProdMeas}_2(\Omega_2^{**}(\delta)) \leq \frac{2 \cdot 10^3 \varepsilon}{\delta}. \tag{17.34}$$

Let

$$\delta = 10^{-50}. \tag{17.35}$$

Summarizing, we have

$$\text{ProdMeas}_3 \{\omega \in \Omega_2 : \omega \in \Omega_2^*(\varepsilon; t) \text{ holds for a set of } t \in [0, T]$$
$$\text{with length} \geq T - \delta T\} \geq 1 - \frac{2 \cdot 10^3 \varepsilon}{\delta}$$

with

$$\delta T = 10^{-50}T \quad \text{and} \quad \frac{2\cdot 10^3\varepsilon}{\delta} < 10^{-46}. \tag{17.36}$$

By (17.29) and (17.30),

$$\frac{4\log N}{N^{1/4}\sqrt{\varepsilon}\sqrt{\text{TiEvFun}_2(t)}}\left(1+\frac{2}{\sqrt{\text{TiEvFun}_2(t)}}\right)+\frac{8}{\sqrt{\varepsilon}\text{TiEvFun}_2(t)}$$

$$\leq \frac{10}{\sqrt{\varepsilon}\text{TiEvFun}_2(t)} \leq \frac{10\cdot 10^{50}}{10^{101}} = 10^{-50}. \tag{17.37}$$

Combining these facts, we obtain a subset $\Omega^{***}\subset\Omega_2$ such that

$$\text{ProdMeas}_2\left(\Omega^{***}\right) \geq 1-\frac{2\cdot 10^3\varepsilon}{\delta} = 1-10^{-46}, \tag{17.38}$$

and for every $\omega\in\Omega^{***}$

$$\text{Imba}_2(\mathcal{Y}_{\mathbf{r};\omega;t}) = \left(1\pm 10^{-50}\right)\text{TiEvFun}_2(t) \tag{17.39}$$

holds for every time t in the interval $0<t<T=\frac{1}{2\pi}10^{-600}$ with the possible exception of a set $t\in[0,T]$ of total length less than $10^{-50}T$. Thus we obtain the following result.

Illustration of Theorem 17.1. *Starting from the cube lattice $\mathcal{Y}=\mathcal{Y}_{m,m,m}$ with $|\mathcal{Y}|=N=10^{3000}$ and $|\mathbf{r}(t)|=t$, there is a subset $\Omega^{***}\subset\Omega_2$ of the initial velocity space of the second kind $\Omega^{***}\subset\Omega_2$ with*

$$\text{ProdMeas}_2\left(\Omega^{***}\right) \geq 1-10^{-46},$$

representing an overwhelming majority, such that for every $\omega\in\Omega^{***}$ the time evolution of the ω-dependent $\text{Imba}_2(\mathcal{Y}_{\mathbf{r};\omega;t})$ is approximated by the explicit ω-independent Time Evolution Function

$$\text{TiEvFun}_2(t) = \left(N^{1/3}\max_{\mathbf{h}\in\mathbb{Z}^3\setminus\mathbf{0}}\left(\frac{\sin(2\pi N^{1/3}t|\mathbf{h}|)}{2\pi N^{1/3}t|\mathbf{h}|^2}\right)^2 \text{ or } 1\right)_{\max}$$

with extreme accuracy. Indeed, we have the time-quantitative result that

$$\text{Imba}_2(\mathcal{Y}_{\mathbf{r};\omega;t}) = \left(1\pm 10^{-50}\right)\text{TiEvFun}_2(t)$$

holds for every time t in the (extremely short) time interval $0<t<T=\frac{1}{2\pi}10^{-600}$ with the possible exception of a totally negligible set of violators $t\in[0,T]$ of total length less than $10^{-50}T$. Note that the time interval $[0,T]$ corresponds to the decrease of the Time Evolution Function of the second kind from the maximum $N=10^{1000}$ to 10^{200}. □

This version of the Second Law — Theorem 17.1 for the second kind — describes the typical time evolution of an individual system strikingly well. As usual, we obtain a strikingly good approximation of the ω-dependent Imbalance by the explicit ω-independent Time Evolution Function as long as the system is far from equilibrium.

Majority behavior vs minority behavior: Paradoxes. The cube lattice $\mathcal{Y} = \mathcal{Y}_{m,m,m}$ leads us to a very interesting "entropy paradox". In the case of the cube lattice we have the explicit formula for the Imbalance

$$\text{Imba}_2(\mathcal{Y}_{m,m,m}) = \max_{\mathbf{n}\in\mathbb{Z}^3\setminus\mathbf{0}} \frac{1}{|\mathbf{n}|^2} \left(\frac{\left|\sum_{0\le j,k,\ell<m} e^{2\pi i\mathbf{n}\cdot\mathbf{y}_{j,k,\ell}}\right|^2}{N} \text{ or } 1 \right)_{\max}$$

$$= \left(\frac{N^2}{Nm^2} \text{ or } 1 \right)_{\max} = m = N^{1/3}. \tag{17.40}$$

We recall that $\text{Imbalance}_2(\mathcal{Y}_{m,m,m})$ represents the square of the Relaxation Drifting Distance for the typical time evolution of $(\mathcal{Y}_{m,m,m};\mathbf{r};\Omega_2)$ (Relaxation Time for $|\mathbf{r}(t)| = t$), so by (17.40) the Relaxation Drifting Distance (= required drifting distance to reach LA-equilibrium) for the typical time evolution is around $N^{1/6}$.

On the other hand, the typical time evolution of the ω-dependent Imbalance $\text{Imba}_2(\mathcal{Y}_{\mathbf{r};\omega;t})$ is well-approximated by the explicit ω-independent Time Evolution Function

$$\text{TiEvFun}_2(t) = \left(N^{1/3} \max_{\mathbf{h}\in\mathbb{Z}^3\setminus\mathbf{0}} \left(\frac{\sin(2\pi N^{1/3}t|\mathbf{h}|)}{2\pi N^{1/3}t|\mathbf{h}|^2} \right)^2 \text{ or } 1 \right)_{\max},$$

and $\text{TiEvFun}_2(t) = O(1)$ for $t \ge N^{-1/6}$.

(Note that in the Illustration above we picked the time interval $0 < t < 10^{-600} = 10^{-3000/5} = N^{-1/5}$, which is substantially shorter than $0 < t < N^{-1/6}$; so here $\text{TiEvFun}_2(t)$ is large, implying a very precise approximation of the Imbalance by the Time Evolution Function.)

The fact $\text{TiEvFun}_2(t) = O(1)$ for $t \ge N^{-1/6}$ means that, starting from the cube lattice $\mathcal{Y} = \mathcal{Y}_{m,m,m}$, the typical time evolution of the Imbalance extremely quickly (in time less than $N^{-1/6}$) reaches $\text{Imba}_2(\mathcal{Y}_{\mathbf{r};\omega;t}) = O(1)$, i.e., $\mathcal{Y}_{\mathbf{r};\omega;t}$ is an Ω_2-equilibrium configuration, or equivalently, the Relaxation Drifting Distance for the typical time evolution starting from $\mathcal{Y}_{\mathbf{r};\omega;t}$ is $O(1)$. This argument seems to imply the contradiction

$$N^{-1/6} + O(1) = N^{1/6}, \tag{17.41}$$

since $N^{1/6}$ is the Relaxation Drifting Distance for the typical time evolution of the cube lattice $(\mathcal{Y}_{m,m,m}; \mathbf{r}; \Omega_2)$ with $|\mathbf{r}(t)| = t$.

The resolution of "paradox (17.41)" is based on a careful analysis of the term *typical*. Just because the "Relaxation Drifting Distance for the typical time evolution starting from $\mathcal{Y}_{\mathbf{r};\omega;t}$ is $O(1)$", it does not mean anything for the Relaxation Drifting Distance for the typical time evolution starting from $\mathcal{Y}_{m,m,m}$. Indeed, for $t \geq N^{-1/6}$ the Micro-Entropy of $\mathcal{Y}_{\mathbf{r};\omega;t}$ is $N - O(1)$, which is substantially larger than the Micro-Entropy $N - N^{1/3}$ of $\mathcal{Y}_{m,m,m}$. Larger entropy states represent a *much* larger volume in the configuration space (since the dynamics is divergence-free, there is a time-invariant configuration space volume; see the proof below); for a simple volume-computation proof see e.g. the Remarks after the Metatheorem. So, the typical time evolution starting from $\mathcal{Y}_{m,m,m}$ at time $t \geq N^{-1/6}$ represents a negligible minority, and this *minority* has nothing to do with the *majority* time evolution starting from the larger entropy state $\mathcal{Y}_{\mathbf{r};\omega;t}$.

In the argument above we used the fact that the time evolution of our dynamical system preserves the volume in the configuration space. The reason is that the dynamics of our torus model represents a divergence-free vector field, and so — by using the generalized Liouville's theorem — the corresponding time flow is volume-preserving. The details go as follows. Consider a general system of ordinary differential equations

$$\frac{d\mathbf{x}}{dt} = \mathbf{f}(t;\mathbf{x}), \quad \mathbf{x} \in \mathbf{R}^n \tag{17.42}$$

with

$$\mathbf{f}(t;\mathbf{x}) = (f_1(t;x_1,\ldots,x_n),\ldots,f_n(t;x_1,\ldots,x_n)).$$

Let $\Omega(0) \subset \mathbf{R}^n$ be a set of finite n-dimensional volume $\mu(0)$, and let $\Omega(\tau)$ denote the image of the set $\Omega(0)$ under the time evolution from $t = 0$ to $t = \tau$, i.e., for every solution $\mathbf{x}$ of (17.42), we have the mapping $\mathbf{x}(0) \to \mathbf{x}(\tau)$ — it is called the time flow induced by the system (17.42). Let $\mu(\tau)$ denote the n-dimensional volume of $\Omega(\tau)$; then we have the well known formula

$$\frac{d\mu(t)}{dt} = \int_{\Omega(t)} \operatorname{div} \mathbf{f}(t;\mathbf{x})\, d\mathbf{x},$$

where

$$\operatorname{div} \mathbf{f}(t;\mathbf{x}) = \sum_{j=1}^{n} \frac{\partial f_j(t;x_1,\ldots,x_n)}{\partial x_j} \quad \text{is the divergence.}$$

It follows that, if $\operatorname{div} \mathbf{f}(t;\mathbf{x}) = 0$ for all $\mathbf{x} \in \mathbf{R}^n$ then the volume is conserved by the time flow. This result is usually called generalized Liouville's theorem (see e.g. P. Hartman, Ordinary Differential Equations, Birkhäuser 1982, pp. 220–227); the proof is based on the n-dimensional divergence theorem.

For a fixed parametrized curve $\mathbf{r}(t)$ and $\omega = (\vartheta_1, \vartheta_2, \ldots, \vartheta_N) \in (SO(3))^N$, our torus model of the second kind has the concrete form

$$\mathcal{Y}_t = \mathcal{Y}_{\mathbf{r};\omega;t} = \{\mathbf{z}_k = \mathbf{y}_k + \vartheta_k(\mathbf{r}(t)) \text{ modulo one} : \ 1 \le k \le N\}, \quad (17.43)$$

where every $\mathbf{z}_k = (z_{k,1}, z_{k,2}, z_{k,3})$ has 3 coordinates. So (17.43) gives the system

$$\frac{d\mathbf{z}_k}{dt} = \mathbf{g}_k(t) = \mathbf{g}_k(\mathbf{r}; \vartheta_k; t), \quad 1 \le k \le N, \quad (17.44)$$

which is a special case of (17.42) with $n = 3N$,

$$\mathbf{f} = (\mathbf{g}_1(t), \ldots, \mathbf{g}_N(t)),$$

and $x_{3k-2} = z_{k,1}, x_{3k-1} = z_{k,2}, x_{3k} = z_{k,3}$, $1 \le k \le N$. By (17.44) the partial derivatives are all zero

$$\frac{\partial f_i(t; x_1, \ldots, x_n)}{\partial x_i} = 0 \ \text{ for every } 1 \le i \le n = 3N,$$

so the divergence is zero $\operatorname{div} \mathbf{f}(t; \mathbf{x}) = 0$, which implies that the generalized Liouville's theorem applies.

Since the time-flow of the dynamics of our torus model is volume-preserving in the configuration space, it may seem a very natural idea to apply the results of ergodic theory. Unfortunately, (traditional measure-theoretic) ergodic theory ignores zero measure sets, i.e., it does not work for non-typical off-equilibrium initial configurations; moreover, Birkhoff's ergodic theorem does not give any error term. These limitations explain why we do not use ergodic theory.

Here is a second example of the "entropy paradox"; this time for a system of the first kind. In the interesting special case $|\mathbf{r}(t)| = t$, by (6.9′) we have

$$t_0 = \text{Relaxation Time}$$

$$= \frac{1}{2\pi} \max_{\mathbf{n} \in \mathbb{Z}^3 \setminus \mathbf{0}} \frac{1}{|\mathbf{n}|} \left(\log \left(\frac{\left| \sum_{k=1}^{N} \mu_k e^{2\pi i \mathbf{n} \cdot \mathbf{y}_k} \right|^2}{\sum_{k=1}^{N} \mu_k^2} \ \text{ or } \ 1 \right)_{\max} \right)^{1/2}. \quad (17.45)$$

That is, (17.45) represents the time the system needs to reach square-root equilibrium in a typical time evolution of the first kind $(\mathcal{Y}, \mathcal{M}; \mathbf{r}; \Omega_1)$. Let

$0 < t < t_0 =$ Relaxation Time. Under the condition of *Illustration of Theorem 13.1* in Section 16, we have

$$\text{Imbalance}_1(\mathcal{Y}_{\mathbf{r};\omega;t}, \mathcal{M}) = (1+o(1))\text{TiEvFun}_1(\mathcal{Y}, \mathcal{M}; \mathbf{r}; t)$$

$$= (1+o(1)) \left(e^{-(2\pi t)^2} \max_{\mathbf{n}\in\mathbb{Z}^3\setminus\mathbf{0}} \left(\frac{\left|\sum_{k=1}^N \mu_k e^{2\pi i \mathbf{n}\cdot\mathbf{y}_k}\right|^2}{\sum_{k=1}^N \mu_k^2} \right)^{1/|\mathbf{n}|^2} \text{ or } 1 \right)_{\max}. \tag{17.46}$$

Let t_1 denote the Relaxation Time for $(\mathcal{Y}_{\mathbf{r};\omega;t}, \mathcal{M}; \mathbf{r}; \Omega_1)$. By (6.9)–(6.10) we have with $|\mathbf{r}(t)| = t$

$$\text{Imbalance} = \exp\left(2\pi \cdot \text{RelaxationTime}\right)^2,$$

or, more precisely

$$\text{Imbalance}_1(\mathcal{Y}_{\mathbf{r};\omega;t}, \mathcal{M}) = e^{(2\pi t_1)^2}. \tag{17.47}$$

Thus by (17.45)–(17.47),

$$t_1 = \frac{1}{2\pi}\left(\log \text{Imbalance}_1(\mathcal{Y}_{\mathbf{r};\omega;t}, \mathcal{M})\right)^{1/2} = \frac{1}{2\pi}$$

$$\left(\left(\log(1+o(1)) - (2\pi t)^2 + \log \max_{\mathbf{n}\in\mathbb{Z}^3\setminus\mathbf{0}} \left(\frac{\left|\sum_{k=1}^N \mu_k e^{2\pi i \mathbf{n}\cdot\mathbf{y}_k}\right|^2}{\sum_{k=1}^N \mu_k^2} \right)^{1/|\mathbf{n}|^2} \right) \text{ or } 0\right)_{\max}^{1/2}$$

$$= \frac{1}{2\pi}\left(\left(-(2\pi t)^2 + \max_{\mathbf{n}\in\mathbb{Z}^3\setminus\mathbf{0}} \frac{1}{|\mathbf{n}|^2}\log\left(\frac{\left|\sum_{k=1}^N \mu_k e^{2\pi i \mathbf{n}\cdot\mathbf{y}_k}\right|^2}{\sum_{k=1}^N \mu_k^2} \right)\right) \text{ or } 0\right)_{\max}^{1/2} + o(1)$$

$$= \frac{1}{2\pi}\sqrt{-(2\pi t)^2 + (2\pi t_0)^2} + o(1) = \sqrt{t_0^2 - t^2} + o(1). \tag{17.48}$$

Clearly

$$\sqrt{t_0^2 - t^2} = (t_0 - t)\sqrt{\frac{t_0 + t}{t_0 - t}} > (t_0 - t), \tag{17.49}$$

so by (17.48)–(17.49),

$$t_1 > t_0 - t. \tag{17.50}$$

The message of (17.50) is that, the typical time evolution of a first kind system $(\mathcal{Y}, \mathcal{M}; \mathbf{r}; \Omega_1)$ at a state $t > 0$ — before reaching equilibrium —

exhibits a tendency of "speeding up" compared to the typical time evolution starting from the later stage $(\mathcal{Y}_{\mathbf{r};\omega;t}, \mathcal{M}; \mathbf{r}; \Omega_1)$. This "speeding up" explains why the *remaining time* $t_0 - t$ (to reach equilibrium) is shorter than $t_1 =$ the Relaxation Time starting from $(\mathcal{Y}_{\mathbf{r};\omega;t}, \mathcal{M}; \mathbf{r}; \Omega_1)$.

This "speeding up" is surprising, but it is not really a paradox. It is just one example where the minority behavior is different from the majority behavior.

In sharp contrast with the "speeding up" of the first kind, for the second and third kinds we have a tendency of "slowing down". In the next example of the "entropy paradox", we study the time evolution of a system of the third kind. In the special case $|\mathbf{r}(t)| = t$, by (6.15′) we have

$$t_0 = \text{Relaxation Time}$$

$$= \frac{1}{2\pi} \max_{\mathbf{n}\in\mathbb{Z}^3\setminus\mathbf{0}} \frac{1}{|\mathbf{n}|} \left(\frac{9\left|\sum_{k=1}^N \mu_k e^{2\pi i \mathbf{n}\cdot\mathbf{y}_k}\right|^2}{\sum_{k=1}^N \mu_k^2} \text{ or } 1 \right)_{\max}^{1/4}. \tag{17.51}$$

That is, (17.51) represents the time the system needs to reach square-root equilibrium in a typical time evolution of the third kind $(\mathcal{Y}, \mathcal{M}; \mathbf{r}; \Omega_3)$. Let $0 < t < t_0 =$ Relaxation Time. Under the condition of the *Second Illustration of Theorem 13.2* in Section 15, we have

$$\text{Imbalance}_3(\mathcal{Y}_{\mathbf{r};\omega;t}, \mathcal{M}) = (1+o(1))\text{TiEvFun}_3(\mathcal{Y}, \mathcal{M}; \mathbf{r}; t) = (1+o(1))$$

$$\max_{\mathbf{n}\in\mathbb{Z}^3\setminus\mathbf{0}} \frac{1}{|\mathbf{n}|^4} \left(\frac{\left|\sum_{k=1}^N \mu_k e^{2\pi i \mathbf{n}\cdot\mathbf{y}_k}\right|^2 9\left(\frac{\cos(2\pi t\ |\mathbf{n}|)}{(2\pi t\ |\mathbf{n}|)^2} - \frac{\sin(2\pi t\ |\mathbf{n}|)}{(2\pi t\ |\mathbf{n}|)^3}\right)^2}{\sum_{k=1}^N \mu_k^2} \text{ or } 1 \right)_{\max}. \tag{17.52}$$

Let t_1 denote the Relaxation Time for $(\mathcal{Y}_{\mathbf{r};\omega;t}, \mathcal{M}; \mathbf{r}; \Omega_3)$. By (6.15)–(6.16) we have with $|\mathbf{r}(t)| = t$

$$\text{Imbalance} = (2\pi \cdot \text{RelaxationTime})^4,$$

or, more precisely

$$\text{Imbalance}_3(\mathcal{Y}_{\mathbf{r};\omega;t}, \mathcal{M}) = (2\pi t_1)^4. \tag{17.53}$$

Thus by (17.51)–(17.53),

$$t_1 = \frac{1}{2\pi} \left(\text{Imbalance}_3(\mathcal{Y}_{\mathbf{r};\omega;t}, \mathcal{M})\right)^{1/4} \leq \frac{1+o(1)}{2\pi}$$

$$\max_{\mathbf{n}\in\mathbb{Z}^3\setminus\mathbf{0}} \frac{1}{|\mathbf{n}|}\left(\frac{\left|\sum_{k=1}^N \mu_k e^{2\pi i\mathbf{n}\cdot\mathbf{y}_k}\right|^2 9\left(\frac{1}{(2\pi t\ |\mathbf{n}|)^2}+\frac{1}{(2\pi t\ |\mathbf{n}|)^3}\right)^2}{\sum_{k=1}^N \mu_k^2} \text{ or } 1\right)_{\max}^{1/4}$$

$$\leq \frac{1+o(1)}{2\pi}\max_{\mathbf{n}\in\mathbb{Z}^3\setminus\mathbf{0}} \frac{1}{|\mathbf{n}|}\left(\frac{9\left|\sum_{k=1}^N \mu_k e^{2\pi i\mathbf{n}\cdot\mathbf{y}_k}\right|^2 \left(\frac{1}{(2\pi t)^2}+\frac{1}{(2\pi t)^3}\right)^2}{\sum_{k=1}^N \mu_k^2} \text{ or } 1\right)_{\max}^{1/4}$$

$$\leq (1+o(1))t_0\left(\frac{1}{(2\pi t)^2}+\frac{1}{(2\pi t)^3}\right)^{1/2}. \tag{17.54}$$

Clearly the upper bound

$$(1+o(1))t_0\left(\frac{1}{(2\pi t)^2}+\frac{1}{(2\pi t)^3}\right)^{1/2}$$

of t_1 (see (17.54)) can be much smaller than $t_0 - t$. For example, if $t = \sqrt{t_0}$ than by (17.54)

$$t_1 = \text{Relaxation Time for } (\mathcal{Y}_{\mathbf{r};\omega;t}, \mathcal{M}; \mathbf{r}; \Omega_3) = O(\sqrt{t_0}),$$

that is, at time t the system is already $O(\sqrt{t_0})$-close to Ω_3-equilibrium, despite the fact that the *remaining time* $t_0 - t = t_0 - \sqrt{t_0}$ (for the system $(\mathcal{Y}, \mathcal{M}; \mathbf{r}; \Omega_3)$ to reach equilibrium) is much larger (basically the square). The message is that, the typical time evolution of a third kind system $(\mathcal{Y}, \mathcal{M}; \mathbf{r}; \Omega_3)$ at a state $t > 0$ — still being very far from equilibrium — exhibits a tendency of "slowing down" compared to the typical time evolution starting from the later stage $(\mathcal{Y}_{\mathbf{r};\omega;t}, \mathcal{M}; \mathbf{r}; \Omega_3)$. This "slowing down" is surprising, but it is not a paradox. It is just another example where the minority behavior is different from the majority behavior.

We have a very similar "slowing down" phenomenon for the second kind.

We can summarize the "entropy paradox" as follows. Starting from a low entropy state at $t = 0$, and reaching a higher entropy state in a typical time evolution later at $t = t'$, the rest of the time evolution of the system (from t' till equilibrium) represents a *minority* behavior, and it has nothing to do with the *majority* behavior starting from $t = t'$.

Finally, note that the so-called Loschmidt paradox (and the similar irreversibility-paradox) can be explained the same way: the minority behavior is different from the majority behavior. We recall the Loschmidt paradox: for every motion of a large system $t > 0$, we can associate with the reverse motion where every velocity is reversed at $t = 0$, and the time evolution of the reverse motion contradicts the law of increasing entropy (Second Law).

Boltzmann gave a statistical explanation to the Loschmidt-paradox/ irreversibility-paradox, and it can be summarized as follows. Given the energy of a many-particle system, the overwhelming majority of its phase points on the corresponding energy surface are "equilibrium points", all of which look macroscopically the same. Since there are many more "equilibrium points" than "nonequilibrium points" in the phase space, it is plausible to expect that the time evolution of a typical trajectory (of the system) in the phase space exhibits a rapid transition from "nonequilibrium" to "equilibrium" as the time $t \to \infty$.

We can easily justify Boltzmann's intuitive explanation with a rigorous argument using our precise Second Law. According to our Second Law, a typical backward time evolution $t < 0$ (i.e., every velocity is reversed at $t = 0$) also exhibits (spatial) entropy increase, exactly like a typical forward time evolution $t > 0$. That is, at $t < 0$ the entropy of the system is (usually) larger than the entropy of the system at $t = 0$. This fact, however, does not contradict our Second Law (= typical entropy increase). Indeed, it is just another example where the minority behavior (this particular time evolution from negative t to 0 exhibiting entropy decrease) is different from the majority behavior (i.e., entropy increase) starting from this negative time point t.

Chapter 4

More on the Second Law

18. The case of general speed distribution (I)

What is in Chapter 4? *So far we formulated our Second Law for the three basic torus models: the torus model of the first kind (Gaussian), the second kind ("like photons") and the third kind ("uniform in the unit ball"). Here we discuss some far-reaching generalizations.*

We recall that in the general 3-dimensional isotropic torus model with $\mathbf{r}_1 = \ldots = \mathbf{r}_N = \mathbf{r}$, the trajectory of the k-th particle is

$$\mathbf{y}_k + \vartheta_k(\rho_k \mathbf{r}(t)) \text{ modulo one}, \quad 1 \le k \le N, \tag{18.1}$$

where ϑ_k is uniformly distributed in $SO(3)$, and the distribution of $\rho_j \in [0, \infty)$ is described by a probability measure; for simplicity assume that the measure has a density function

$$\Pr[\rho_j \le u] = \int_0^u g(y)\, dy \tag{18.2}$$

and

$$\text{Expectation} = \int_0^\infty u g(u)\, du < \infty,$$

where the latter means that the "mean speed is finite". We allow g to be a generalized function like the Dirac delta function concentrated on 1 representing the second kind (but of course we exclude the trivial case of Dirac delta concentrated on 0). In fact, we want the Expectation (= "mean speed") to be one, or at least reasonably close to one. (The reason is that we compare how fast systems approach equilibrium with different probability density functions g; so, it is natural to assume that they have the same Expectation (= "mean speed"), preferably one. But, for notational convenience, we allow probability density functions for which the Expectation is "close to one"; like the Expectation for the first kind is $\sqrt{8/\pi}$, and for the third kind it is 3/4.)

The corresponding initial velocity space is

$$\omega = (\rho_1, \vartheta_1, \rho_2, \vartheta_2, \ldots, \rho_N, \vartheta_N) \in \Omega_g = ([0,\infty) \times SO(3))^N$$

equipped with the product measure (Haar measure for $SO(3)$ and (18.2) for $[0,\infty)$).

As we explained in Theorem 4.2(b), the key functions

$$e^{-w^2/2}, \quad \frac{\sin w}{w} \quad \text{and} \quad -3\frac{\cos w}{w^2} + 3\frac{\sin w}{w^3}$$

in Theorems 3.1, 4.1 and 4.2(a) are replaced by the corresponding isotropic Fourier transform

$$\mathbf{F}_g(\mathbf{w}) = \int_0^\infty \frac{1}{4\pi} \int_{\mathbf{v}\in\mathbf{S}^2} e^{\mathrm{i}u\mathbf{v}\cdot\mathbf{w}} g(u)\, du\, d\mathrm{SA}(\mathbf{v})$$

$$= \int_0^\infty \frac{1}{4\pi} \int_{\mathbf{v}\in\mathbf{S}^2} \cos(u\mathbf{v}\cdot\mathbf{w}) g(u)\, du\, d\mathbf{v}, \tag{18.3}$$

where $\mathbf{w}$ is an arbitrary 3-dimensional vector and $d\mathrm{SA}(\mathbf{v})$ represents the surface area (notice that $\mathbf{F}_g(\mathbf{w})$ defines a real-valued function depending only on the length $w = |\mathbf{w}|$ of the input vector). The definition implies the trivial but important facts

$$\mathbf{F}_g(0) = \mathbf{F}_g(\mathbf{0}) = 1 \quad \text{and} \quad |\mathbf{F}_g(w)| = |\mathbf{F}_g(\mathbf{w})| < 1 \quad \text{for} \quad w = |\mathbf{w}| > 0. \tag{18.4}$$

For the convenience of the reader we summarize some of the results of Theorem 4.2(b) in the following lemma.

Lemma 18.1. *Let $g(u)$ be a probability density function in $[0,\infty)$, let us extend g over the whole real line as an even function $g(-u) = g(u)$, and let $\widehat{g}(x)$ denote the Fourier transform of g. The isotropic Fourier transform $\mathbf{F}_g(\mathbf{w})$ (see (18.3)) can be expressed in terms of the Fourier transform $\widehat{g}(x)$ as follows*

$$\mathbf{F}_g(w) = \mathbf{F}_g(\mathbf{w}) = \frac{\sqrt{\pi}}{2\sqrt{2}w} \int_{-w}^{w} \widehat{g}(x)\, dx = \sqrt{\frac{\pi}{2}} \frac{1}{w} \int_0^w \widehat{g}(x)\, dx.$$

Moreover, if $g(y)$ is continuous at $y = 0$ then

$$g(0) = \frac{1}{\sqrt{2\pi}} \int_{-\infty}^{\infty} \widehat{g}(x)\, dx,$$

and so we can rewrite $\mathbf{F}_g(w)$ *as*

$$\mathbf{F}_g(w) = \frac{\sqrt{\pi}}{2\sqrt{2}w}\left(g(0)\sqrt{2\pi} - \int_{|x|\geq w} \widehat{g}(x)\, dx\right).$$

□

We show several applications of this simple but very useful lemma.

We already introduced the torus model of the fourth kind in (4.32). We show how Lemma 18.1 applies in this case.

Torus model of the fourth kind: the speed density function is $g_4(y) = \sqrt{2/\pi}e^{-y^2/2}$, that is,

$$\Pr[\rho_j \leq u] = \sqrt{\frac{2}{\pi}}\int_0^u e^{-y^2/2}\, dy \quad \text{for all positive } u. \tag{18.5}$$

Since the Fourier transform of $g_4(y)$ is itself, Lemma 18.1 gives

$$\mathbf{F}_{g_4}(w) = \frac{\sqrt{\pi}}{2\sqrt{2}w}\left(g(0)\sqrt{2\pi} - \int_{|x|\geq w} \widehat{g}(x)\, dx\right)$$

$$= \frac{\sqrt{\pi}}{2\sqrt{2}w}\left(\sqrt{2/\pi}\sqrt{2\pi} - \int_{|x|\geq w} e^{-x^2/2}\, dx\right) = \frac{\sqrt{\pi}}{2\sqrt{2}w}\left(2 - \int_{|x|\geq w} e^{-x^2/2}\, dx\right). \tag{18.6}$$

By (18.6) we obtain the asymptotic behavior of $\mathbf{F}_{g_4}(w)$ as $w \to \infty$:

$$\mathbf{F}_{g_4}(w) = \frac{C + o(1)}{w} \quad \text{where } C = \sqrt{\pi/2}. \tag{18.7}$$

It follows from (18.7) that the typical time evolution of the torus model of the fourth kind is basically the same as that of the second kind. Repeating the arguments of (6.11)–(6.13) with (18.7) instead of $\frac{\sin w}{w}$, we obtain the Imbalance for the fourth kind

$$\text{Imbalance}_{g_4}(\mathcal{Y}, \mathcal{M}) = \max_{\mathbf{n}\in\mathbb{Z}^3\setminus\mathbf{0}} \frac{1}{|\mathbf{n}|^2}\left(\frac{\left|\sum_{k=1}^N \mu_k e^{2\pi i\mathbf{n}\cdot\mathbf{y}_k}\right|^2}{\sum_{k=1}^N \mu_k^2} \text{ or } 1\right)_{\max},$$

which is exactly the same as (6.13) ("second kind").

As in the Metatheorem, the corresponding entropy has a complementary relation with the Imbalance:

$$\text{MicroEntropy}_{g_4}(\mathcal{Y}, \mathcal{M}) = \frac{N^2}{\sum_{k=1}^N \mu_k^2} - \text{Imbalance}_{g_4}(\mathcal{Y}, \mathcal{M}).$$

In particular, the fourth kind and the second kind exhibit the same very slow approach to LA-equilibrium, assuming they start from a far-from-equilibrium initial configuration. For example, starting from close to Big Bang, for a typical initial velocity vector $\omega \in \Omega_4$, the drifting distance $|\mathbf{r}(t)|$ has to be in the range of $\sqrt{N}$ to reach equilibrium (exactly like for the second kind).

If we modify the speed density function $g_4(y) = \sqrt{2/\pi}e^{-y^2/2}$ by a factor of y^2, then of course we get back the familiar first kind (Gaussian distribution), which is extremely fast. Indeed, as we explained in Section 3, starting from Big Bang, or from any other initial point configuration, for a typical initial velocity vector $\omega \in \Omega_1$, the system reaches L-equilibrium when the drifting distance $|\mathbf{r}(t)|$ is in the range of $\sqrt{\log N}$.

Next we study what happens when we modify the speed density function $g_4(y) = \sqrt{2/\pi}e^{-y^2/2}$ by a factor of y^4.

Torus model of the fifth kind: the speed density function is $g_5(y) = \frac{\sqrt{2}}{3\sqrt{\pi}}y^4e^{-y^2/2}$, that is,

$$\Pr[\rho_j \leq u] = \frac{\sqrt{2}}{3\sqrt{\pi}}\int_0^u y^4e^{-y^2/2}\,dy \quad \text{for all positive } u. \tag{18.8}$$

To compute the Fourier transform of $g_5(y)$ we use the following general fact $(\mathrm{i} = \sqrt{-1})$:

$$\text{the Fourier transform of } y^n f(y) \text{ is } \mathrm{i}^n \text{ times the } n^{\text{th}} \text{ derivative of } \widehat{f}. \tag{18.9}$$

Applying (18.9) for

$$f(y) = \frac{\sqrt{2}}{3\sqrt{\pi}}e^{-y^2/2} \quad \text{and} \quad n = 4,$$

we obtain

$$\widehat{g_5}(x) = \frac{\sqrt{2}}{3\sqrt{\pi}}\left(\frac{d}{dx}\right)^4 e^{-x^2/2} = \frac{\sqrt{2}}{3\sqrt{\pi}}(x^4 - 6x^2 + 3)e^{-x^2/2}.$$

Using this in Lemma 18.1, we have

$$\mathbf{F}_{g_5}(w) = \frac{\sqrt{\pi}}{\sqrt{2}w}\int_0^w \widehat{g_5}(x)\,dx$$

$$= \frac{\sqrt{\pi}}{\sqrt{2}w}\int_0^w \frac{1}{3}\sqrt{\frac{2}{\pi}}(x^4 - 6x^2 + 3)e^{-x^2/2}\,dx = \frac{1}{3w}\int_0^w (x^4 - 6x^2 + 3)e^{-x^2/2}\,dx. \tag{18.10}$$

Using integration by parts,

$$\int_0^w x^2 e^{-x^2/2}\,dx = \int_0^w x\cdot xe^{-x^2/2}\,dx = -we^{-w^2/2} + \int_0^w e^{-x^2/2}\,dx, \quad (18.11)$$

and similarly,

$$\int_0^w x^4 e^{-x^2/2}\,dx = \int_0^w x^3 \cdot xe^{-x^2/2}\,dx = -w^3 e^{-w^2/2} + 3\int_0^w x^2 e^{-x^2/2}\,dx$$

$$= -w^3 e^{-w^2/2} + 3\left(-we^{-w^2/2} + \int_0^w e^{-x^2/2}\,dx\right)$$

$$= -w^3 e^{-w^2/2} - 3we^{-w^2/2} + 3\int_0^w e^{-x^2/2}\,dx, \quad (18.12)$$

where in the last step we used (18.11).

Using (18.11)–(18.12) in (18.10), we obtain

$$\mathbf{F}_{g_5}(w) = \frac{\sqrt{\pi}}{\sqrt{2}w}\int_0^w \widehat{g_5}(x)\,dx = \frac{1}{3w}\int_0^w (x^4 - 6x^2 + 3)e^{-x^2/2}\,dx$$

$$= \frac{1}{3w}\left(-w^3 e^{-w^2/2} + 3we^{-w^2/2}\right) = \frac{1}{3}\left(-w^2 + 3\right)e^{-w^2/2}. \quad (18.13)$$

Since in the application we square $\mathbf{F}_{g_5}(w)$ anyway, it suffices to study the monotonicity of

$$-\mathbf{F}_{g_5}(w) = \frac{1}{3}\left(w^2 - 3\right)e^{-w^2/2}. \quad (18.14)$$

Since the derivative of (18.14) is

$$-\frac{1}{3}\left(w^3 - 5w\right)e^{-w^2/2},$$

we see that (18.14) is monotone decreasing for all $w > \sqrt{5}$. Moreover, (18.14) tends to zero superexponentially fast as $w \to \infty$.

It follows that the typical time evolution of the torus model of the fifth kind is almost exactly the same as that of the first kind. Both exhibit extremely fast approach to L-equilibrium: even starting from Big Bang, or any other initial point configuration, for a typical initial velocity vector $\omega \in \Omega_i$, $i = 1$ or 5, the system reaches L-equilibrium when the drifting distance $|\mathbf{r}(t)|$ is in the range of $\sqrt{\log N}$.

Also the Imbalance (and the Micro-Entropy) of the fifth kind is basically the same as that of the first kind.

Similarly, we can replace the function $y^4 e^{-y^2/2}$ with $y^6 e^{-y^2/2}$, or in general, with any probability density function of the form

$$g(y) = (c_1 y^2 + c_2 y^4 + c_3 y^6 + \ldots + c_k y^{2k}) e^{-y^2/2} \tag{18.15}$$

(of course we need $g(y) \geq 0$ which can be guaranteed e.g. if all coefficients c_i are nonnegative). The Imbalance (and the Micro-Entropy) for the whole class (18.15) is basically the same as that of the first kind. In particular, all cases in (18.15) exhibit superexponentially fast approach to L-equilibrium exactly like the first kind.

Note that $y^2 e^{-y^2/2}$ and $y^4 e^{-y^2/2}$ are both "infinitely smooth", in fact, analytic over the complex plane (and also both have zero value at $y = 0$). This is why the extra factor y^2 has no effect on the threshold magnitudes: it is $\sqrt{\log N}$ for both.

Besides the "Gaussian class" (18.15), there are many other even analytic functions; for example, consider the class

$$y^2 e^{-y^4},\ y^4 e^{-y^4},\ y^6 e^{-y^4}, \ldots,\ y^2 e^{-y^6},\ y^4 e^{-y^6},\ y^6 e^{-y^6}, \ldots,$$

all finite products formed from these functions, and finally,

all finite sums of these products taken with positive coefficients. (18.16)

The corresponding systems behave quite similarly to the first kind: the typical time evolution of the system converges to equilibrium superexponentially fast in terms of the drifting distance $|\mathbf{r}(t)|$. This follows from the combination of Lemma 18.1 with the following "Paley–Wiener type" result in complex variables. Let $f(z)$ be a function analytic in the horizontal strip on the complex plane where the imaginary part of z has absolute value $\leq a$, and let

$$\sup_{|y| \leq a} \int_{-\infty}^{\infty} |f(x + \mathrm{i}y)|^2 \, dx < \infty.$$

Then the tail of the Fourier transform has the decay

$$\left| \widehat{f}(z) \right| \leq C_b e^{-b|z|} \text{ for every } z \in \mathbb{C} \text{ and } 0 < b < a. \tag{18.17}$$

For a proof, see e.g. Reed and Simon, Methods of Modern Mathematical Physics, Vol. 2, Theorem IX.14.

For example, the function $f(z) = e^{-z^4}$ is analytic on the whole complex plane ($z = x + \mathrm{i}y$),

$$f(z) = e^{-z^4} \text{ has absolute value } |f(z)| = e^{-x^4 + 6x^2 y^2 - y^4},$$

and so

$$\sup_{|y|\le a}\int_{-\infty}^{\infty}|f(x+\mathrm{i}y)|^2\,dx=\sup_{|y|\le a}e^{-2a^4}\int_{-\infty}^{\infty}e^{-2x^4+12a^2x^2}\,dx=C(a)<\infty$$

for every $a>0$. Since (18.17) applies for arbitrarily large a, it follows that $\left|\widehat{f}(z)\right|$ tends to zero superexponentially fast as $|z|\to\infty$ on the complex plane. The same argument — applying (18.17) — works for all members of the class (18.16).

The only difference between the classes (18.15) and (18.16) is that for the latter we do not have an explicit formula for the Fourier transform. Nevertheless, the infinite smoothness of the density function makes it possible to carry out an unlimited number of integrations by parts. This is how we demonstrate the extremely rapid approach to equilibrium. Unfortunately, unlimited number of integrations by parts is an unpleasant technical business, but it is certainly doable. Here is an illustration. We consider the first function

$$f(u)=u^2e^{-u^4/4}$$

in the list (18.16), and study the decay of the tail of its Fourier transform (the same argument works for the whole class (18.6)). Note that the Paley–Wiener type result (18.17) gives

$$\widehat{f}(z)=O\left(e^{a|z|}\right)\quad\text{for every } a>0,$$

which can be interpreted as a *superexponentially fast decay* in the weakest quantitative sense. In other words, (18.17) implies that the Relaxation Drifting Distance is $o(\log N)$. The following result, Lemma 18.2 below, gives a substantially better upper bound on the speed of the decay: it implies that the Relaxation Drifting Distance is $O((\log N)^{3/4})$.

For notational convenience, write

$$A(x)=\sqrt{2\pi}\widehat{f}(x)=\int_{-\infty}^{\infty}e^{\mathrm{i}xu}u^2e^{-u^4/4}\,du=2\int_0^{\infty}\cos(xu)u^2e^{-u^4/4}\,du. \tag{18.18}$$

Lemma 18.2. *There are positive absolute constants C_1 and c_2 such that*

$$|A(x)|\le C_1e^{-c_2|x|^{4/3}} \tag{18.19}$$

holds for all x.

We postpone the proof to the end of the section.

Next we show a general method for finding probability density functions, far beyond the explicit class (18.15), for which we can guarantee Relaxation

Drifting Distance $O\left(\sqrt{\log N}\right)$ (like the first kind). This method works in every dimension $d \geq 2$. Let $\mathbf{h}(\mathbf{x})$, $\mathbf{x} \in \mathbf{R}^d$ be a non-negative rotation-invariant function in d variables, i.e., $\mathbf{h}(\mathbf{x}) = \mathbf{h}(x, 0, \ldots, 0) \geq 0$ for all $|\mathbf{x}| = x$, and we also require the polynomial decay condition

$$\mathbf{h}(x, 0, \ldots, 0) = O\left(\min\left\{1, x^{-3d}\right\}\right) \quad \text{for all} \quad x > 0. \tag{18.20}$$

Let $\mathbf{G}(\mathbf{x}) = \mathbf{Gauss}(\mathbf{x}) = e^{-|\mathbf{x}|^2/2}$. The convolution $\mathbf{h}_2 = \mathbf{h} * \mathbf{G}$ defines another non-negative rotation-invariant function in d variables.

Applying the well known fact that the Fourier transform of a convolution is the product of the Fourier transforms, we have

$$\widehat{\mathbf{h}_2} = \widehat{\mathbf{h} * \mathbf{G}} = \widehat{\mathbf{h}}\mathbf{G}, \tag{18.21}$$

where we used that $\mathbf{G}(\mathbf{x}) = e^{-|\mathbf{x}|^2/2}$ is a fixed point of the Fourier transform.

Since $\mathbf{h}_2$ is a non-negative rotation-invariant function in d variables, the equation

$$g(u) = u^{d-1}\mathbf{h}_2(\mathbf{u}) \quad \text{with} \quad u = |\mathbf{u}| \tag{18.22}$$

defines a non-negative function in one variable $u \geq 0$. Condition (18.20) implies that $g(u)$ has a finite integral

$$\int_0^\infty g(u)\, du < \infty.$$

What is more, multiplying the initial function $\mathbf{h}(\mathbf{x})$, $\mathbf{x} \in \mathbf{R}^d$ with some constant, we can also guarantee that $g(u)$, $u \geq 0$ is a probability density function

$$\int_0^\infty g(u)\, du = 1.$$

By modifying the polynomial decay condition in (18.20), we can even guarantee that the corresponding random variable has finite expectation, finite variance, finite third moment, finite fourth moment, and so on.

We claim that the isotropic Fourier transform $\mathbf{F}_g$ of this g has the superexponential decay

$$|\mathbf{F}_g(w)| \leq O\left(e^{-w^2/2}\right). \tag{18.23}$$

Inequality (18.23) clearly guarantees that the Relaxation Drifting Distance is in the range $O\left(\sqrt{\log N}\right)$ (like the first kind).

Note that (18.23) is a simple application of the last formula in Theorem 4.2:

$$\mathbf{F}_g(\mathbf{w}) = \int_{\mathbf{R}^d} e^{i\mathbf{x}\cdot\mathbf{w}} \frac{g(|\mathbf{x}|)}{|\mathbf{x}|^{d-1}}\, d\mathbf{x} = (2\pi)^{d/2}\widehat{\mathbf{g}_d}(-\mathbf{w}), \tag{18.24}$$

where

$$\mathbf{g}_d(\mathbf{x}) = \frac{g(|\mathbf{x}|)}{|\mathbf{x}|^{d-1}}$$

and $\widehat{\mathbf{g}_d}(\mathbf{w})$ is its Fourier transform. Indeed, in this special case (18.22) yields

$$\mathbf{g}_d(\mathbf{x}) = \frac{g(|\mathbf{x}|)}{|\mathbf{x}|^{d-1}} = \mathbf{h}_2(\mathbf{x}),$$

and so

$$\widehat{\mathbf{g}_d}(\mathbf{w}) = \widehat{\mathbf{h}_2}(\mathbf{w}). \tag{18.25}$$

Using (18.21) and (18.25) in (18.24), we have

$$\begin{aligned}\mathbf{F}_g(w) = \mathbf{F}_g(\mathbf{w}) &= (2\pi)^{d/2}\widehat{\mathbf{g}_d}(-\mathbf{w}) \\ &= (2\pi)^{d/2}\widehat{\mathbf{h}_2}(\mathbf{w}) = (2\pi)^{d/2}\mathbf{G}(\mathbf{w})\widehat{\mathbf{h}}(\mathbf{w}),\end{aligned}$$

which implies

$$|\mathbf{F}_g(w)| \leq e^{-w^2/2} \int_{\mathbf{R}^d} \mathbf{h}_2(\mathbf{x})\, d\mathbf{x} = O\left(e^{-w^2/2}\right), \tag{18.26}$$

since the upper bound

$$\max_{\mathbf{w}} |\widehat{\mathbf{h}}(\mathbf{w})| \leq (2\pi)^{-d/2} \int_{\mathbf{R}^d} \mathbf{h}_2(\mathbf{x})\, d\mathbf{x} = O\left(1\right)$$

is trivial from the decay condition (18.20). Finally, (18.26) gives (18.23)

As we promised, we conclude the section with the

Proof of Lemma 18.2. Since $A(-x) = A(x)$, it suffices to estimate $|A(x)|$ for $x \geq 0$. The lemma is trivial for $0 \leq x \leq 10^3$. Let $x > 10^3$ be fixed, and let $k \geq 1$ be the nearest integer to $x^{4/3}/c_3$, where the constant c_3 will be specified later (note that $c_3 = 100$ is a good choice). For notational convenience we just write $k = x^{4/3}/c_3$. Write $F_0(u) = u^2 e^{-u^4/4}$, and let $F_j(u)$ be the j-th derivative of $F_0(u)$. Applying integration by parts k times, we have

$$\begin{aligned}A(x) &= \int_{-\infty}^{\infty} e^{ixu} u^2 e^{-u^4/4}\, du \\ &= \frac{(-1)^k}{(\mathrm{i}x)^k} \int_{-\infty}^{\infty} e^{ixu} F_k(u)\, du = 2\frac{\mathrm{i}^k}{x^k} \int_0^{\infty} \cos(xu) F_k(u)\, du.\end{aligned} \tag{18.27}$$

We have

$$F_1(u) = (-u^5 + 2u)e^{-u^4/4}, \quad F_2(u) = (u^8 - 7u^4 + 2)e^{-u^4/4},$$

and in general, $F_j(u) = P_j(u)e^{-u^4/4}$, where $P_j(u)$ is a polynomial of the form

$$P_j(u) = a_{j,0}u^{3j+2} + a_{j,1}u^{3j-2} + a_{j,2}u^{3j-6} + \ldots + a_{j,r}u^{3j+2-4r} + \ldots$$
$$\text{with } 3j+2 \geq 4r. \tag{18.28}$$

Since

$$P_{j+1}(u)e^{-u^4/4} = F_{j+1}(u) = F_j'(u) = (P_j'(u) - u^3 P_j(u))e^{-u^4/4},$$

which implies $P_{j+1}(u) = P_j'(u) - u^3 P_j(u)$, or equivalently,

$$a_{j+1,r} = -a_{j,r} + (3j+6-4r)a_{j,r-1}. \tag{18.29}$$

Clearly $a_{n,0} = (-1)^n$. For the sequence $a_{n,1}$ equation (18.29) gives the recurrence formula

$$a_{j+1,1} = -a_{j,1} + (3j+2)a_{j,0},$$

which implies the inequality

$$|a_{j+1,1}| \leq |a_{j,1}| + (3j+2). \tag{18.30}$$

By (18.30),

$$|a_{n,1}| \leq \sum_{j=0}^{n-1} (3j+2) \leq 2n^2. \tag{18.31}$$

For the sequence $a_{n,2}$ equation (18.29) gives the recurrence formula

$$a_{j+1,2} = -a_{j,2} + (3j-2)a_{j,1},$$

which implies the inequality

$$|a_{j+1,2}| \leq |a_{j,2}| + (3j-2) \cdot 2j^2. \tag{18.32}$$

By (18.31) and (18.32),

$$|a_{n,2}| \leq \sum_{j=0}^{n-1} (3j-2) \cdot 2j^2 \leq \frac{2^2 n^4}{2}. \tag{18.33}$$

For the sequence $a_{n,3}$ equation (18.29) gives the recurrence formula

$$a_{j+1,3} = -a_{j,3} + (3j-6)a_{j,2},$$

which implies the inequality

$$|a_{j+1,3}| \leq |a_{j,3}| + (3j-6) \cdot \frac{2^2 j^4}{2}. \tag{18.34}$$

By (18.33) and (18.34),

$$|a_{n,3}| \le \sum_{j=0}^{n-1}(3j-6)\cdot\frac{2^2 j^4}{2} \le \frac{2^3 n^6}{3!}.$$

Repeating the same argument, in general we have

$$|a_{n,r}| \le \frac{2^r n^{2r}}{r!} \text{for all } 3n+2-4r \ge 0. \tag{18.35}$$

Using (18.35) in (18.28), we have

$$|P_k(u)| \le \sum_{0\le r\le \frac{3k+2}{4}} \frac{2^r k^{2r}}{r!} u^{3k+2-4r} = u^{3k+2} \sum_{0\le r\le \frac{3k+2}{4}} \frac{1}{r!}\left(\frac{2k^2}{u^4}\right)^r. \tag{18.36}$$

Next we study the end-sum in (18.36)

$$\sum_{0\le r\le \frac{3k+2}{4}} \frac{1}{r!}\left(\frac{2k^2}{u^4}\right)^r. \tag{18.37}$$

Of course we have the trivial upper bound

$$\sum_{0\le r\le \frac{3k+2}{4}} \frac{1}{r!}\left(\frac{2k^2}{u^4}\right)^r \le \sum_{r=0}^{\infty} \frac{1}{r!}\left(\frac{2k^2}{u^4}\right)^r = \exp\left(\frac{2k^2}{u^4}\right), \tag{18.38}$$

which always applies.

In the following special case, that we call Case 0, we do not use (18.38).

Case 0: $\frac{2k^2}{u^4} \ge 2\cdot\frac{3k+2}{4}$

In Case 0 (18.37) is dominated by the last term of the sum. Indeed, the terms are increasing, and the next one is at least twice as large as the previous one. So the whole sum (18.37) is less than twice the last term:

$$\sum_{0\le r\le \frac{3k+2}{4}} \frac{1}{r!}\left(\frac{2k^2}{u^4}\right)^r \le 2\frac{1}{r_0!}\left(\frac{2k^2}{u^4}\right)^{r_0}, \tag{18.39}$$

where $r_0 = \frac{3k+2}{4}$ (for notational convenience assume that r_0 is an integer). Using the weak form of the Stirling formula $n! \ge (n/e)^n$ with $n = r_0$ in (18.39), we have

$$\sum_{0\le r\le \frac{3k+2}{4}} \frac{1}{r!}\left(\frac{2k^2}{u^4}\right)^r \le 2\left(\frac{8ek^2}{(3k+2)u^4}\right)^{r_0}. \tag{18.40}$$

Using (18.40) and $r_0 = \frac{3k+2}{4}$ in (18.36),

$$|P_k(u)| \le u^{3k+2}\cdot 2\left(\frac{8ek^2}{(3k+2)u^4}\right)^{r_0} = 2\left(\frac{8ek^2}{3k+2}\right)^{r_0}. \tag{18.41}$$

We divide the integral (18.27) into 2 parts

$$A(x) = 2\frac{\mathrm{i}^k}{x^k}\int_0^\infty \cos(xu)F_k(u)\,du = 2\sum_{j=1}^{2}\frac{\mathrm{i}^k}{x^k}\int_{u\in I_j}\cos(xu)F_k(u)\,du, \tag{18.42}$$

where

$$I_1 = \left[0, k^{1/4}\right] \quad \text{and} \quad I_2 = [0,\infty)\setminus I_1.$$

Notice that Case 0 above corresponds to the interval

$$I_0 = \left[0, \left(\frac{4k^2}{3k+2}\right)^{1/4}\right],$$

and $I_1 \subset I_0$. It follows that, to estimate the contribution of I_1 in (18.42), we can use (18.41). Thus we have

$$\left|2\frac{\mathrm{i}^k}{x^k}\int_{u\in I_1}\cos(xu)F_k(u)\,du\right| \le 4\int_{u\in I_1} x^{-k}\left(\frac{8ek^2}{3k+2}\right)^{(3k+2)/4} e^{-u^4/4}\,du. \tag{18.43}$$

Clearly

$$x^{-k}\left(\frac{8ek^2}{3k+2}\right)^{(3k+2)/4} = \left(\frac{1}{(c_3k)^{3/4}}\cdot\left(\frac{8ek^2}{3k+2}\right)^{\frac{3}{4}+\frac{1}{2k}}\right)^k \le e^{-k} = e^{-x^{4/3}/c_3} \tag{18.44}$$

if c_3 is a sufficiently large constant; for example, $c_3 = 100$ is a good choice. Since

$$\int_0^\infty e^{-u^4/4}\,du = c_4 < \infty,$$

using (18.44) in (18.43), we obtain

$$\left|2\frac{\mathrm{i}^k}{x^k}\int_{u\in I_1}\cos(xu)F_k(u)\,du\right| \le 4c_4e^{-x^{4/3}/c_3}. \tag{18.45}$$

To estimate the contribution of I_2 in (18.35), we use the trivial upper bound (18.31). Since

$$\exp\left(\frac{2k^2}{u^4}\right) \le e^{2k} \quad \text{for all } u > k^{1/4},$$

by (18.29) we have

$$\left|2\frac{\mathrm{i}^k}{x^k}\int_{u\in I_2}\cos(xu)F_k(u)\,du\right| \le 2\int_{u\ge k^{1/4}} h(u)\,du \tag{18.46}$$

with

$$h(u) = x^{-k} u^{3k+2} e^{2k} e^{-u^4/4}. \tag{18.47}$$

At the left endpoint of the integral $u = k^{1/4}$ the function $h(u)$ has a very small value if c_3 is large enough:

$$h\left(k^{1/4}\right) = (c_3 k)^{-3k/4} k^{(3k+2)/4} e^{2k} e^{-k/4} = c_3^{-3k/4} \sqrt{k} e^{7k/4}$$

$$= e^{-c_5 k} = e^{-c_5 x^{4/3}/c_3} \tag{18.48}$$

for some positive constant $c_5 > 0$. Using routine calculus, it is easy to check that $h(u)$ is rapidly decreasing as u increases in the interval $u \in I_2$, assuming c_3 is large enough. With the choice $c_3 = 100$ we have (see (18.46)–(18.48))

$$\left| 2 \frac{\mathrm{i}^k}{x^k} \int_{u \in I_2} \cos(xu) F_k(u)\, du \right| \leq 2 \int_{u \geq k^{1/4}} h(u)\, du \leq e^{-c_5 x^{4/3}/c_3}. \tag{18.49}$$

Combining (18.27), (18.42), (18.46) and (18.49), (18.19) follows, completing the proof of Lemma 18.2. □

19. The case of general speed distribution (II): A general Second Law

We switch from the class of probability density functions with "at least exponential decay for the Fourier transform" — studied in Section 18 — to the class of "polynomial decay for the Fourier transform". Note that the graph of the probability density function $\frac{1}{2}u^2 e^{-u}$, $u \geq 0$ looks very much the same as that of the first kind (Gaussian distribution) or the fifth kind. Does the strong similarity of the graphs of the functions imply similarity in the speed of reaching equilibrium for the typical time evolutions of the corresponding systems? Well, the answer is no.

Torus model of the sixth kind: the speed density function is $g_6(y) = \frac{1}{2} y^2 e^{-y}$, that is,

$$\Pr[\rho_j \leq u] = \frac{1}{2} \int_0^u y^2 e^{-y}\, dy \quad \text{for all positive } u. \tag{19.1}$$

To compute the Fourier transform of $g_6(y)$ we apply (18.9) for

$$f(y) = \frac{1}{2} e^{-|y|} \quad \text{and} \quad n = 2.$$

Using the well known fact that the Fourier transform of $f(y) = \frac{1}{2} e^{-|y|}$ is $\left(\sqrt{2\pi}(1+x^2)\right)^{-1}$, we obtain

$$\widehat{g_6}(x) = -\frac{1}{\sqrt{2\pi}} \left(\frac{d}{dx}\right)^2 \frac{1}{1+x^2}$$

$$= \sqrt{\frac{2}{\pi}}\left(\frac{1}{(1+x^2)^2} - \frac{4x^2}{(1+x^2)^3}\right). \tag{19.2}$$

Using (19.2) in Lemma 18.1, we have

$$\mathbf{F}_{g_6}(w) = \frac{\sqrt{\pi}}{\sqrt{2}w}\int_0^w \widehat{g_6}(x)\,dx$$

$$= \frac{\sqrt{\pi}}{\sqrt{2}w}\int_0^w \sqrt{\frac{2}{\pi}}\left(\frac{1}{(1+x^2)^2} - \frac{4x^2}{(1+x^2)^3}\right)dx$$

$$= \frac{1}{w}\int_0^w \left(\frac{1}{(1+x^2)^2} - \frac{4x^2}{(1+x^2)^3}\right)dx. \tag{19.3}$$

Using integration by parts,

$$\int_0^w \frac{4x^2}{(1+x^2)^3}\,dx = \int_0^w x\cdot 2\frac{2x}{(1+x^2)^3}\,dx = -\frac{w}{(1+w^2)^2} + \int_0^w \frac{1}{(1+x^2)^2}\,dx. \tag{19.4}$$

Using (19.4) in (19.3), we obtain

$$\mathbf{F}_{g_6}(w) = \frac{1}{w}\int_0^w \widehat{g_6}(x)\,dx = \frac{1}{w}\frac{w}{(1+w^2)^2} = \frac{1}{(1+w^2)^2}. \tag{19.5}$$

Clearly (19.5) is monotone decreasing for all $w \ge 0$.

It follows that the typical time evolution of the torus model of the sixth kind is between the first kind and the third kind. It exhibits medium fast approach to LA-equilibrium: starting from close to Big Bang, for a typical initial velocity vector $\omega \in \Omega_6$, the drifting distance $|\mathbf{r}(t)|$ has to be in the range of $N^{1/8}$ to reach equilibrium.

Therefore, despite the strong similarity of the graphs of the functions $g_1(y) = \sqrt{2/\pi}y^2e^{-y^2/2}$ and $g_6(y) = \frac{1}{2}y^2e^{-y}$ for $y \ge 0$, the corresponding threshold magnitudes $\sqrt{\log N}$ and $N^{1/8}$ are very different if N is very large.

Next we replace the probability density function $\frac{1}{2}u^2e^{-u}$ with $\frac{1}{24}u^4e^{-u}$ ($u \ge 0$). We recall that in the Gaussian class the extra factor u^2 did not make any difference: for the first and fifth kinds the threshold is the same $\sqrt{\log N}$. Does the extra factor u^2 make a difference here? Well, the answer is yes.

Torus model of the seventh kind: the speed density function is $g_7(y) = \frac{1}{24}y^4e^{-y}$, that is,

$$\Pr[\rho_j \le u] = \frac{1}{24}\int_0^u y^4e^{-y}\,dy \quad \text{for all positive } u. \tag{19.6}$$

To compute the Fourier transform of $g_6(y)$ we use (18.9) for

$$f(y) = \frac{1}{24}e^{-|y|} \quad \text{and} \quad n = 4.$$

Using the well known fact that the Fourier transform of $f(y) = \frac{1}{2}e^{-|y|}$ is $\left(\sqrt{2\pi}(1+x^2)\right)^{-1}$, we obtain

$$\widehat{g_6}(x) = \frac{1}{12}\frac{1}{\sqrt{2\pi}}\left(\frac{d}{dx}\right)^4 \frac{1}{1+x^2}$$

$$= -\frac{1}{12}\sqrt{\frac{2}{\pi}}\left(\frac{d}{dx}\right)^2\left(\frac{1}{(1+x^2)^2} - \frac{4x^2}{(1+x^2)^3}\right). \tag{19.7}$$

Repeating the same routine calculations what we did above for the sixth kind, we conclude that the switch from the sixth kind to the seventh kind means that the key function $(1+w^2)^{-2}$ in (19.5) is replaced by $(1+w^2)^{-3}$.

It follows that the typical time evolution of the torus model of the seventh kind is between the first kind and the sixth kind. It exhibits medium fast approach to LA-equilibrium: starting from close to Big Bang, for a typical initial velocity vector $\omega \in \Omega_7$, the drifting distance $|\mathbf{r}(t)|$ has to be in the range of $N^{\frac{1}{12}}$ to reach equilibrium.

Therefore, despite the strong similarity of the graphs of the probability density functions $g_6(y) = \frac{1}{2}y^2e^{-y}$ and $g_7(y) = \frac{1}{24}y^4e^{-y}$ $(y \geq 0)$, the corresponding threshold magnitudes $N^{1/8}$ and $N^{\frac{1}{12}}$ are very different if N is very large.

Similarly, if we switch to

$$\frac{1}{(2k)!}y^{2k}e^{-y}, \quad y \geq 0, \quad k = 1, 2, 3, \ldots \tag{19.8}$$

then the the corresponding threshold magnitude is $N^{1/(4k+4)}$.

Similarly to the fourth kind, the case $k = 0$ in (19.8) is exceptional: the threshold magnitude is $N^{1/2}$ (and not $N^{1/4}$, which comes from $N^{1/(4k+4)}$ at $k = 0$).

This "violation of the rule" is explained by the second part of Lemma 18.1. Indeed, if $g(0) \neq 0$ then the asymptotic behavior of $\mathbf{F}_g(w)$ as $w \to \infty$ is simply

$$\mathbf{F}_g(w) = \frac{\pi g(0) + o(1)}{2w},$$

which gives the threshold magnitude $N^{1/2}$.

Once we have $g(0) = 0$, all what matters is the "smoothness" of the density function $g(y)$. Indeed, it is a basic principle of Fourier analysis that the asymptotic behavior

$$\lim_{x\to\infty} \widehat{g}(x) = 0 \tag{19.9}$$

is strongly related to the "smoothness" of $g(y)$ on the *whole* real line (extended as an even function). In other words, the smoother $g(y)$, $-\infty < y < \infty$, the faster the limit (19.9) tends to zero as $x \to \infty$. And of course if the the limit (19.9) tends to zero very fast, then the speed of convergence of the tail integral

$$\lim_{w\to\infty} \int_{x\geq w} \widehat{g}(x)\, dx = 0$$

is also very fast. For example, consider the density function $g_6(y) = \frac{1}{2}y^2e^{-y}$ of the sixth kind: its even extension is $g_6^*(y) = \frac{1}{2}y^2e^{-|y|}$, and its third derivative does not exist at $y = 0$ (due to the fact that the absolute value function $|y|$ is not differentiable at $y = 0$).

On the other hand, for the density function $g_7(y) = \frac{1}{24}y^4e^{-y}$ of the seventh kind the even extension is $g_7^*(y) = \frac{1}{24}y^4e^{-|y|}$, and the third derivative is perfectly fine. The first violation of "smoothness" occurs at the fifth derivative: it does not exist at $y = 0$. We can say, therefore, that $g_7^*(y)$ is "smoother" than $g_6^*(y)$, and this qualitative difference is manifested in the substantial quantitative difference in the threshold magnitudes: $N^{\frac{1}{12}}$ is substantially smaller than $N^{1/8}$, i.e, the typical time evolution of the seventh kind converges to equilibrium faster than that of the sixth kind.

We put this in the form of a lemma.

Lemma 19.1. *Let $g(u)$ be a probability density function in $[0,\infty)$ with $g(0) = 0$, and extend it over the whole real line as an even function: $g(-u) = g(u)$. Let $k \geq 2$, and assume that $g(u)$, $-\infty < u < \infty$ is k-times continuously differentiable, also assume that for every integer $1 \leq j \leq k$, the j-th derivative $g^{(j)}(u) \to 0$ as $u \to \infty$, and finally assume*

$$\int_{-\infty}^{\infty} |g^{(j)}(u)|\, du \leq C_0 < \infty.$$

Then the isotropic Fourier transform

$$\mathbf{F}_g(w) = \int_0^\infty \frac{1}{4\pi}\int_{\mathbf{v}\in\mathbf{S}^2} e^{\mathrm{i}u\mathbf{v}\cdot\mathbf{w}} g(u)\, du\, d\mathrm{SA}(\mathbf{v})$$

with $w = |\mathbf{w}|$ satisfies the upper bound

$$|\mathbf{F}_g(w)| \leq \frac{C_0}{2(k-1)w^k}.$$

Note that $d\mathrm{SA}(\mathbf{v})$ *represents the surface area.*

Proof. Applying integration by parts k times in the definition of the Fourier transform, we have

$$\widehat{g}(x) = \frac{1}{\sqrt{2\pi}} \int_{-\infty}^{\infty} e^{-\mathrm{i}xu} g(u)\, du = \frac{1}{\sqrt{2\pi}} \int_{-\infty}^{\infty} e^{-\mathrm{i}xu} \frac{(-1)^k g^{(k)}(u)}{(-\mathrm{i}x)^k}\, du$$

$$= \frac{1}{\sqrt{2\pi}} \frac{1}{(\mathrm{i}x)^k} \int_{-\infty}^{\infty} e^{-\mathrm{i}xu} g^{(k)}(u)\, du,$$

and so

$$|\widehat{g}(x)| \le \frac{1}{\sqrt{2\pi}|x|^k} \int_{-\infty}^{\infty} \left| g^{(k)}(u) \right| du \le \frac{C_0}{\sqrt{2\pi}|x|^k}. \tag{19.10}$$

By Lemma 18.1,

$$\mathbf{F}_g(w) = -\frac{\sqrt{\pi}}{\sqrt{2}w} \int_{x \ge w} \widehat{g}(x)\, dx,$$

and by using (19.10) here, we conclude

$$|\mathbf{F}_g(w)| \le \frac{\sqrt{\pi}}{\sqrt{2}w} \int_{x \ge w} |\widehat{g}(x)|\, dx \le \frac{\sqrt{\pi}}{\sqrt{2}w} \int_{x \ge w} \frac{C_0}{\sqrt{2\pi} x^k}\, dx \le \frac{C_0}{2(k-1)w^k},$$

completing the proof of Lemma 19.1. □

It is interesting to point out that Lemma 19.1 is not the only way to generate probability density functions g for which the tail of $|\mathbf{F}_g(w)|$ tends to zero polynomially fast. An alternative way is to adapt/modify the "convolution method" at the end of Section 18 (before the proof of Lemma 18.2). The only change is to replace

$$\mathbf{G}(\mathbf{x}) = \mathbf{Gauss}(\mathbf{x}) = e^{-|\mathbf{x}|^2/2}$$

with another *sufficiently smooth* non-negative rotation-invariant function.

Let us return now to the sixth kind and the seventh kind. Notice that Lemma 19.1 applies for the sixth kind $g_6(y) = \frac{1}{2}y^3 e^{-y}$ with $k = 2$, and the lemma applies for the seventh kind $g_7(y) = \frac{1}{24}y^4 e^{-y}$ with $k = 4$. So Lemma 19.1 implies the upper bounds $O(w^{-2})$ and $O(w^{-4})$, but they are not sharp, because the explicit forms $(1+w^2)^{-2}$ and $(1+w^2)^{-3}$ give the true orders $O(w^{-4})$ and $O(w^{-6})$, respectively.

Here is another application of Lemma 19.1. For every $\ell \ge 1$ consider the probability density function

$$g(y) = c_k y^{2k+2}(1-y^2)^{2k+2} \text{ for } 0 \le y \le 1 \text{ and zero otherwise,}$$

where $c_k > 0$ is the norming factor to make the integral equal to one. Notice that $g(y)$, $-\infty < y < \infty$ is k-times continuously differentiable, and Lemma 19.1 gives

$$|\mathbf{F}_g(w)| \le \frac{C_k}{w^k}.$$

In Section 18 we studied the analytic probability density functions g, for which the tail of $|\mathbf{F}_g(w)|$ tends to zero at least exponentially fast (see (18.17)), assuming $g(0) = 0$. In this section we studied density functions g, for which the tail of $|\mathbf{F}_g(w)|$ tends to zero polynomially fast. An important subclass is the family of probability density functions g that satisfy the following extra property: for some integer $k \ge 2$ and some positive constants C and C_1,

$$\frac{C}{\max\{w^k,1\}} - \frac{C_1}{\max\{w^{k+1},1\}} \le \sup_{\tau \ge w} |\mathbf{F}_g(\tau)| \le \frac{C}{\max\{w^k,1\}} + \frac{C_1}{\max\{w^{k+1},1\}} \tag{19.11}$$

holds for every $0 \le w < \infty$.

For example, the third kind belongs to class (19.11). Not surprisingly, we can easily adapt/extend the proof of Theorem 13.2 (= Second Law for the third kind) for the whole class (19.11).

Given a general probability density function g, we define the corresponding Unevenness Function

$$\text{UnevenFun}_g(t) = \text{UnevenFun}_g(\mathcal{Y}, \mathcal{M}; \mathbf{r}; t)$$

in terms of the isotropic Fourier transform $\mathbf{F}_g$ as follows:

$$\text{UnevenFun}_g(\mathcal{Y}, \mathcal{M}; \mathbf{r}; t)$$

$$= \max\left\{ \max_{\mathbf{n}\in\mathbb{Z}^3\setminus\mathbf{0}} \left| \sum_{k=1}^{N} \mu_k e^{2\pi i \mathbf{n}\cdot\mathbf{y}_k} \right|^2 \cdot \sup_{\tau\ge t} \mathbf{F}_g^2\left(2\pi|\mathbf{r}(\tau)|\,|\mathbf{n}|\right), \sum_{k=1}^{N} \mu_k^2 \right\}. \tag{19.12}$$

The motivation for (19.12) comes from two parts: (1) the key formula (4.19′)

$$\overline{\mathbf{D}^2}_{\Omega_g}(\mathcal{Y}, \mathcal{M}; \mathbf{r}; f; t) = \sum_{k=1}^{N} \mu_k^2 \sum_{\substack{\mathbf{n}\in\mathbb{Z}^3: \\ \mathbf{n}\ne\mathbf{0}}} |a_{\mathbf{n}}|^2 \left(1 - \mathbf{F}_g^2(2\pi|\mathbf{r}(t)|\,|\mathbf{n}|)\right)$$

$$+\sum_{\substack{\mathbf{n}\in\mathbb{Z}^3:\\ \mathbf{n}\neq\mathbf{0}}} |a_{\mathbf{n}}|^2 \left|\sum_{k=1}^{N} \mu_k e^{2\pi i\mathbf{n}\cdot\mathbf{y}_k}\right|^2 \cdot \mathbf{F}_g^2(2\pi|\mathbf{r}(t)|\ |\mathbf{n}|)$$

$$= \sum_{\substack{\mathbf{n}\in\mathbb{Z}^3:\\ \mathbf{n}\neq\mathbf{0}}} |a_{\mathbf{n}}|^2 \left|\sum_{k=1}^{N} \mu_k e^{2\pi i\mathbf{n}\cdot\mathbf{y}_k}\right|^2 \cdot \mathbf{F}_g^2(2\pi|\mathbf{r}(t)|\ |\mathbf{n}|)$$

$$+\sigma_0^2(f)\sum_{k=1}^{N}\mu_k^2 + \text{ negligible,}$$

and the second part of the motivation is (2) to take maximum over the whole space $f \in L^2$ (as we explained it in (12.12)–(12.16) in the special case of the first kind).

Notice that (19.12) is the straightforward generalization of (12.16), (12.21) and (12.22), which define the Unevenness Function in the 3 basic special cases.

To clarify the meaning of (19.12) note that the right-hand side of (19.12) describes the typical time evolution of the system toward LA-equilibrium for the majority of $\omega \in \Omega_g$. This follows from Theorem 4.2(b) and Theorem 5.1(c) exactly the same way as we explained/proved it in the 3 basic special cases (first, second and third kinds) in Sections 11–12.

Using (19.11) in (19.12), we have

$$\text{UnevenFun}_g(\mathcal{Y},\mathcal{M};\mathbf{r};t)$$

$$= \max\left\{ \max_{\mathbf{n}\in\mathbb{Z}^3\setminus\mathbf{0}} \left|\sum_{k=1}^{N}\mu_k e^{2\pi i\mathbf{n}\cdot\mathbf{y}_k}\right|^2 \min\left\{\sup_{\tau\geq t}\frac{C^2}{(2\pi|\mathbf{r}(\tau)|\ |\mathbf{n}|)^{2k}}, 1\right\}, \sum_{k=1}^{N}\mu_k^2\right\}$$

$$+ \text{negligible.} \tag{19.13}$$

It follows from (19.13) that the corresponding torus model reaches LA-equilibrium at time t_0 when

$$2\pi \inf_{\tau\geq t_0}|\mathbf{r}(\tau)| = \max_{\mathbf{n}\in\mathbb{Z}^3\setminus\mathbf{0}} \frac{1}{|\mathbf{n}|}\left(\frac{C^2\left|\sum_{k=1}^{N}\mu_k e^{2\pi i\mathbf{n}\cdot\mathbf{y}_k}\right|^2}{\sum_{k=1}^{N}\mu_k^2} \text{ or } 1\right)_{\max}^{\frac{1}{2k}}. \tag{19.14}$$

We refer to the time t_0 and the euclidean distance $\inf_{\tau\geq t_0}|\mathbf{r}(\tau)|$ defined by the equality (19.14) as the Relaxation Time and the Relaxation Drifting Distance, respectively, of the system $(\mathcal{Y},\mathcal{M};\mathbf{r};\Omega_g)$.

Note that in the interesting special case $|\mathbf{r}(t)| = t$ (which includes, among others, the Bernoulli billiard/torus model)

$$\max_{\mathbf{n}\in\mathbb{Z}^3\setminus\mathbf{0}} \frac{1}{|\mathbf{n}|}\left(\frac{C^2\left|\sum_{k=1}^N \mu_k e^{2\pi i \mathbf{n}\cdot\mathbf{y}_k}\right|^2}{\sum_{k=1}^N \mu_k^2} \text{ or } 1\right)_{\max}^{\frac{1}{2k}}$$

represents the true order of magnitude of the Relaxation Time (i.e., the time to reach LA-equilibrium) for the typical time evolution of the system $(\mathcal{Y},\mathcal{M};\mathbf{r};\Omega_g)$.

In (19.14) we basically used the function $\left(C^2x\right)^{1/2k}$. To define the Imbalance, we apply the inverse function $C^{-2}x^{2k}$ in (19.14): we have

$$\text{Imbalance}_g(\mathcal{Y},\mathcal{M}) = \max_{\mathbf{n}\in\mathbb{Z}^3\setminus\mathbf{0}} \frac{1}{|\mathbf{n}|^{2k}}\left(\frac{\left|\sum_{k=1}^N \mu_k e^{2\pi i \mathbf{n}\cdot\mathbf{y}_k}\right|^2}{\sum_{k=1}^N \mu_k^2} \text{ or } 1\right)_{\max}. \tag{19.15}$$

We may say, therefore, that $\text{Imbalance}_g(\mathcal{Y},\mathcal{M})$ represents the $2k$-th power of the Relaxation Drifting Distance for the typical time evolution of $(\mathcal{Y},\mathcal{M};\mathbf{r};\Omega_g)$.

The corresponding entropy has the usual complementary relation with the Imbalance:

$$\text{MicroEntropy}_g(\mathcal{Y},\mathcal{M}) = \frac{N^2}{\sum_{k=1}^N \mu_k^2} - \text{Imbalance}_g(\mathcal{Y},\mathcal{M}). \tag{19.16}$$

We are ready now to prove an analog of Theorem 13.2 for the whole class (19.11) of density functions g. Note that the condition $k \geq 2$ is important to avoid the "divergence problem" of the second kind.

As usual,

$$\mathcal{Y}_{\mathbf{r};\omega;t} = \{\mathbf{z}_k = \mathbf{y}_k + \vartheta_k(\rho_k\mathbf{r}(t)) \text{ modulo one}: \ 1 \leq k \leq N\},$$

where

$$\omega = (\rho_1,\vartheta_1,\rho_2,\vartheta_2,\ldots,\rho_N,\vartheta_N) \in \Omega_g = \left([0,\infty)\times SO(3)\right)^N$$

with

$$\Pr[\rho_k \leq u] = \int_0^u g(y)\,dy.$$

Because of the complementary relation between our entropy and the Imbalance, it suffices to study

$$\text{Imba}_g(\mathcal{Y}_{\mathbf{r};\omega;t},\mathcal{M}) = \text{Imbalance}_g(\mathcal{Y}_{\mathbf{r};\omega;t},\mathcal{M}).$$

Again we want to find the relation between $\text{Imba}_g(\mathcal{Y}_{\mathbf{r};\omega;t}, \mathcal{M})$ for the majority of $\omega \in \Omega_g$ and the corresponding ω-independent Time Evolution Function (t is the time variable)

$$\text{TiEvFun}_g(t) = \text{TiEvFun}_g(\mathcal{Y}, \mathcal{M}; \mathbf{r}; t)$$

$$= \max_{\mathbf{n}\in\mathbb{Z}^3\setminus\mathbf{0}} \frac{1}{|\mathbf{n}|^{2k}} \left(\frac{\left|\sum_{k=1}^N \mu_k e^{2\pi i \mathbf{n}\cdot\mathbf{y}_k}\right|^2 \mathbf{F}_g^2\left(2\pi|\mathbf{r}(t)|\ |\mathbf{n}|\right)}{\sum_{k=1}^N \mu_k^2} \text{ or } 1 \right)_{\max} . \quad (19.17)$$

Notice that (19.17) is a straightforward analog of the Time Evolution Function of the second kind in (14.3).

By a simple adaptation of the proof argument of Theorem 13.1 up to (13.22), we obtain that there is a measurable subset $\Omega_g^* \subset \Omega_g$ such that

$$\text{ProdMeas}_g(\Omega_g^*) > 1 - 10^2\varepsilon, \quad (19.18)$$

and for every $\omega \in \Omega_g^*$ and every nonzero lattice point $\mathbf{n} \in \mathbb{Z}^3 \setminus \mathbf{0}$ we have the analog of (13.22):

$$\left| \frac{\left|\sum_{\mathbf{z}_k\in\mathcal{Y}_{\mathbf{r};\omega;t}} \mu_k e^{2\pi i \mathbf{n}\cdot\mathbf{z}_k}\right|^2}{\sum_{k=1}^N \mu_k^2} - \frac{\overline{\mathbf{D}^2}_g(\mathbf{n};t)}{\sum_{k=1}^N \mu_k^2} \right|$$

$$\leq \frac{4|\mathbf{n}|^2}{\sqrt{\varepsilon}} \left(\left(\frac{\overline{\mathbf{D}^2}_g(\mathbf{n};t)}{\sum_{k=1}^N \mu_k^2} \right)^{1/2} + 2 \right), \quad (19.19)$$

where

$$\overline{\mathbf{D}^2}_g(\mathbf{n};t) = \left| \sum_{k=1}^N \mu_k e^{2\pi i \mathbf{n}\cdot\mathbf{y}_k} \right|^2 \mathbf{F}_g^2\left(2\pi|\mathbf{r}(t)|\ |\mathbf{n}|\right). \quad (19.20)$$

By using the notation (19.20), we can rewrite the definition (19.17) of the Time Evolution Function as follows:

$$\text{TiEvFun}_g(t) = \max_{\mathbf{n}\in\mathbb{Z}^3\setminus\mathbf{0}} \frac{1}{|\mathbf{n}|^{2k}} \left(\frac{\overline{\mathbf{D}^2}_g(\mathbf{n};t)}{\sum_{k=1}^N \mu_k^2} \text{ or } 1 \right)_{\max} . \quad (19.21)$$

Combining (19.19) and (19.21), we have

$$\left| \frac{1}{|\mathbf{n}|^{2k}} \left(\frac{\left|\sum_{\mathbf{z}_k\in\mathcal{Y}_{\mathbf{r};\omega;t}} \mu_k e^{2\pi i \mathbf{n}\cdot\mathbf{z}_k}\right|^2}{\sum_{k=1}^N \mu_k^2} \text{ or } 1 \right)_{\max} - \frac{1}{|\mathbf{n}|^{2k}} \left(\frac{\overline{\mathbf{D}^2}_g(\mathbf{n};t)}{\sum_{k=1}^N \mu_k^2} \text{ or } 1 \right)_{\max} \right|$$

$$\leq \frac{1}{|\mathbf{n}|^{2k}} \frac{4|\mathbf{n}|^2}{\sqrt{\varepsilon}} \left(\left(\frac{\overline{\mathbf{D}^2}_g(\mathbf{n};t)}{\sum_{k=1}^N \mu_k^2} \right)^{1/2} + 2 \right)$$

$$\leq \frac{4|\mathbf{n}|^{2-k}}{\sqrt{\varepsilon}} \left(\left(\frac{1}{|\mathbf{n}|^{2k}} \left(\frac{\overline{\mathbf{D}^2}_g(\mathbf{n};t)}{\sum_{k=1}^N \mu_k^2} \text{ or } 1 \right)_{\max} \right)^{1/2} + 2 \right)$$

$$\leq \frac{2|\mathbf{n}|^{2-k}}{\sqrt{\varepsilon}} \left((\text{TiEvFun}_g(t))^{1/2} + 2 \right)$$

$$= \frac{4}{\sqrt{\varepsilon}\sqrt{\text{TiEvFun}_g(t)}} \left(1 + \frac{2}{\sqrt{\text{TiEvFun}_g(t)}} \right) \cdot \text{TiEvFun}_g(t), \qquad (19.22)$$

where in the last step we used the condition $k \geq 2$ in (19.11).

Let $\mathbf{n}_1 \in \mathbb{Z}^3 \setminus \mathbf{0}$ denote the nonzero lattice point for which $\text{Imba}_g(\mathcal{Y}_{\mathbf{r};\omega;t}, \mathcal{M})$ attains its maximum, and let $\mathbf{n}_2 \in \mathbb{Z}^3 \setminus \mathbf{0}$ denote the nonzero lattice point for which $\text{TiEvFun}_g(t)$ attains its maximum.

By (19.22) and the triangle inequality,

$$\text{Imba}_g(\mathcal{Y}_{\mathbf{r};\omega;t}, \mathcal{M}) = \frac{1}{|\mathbf{n}_1|^{2k}} \left(\frac{\left| \sum_{\mathbf{z}_k \in \mathcal{Y}_{\mathbf{r};\omega;t}} \mu_k e^{2\pi i \mathbf{n}_1 \cdot \mathbf{z}_k} \right|^2}{\sum_{k=1}^N \mu_k^2} \text{ or } 1 \right)_{\max}$$

$$\leq \frac{1}{|\mathbf{n}_1|^{2k}} \left(\frac{\overline{\mathbf{D}^2}_g(\mathbf{n}_1;t)}{\sum_{k=1}^N \mu_k^2} \text{ or } 1 \right)_{\max}$$

$$+ \frac{4}{\sqrt{\varepsilon}\sqrt{\text{TiEvFun}_g(t)}} \left(1 + \frac{2}{\sqrt{\text{TiEvFun}_g(t)}} \right) \cdot \text{TiEvFun}_g(t)$$

$$\leq \text{TiEvFun}_g(t) + \frac{4}{\sqrt{\varepsilon}\sqrt{\text{TiEvFun}_g(t)}} \left(1 + \frac{2}{\sqrt{\text{TiEvFun}_g(t)}} \right) \cdot \text{TiEvFun}_g(t), \qquad (19.23)$$

and similarly,

$$\text{Imba}_g(\mathcal{Y}_{\mathbf{r};\omega;t}, \mathcal{M}) \geq \frac{1}{|\mathbf{n}_2|^{2k}} \left(\frac{\left| \sum_{\mathbf{z}_k \in \mathcal{Y}_{\mathbf{r};\omega;t}} \mu_k e^{2\pi i \mathbf{n}_2 \cdot \mathbf{z}_k} \right|^2}{\sum_{k=1}^N \mu_k^2} \text{ or } 1 \right)_{\max}$$

$$\geq \frac{1}{|\mathbf{n}_2|^{2k}} \left(\frac{\overline{\mathbf{D}^2}_g(\mathbf{n}_2;t)}{\sum_{k=1}^N \mu_k^2} \text{ or } 1 \right)_{\max} - \frac{4}{\sqrt{\varepsilon}\sqrt{\text{TiEvFun}_g(t)}}$$

$$\left(1+\frac{2}{\sqrt{\text{TiEvFun}_g(t)}}\right)\text{TiEvFun}_g(t)$$

$$= \text{TiEvFun}_g(t) - \frac{4}{\sqrt{\varepsilon}\sqrt{\text{TiEvFun}_g(t)}}\left(1+\frac{2}{\sqrt{\text{TiEvFun}_g(t)}}\right)\cdot \text{TiEvFun}_g(t). \tag{19.24}$$

Combining (19.23) and (19.24), we have

$$|\text{Imba}_g(\mathcal{Y}_{\mathbf{r};\omega;t},\mathcal{M}) - \text{TiEvFun}_g(t)|$$

$$\le \frac{4}{\sqrt{\varepsilon}\sqrt{\text{TiEvFun}_g(t)}}\left(1+\frac{2}{\sqrt{\text{TiEvFun}_g(t)}}\right)\cdot \text{TiEvFun}_g(t)$$

for all $\omega \in \Omega_g^*$, where $\Omega_g^* \subset \Omega_g$ is a measurable subset such that

$$\text{ProdMeas}_g(\Omega_g^*) > 1 - 10^2\varepsilon.$$

Thus we obtain the following analog of Theorem 13.2: a Second Law for the class (19.11).

Theorem 19.1. *Let $g(u)$, $0 \le u < \infty$ be a probability density function such that the isotropic Fourier transform $\mathbf{F}_g$ satisfies (19.11). Consider the corresponding 3-dimensional isotropic torus model $(\mathcal{Y},\mathcal{M};\mathbf{r};\Omega_g)$. Given a time $t > 0$ and $0 < \varepsilon < 10^{-2}$, there exists a measurable subset $\Omega_g^* = \Omega_g^*(\varepsilon;t) \subset \Omega_g$ of the corresponding initial velocity space such that*

$$\text{ProdMeas}_g(\Omega_g^*) > 1 - 10^2\varepsilon$$

and

$$|\text{Imbalance}_g(\mathcal{Y}_{\mathbf{r};\omega;t},\mathcal{M}) - \text{TiEvFun}_g(\mathcal{Y},\mathcal{M};\mathbf{r};t)|$$

$$\le \frac{4}{\sqrt{\varepsilon}\sqrt{\text{TiEvFun}_g(t)}}\left(1+\frac{2}{\sqrt{\text{TiEvFun}_g(t)}}\right)\cdot \text{TiEvFun}_g(t)$$

holds for all $\omega \in \Omega_g^*$. Here $\text{TiEvFun}_g(t) = \text{TiEvFun}_g(\mathcal{Y},\mathcal{M};\mathbf{r};t)$ is defined in (19.17). □

Similarly, it is not too difficult to prove a Second Law for the class of probability density functions g with the property that the tail of $|\mathbf{F}_g(w)|$ tends to zero at least exponentially fast (see Section 18). Indeed, one can simply adapt the proof of Theorem 13.1.

Next we give the formal definition of the Micro-Entropy in the general case. It is just a straightforward generalization of the special cases discussed above. The first step is to recall (19.12): given a probability density function g with its isotropic Fourier transform $\mathbf{F}_g$, the corresponding Unevenness Function is defined as

$$\text{UnevenFun}_g(\mathcal{Y}, \mathcal{M}; \mathbf{r}; t)$$

$$= \max\left\{ \max_{\mathbf{n}\in\mathbb{Z}^3\setminus\mathbf{0}} \left|\sum_{k=1}^{N} \mu_k e^{2\pi i \mathbf{n}\cdot\mathbf{y}_k}\right|^2 \sup_{\tau\geq t} \mathbf{F}_g^2\left(2\pi|\mathbf{r}(\tau)|\,|\mathbf{n}|\right), \sum_{k=1}^{N} \mu_k^2 \right\}. \tag{19.25}$$

Write

$$\mathcal{F}_g(z) = \sup_{u\geq z} |\mathbf{F}_g(u)|, \tag{19.26}$$

noting that $\mathcal{F}_g(z)$ is a positive continuous decreasing function.

We have the simple equality

$$\sup_{\tau\geq t} \mathbf{F}_g^2\left(2\pi|\mathbf{r}(\tau)|\,|\mathbf{n}|\right) = \mathcal{F}_g^2\left(\inf_{\tau\geq t} 2\pi|\mathbf{r}(\tau)|\,|\mathbf{n}|\right). \tag{19.27}$$

Indeed, by (19.26)

$$\mathcal{F}_g^2\left(\inf_{\tau\geq t} 2\pi|\mathbf{r}(\tau)|\,|\mathbf{n}|\right) = \sup_{u\geq\inf_{\tau\geq t} 2\pi|\mathbf{r}(\tau)|\,|\mathbf{n}|} \mathbf{F}_g^2(u)$$

$$= \sup_{\tau\geq t} \mathbf{F}_g^2\left(2\pi|\mathbf{r}(\tau)|\,|\mathbf{n}|\right),$$

since $\mathbf{F}_g^2(z)$ and $|\mathbf{r}(\tau)|$ are continuous and $|\mathbf{r}(\tau)| \to \infty$ as $\tau \to \infty$.

For simplicity assume that $\mathcal{F}_g(z)$ is *strictly* decreasing; then

$$\mathcal{H}_g(z) = \frac{1}{\mathcal{F}_g^2(z)} \tag{19.28}$$

is a *strictly* increasing function, and it has a well-defined inverse $\mathcal{H}_g^{-1}$.

It follows from (19.25), (19.27) and (19.28) that the corresponding torus model reaches LA-equilibrium at time t_0 when

$$2\pi \inf_{\tau\geq t_0} |\mathbf{r}(\tau)| = \max_{\mathbf{n}\in\mathbb{Z}^3\setminus\mathbf{0}} \left(\frac{1}{|\mathbf{n}|}\mathcal{H}_g^{-1}\left(\frac{\left|\sum_{k=1}^{N} \mu_k e^{2\pi i \mathbf{n}\cdot\mathbf{y}_k}\right|^2}{\sum_{k=1}^{N}\mu_k^2} \right) \text{ or } 1 \right)_{\max}. \tag{19.29}$$

We refer to the time t_0 and the euclidean distance $\inf_{\tau\geq t_0} |\mathbf{r}(\tau)|$ defined by the equality (19.29) as the Relaxation Time and the Relaxation Drifting Distance, respectively, of the system $(\mathcal{Y}, \mathcal{M}; \mathbf{r}; \Omega_g)$.

Note that in the special case $|\mathbf{r}(t)| = t$

$$\left(\max_{\mathbf{n}\in\mathbb{Z}^3\setminus\mathbf{0}} \frac{1}{|\mathbf{n}|}\mathcal{H}_g^{-1}\left(\frac{\left|\sum_{k=1}^{N}\mu_k e^{2\pi i\mathbf{n}\cdot\mathbf{y}_k}\right|^2}{\sum_{k=1}^{N}\mu_k^2}\right) \quad \text{or} \quad 1\right)_{\max}$$

represents the true order of magnitude of the Relaxation Time (i.e., the time to reach LA-equilibrium) for the typical time evolution of the system $(\mathcal{Y}, \mathcal{M}; \mathbf{r}; \Omega_g)$.

In (19.29) we used the inverse function $\mathcal{H}_g^{-1}$. To define the Imbalance, we apply the inverse of $\mathcal{H}_g^{-1}$, i.e., we apply $\mathcal{H}_g$ itself (see (19.28)):

$$\text{Imbalance}_g(\mathcal{Y}, \mathcal{M})$$

$$= \max_{\mathbf{n}\in\mathbb{Z}^3\setminus\mathbf{0}}\left(\mathcal{H}_g\left(\frac{1}{|\mathbf{n}|}\mathcal{H}_g^{-1}\left(\frac{\left|\sum_{k=1}^{N}\mu_k e^{2\pi i\mathbf{n}\cdot\mathbf{y}_k}\right|^2}{\sum_{k=1}^{N}\mu_k^2}\right)\right) \quad \text{or} \quad 1\right)_{\max}. \tag{19.30}$$

Notice that (19.30) is the straightforward generalization of how we defined the Imbalance in the 3 basic special cases in the Metatheorem (see Section 6), and also in (19.15).

Finally, the corresponding entropy has the usual complementary relation with the Imbalance:

$$\text{MicroEntropy}_g(\mathcal{Y}, \mathcal{M}) = \frac{N^2}{\sum_{k=1}^{N}\mu_k^2} - \text{Imbalance}_g(\mathcal{Y}, \mathcal{M}). \tag{19.31}$$

We define the corresponding ω-independent Time Evolution Function as (t is the time variable)

$$\text{TiEvFun}_g(t) = \text{TiEvFun}_g(\mathcal{Y}, \mathcal{M}; \mathbf{r}; t)$$

$$= \max_{\mathbf{n}\in\mathbb{Z}^3\setminus\mathbf{0}}\left(\mathcal{H}_g\left(\frac{1}{|\mathbf{n}|}\mathcal{H}_g^{-1}\left(\frac{\left|\sum_{k=1}^{N}\mu_k e^{2\pi i\mathbf{n}\cdot\mathbf{y}_k}\right|^2 \mathbf{F}_g^2\left(2\pi|\mathbf{r}(t)|\;|\mathbf{n}|\right)}{\sum_{k=1}^{N}\mu_k^2}\right)\right) \text{or } 1\right)_{\max}. \tag{19.32}$$

Notice that (19.32) is a straightforward generalization of how we defined the Imbalance in the 3 basic special cases and also in (19.17).

Repeating the proof of Theorem 19.1 it is not too difficult to prove an analog Second Law in the general case of (19.30)–(19.32).

20. Returning to the torus-via-unfolding model

At the end of Section 1 we explained why a particles-in-a-box model — both the simplest linear constant speed case in (1.19), and also the general (not necessarily linear) case in (1.22) — is equivalent to a *torus-via-unfolding* model (see (1.23) and (1.27)). Note that this torus-via-unfolding model is *not* a special case of the general torus model. Nevertheless, there is a rather simple way to save/adapt the results in Sections 3–19 for the torus-via-unfolding model.

We use the notation introduced in (1.19)–(1.27) and (2.33)–(2.38). Let $S \subset I^3 = [0,1)^3$ be a measurable test set, and write $f = \chi_{S/2}$, i.e., f is the 0-1 valued characteristic function of the "half set" $\frac{1}{2}S$. By (1.23) we have

$$N(\text{TorViaUnfold}; \mathcal{Y}, \mathcal{M}; \mathcal{R}; \omega; f; t) - N8\int_{I^3} f\, dV = \sum_{\mathbf{n}=(n_1,n_2,n_3)\in\mathbb{Z}^3\setminus\mathbf{0}} a_{\mathbf{n}} \cdot$$

$$\cdot \sum_{(\varepsilon_1,\varepsilon_2,\varepsilon_3)\in\{1,-1\}^3} \sum_{k=1}^{N} \mu_k \exp\left(2\pi\mathrm{i}\left(\sum_{j=1}^{3} n_j \frac{1}{2}\varepsilon_j c_{k,j}(t)\right)\right), \tag{20.1}$$

where

$$\mathbf{y}_k + \vartheta_k(\rho_k \mathbf{r}_k(t)) = \mathbf{c}_k(t) = (c_{k,1}(t), c_{k,2}(t), c_{k,3}(t)), \tag{20.2}$$

and, as usual, $a_{\mathbf{n}}$ denote the Fourier coefficients of f.

By using the notation $\mathbf{n}(\varepsilon_1,\varepsilon_2,\varepsilon_3) = (\varepsilon_1 n_1, \varepsilon_2 n_2, \varepsilon_3 n_3)$, we can rewrite (20.1)–(20.2) as follows:

$$N(\text{TorViaUnfold}; \mathcal{Y}, \mathcal{M}; \mathcal{R}; \omega; f; t) - N8\int_{I^3} f\, dV$$

$$= \sum_{\mathbf{n}\in\mathbb{Z}^3\setminus\mathbf{0}} a_{\mathbf{n}} \sum_{(\varepsilon_1,\varepsilon_2,\varepsilon_3)\in\{1,-1\}^3} \sum_{k=1}^{N} \mu_k e^{\pi\mathrm{i}\mathbf{n}(\varepsilon_1,\varepsilon_2,\varepsilon_3)\cdot\mathbf{c}_k(t)}$$

$$= \sum_{\mathbf{n}\in\mathbb{Z}^3\setminus\mathbf{0}} a_{\mathbf{n}} \sum_{(\varepsilon_1,\varepsilon_2,\varepsilon_3)\in\{1,-1\}^3} \sum_{k=1}^{N} \mu_k e^{\pi\mathrm{i}\mathbf{n}(\varepsilon_1,\varepsilon_2,\varepsilon_3)\cdot\mathbf{y}_k} e^{\pi\mathrm{i}\mathbf{n}(\varepsilon_1,\varepsilon_2,\varepsilon_3)\cdot\vartheta_k(\rho_k\mathbf{r}_k(t))}. \tag{20.3}$$

If f is replaced with a translation $f_{\mathbf{w}}$ (see (2.30)), then (20.3) changes to

$$N(\text{TorViaUnfold}; \mathcal{Y}, \mathcal{M}; \mathcal{R}; \omega; f_{\mathbf{w}}; t) - N8\int_{I^3} f\, dV = \sum_{\mathbf{n}\in\mathbb{Z}^3\setminus\mathbf{0}} a_{\mathbf{n}} e^{-2\pi\mathrm{i}\mathbf{n}\cdot\mathbf{w}}$$

$$\sum_{(\varepsilon_1,\varepsilon_2,\varepsilon_3)\in\{1,-1\}^3} \sum_{k=1}^{N} \mu_k e^{\pi \mathrm{i}\mathbf{n}(\varepsilon_1,\varepsilon_2,\varepsilon_3)\cdot \mathbf{y}_k} e^{\pi \mathrm{i}\mathbf{n}(\varepsilon_1,\varepsilon_2,\varepsilon_3)\cdot \vartheta_k(\rho_k \mathbf{r}_k(t))}. \tag{20.4}$$

Notice that (20.3) is an analog of (7.3).

By (20.4) we have

$$\mathbf{D}^2(\text{TorViaUnfold}; \mathcal{Y}, \mathcal{M}; \mathcal{R}; \omega; f_{\mathbf{w}}; t)$$

$$= \left| N(\text{TorViaUnfold}; \mathcal{Y}, \mathcal{M}; \mathcal{R}; \omega; f_{\mathbf{w}}; t) - N8 \int_{I^3} f \, dV \right|^2$$

$$= \sum_{\mathbf{n}_1 \in \mathbb{Z}^3 \setminus \mathbf{0}} a_{\mathbf{n}_1} e^{-2\pi \mathrm{i}\mathbf{n}_1 \cdot \mathbf{w}} \sum_{(\varepsilon_1,\varepsilon_2,\varepsilon_3)\in\{1,-1\}^3} \sum_{k_1=1}^{N} \mu_{k_1} e^{\pi \mathrm{i}\mathbf{n}_1(\varepsilon_1,\varepsilon_2,\varepsilon_3)\cdot \mathbf{y}_{k_1}}$$

$$\cdot e^{\pi \mathrm{i}\mathbf{n}_1(\varepsilon_1,\varepsilon_2,\varepsilon_3)\cdot \vartheta_{k_1}(\rho_{k_1} \mathbf{r}_{k_1}(t))}$$

$$\cdot \sum_{\mathbf{n}_2 \in \mathbb{Z}^3 \setminus \mathbf{0}} \overline{a_{\mathbf{n}_2}} e^{-2\pi \mathrm{i}\mathbf{n}_2 \cdot \mathbf{w}} \sum_{(\varepsilon_4,\varepsilon_5,\varepsilon_6)\in\{1,-1\}^3} \sum_{k_2=1}^{N} \mu_{k_2} e^{-\pi \mathrm{i}\mathbf{n}_2(\varepsilon_4,\varepsilon_5,\varepsilon_6)\cdot \mathbf{y}_{k_2}}$$

$$\cdot e^{-\pi \mathrm{i}\mathbf{n}_2(\varepsilon_4,\varepsilon_5,\varepsilon_6)\cdot \vartheta_{k_2}(\rho_{k_2} \mathbf{r}_{k_2}(t))}, \tag{20.5}$$

where $\overline{z} = x - iy$ denotes the complex conjugate of a complex number $z = x + iy$.

Since

$$\int_{I^3} e^{-2\pi \mathrm{i}(\mathbf{n}_1 - \mathbf{n}_2)\cdot \mathbf{w}} \, d\mathbf{w} = 0 \tag{20.6}$$

except the trivial case $\mathbf{n}_1 = \mathbf{n}_2$, by (20.5) and (20.6) we have

$$\overline{\mathbf{D}^2}(\text{TorViaUnfold}; \mathcal{Y}, \mathcal{M}; \mathcal{R}; \omega; f; t)$$

$$= \int_{I^3} \mathbf{D}^2(\text{TorViaUnfold}; \mathcal{Y}, \mathcal{M}; \mathcal{R}; \omega; f_{\mathbf{w}}; t) \, d\mathbf{w}$$

$$= \int_{I^3} \left| N(\text{TorViaUnfold}; \mathcal{Y}, \mathcal{M}; \mathcal{R}; \omega; f_{\mathbf{w}}; t) - N8 \int_{I^3} f \, dV \right|^2 d\mathbf{w} = \sum_{\mathbf{n} \in \mathbb{Z}^3 \setminus \mathbf{0}} |a_{\mathbf{n}}|^2$$

$$\sum_{(\varepsilon_1,\varepsilon_2,\varepsilon_3)\in\{1,-1\}^3} \sum_{k_1=1}^{N} \mu_{k_1} e^{\pi \mathrm{i}\mathbf{n}(\varepsilon_1,\varepsilon_2,\varepsilon_3)\cdot \mathbf{y}_{k_1}} e^{\pi \mathrm{i}\mathbf{n}(\varepsilon_1,\varepsilon_2,\varepsilon_3)\cdot \vartheta_{k_1}(\rho_{k_1} \mathbf{r}_{k_1}(t))}$$

$$\cdot \sum_{(\varepsilon_4,\varepsilon_5,\varepsilon_6)\in\{1,-1\}^3} \sum_{k_2=1}^{N} \mu_{k_2} e^{-\pi i \mathbf{n}(\varepsilon_4,\varepsilon_5,\varepsilon_6)\cdot \mathbf{y}_{k_2}} e^{-\pi i \mathbf{n}(\varepsilon_4,\varepsilon_5,\varepsilon_6)\cdot \vartheta_{k_2}(\rho_{k_2}\mathbf{r}_{k_2}(t))}$$

$$= \sum_{\mathbf{n}\in\mathbb{Z}^3\setminus\mathbf{0}} |a_{\mathbf{n}}|^2 \sum_{k=1}^{N} \mu_k^2 \sum_{(\varepsilon_1,\varepsilon_2,\varepsilon_3)\in\{1,-1\}^3} \sum_{(\varepsilon_4,\varepsilon_5,\varepsilon_6)\in\{1,-1\}^3}$$

$$\cdot e^{\pi i(\mathbf{n}(\varepsilon_1,\varepsilon_2,\varepsilon_3)-\mathbf{n}(\varepsilon_4,\varepsilon_5,\varepsilon_6))\cdot \mathbf{y}_k} e^{\pi i(\mathbf{n}(\varepsilon_1,\varepsilon_2,\varepsilon_3)-\mathbf{n}(\varepsilon_4,\varepsilon_5,\varepsilon_6))\cdot \vartheta_k(\rho_k\mathbf{r}_k(t))}$$

$$+ \sum_{\mathbf{n}\in\mathbb{Z}^3\setminus\mathbf{0}} |a_{\mathbf{n}}|^2 \sum_{k_1=1}^{N} \sum_{\substack{1\le k_2\le N:\\ k_2\neq k_1}} \mu_{k_1}\mu_{k_2} \sum_{(\varepsilon_1,\varepsilon_2,\varepsilon_3)\in\{1,-1\}^3} \sum_{(\varepsilon_4,\varepsilon_5,\varepsilon_6)\in\{1,-1\}^3}$$

$$\cdot e^{\pi i \mathbf{n}(\varepsilon_1,\varepsilon_2,\varepsilon_3)\cdot \mathbf{y}_{k_1}} e^{\pi i \mathbf{n}(\varepsilon_1,\varepsilon_2,\varepsilon_3)\cdot \vartheta_{k_1}(\rho_{k_1}\mathbf{r}_{k_1}(t))} e^{-\pi i \mathbf{n}(\varepsilon_4,\varepsilon_5,\varepsilon_6)\cdot \mathbf{y}_{k_2}}$$

$$e^{-\pi i \mathbf{n}(\varepsilon_4,\varepsilon_5,\varepsilon_6)\cdot \vartheta_{k_2}(\rho_{k_2}\mathbf{r}_{k_2}(t))}. \tag{20.7}$$

Next we integrate (20.7) over the big "initial velocity space" $\Omega_1 = ([0,\infty)\times SO(3))^N$ (see (2.13)), and apply Lemma 7.1. Thus we obtain the following analog of (7.8):

$$\overline{\mathbf{D}^2}_{\Omega_1}(\text{TorViaUnfold}; \mathcal{Y}, \mathcal{M}; \mathcal{R}; f; t)$$

$$= \int_{\Omega_1} \overline{\mathbf{D}^2}(\text{TorViaUnfold}; \mathcal{Y}, \mathcal{M}; \mathcal{R}; \omega; f; t)\, d\omega$$

$$= \sqrt{\frac{2}{\pi}} \int_{\rho_1=0}^{\infty} \int_{\vartheta_1\in SO(3)} \cdots \sqrt{\frac{2}{\pi}} \int_{\rho_N=0}^{\infty} \int_{\vartheta_N\in SO(3)}$$

$$\mathbf{D}^2(\text{TorViaUnfold}; \mathcal{Y}, \mathcal{M}; \mathcal{R}; \omega; f; t)\rho_1^2$$

$$\cdot e^{-\rho_1^2/2}\, d\rho_1\, d\vartheta_1 \cdots \rho_N^2 e^{-\rho_N^2/2}\, d\rho_N\, d\vartheta_N$$

$$= \sum_{\mathbf{n}\in\mathbb{Z}^3\setminus\mathbf{0}} |a_{\mathbf{n}}|^2 \sum_{k=1}^{N} \mu_k^2 \sum_{(\varepsilon_1,\varepsilon_2,\varepsilon_3)\in\{1,-1\}^3}$$

$$+ \sum_{\mathbf{n}\in\mathbb{Z}^3\setminus\mathbf{0}} |a_{\mathbf{n}}|^2 \sum_{k=1}^{N} \mu_k^2 \sum_{(\varepsilon_1,\varepsilon_2,\varepsilon_3)\in\{1,-1\}^3} \sum_{\substack{(\varepsilon_4,\varepsilon_5,\varepsilon_6)\in\{1,-1\}^3\\ (\varepsilon_4,\varepsilon_5,\varepsilon_6)\neq(\varepsilon_1,\varepsilon_2,\varepsilon_3)}}$$

$$\cdot e^{\pi i(\mathbf{n}(\varepsilon_1,\varepsilon_2,\varepsilon_3)-\mathbf{n}(\varepsilon_4,\varepsilon_5,\varepsilon_6))\cdot \mathbf{y}_k} e^{-(\pi|\mathbf{n}(\varepsilon_1,\varepsilon_2,\varepsilon_3)-\mathbf{n}(\varepsilon_4,\varepsilon_5,\varepsilon_6)|\,|\mathbf{r}_k(t)|)^2/2}$$

$$+\sum_{\mathbf{n}\in\mathbb{Z}^3\setminus\mathbf{0}}|a_{\mathbf{n}}|^2\sum_{k_1=1}^{N}\sum_{\substack{1\le k_2\le N:\\ k_2\ne k_1}}\mu_{k_1}\mu_{k_2}\sum_{(\varepsilon_1,\varepsilon_2,\varepsilon_3)\in\{1,-1\}^3}\ \sum_{(\varepsilon_4,\varepsilon_5,\varepsilon_6)\in\{1,-1\}^3}$$

$$\cdot\, e^{\pi i\mathbf{n}(\varepsilon_1,\varepsilon_2,\varepsilon_3)\cdot\mathbf{y}_{k_1}}e^{-\pi i\mathbf{n}(\varepsilon_4,\varepsilon_5,\varepsilon_6)\cdot\mathbf{y}_{k_2}}e^{-(\pi|\mathbf{n}|\ |\mathbf{r}_{k_1}(t)|)^2/2}e^{-(\pi|\mathbf{n}|\ |\mathbf{r}_{k_2}(t)|)^2/2}. \tag{20.8}$$

We can rewrite (20.8) in the alternative form

$$\overline{\mathbf{D}^2}_{\Omega_1}(\text{TorViaUnfold};\mathcal{Y},\mathcal{M};\mathcal{R};f;t)$$

$$=8\sum_{k=1}^{N}\mu_k^2\sum_{\substack{\mathbf{n}\in\mathbb{Z}^3:\\ \mathbf{n}\ne\mathbf{0}}}|a_{\mathbf{n}}|^2\left(1-\mathbf{F}^2(\pi|\mathbf{r}_k(t)|\ |\mathbf{n}|)\right)$$

$$+\sum_{\substack{\mathbf{n}\in\mathbb{Z}^3:\\ \mathbf{n}\ne\mathbf{0}}}|a_{\mathbf{n}}|^2\left|\sum_{k=1}^{N}\mu_k\mathbf{F}(\pi|\mathbf{r}_k(t)|\ |\mathbf{n}|)\sum_{(\varepsilon_1,\varepsilon_2,\varepsilon_3)\in\{1,-1\}^3}e^{\pi i\mathbf{n}(\varepsilon_1,\varepsilon_2,\varepsilon_3)\cdot\mathbf{y}_k}\right|^2, \tag{20.9}$$

where $\mathbf{F}(x)=e^{-x^2/2}$. The evaluation of the end-sum in (20.9) is easy:

$$\sum_{(\varepsilon_1,\varepsilon_2,\varepsilon_3)\in\{1,-1\}^3}e^{\pi i\mathbf{n}(\varepsilon_1,\varepsilon_2,\varepsilon_3)\cdot\mathbf{y}_k}=\prod_{j=1}^{3}\left(e^{\pi i n_j y_{k,j}}+e^{-\pi i n_j y_{k,j}}\right)$$

$$=8\prod_{j=1}^{3}\cos\left(\pi n_j y_{k,j}\right). \tag{20.10}$$

Using (20.10) in (20.9), we have

$$\overline{\mathbf{D}^2}_{\Omega_1}(\text{TorViaUnfold};\mathcal{Y},\mathcal{M};\mathcal{R};f;t)$$

$$=8\sum_{k=1}^{N}\mu_k^2\sum_{\substack{\mathbf{n}\in\mathbb{Z}^3:\\ \mathbf{n}\ne\mathbf{0}}}|a_{\mathbf{n}}|^2\left(1-\mathbf{F}^2(\pi|\mathbf{r}_k(t)|\ |\mathbf{n}|)\right)$$

$$+64\sum_{\substack{\mathbf{n}\in\mathbb{Z}^3:\\ \mathbf{n}\ne\mathbf{0}}}|a_{\mathbf{n}}|^2\left|\sum_{k=1}^{N}\mu_k\mathbf{F}(\pi|\mathbf{r}_k(t)|\ |\mathbf{n}|)\prod_{j=1}^{3}\cos\left(\pi n_j y_{k,j}\right)\right|^2. \tag{20.11}$$

Notice that (20.11) is an analog of (7.10).

Consider now the special case $\mathbf{r}_1(t) = \mathbf{r}_2(t) = \ldots = \mathbf{r}_N(t) = \mathbf{r}(t)$. Then, using the notation

$$\Phi(\text{TorViaUnfold}; \mathcal{Y}, \mathcal{M}; \mathbf{n}) = \sum_{k=1}^{N} \mu_k \prod_{j=1}^{3} \cos\left(\pi n_j y_{k,j}\right), \tag{20.12}$$

we can rewrite (20.11) in the form

$$\begin{aligned}
&\overline{\mathbf{D}^2}_{\Omega_1}(\text{TorViaUnfold}; \mathcal{Y}, \mathcal{M}; \mathbf{r}; f; t) \\
&= \int_{\Omega_1} \overline{\mathbf{D}^2}(\text{TorViaUnfold}; \mathcal{Y}, \mathcal{M}; \mathbf{r}; \omega; f; t)\, d\omega \\
&= 8 \sum_{k=1}^{N} \mu_k^2 \sum_{\substack{\mathbf{n} \in \mathbb{Z}^3: \\ \mathbf{n} \neq \mathbf{0}}} |a_{\mathbf{n}}|^2 \left(1 - \mathbf{F}^2(\pi |\mathbf{r}(t)|\, |\mathbf{n}|)\right) \\
&+ 64 \sum_{\substack{\mathbf{n} \in \mathbb{Z}^3: \\ \mathbf{n} \neq \mathbf{0}}} |a_{\mathbf{n}}|^2 \left|\Phi(\text{TorViaUnfold}; \mathcal{Y}, \mathcal{M}; \mathbf{n})\right|^2 \mathbf{F}^2(\pi |\mathbf{r}(t)|\, |\mathbf{n}|).
\end{aligned} \tag{20.13}$$

Notice that (20.13) is an analog of (3.4), which is a key statement in Theorem 3.1. Thus we obtain the following analog of (3.9):

$$\begin{aligned}
&\overline{\mathbf{D}^2}_{\Omega_1, uneven}(\text{TorViaUnfold}; \mathcal{Y}, \mathcal{M}; \mathbf{r}; f; t) \\
&= \sum_{\substack{\mathbf{n} \in \mathbb{Z}^3: \\ \mathbf{n} \neq \mathbf{0}}} |a_{\mathbf{n}}|^2 \left|\Phi(\text{TorViaUnfold}; \mathcal{Y}, \mathcal{M}; \mathbf{n})\right|^2 \mathbf{F}^2(\pi |\mathbf{r}(t)|\, |\mathbf{n}|),
\end{aligned} \tag{20.14}$$

where $\mathbf{F}(x) = e^{-x^2/2}$. Similarly to (3.9), *unevenness* (20.14) represents the dominating part in

$$\overline{\mathbf{D}^2}_{\Omega_1}(\text{TorViaUnfold}; \mathcal{Y}, \mathcal{M}; \mathbf{r}; f; t)$$

as long as the system stays in off-equilibrium.

In view of (20.14), the corresponding analog of the Unevenness Function (12.16) goes as follows: for $i = 1, 2, 3$ we have

$$\begin{aligned}
&\text{UnevenFun}_{\Omega_i}(\text{TorViaUnfold}; \mathcal{Y}, \mathcal{M}; \mathbf{r}; t) \\
&= \sup_{f \in L^2:\ \sigma_0^2(f) > 0} \sup_{\tau \geq t} \frac{1}{\sigma_0^2(f)} \overline{\mathbf{D}^2}_{\Omega_i, uneven}(\text{TorViaUnfold}; \mathcal{Y}, \mathcal{M}; \mathbf{r}; f; \tau),
\end{aligned} \tag{20.15}$$

where

$$\overline{\mathbf{D}^2}_{\Omega_i,uneven}(\text{TorViaUnfold}; \mathcal{Y}, \mathcal{M}; \mathbf{r}; f; t)$$

$$= \sum_{\substack{\mathbf{n}\in\mathbb{Z}^3:\\ \mathbf{n}\neq\mathbf{0}}} |a_{\mathbf{n}}|^2 \left|\Phi(\text{TorViaUnfold}; \mathcal{Y}, \mathcal{M}; \mathbf{n})\right|^2 \mathbf{F}_i^2(\pi|\mathbf{r}(t)|\ |\mathbf{n}|) \tag{20.16}$$

with

$$\mathbf{F}_1(x) = e^{-x^2/2},\ \mathbf{F}_2(x) = \frac{\sin x}{x} \text{ and } \mathbf{F}_3(x) = -3\frac{\cos x}{x^2} + 3\frac{\sin x}{x^3}. \tag{20.17}$$

Following Section 6 (and Sections 18–19 in the general isotropic case) it is a routine straightforward generalization to define the Relaxation Time, Relaxation Drifting Distance, Imbalance, Time Evolution Function and Micro-Entropy for the billiard system.

Theorem 20.1. *The typical time evolution of the 3-dimensional torus-via-unfolding model behaves exactly the same way — approach to equilibrium, Second Law — as the ordinary 3-dimensional torus model with the following natural change. The key exponential sum*

$$\Phi(\mathcal{Y}, \mathcal{M}; \mathbf{n}) = \sum_{k=1}^{N} \mu_k e^{2\pi i \mathbf{n}\cdot\mathbf{y}_k} \tag{20.18}$$

for the ordinary general torus model is replaced by the sum defined in (20.12) for the torus-via-unfolding model:

$$\Phi(\text{TorViaUnfold}; \mathcal{Y}, \mathcal{M}; \mathbf{n}) = \sum_{k=1}^{N} \mu_k \prod_{j=1}^{3} \cos\left(\pi n_j y_{k,j}\right), \tag{20.19}$$

where $\mathbf{n} = (n_1, n_2, n_3)$ *and* $\mathcal{Y} = \{\mathbf{y}_1, \dots, \mathbf{y}_N\}$ *with* $\mathbf{y}_k = (y_{k,1}, y_{k,2}, y_{k,3})$. *The 3-dimensional case has a straightforward generalization for every dimension.* □

What we did so far in Section 20 was to emphasize the strong similarity between the ordinary general torus model and the torus-via-unfolding model. Nevertheless, there are some minor differences. An interesting minor difference between (20.18) and (20.19) shows up in the special case of Big Bang $\mathcal{Y} = \mathcal{Y}_{BB}$ with the extra condition that every mass is one (so we omit $\mathcal{M}$) and $|\mathbf{r}(t)| = t$. Clearly

$$|\Phi(\mathcal{Y}_{BB}; \mathbf{n})| = N$$

independently of the choice of the lattice point $\mathbf{n}$; on the other hand, the value of

$$|\Phi(\text{TorViaUnfold}; \mathcal{Y}_{BB}; \mathbf{n})| = N \prod_{j=1}^{3} |\cos(\pi n_j y_j)|$$

heavily depends on the choice of the lattice point $\mathbf{n}$ and on the choice of the single point $\mathbf{y} = (y_1, y_2, y_3)$ representing the Big Bang (i.e., $\mathcal{Y}_{BB}$ is the single point $\mathbf{y} = \mathbf{y}_1 = \ldots = \mathbf{y}_N$ with multiplicity N).

As we demonstrated for the ordinary general torus model, the Second Law describes the typical time evolution of an individual system. As a rule, we obtain a good approximation of the ω-dependent Imbalance by the explicit ω-independent Time Evolution Function as long as the system is far from equilibrium. The same holds for the torus-via-unfolding model.

In our special case the ω-independent Time Evolution Function of the first kind for the ordinary torus model simplifies to (see (17.1))

$$\text{TiEvFun}_1(\mathcal{Y}_{BB}; \mathbf{r}; t) = \left(e^{-(2\pi t)^2} N \quad \text{or} \quad 1\right)_{\max}. \tag{20.20}$$

On the other hand, the torus-via-unfolding analog of (20.20) is

$$\text{TiEvFun}_1(\text{TorViaUnfold}; \mathcal{Y}_{BB}; \mathbf{r}; t)$$

$$= \left(e^{-(\pi t)^2} \max_{\mathbf{n} \in \mathbb{Z}^3 \setminus \mathbf{0}} \left(N \prod_{j=1}^{3} \cos^2(\pi n_j y_j) \right)^{1/|\mathbf{n}|^2} \quad \text{or} \quad 1 \right)_{\max}, \tag{20.21}$$

where $\mathcal{Y}_{BB}$ is the single point $\mathbf{y} = (y_1, y_2, y_3)$ with multiplicity N.

As we explained in Section 5 (see (5.12)), in our special case, starting from Big Bang, for the overwhelming majority of the initial conditions $\omega \in \Omega_1$, the relaxation time (= required time to reach LA-equilibrium or L-equilibrium) in the ordinary torus model of the first kind is precisely

$$\frac{\sqrt{\log N}}{2\pi} + O(1), \tag{20.22}$$

which comes from (20.20) via the equation

$$e^{-(2\pi t)^2} N = 1 \iff t = \frac{\sqrt{\log N}}{2\pi}. \tag{20.23}$$

Similarly, in the Bernoulli torus model of the first kind, starting from Big Bang at $\mathbf{y}$, the analog majority relaxation time is

$$\frac{1}{\pi} \left(\log \max_{\mathbf{n} \in \mathbb{Z}^3 \setminus \mathbf{0}} \left(N \prod_{j=1}^{3} \cos^2(\pi n_j y_j) \right)^{1/|\mathbf{n}|^2} \right)^{1/2} + O(1), \tag{20.24}$$

which comes from (20.21) via the equation

$$e^{-(\pi t)^2} \max_{\mathbf{n}\in\mathbb{Z}^3\setminus\mathbf{0}} \left(N \prod_{j=1}^{3} \cos^2(\pi n_j y_j) \right)^{1/|\mathbf{n}|^2} = 1. \tag{20.25}$$

In view of (20.25) it is natural to study the function

$$h(\mathbf{y}) = \max_{\mathbf{n}\in\mathbb{Z}^3\setminus\mathbf{0}} \left(N \prod_{j=1}^{3} \cos^2(\pi n_j y_j) \right)^{1/|\mathbf{n}|^2} . \tag{20.26}$$

Notice that the function $h(\mathbf{y})$ in (20.26) exhibits a dramatic change as $\mathbf{y} \in [0,1]^3$ (the single point representing the Big Bang) converges to the center $(1/2, 1/2, 1/2)$ of the unit cube $I^3[0,1)^3$. Indeed, if $\mathbf{y}$ starts from a corner of the cube $(0,0,0)$ (say), and moves to the center, then $h(\mathbf{y})$ decreases from N to $N^{1/4}$ (due to the fact that $\cos^2(\pi/2) = 0$ and $\cos^2(\pi) = 1$). It is easy to see that the function $h(\mathbf{y})$ in (20.26) attains its minimum at the center $\mathbf{y} = (1/2, 1/2, 1/2)$ of the unit cube.

Using this fact in (20.24), we obtain that the majority relaxation time in the torus-via-unfolding model of the first kind is between

$$\frac{\sqrt{\log N}}{\pi} + O(1) \quad \text{and} \quad \frac{\sqrt{\log N^{1/4}}}{\pi} + O(1) = \frac{\sqrt{\log N}}{2\pi} + O(1),$$

depending on the location $\mathbf{y}$ of the Big Bang. The minimum is

$$\frac{\sqrt{\log N}}{2\pi} + O(1), \tag{20.27}$$

which is achieved at the center $\mathbf{y} = (1/2, 1/2, 1/2)$. Notice that (20.27) is exactly the same as the threshold (20.22) in the ordinary torus model (we may say that in the torus every point is a center).

The dependence on the location $\mathbf{y}$ of the Big Bang in the billiard model is not surprising, since the unit cube $[0,1]^3$, the underlying set of the billiard, is not translation invariant. It is plausible by symmetry that the center is the best place for the Big Bang in the torus-via-unfolding model.

Finally, note that Theorem 20.1 holds in every dimension.

21. Billiards in other shapes

For simplicity we consider the 2-dimensional case. What happens if we change the rectangular billiard table to another shape? For which shapes can we save/adapt our Fourier method developed in Sections 3–20? We can certainly save/adapt the proofs as long as the process of *unfolding* gives a perfect tiling of the plane (see Section 1). Let $\mathcal{P}$ be a polygon; *unfolding* of $\mathcal{P}$ means the following infinite process. We select a side of $\mathcal{P}$, and reflect $\mathcal{P}$ across this side; we obtain a new polygon $\mathcal{P}_1$; next we select a side of $\mathcal{P}_1$, and reflect $\mathcal{P}_1$ across this side; we obtain another polygon $\mathcal{P}_2$; and so on — we keep doing this in all possible ways. Does the complete sequence $\mathcal{P}, \mathcal{P}_1, \mathcal{P}_2, \ldots$ tile the plane? Unfortunately, the list of polygons $\mathcal{P}$ for which unfolding of $\mathcal{P}$ gives a tiling of the plane is very short: (1) rectangles, (2) regular hexagon, (3) equilateral triangle, (4) isosceles right triangle (= half of a square), (5) right triangle with angle $\pi/3$ (= half of an equilateral triangle).

For these 5 shapes — visualizing them as unconventional billiard tables — we can save our Fourier method, and we can *prove* that the corresponding billiard systems exhibit the same behavior — approach to equilibrium, Second Law — as the standard rectangular billiard, or the 2-dimensional torus model. Unfortunately, beyond these 5 shapes, we know very little. Even the simplest problems are wide open. For example, it is well known that a single rectangular billiard is uniformly distributed for almost every initial angle (König–Szűcs–Weyl), explaining why we used volume (= uniform Lebesgue measure) in our definition of equilibrium; see Definitions 2.3 and 2.4. Unfortunately, beyond the 5 shapes, we know very little about the uniformity of the corresponding single billiard path. For example, let Δ be an arbitrary triangle: is it true that the corresponding billiard is uniform for almost every initial direction? Shockingly, this innocent-looking question remains a long-standing unsolved problem. We know that the answer to this question is affirmative in the special case where the triangle is "rational", meaning that the angles are rational multiples of π. The proof of this narrow special case is already very difficult; see [Ke-M-S86].

For many natural shapes the 2-dimensional billiard is far from being uniform. For example, if the billiard table is a circular disk (= solid circle), then *no* billiard path is uniformly distributed inside the circular disk. (Here we are talking about uniformity in the traditional sense of Weyl, i.e., uniform with respect to all Riemann integrable test functions, or equivalently, for all Jordan measurable test sets.)

Just because we know so little about the (single) billiard paths beyond the 5 special shapes listed above, it does not mean that there is no other example of systems for which we can save our Fourier method. Here is a new example.

The geodesics flow on an equifacial tetrahedron surface. As we said above, there are only 3 special triangles that provide a perfect tiling of the plane via *unfolding*. Unfolding is based on reflection across a line. But we also have a different kind of reflection called "reflection in a point". Reflection in a point is much more flexible than reflection across a line: by using reflection in a point, we can tile the plane with an arbitrary triangle. This is of course the same thing as to start with a parallelogram lattice, and split every parallelogram into two congruent triangles by means of a diagonal (we always use the same type of diagonal, i.e., they are all parallel). The neighboring triangles (which share a side) are congruent via reflection in the midpoint of the common side.

Let Δ be one of these congruent triangles in the plane; assume that Δ is acute-angled. We can fold Δ into an equifacial tetrahedron surface in the 3-space, where equifacial means that the 4 faces of the tetrahedron are congruent triangles. To obtain 4 congruent triangles, we consider the midpoints; formally, if A, B, C are the vertices of Δ, then let C_1, A_1, B_1 be the midpoints of the sides AB, BC, CA of Δ in this order (we recommend the reader to draw a picture). We simply fold along the three line segments in Δ which join the midpoints A_1, B_1, C_1 of two of the three sides. Then the three vertices A, B, C come together in a point D in the 3-space, and we obtain an equifacial tetrahedron surface $T = T(\Delta)$ with vertices A_1, B_1, C_1, D. Note that this way the boundary points of the triangle Δ glue together in pairs; for example, B_1A becomes identical to B_1C via the reflection in the midpoint B_1.

Note that the simplest equifacial tetrahedron is the regular tetrahedron (or symplex) in the 3-space.

Let L be a straight line in the plane that contains the triangle Δ. For simplicity, assume that L intersects the inside of Δ; let $Q \in L \cap \Delta$ be such a common point. It is easy to see that the whole straight line L in the plane is wrapped up as a *geodesic* on the equifacial tetrahedron surface $T = T(\Delta) = A_1B_1C_1D$. Note that in general, a geodesic on a polyhedron surface consists of straight line segments that fit together to a straight line at the boundary, where the line segment hits an edge: the consecutive line segments become collinear when we turn the two neighboring faces (that share the edge) into coplanar position.

The geometric explanation why a straight line L in the plane is wrapped up as a *geodesic* on the equifacial tetrahedron surface $T = T(\Delta) = A_1B_1C_1D$ is based on the effect of reflection in a point. Indeed, moving from point Q along the line L in one direction, we reach a congruent neighbor Δ_1 of Δ in the triangle lattice, and the fact that Δ_1 is congruent to Δ via reflection in the midpoint of the common side means that the part of L in Δ_1 corresponds exactly to the next part of the geodesic on the tetrahedron surface $T = T(\Delta)$. Similarly, when moving further along L we reach a neighbor Δ_2 of Δ_1 in the triangle lattice, then the part of L in Δ_2 corresponds exactly to the next part of the geodesic on the tetrahedron surface $T = T(\Delta)$, and so on.

An alternative way to say the same thing in terms of the smaller midpoint triangles goes as follows. We label the midpoint triangle $A_1B_1C_1$ with 1, label AC_1B_1 with 2, label BA_1C_1 with 3, and label CB_1A_1 with 4; this way the four faces of tetrahedron $T = T(\Delta)$ are labeled with 1, 2, 3, 4. Next we divide every congruent copy of Δ in the triangle lattice into four congruent smaller midpoint triangles ("halves"), and extend the labeling of the four halves of Δ to the whole plane by using the following simple rule: if two midpoint triangles have exactly one common vertex then they have the same label (1 or 2 or 3 or 4). Now wrapping-up-a-line means that, if a motion on a straight line L in the plane is in a midpoint triangle labeled i, then the corresponding motion on the geodesic (= wrapped up line) is on the face of the tetrahedron $T = T(\Delta) = A_1B_1C_1D$ with the same label i $(i = 1, 2, 3, 4)$.

Let S be an arbitrary measurable subset of the triangle Δ with vertices A, B, C. Again let A_1 denote the midpoint of BC, and let Δ_1 denote the reflection of Δ in the midpoint A_1. Similarly, let S_1 denote the reflection of S in the midpoint A_1, and let A_2 denote the reflection of A in the midpoint A_1. Let $\mathcal{P}$ denote the parallelogram with vertices A, B, A_2, C, and let $\mathcal{L} = \mathcal{L}(\Delta)$ denote the parallelogram lattice on the plane with fundamental parallelogram $\mathcal{P}$. The geometric method of "wrapping up a straight line to a geodesic on an equifacial tetrahedron surface" explains why the uniformity of a straight line L modulo $\mathcal{P}$ in the plane with respect to the test set $\widetilde{S} = S \cup S_1$ is equivalent to the uniformity of the corresponding geodesic (= wrapped up copy of line L) on the equifacial tetrahedron surface $T = T(\Delta)$ with respect to the test set S.

The uniformity of a straight line L modulo $\mathcal{P}$ is analogous to the problem of the uniformity of a torus line modulo $[0, 1)^2$, since every parallelogram lattice $\mathcal{L}$ in the plane can be mapped to the integer square lattice $\mathbb{Z}^2$ via some linear transformation of the plane. This means, to describe the time evolution of the corresponding *dynamical system*, we can save/adapt the 2-dimensional results of Sections 3–19. (Note that a linear transformation may convert an isotropic initial velocity distribution into a non-isotropic initial velocity distribution. However, this change is basically irrelevant. For example, the linear transformation of a multivariable normal — i.e., Gaussian — distribution is another multivariable normal distribution.)

Here the *dynamical system* simply means that we have N points on a given equifacial tetrahedron surface (= initial point configuration), we choose an initial velocity distribution (e.g., Gaussian), and study the time evolution of the system.

Theorem 21.1. *The typical time evolution of the geodesic flow on an equifacial tetrahedron surface behaves exactly the same way — approach to equilibrium, Second Law — as the simplest 2-dimensional torus model on* $I^2 = [0,1)^2$ *with torus lines*

$$\mathbf{r}_1(t) = \mathbf{r}_2(t) = \ldots = \mathbf{r}_N(t) = (t, 0).$$

□

It is very surprising that this argument (wrapping up a straight line via reflection in a point), which works so well for the regular tetrahedron surface, does not seem to work for the seemingly equally simple cube surface. Thus we face the following questions.

Open Problems. Can one extend Theorem 21.1 for the geodesic flow on the cube surface? What about the surfaces of the other three Platonic solids (octahedron, dodecahedron, icosahedron)? What about an arbitrary convex polyhedron? Finally, can one extend Theorem 21.1 for the geodesic flow on an arbitrary convex surface?

The simplest smooth convex surface is the sphere. The geodesics on the sphere are the great circles; they are all closed curves (completely different from the case of the torus). In Chapter 5 we describe how the typical time evolution of a geodesic flow system on the sphere (the particles move on great circles) approaches equilibrium.

22. Equivalent dynamical systems (I)

Discrepancy is the basic concept. We emphasize the fact that everything we did so far was based on the elementary concept of *discrepancy,* which is

$$\text{Point Counting in a test set } S - \text{ExpectedValue}$$

$$= \text{Point Counting in a test set } S - N \cdot \text{vol}(S)$$

in the simplest special case ($S \subset [0,1]^d$), and in the general case (see (2.27))

$$N(\mathcal{Y}, \mathcal{M}; \mathcal{R}; \omega; f; t) - N \int_{I^d} f \, d\mathrm{V}. \tag{22.1}$$

Indeed, we started Section 6 with the concept of Disparity Function (see (6.2); note that (6.1) represents a failed attempt)

$$\mathrm{DispFun}_i(t) = \mathrm{DispFun}_i(\mathcal{Y}, \mathcal{M}; \mathbf{r}; t)$$

$$= \max\left\{ \sup_{f \in L^2:\ \sigma_0^2(f) > 0} \frac{1}{\sigma_0^2(f)} \overline{\mathbf{D}^2}_{\Omega_i}(\mathcal{Y}, \mathcal{M}; \mathbf{r}; f; t), \sum_{k=1}^{N} \mu_k^2 \right\}, \tag{22.2}$$

where

$$\overline{\mathbf{D}^2}_{\Omega_i}(\mathcal{Y}, \mathcal{M}; \mathcal{R}; f; t) = \int_{\Omega_i} \overline{\mathbf{D}^2}(\mathcal{Y}, \mathcal{M}; \mathcal{R}; \omega; f; t) \, d\omega, \tag{22.3}$$

and

$$\overline{\mathbf{D}^2}(\mathcal{Y}, \mathcal{M}; \mathcal{R}; \omega; f; t) = \int_{I^d} \mathbf{D}^2(\mathcal{Y}, \mathcal{M}; \mathcal{R}; \omega; f_{\mathbf{w}}; t) \, d\mathbf{w}$$

$$= \int_{I^d} \left| N(\mathcal{Y}, \mathcal{M}; \mathcal{R}; \omega; f_{\mathbf{w}}; t) - N \int_{I^d} f \, d\mathrm{V} \right|^2 d\mathbf{w} \tag{22.4}$$

with

$$f_{\mathbf{w}}(\mathbf{x}) = f(\mathbf{x} - \mathbf{w}). \tag{22.5}$$

We refer to (22.3) as the Big Average Square-Discrepancy (relative to f) of the first or second or third kind for $i = 1$ or 2 or 3, respectively. In the general isotropic case we replace Ω_i with Ω_g (see Section 18).

Moreover, from the Disparity Function we derive the Sup-Disparity Function simply by taking supremum; see (6.3). Also, the Sup-Disparity Function is practically equal to the Unevenness Function (the difference is the negligible "white noise of the equilibrium", see (6.4)), and the rest of the key concepts, Relaxation Time and Drifting Distance, Imbalance, Micro-Entropy, are all expressed in terms of the Unevenness Function; see Section 6.

The reason why we emphasize the central role of the discrepancy (6.1) is that the formulas in the Metatheorem (see Section 6) are all based on exponential sums like

$$\left| \sum_{k=1}^{N} \mu_k e^{2\pi i \mathbf{n} \cdot \mathbf{y}_k} \right|^2, \quad \mathbf{n} \in \mathbb{Z}^d,$$

which are torus-specific expressions, coming from (classical) Fourier analysis on the torus. This raises the question: how to extend our results beyond the case of the torus? For example, what can we say about the typical time evolution of large dynamical systems on the sphere? What about other surfaces? What about the unit disk?

The torus is invariant under the group of translations in the Euclidean space — this group is commutative. The sphere $\mathbf{S}^2$, on the other hand, is invariant under the (special) orthogonal group $SO(3)$, which is noncommutative. Spherical Fourier analysis is the simplest special case of noncommutative Fourier analysis; nevertheless, it is already much more complicated than classical Fourier analysis (on the torus). Moreover, nobody knows how to create an effective theory of "Fourier analysis on the unit disk, or in general, on an arbitrary bounded surface". We may say that "Fourier analysis is very rigid".

There is, however, a straightforward way to extend our results on torus systems to dynamical systems on arbitrary surfaces, and beyond. The basic idea is to define an *equivalence relation* between dynamical systems. Roughly speaking, two dynamical systems are equivalent if there is a *discrepancy-preserving one-to-one correspondence* between them. Instead of starting with a general formal definition, we show some examples.

Examples of induced systems. First consider the following explicit mapping from a rectangle to the unit circle (= circular disk) $x^2 + y^2 \leq 1$ using modified polar coordinates

$$\Psi(w, \theta) = \Psi_1(w, \theta) = (\sqrt{2w}\cos\theta, \sqrt{2w}\sin\theta). \tag{22.6}$$

$\Psi = \Psi_1$ in (22.6) maps the rectangle $0 < w \leq 1/2$, $0 \leq \theta < 2\pi$ to the unit disk in a one-to-one way (except the origin). It is area-preserving: the corresponding Jacobian is one

$$\text{Jacobian} = J_\Psi = \begin{vmatrix} \frac{\cos\theta}{\sqrt{2w}} & -\sqrt{2w}\sin\theta \\ \frac{\sin\theta}{\sqrt{2w}} & \sqrt{2w}\cos\theta \end{vmatrix} = 1. \tag{22.7}$$

Let S be an arbitray Lebesgue measurable subset of the unit disk $x^2 + y^2 \leq 1$. By (22.7), $\text{area}(S) = \text{area}(\Psi^{-1}(S))$, where of course $\Psi^{-1}(S)$ is a measurable subset of the rectangle $0 < w \leq 1/2$, $0 \leq \theta < 2\pi$.

Given an arbitrary parametrized curve Γ in the rectangle $0 < w \leq 1/2$, $0 \leq \theta < 2\pi$ (representing the motion of a particle in the rectangle), $\Psi_1(\Gamma)$ defines a parametrized curve in unit disk (representing the motion of a particle in the disk). This is how Ψ defines the *induced motion.*

Given an arbitrary system of particles in motion in the rectangle $0 < w \le 1/2$, $0 \le \theta < 2\pi$, Ψ_1 defines the corresponding induced motion of particles in the disk.

For example, let $0 < T \le \infty$ be arbitrary, and let $\alpha \ne 0$ be any real ("slope"). For every "starting point" $\mathbf{s} = (s_1, s_2)$, consider the following HalfBilliardpath-HalfTorusline $\Gamma[\mathbf{s}; \alpha; T]$ in the rectangle $0 < w \le 1/2$, $0 \le \theta < 2\pi$:

$$(\|s_1 + t\|, s_2 + \alpha t \pmod{2\pi}), \quad 0 \le t \le T, \tag{22.8}$$

where $\|x\|$ denotes the distance of a real number x from the nearest integer (so clearly $0 \le \|x\| \le 1/2$), and the somewhat informal $s_2 + \alpha t \pmod{2\pi}$ has the precise definition

$$2\pi \left\{ \frac{s_2 + \alpha t}{2\pi} \right\},$$

where $\{x\}$ denotes the fractional part.

Equation (22.8) defines a billiard path in the direction of variable w (bouncing back on the vertical sides of the rectangle, if w represents the horizontal axis), and it defines a torus line in the direction of θ (since sine and cosine are 2π-periodic; see (21.1)). Write $\Gamma = \Gamma[\mathbf{s}; \alpha; T]$ and

$$\Gamma(t) = (\|s_1 + t\|, s_2 + \alpha t \pmod{2\pi}) \tag{22.9}$$

for $0 \le t \le T$ (see (22.8)). $\Gamma(t)$ describes a linear motion of a particle with constant speed $\sqrt{1+\alpha^2}$.

The induced motion $\Psi_1(\Gamma)$, defined by (22.6) and (22.9), does not have constant speed. Indeed, its speed is the magnitude of the velocity, that is, the magnitude of the derivative:

$$\text{speed of } \Psi_1(\Gamma) \text{ at time } t = \left((x_w w'(t) + x_\theta \theta'(t))^2 + (y_w w'(t) + y_\theta \theta'(t))^2 \right)^{1/2},$$

where $x = x(w,\theta) = \sqrt{2w}\cos\theta$, $y = y(w,\theta) = \sqrt{2w}\sin\theta$ (see (22.6)), $x_w, x_\theta, y_w, y_\theta$ are the partial derivatives, and $w(t) = \|s_1 + t\|$, $\theta(t) = s_2 + \alpha t \pmod{2\pi}$ (see (22.6) and (22.8)–(22.9)). Clearly

$$(x_w \cdot w'(t) + x_\theta \cdot \theta'(t))^2 + (y_w \cdot w'(t) + y_\theta \cdot \theta'(t))^2$$

$$= \left(\pm \frac{\cos\theta}{\sqrt{2w}} - \alpha\sqrt{2w}\sin\theta \right)^2 + \left(\pm \frac{\sin\theta}{\sqrt{2w}} + \alpha\sqrt{2w}\cos\theta \right)^2 = \frac{1}{2w} + 2\alpha^2 w,$$

which implies that

$$\text{speed of } \Psi_1(\Gamma) \text{ at } t = \sqrt{\frac{1}{2w(t)} + 2\alpha^2 w(t)} = \frac{\sqrt{1 + 4\alpha^2 r^4}}{r}, \tag{22.10}$$

where $r = \sqrt{2w}$ is the distance of the particle from the origin = center of the unit disk $x^2 + y^2 \leq 1$, see (22.6).

Note that the curve $\Psi_1(\Gamma)$ here is a union of smooth pieces called the Fermat's spiral (also known as a parabolic spiral). The Fermat's spiral has the equation $r = a\sqrt{\theta} + b$ in polar coordinates (r, θ). (If T is infinite, then $\Psi_1(\Gamma)$ passes through the origin infinitely many times with infinite speed in a nearly periodic fashion.) Note that $\Psi_1(\Gamma)$ is smooth at the origin, but it is nonsmooth where $\Psi_1(\Gamma)$ hits the boundary circle, and bounces back (exactly the same way as a billiard path is nonsmooth where it hits the wall of a rectangle). We may call $\Psi_1(\Gamma)$ (see (22.6) and (22.9) with speed (22.10)) the *Fermat spiral billiard.* (Note that the average speed of a Fermat spiral billiard motion on a long run is $\sqrt{1+\alpha^2}$ plus negligible.)

Thus we obtain

Theorem 22.1. *The typical time evolution of the Fermat spiral billiard system in the unit disk, induced by the mapping Ψ_1 defined in (22.6), behaves exactly the same way — approach to equilibrium, Second Law — as the simplest 2-dimensional torus model on $I^2 = [0,1)^2$ with torus lines*

$$\mathbf{r}_1(t) = \mathbf{r}_2(t) = \ldots = \mathbf{r}_N(t) = (t, 0).$$

□

Indeed, since $\Psi = \Psi_1$ in (22.6) is measure-preserving, the two systems have the same discrepancy (22.1), and as we explained it at the beginning of the section, that makes the two systems equivalent.

We have to clarify a slight formal/technical problem in Theorem 22.1: equation (22.8) is neither a torus line, nor a billiard path. However, we can use *unfolding* in one dimension (see Sections 1 and 20), and thus the Half Billiardpath-Half Torusline becomes a torus line.

The choice of the simplest torus model with torus line trajectories in Theorem 22.1 was accidental. Of course, Theorem 22.1 remains true if we start with a general torus model having arbitrary trajectories $\mathbf{r}_1(t), \ldots, \mathbf{r}_N(t)$ in $I^2 = [0,1)^2$. Then again the induced system in the unit disk, induced by the mapping Ψ_1 in (22.6), behaves exactly the same way — approach to equilibrium, Second Law — as the general 2-dimensional torus model.

Warning: when we choose a particular initial velocity distribution in the torus model — e.g., the Gaussian or the second kind — then the induced initial velocity distribution in the unit disk is not the same: the induced distribution is not the Gaussian, and similarly, the induced distribution is not the second kind. This is a trivial consequence of the fact that mapping Ψ_1 in (22.6) is neither angle-preserving, nor distance-preserving.

Next we explain why the area-preserving property of mapping (22.6) was not really important — it was just a convenience. The following mapping, $\Psi = \Psi_2$ uses the usual polar coordinates

$$\Psi(r,\theta) = \Psi_2(r,\theta) = (x,y) = (r\cos\theta, r\sin\theta). \tag{22.11}$$

By using (22.11), we obtain another induced system in the circular disk. Mapping Ψ_2 in (22.11) maps the rectangle $0 < r \le 1/2$, $0 \le \theta < 2\pi$ to the disk $x^2 + y^2 \le 1/4$ of radius $1/2$ (except the origin; the choice of radius $1/2$ is motivated by the convenience of working with the function $\|x\|$; see below). This mapping is not measure-preserving: the Jacobian of the polar coordinates is r. So, computing the discrepancy (22.1) for the induced system in the disk of radius $1/2$, in the definition of the *expected value* of a measurable subset S we use

$$\frac{1}{\pi}\text{area}(\Psi^{-1}(S)) \text{ instead of } \frac{4}{\pi}\text{area}(S) = \frac{4}{\pi}\iint_{\Psi^{-1}(S)} r\,dr\,d\theta \tag{22.12}$$

(in different examples Ψ means different mappings).

Similarly to the Fermat spiral billiard, for every $0 < T \le \infty$, $\alpha \ne 0$ ("slope"), and $\mathbf{s} = (s_1, s_2)$ ("starting point"), consider the following HalfBilliardpath-HalfTorusline $\Gamma(\mathbf{s};\alpha;T)$ in the rectangle $0 < r \le 1/2$, $0 \le \theta < 2\pi$:

$$(\|s_1 + t\|, s_2 + \alpha t \pmod{2\pi}), \quad 0 \le t \le T, \tag{22.13}$$

where $\|x\|$ denotes the distance of a real number x from the nearest integer. (Equation (22.13) means a billiard path in the direction of r, bouncing back on the vertical sides of the rectangle, if r represents the horizontal axis, and it means a torus line in the direction of θ, since sine and cosine are 2π-periodic.) Write $\Gamma = \Gamma(\mathbf{s};\alpha;T)$ and

$$\Gamma(t) = (\|s_1 + t\|, s_2 + \alpha t \pmod{2\pi}) \tag{22.14}$$

for $0 \le t \le T$.

The speed of the parametrized curve $\Psi_2(\Gamma)$ is

$$\text{speed} = \left((x_r \cdot r'(u) + x_\theta \cdot \theta'(u))^2 + (y_r \cdot r'(u) + y_\theta \cdot \theta'(u))^2\right)^{1/2} du,$$

where $x = x(r,\theta) = r\cos\theta$, $y = y(r,\theta) = r\sin\theta$ (see (2.15)), and $r(u) = \|s_1 + u\|$, $\theta(u) = s_2 + \alpha u \pmod{2\pi}$. Routine calculation gives

$$\text{speed} = \left((\pm\cos\theta - \alpha r\sin\theta)^2 + (\pm\sin\theta + \alpha r\cos\theta)^2\right)^{1/2} = \sqrt{1+\alpha^2 r^2}. \tag{22.15}$$

The parametrized curve $\Psi_2(\Gamma)$, defined in (22.11)–(22.14), is in the disk $x^2 + y^2 \leq 1/4$ of radius $1/2$, and just like a billiard path is a union of some *straight line segments*, the curve $\Psi_2(\Gamma)$ is a union of smooth pieces called the *Archimedean spiral* (also known as the arithmetical spiral). The Archimedean spiral has the simplest linear equation $r = a\theta + b$ in polar coordinates (r, θ). The alternative name "arithmetical spiral" comes from the fact that successive turnings of the spiral have a constant separation distance $2\pi a$ (like an arithmetic progression). If T is large, then again $\Psi_2(\Gamma)$ is passing through the origin several times with infinite speed (see (22.15)). Again $\Psi_2(\Gamma)$ is smooth at the origin, but it is nonsmooth where $\Psi_2(\Gamma)$ hits the boundary circle, and bounces back (exactly the same way as the billiard path is nonsmooth where it hits the wall). We may call $\Psi_2(\Gamma)$ the *Archimedean spiral billiard*, for which we have the perfect analog of Theorem 22.1.

Theorem 22.2. *The typical time evolution of the Archimedean spiral billiard system in the disk $x^2 + y^2 \leq 1/4$, induced by the mapping Ψ_2 defined in (22.11), behaves exactly the same way — approach to equilibrium, Second Law — as the simplest 2-dimensional torus model on $I^2 = [0,1)^2$ with torus lines*

$$\mathbf{r}_1(t) = \mathbf{r}_2(t) = \ldots = \mathbf{r}_N(t) = (t, 0).$$

Since Ψ_2 is not area-preserving, we have to modify the expected value according to (22.12). □

Next we study induced motions on the geometric torus surface. The standard way to represent a torus as a surface in the 3-space is to rotate the circle $(x-b)^2 + z^2 = a^2$ of radius a and center $(b, 0, 0)$ with $0 < a < b$ around the z-axis. We denote this geometric torus by $Tor(a,b)$, and refer to it as the donut surface. It is a closed surface of surface area $4\pi^2 ab$. We use the angular coordinates

$$\Psi(\varphi, \theta) = \Psi_3(\varphi, \theta) = (x, y, z) = ((b + a\cos\varphi)\cos\theta, (b + a\cos\varphi)\sin\theta, a\sin\varphi), \tag{22.16}$$

where the mapping $\Psi = \Psi_3$ in (22.16) maps the square $0 \leq \varphi < 2\pi$, $0 \leq \theta < 2\pi$ to the donut surface $Tor(a,b)$ in a one-to-one way.

Note that the mapping $\Psi = \Psi_3$ is not area-preserving. Indeed, let S be an arbitrary measurable subset of the donut surface $Tor(a,b)$ with surface area $\text{SuArea}(S)$. We apply the well known surface area formula involving the cross product of vectors:

$$\text{SuArea}(S) = \iint_{\Psi^{-1}(S)} |\mathbf{r}_\varphi \times \mathbf{r}_\theta| \, d\varphi \, d\theta,$$

where

$$\mathbf{r}_\varphi = (x_\varphi, y_\varphi, z_\varphi) = (-a\sin\varphi\cos\theta, -a\sin\varphi\sin\theta, a\cos\varphi)$$

and

$$\mathbf{r}_\theta = (x_\theta, y_\theta, z_\theta) = (-(b + a\cos\varphi)\sin\theta, (b + a\cos\varphi)\cos\theta, 0).$$

An easy calculation shows that the magnitude of the cross product $|\mathbf{r}_\varphi \times \mathbf{r}_\theta|$ equals $a(b + a\cos\varphi)$, implying that

$$\text{SuArea}(S) = \iint_{\Psi^{-1}(S)} a(b + a\cos\varphi)\, d\varphi\, d\theta,$$

where the factor $b + a\cos\varphi$ depends on φ.

So, computing the discrepancy (22.1) for the induced system on the donut surface, in the definition of the *expected value* of a measurable subset $S \subset Tor(a, b)$ we use

$$\frac{1}{4\pi^2}\text{area}(\Psi^{-1}(S)) \text{ instead of } \frac{1}{4\pi^2 ab}\text{SuArea}(S) \tag{22.17}$$

(in different examples Ψ means different mappings).

As usual, for every $0 < T \leq \infty$, $\alpha \neq 0$ ("slope"), and $\mathbf{s} = (s_1, s_2)$ ("starting point"), consider the following torus line $\Gamma[\mathbf{s}; \alpha; T]$ in the square $0 \leq \varphi < 2\pi$, $0 \leq \theta < 2\pi$:

$$(s_1 + t \pmod{2\pi}), s_2 + \alpha t \pmod{2\pi})), \quad 0 \leq t \leq T. \tag{22.18}$$

Write $\Gamma = \Gamma[\mathbf{s}; \alpha; T]$ and

$$\Gamma(t) = (s_1 + t \pmod{2\pi}, s_2 + \alpha t \pmod{2\pi}) \tag{22.19}$$

for $0 \leq t \leq T$.

The speed of the induced parametrized curve $\Psi_3(\Gamma)$ is

$$\text{speed} = \big((x_\varphi \cdot \varphi'(t) + x_\theta \cdot \theta'(t))^2 + (y_\varphi \cdot \varphi'(t) + y_\theta \cdot \theta'(t))^2 + (z_\varphi \cdot \varphi'(t)$$

$$+ z_\theta \cdot \theta'(t))^2\big)^{1/2},$$

where $x = x(\varphi, \theta) = (b + a\cos\varphi)\cos\theta$, $y = y(\varphi, \theta) = (b + a\cos\varphi)\sin\theta$, $z = z(\varphi, \theta) = a\sin\varphi$, and $\varphi(t) = s_1 + t \pmod{2\pi}$, $\theta(t) = s_2 + \alpha t \pmod{2\pi}$.

Routine calculation gives

$$\text{speed} = \big(1 + \alpha^2(b + a\sin\varphi)^2\big)^{1/2} \tag{22.20}$$

at a point with angular coordinates (φ, θ).

Comparing the case of the circular disk with the case of the donut surface, we see that the donut surface has three advantages: (1) in view of

(22.20) the speed is bounded, (2) the motion is everywhere smooth, (3) the motion does not have any "fixed point" like the center of the disk. Another minor advantage is that we work with the torus line

$$(s_1 + t \pmod{2\pi}, s_2 + \alpha t (\text{mod } 2\pi)), \quad 0 < t < T,$$

so we do not need the trick of unfolding.

We refer to the induced motion $\Psi_3(\Gamma)$ (see (22.18)–(22.19) with speed (22.20)) as a *donut spiral billiard.*

Thus we obtain

Theorem 22.3. *The typical time evolution of the donut spiral billiard system, induced by Ψ_3 defined in (22.16), behaves exactly the same way — approach to equilibrium, Second Law — as the simplest 2-dimensional torus model on $I^2 = [0,1)^2$ with torus lines*

$$\mathbf{r}_1(t) = \mathbf{r}_2(t) = \ldots = \mathbf{r}_N(t) = (t, 0).$$

Since Ψ_3 is not area-preserving, we have to modify the expected value according to (22.17). □

23. Equivalent dynamical systems (II)

Spiral motions on the sphere. Like in Section 22, we start with an area-preserving parametrization of the sphere. The mapping

$$\Psi(z,\theta) = \Psi_4(z,\theta) = (x,y,z) = \left(\sqrt{1-z^2}\cos\theta, \sqrt{1-z^2}\sin\theta, z\right) \quad (23.1)$$

maps the rectangle $-1 < z < 1$, $0 \le \theta < 2\pi$ to the unit sphere $\mathbf{S}^2$: $x^2 + y^2 + z^2 = 1$ in a one-to-one way (except the two poles $(0,0,\pm 1)$). Let $S \subset \mathbf{S}^2$ be an arbitrary measurable subset of the unit sphere with surface area SuArea(S). We use the well known surface area formula

$$\text{SuArea}(S) = \iint_{\Psi^{-1}(S)} |\mathbf{r}_z \times \mathbf{r}_\theta| \, dz \, d\theta,$$

where

$$\mathbf{r}_z = (x_z, y_z, z_z) = \left(\frac{-z}{\sqrt{1-z^2}}\cos\theta, \frac{-z}{\sqrt{1-z^2}}\sin\theta, 1\right)$$

and

$$\mathbf{r}_\theta = (x_\theta, y_\theta, z_\theta) = \left(-\sqrt{1-z^2}\sin\theta, \sqrt{1-z^2}\cos\theta, 0\right).$$

The cross product $\mathbf{r}_z \times \mathbf{r}_\theta$ of the vectors equals

$$\left(\sqrt{1-z^2}\cos\theta, \sqrt{1-z^2}\sin\theta, z\right) = (x,y,z) \in \mathbf{S}^2,$$

so clearly $|\mathbf{r}_z \times \mathbf{r}_\theta| = 1$. Using this fact we have

$$\text{SuArea}(S) = \iint_{\Psi^{-1}(S)} 1\, dz\, d\theta = \text{area}(\Psi^{-1}(S)), \tag{23.2}$$

justifying the claim that the mapping $\Psi = \Psi_4$ in (23.1) is area-preserving.

As usual, for every $0 < T \le \infty$, $\alpha \neq 0$ ("slope"), and $\mathbf{s} = (s_1, s_2)$ ("starting point"), consider the following HalfBilliardpath-HalfTorusline $\Gamma[\mathbf{s}; \alpha; T]$ in the rectangle $-1 \le z \le 1$, $0 \le \theta < 2\pi$:

$$(4\|(s_1 + t + 1)/4\| - 1, s_2 + \alpha t \ (\text{mod } 2\pi)), \quad 0 \le t \le T, \tag{23.3}$$

where $\|y\|$ denotes the distance of a real number y from the nearest integer, and we applied the linear function $x \to (x+1)/4$ that maps the interval $[-1, 1]$ to the shorter interval $[0, 1/2]$ (= the range of the function $\|y\|$). Equation (23.3) represents a billiard path in the direction of the variable z, bouncing back on the vertical sides of the rectangle, if z represents "horizontal", and it represents a torus line in the direction of θ, since sine and cosine are 2π-periodic. Write $\Gamma = \Gamma[\mathbf{s}; \alpha; T]$ and

$$\Gamma(t) = (4\|(s_1 + t + 1)/4\| - 1, s_2 + \alpha t \ (\text{mod } 2\pi)) \tag{23.4}$$

for $0 \le t \le T$.

The speed of the induced motion $\Psi_4(\Gamma)$ is

$$\text{speed} = \big((x_z \cdot z'(t) + x_\theta \cdot \theta'(t))^2 + (y_z \cdot z'(t) + y_\theta \cdot \theta'(t))^2$$
$$+ (z_z \cdot z'(t) + z_\theta \cdot \theta'(t))^2\big)^{1/2}$$

where $x = x(z, \theta) = \sqrt{1 - z^2} \cos\theta$, $y = y(z, \theta) = \sqrt{1 - z^2} \sin\theta$ (see (23.1)), and $z(t) = 4\|(s_1 + t + 1)/4\| - 1$, $\theta(t) = s_2 + \alpha t$ (mod 2π).

Routine calculation gives

$$\text{speed of the induced motion } \Psi_4(\Gamma) = \left(\left(\frac{\mp z}{\sqrt{1 - z^2}} \cos\theta - \alpha\sqrt{1 - z^2} \sin\theta\right)^2\right.$$
$$\left. + \left(\frac{\mp z}{\sqrt{1 - z^2}} \sin\theta + \alpha\sqrt{1 - z^2} \cos\theta\right)^2 + (\pm 1)^2\right)^{\frac{1}{2}}$$
$$= \left(\frac{1}{1 - z^2} + \alpha^2(1 - z^2)\right)^{1/2}. \tag{23.5}$$

In view of (23.5) the speed is not bounded (the speed is infinite at the North and South Poles), but the average speed on the long run is bounded (it is $\sqrt{1 + \alpha^2}$ plus negligible).

Note that the parametrized curve $\Psi_4(\Gamma)$ here is smooth. If $T = \infty$, then it is an infinite smooth spiral that passes through the North and South Poles $(0, 0, \pm 1)$ of the unit sphere $x^2 + y^2 + z^2 = 1$ infinitely many times with infinite speed (see (23.5)). The speed is not bounded, but the average speed on the long run is bounded (it is $\sqrt{1 + \alpha^2}$ plus negligible).

We may call $\Psi_4(\Gamma)$ (see (23.3)–(23.4)) with speed (23.5) the *spherical AP-spiral motion.* Here *AP* refers to *area preserving.* Thus we obtain

Theorem 23.1. *The typical time evolution of the spherical AP-spiral system on the unit sphere $x^2+y^2+z^2 = 1$, induced by the mapping Ψ_4 defined in (23.1), behaves exactly the same way — approach to equilibrium, Second Law — as the simplest 2-dimensional torus model on $I^2 = [0,1)^2$ with torus lines*

$$\mathbf{r}_1(t) = \mathbf{r}_2(t) = \ldots = \mathbf{r}_N(t) = (t, 0).$$

□

Next we replace (23.1) with the usual spherical coordinates

$$\Psi(\varphi, \theta) = \Psi_5(\varphi, \theta) = (x, y, z) = (\cos\theta \sin\varphi, \sin\theta \sin\varphi, \cos\varphi), \qquad (23.6)$$

which will provide a new family of induced motions on the unit sphere. Mapping $\Psi = \Psi_5$ in (23.6) maps the rectangle $0 < \varphi < \pi$, $0 \le \theta < 2\pi$ to the unit sphere $\mathbf{S}^2 : x^2+y^2+z^2 = 1$ (except the poles $(0, 0, \pm 1)$). Mapping (23.6) is not area-preserving: for an arbitrary measurable subset of the unit sphere $S \subset \mathbf{S}^2$ with surface area $\mathrm{SA}(S)$,

$$\mathrm{SA}(S) = \iint_{\Psi^{-1}(S)} \sin\varphi \, d\theta \, d\varphi.$$

So, computing the discrepancy (22.1) for the induced system on the unit sphere, in the definition of the *expected value* of a measurable subset S we use

$$\frac{1}{2\pi^2}\mathrm{area}(\Psi^{-1}(S)) \text{ instead of } \frac{1}{4\pi}\mathrm{SA}(S) = \frac{1}{4\pi}\iint_{\Psi^{-1}(S)} \sin\varphi \, d\theta \, d\varphi \quad (23.7)$$

(in different examples Ψ means different mappings).

As usual, for every $0 < T \le \infty$, $\alpha \ne 0$ ("slope"), and $\mathbf{s} = (s_1, s_2)$ ("starting point"), consider the following "HalfBilliardpath-HalfTorusline" $\Gamma(\mathbf{s}; \alpha; T)$ in the rectangle $0 \le \varphi \le \pi$, $0 \le \theta < 2\pi$:

$$(2\pi\|(s_1 + t)/2\pi\|, s_2 + \alpha t \ (\mathrm{mod}\ 2\pi)), \quad 0 \le t \le T, \qquad (23.8)$$

where $\|y\|$ denotes the distance of a real number y from the nearest integer, and we applied the linear function $x/2\pi$ that maps the interval $[0, \pi]$ to the shorter interval $[0,1/2]$ (= the range of the function $\|y\|$). (Equation (23.8) means a billiard path in the direction of φ, bouncing back on the vertical sides of the rectangle, if φ represents "horizontal", and it means a torus line in the direction of θ, since sine and cosine are 2π-periodic.) Write $\Gamma = \Gamma(\mathbf{s};\alpha;T)$ and

$$\Gamma(t) = (2\pi\|(s_1+t)/2\pi\|, s_2+\alpha t \text{ (mod } 2\pi)) \tag{23.9}$$

for $0 \le t \le T$.

The speed of the induced motion $\Psi_5(\Gamma)$ is

$$\text{speed} = \Big((x_\varphi\varphi'(u)+x_\theta\theta'(u))^2 + (y_\varphi\varphi'(u)+y_\theta\theta'(u))^2 + (z_\varphi\varphi'(u)+z_\theta\theta'(u))^2\Big)^{\frac{1}{2}}, \tag{23.10}$$

where $x = x(\varphi,\theta) = \cos\theta\sin\varphi$, $y = y(\varphi,\theta) = \sin\theta\sin\varphi$, and $\varphi(u) = 2\pi\|(s_1+u)/2\pi\|$, $\theta(u) = s_2+\alpha u$ (mod 2π). Routine calculation gives

$$(x_\varphi\cdot\varphi'(u)+x_\theta\cdot\theta'(u))^2+(y_\varphi\cdot\varphi'(u)+y_\theta\cdot\theta'(u))^2+(z_\varphi\cdot\varphi'(u)+z_\theta\cdot\theta'(u))^2$$

$$=(\pm\cos\varphi\cos\theta-\alpha\sin\varphi\sin\theta)^2+(\pm\cos\varphi\sin\theta+\alpha\sin\varphi\cos\theta)^2+(\mp\sin\varphi)^2$$

$$=1+\alpha^2\sin^2\varphi.$$

Using it in (23.10), we have

$$\text{speed of the induced motion } \Psi_5(\Gamma) = \sqrt{1+\alpha^2\sin^2\varphi}. \tag{23.11}$$

Again the parametrized curve $\Psi_5(\Gamma)$ is smooth; if $T=\infty$, the motion is passing through the poles $(0,0,\pm1)$ infinitely many times.

We may call $\Psi_5(\Gamma)$ (see (23.8)–(23.9)) with speed (23.11) the *spherical NAP-spiral motion.* Here NAP stands for *not area preserving.*

Thus we obtain

Theorem 23.2. *The typical time evolution of the spherical NAP-spiral system on the unit sphere $x^2+y^2+z^2=1$, induced by the mapping Ψ_5 defined in (23.6), behaves exactly the same way — approach to equilibrium, Second Law — as the simplest 2-dimensional torus model on $I^2=[0,1)^2$ with torus lines*

$$\mathbf{r}_1(t) = \mathbf{r}_2(t) = \ldots = \mathbf{r}_N(t) = (t,0).$$

Since Ψ_5 is not area-preserving, we have to modify the expected value according to (23.7). □

Motions with bounded speed. Comparing the results in Sections 22–23, it is safe to say that the case of the donut surface gave the most satisfying result. The donut spiral billiard exhibits everywhere smooth motion with bounded speed. For the circular disk and for the unit sphere the speed was unbounded. This raises the natural question: is it possible to construct induced motions with bounded speed on the circular disk and on the unit sphere? The answer is yes, if we allow piecewise smooth motions.

We start with the unit disk. The crucial change is in the mapping: we do not use the angular coordinates, which lead to fixed points — the center of the disk and the two poles of the sphere — that are visited infinitely many times along the infinite motion. Instead we keep the Cartesian coordinates as follows. Consider the square $[-\frac{1}{2}, \frac{1}{2}]^2$ of area one centered at the origin. Let $\mathcal{D}: x^2+y^2 \le \frac{1}{2}$ denote the circular disk centered at the origin and having radius $1/\sqrt{2}$. Notice that the disk $\mathcal{D}$ contains the square $[-\frac{1}{2}, \frac{1}{2}]^2$, and the boundary circle of $\mathcal{D}$ contains the four vertices of the square $[-\frac{1}{2}, \frac{1}{2}]^2$. The first octant of the square $[-\frac{1}{2}, \frac{1}{2}]^2$ is mapped to the first octant of the disk $\mathcal{D}$ by the bijection

$$\Psi(x, y) = \Psi_6(x, y) = (\sqrt{2}x \cos\theta, \sqrt{2}y \cos\theta), \tag{23.12}$$

where $0 \le y \le x \le \frac{1}{2}$ represents the first octant, and $\theta = \arctan(y/x)$ ("inverse tangent", i.e., θ is the angle of the slope). To extend (23.12) to the whole square, we use the notation

$$\langle z \rangle = \|z\| \text{ if } z \ge 0 \text{ and } \langle z \rangle = -\|-z\| \text{ if } z < 0, \tag{22.13}$$

where (23.13) represents a continuous zig-zag function with period 2 (here, as usual, $\|w\|$ denotes the distance of a real number w from the nearest integer).

Using (23.13), we extend (23.12) to the whole square:

$$\Psi(x, y) = \Psi_6(x, y) = (g(\theta)x, g(\theta)y), \tag{23.14}$$

where

$$g(\theta) = \sqrt{2}\cos\left(\left\|\frac{\theta}{\pi/2}\right\| \frac{\pi}{2}\right) \quad \text{with} \quad \theta = \arctan(y/x). \tag{23.15}$$

We have the obvious bounds $1 \le g(\theta) \le \sqrt{2}$ and $1 \le |g'(\theta)| \le \sqrt{2}$.

By definition,

$$\text{Jacobian} = J_\Psi = \begin{vmatrix} X_x & X_y \\ Y_x & Y_y \end{vmatrix}, \tag{23.16}$$

where $X = g(\theta)x$ and $Y = g(\theta)y$. We have

$$X_x = g(\theta) - \frac{g'(\theta)xy}{x^2+y^2}, \quad X_y = \frac{g'(\theta)x^2}{x^2+y^2}, \tag{23.17}$$

$$Y_x = \frac{-g'(\theta)y^2}{x^2+y^2}, \quad Y_y = g(\theta) + \frac{g'(\theta)xy}{x^2+y^2}. \tag{23.18}$$

Combining (23.16)–(23.18), elementary calculation shows that the Jacobian is $J_\Psi = g^2(\theta)$, so $\Psi = \Psi_6$ is not area-preserving.

So, computing the discrepancy (22.1) for the induced system in the disk $\mathcal{D}$, in the definition of the *expected value* of a measurable subset S we use

$$\text{area}(\Psi^{-1}(S)) \text{ instead of } \frac{2}{\pi}\text{area}(S) = \frac{2}{\pi}\iint_{\Psi^{-1}(S)} g^2(\theta)\,dx\,dy \tag{23.19}$$

(in different examples Ψ means different mappings).

As usual, for every $0 < T \le \infty$, $\alpha \ne 0$ ("slope"), and $\mathbf{s} = (s_1, s_2)$ ("starting point"), consider the following billiard path $\Gamma[\mathbf{s};\alpha;T]$ in the square $[-\frac{1}{2},\frac{1}{2}]^2$:

$$(\langle s_1 + t\rangle, \langle s_2 + \alpha t\rangle), \quad 0 \le t \le T, \tag{23.20}$$

where we use the notation introduced in (23.13). Write $\Gamma = \Gamma[\mathbf{s};\alpha;T]$ and

$$\Gamma(t) = (\langle s_1 + t\rangle, \langle s_2 + \alpha t\rangle) \tag{23.21}$$

for $0 \le t \le T$. Working with billiard paths means that we apply Theorem 20.1.

Notice that the curve $\Psi_6(\Gamma)$ is piecewise smooth.

The speed of the induced motion $\Psi_6(\Gamma)$ is

$$\text{speed} = \left((X_x \cdot x'(t) + X_y \cdot y'(t))^2 + (Y_x \cdot x'(t) + Y_y \cdot y'(t))^2\right)^{1/2}, \tag{23.22}$$

where $x(t) = \langle s_1 + t\rangle$ and $y(t) = \langle s_2 + \alpha t\rangle$.

By using (23.17) and (23.18) in (23.22), routine calculation gives

$$\text{speed} = \left(\left(\left(g(\theta) - \frac{g'(\theta)xy}{x^2+y^2}\right)\varepsilon(t) + \frac{g'(\theta)x^2}{x^2+y^2}\delta(t)\alpha\right)^2\right.$$

$$\left. + \left(\frac{-g'(\theta)y^2}{x^2+y^2}\varepsilon(t) + \left(g(\theta) + \frac{g'(\theta)xy}{x^2+y^2}\right)\delta(t)\alpha\right)^2\right)^{1/2}, \tag{23.23}$$

where $\varepsilon(t) = x'(t) = \pm 1$, $\delta(t) = y'(t) = \pm 1$ (see the definition), and

$$\theta = \arctan(y(t)/x(t)) = \arctan\frac{\langle s_2 + \alpha t\rangle}{\langle s_1 + t\rangle}.$$

It follows from the explicit formula (23.23) that the speed of the induced motion $\Psi_6(\Gamma)$ is bounded, as we wanted.

Next we switch to the unit sphere $\mathbf{S}^2$ centered at the origin $\mathbf{0} \in \mathsf{R}^3$, and consider a regular tetrahedron centered at the origin $\mathbf{0} \in \mathsf{R}^3$ — let $\mathbf{T}$ denote the tetrahedron surface. For simplicity assume that the tetrahedron is inside the sphere $\mathbf{S}^2$. A ray starting from the origin $\mathbf{0} \in \mathsf{R}^3$ first intersects the tetrahedron surface $\mathbf{T}$ at a point $Q = (x, y, z) \in \mathbf{T}$, and then the ray intersects the sphere $\mathbf{S}^2$ at some point $\Psi_7(Q) = \Psi_7(x, y, z) = (X, Y, Z) \in \mathbf{S}^2$, where

$$X = g(x, y, z)x, \ Y = g(x, y, z)y, \ Z = g(x, y, z)z$$

for some function g. So the construction of mapping Ψ_7 is an analog of the construction of mapping Ψ_6. The only difference is that we use Theorem 21.1 instead of Theorem 20.1.

The idea of the construction of Ψ_6 and Ψ_7 clearly works far beyond the circular disk and the sphere. The construction of Ψ_6 certainly generalizes for any convex polygon, and beyond that, it works for any convex region with piecewise smooth boundary. Every nonconvex polygon can be mapped to a convex polygon by using a piecewise linear mapping. Similarly, every nonconvex simply connected region with piecewise linear boundary can be mapped to a convex region with piecewise smooth boundary by using a piecewise smooth mapping. Plane regions that are not simply connected can be turned into simply connected regions by making "cuts", and thus eliminating the "holes".

Similarly, the construction of Ψ_7 can be easily extended far beyond the sphere to large classes of closed (= no boundary curve) oriented surfaces (we assume that the surfaces are connected). It certainly works for any convex polyhedron surface, and beyond that, it works for any convex closed piecewise smooth surface, and so on. Using well known geometric techniques, it is easy to map the regular tetrahedron surface to a closed surface that is homeomorphic to the sphere. Similarly, the donut surface (see Section 22) can be mapped to any closed surface that is homeomorphic to the torus. For closed oriented surfaces with larger genus we can reduce the genus by some well known operations ("drawing circles"), and return to the case of the torus.

24. Extensions of the Second Law

1. Entropy increase = energy dispersal. So far we have been studying *mass balance* (see Case 2 at the beginning of Section 2); and the special case $\mu_k = 1$, $1 \le k \le N$ gave back the case of *particle balance* (see Case 1 at the beginning of Section 2). Now we discuss Case 3: *energy balance*. It reflects the well known intuitive meaning of entropy increase as a form of *energy dispersal*.

The switch from mass balance to (kinetic) energy balance formally means that in the definition of the mass counting function (see (2.25))

$$N(\mathcal{Y}, \mathcal{M}; \mathcal{R}; \omega; f; t) = \sum_{1 \le k \le N} f(\mathbf{x}_k(t))\mu_k$$

we replace the mass μ_k of the k-th particle with its kinetic energy $m_k v_k^2(t)/2$, where in this section m_k denotes the mass and $v_k(t)$ denotes the speed of the k-th particle at time t. That is, we replace (2.25) with

$$N(\mathcal{Y}; \text{Energy}; \mathcal{R}; \omega; f; t) = \sum_{1 \le k \le N} f(\mathbf{x}_k(t)) m_k v_k^2(t)/2. \tag{24.1}$$

Similarly, we replace (2.27′) with

$$\mathbf{D}(\mathcal{Y}; \text{Energy}; \mathcal{R}; \omega; f; t) = \left| N(\mathcal{Y}; \text{Energy}; \mathcal{R}; \omega; f; t) - \text{Energy}(t) \int_{I^3} f \, d\text{V} \right|, \tag{24.2}$$

where

$$\text{Energy}(t) = \sum_{k=1}^{N} m_k v_k^2(t)/2 \tag{24.3}$$

is the energy of the system at time t. (In general we do not even assume that the energy is conserved.)

We recall that in our general 3-dimensional N-particle torus model the trajectory of the k-th particle is

$$\mathbf{x}_k(t) = \mathbf{y}_k + \vartheta_k(\rho_k \mathbf{r}_k(t)) \quad \text{modulo one}, \quad 1 \le k \le N,$$

where the distribution of the scale ρ_k ($1 \le k \le N$) is described by a given initial speed density function $g(y)$ as $\Pr[\rho_j \le u] = \int_0^u g(y)\, dy$.

The corresponding initial velocity space is

$$\omega = (\rho_1, \vartheta_1, \rho_2, \vartheta_2, \ldots, \rho_N, \vartheta_N) \in \Omega_g = ([0, \infty) \times SO(3))^N$$

equipped with the product measure (Haar measure for $SO(3)$ and the given probability density function g for $[0, \infty)$).

By (24.1)–(24.3), we have

$$N(\mathcal{Y}; \text{Energy}; \mathcal{R}; \omega; f; t) = \sum_{1 \le k \le N} f\left(\mathbf{y}_k + \vartheta_k(\rho_k \mathbf{r}_k(t))\right) m_k \rho_k^2 \left|\mathbf{r}_k'(t)\right|^2 /2, \tag{24.4}$$

where $\mathbf{r}_k'(t)$ is the time-derivative of the vector-function $\mathbf{r}_k(t)$.

Now we basically repeat the argument of Section 7. As in Section 7, we make use the Fourier expansion of the (complex valued) test function $f \in L^2(I^3)$, $I^3 = [0,1)^3$:

$$f(\mathbf{z}) = \sum_{\mathbf{n}\in\mathbb{Z}^3} a_{\mathbf{n}} e^{2\pi i \mathbf{n}\cdot\mathbf{z}} \quad \text{with} \quad a_{\mathbf{n}} = a_{\mathbf{n}}(f) = \int_{I^3} f(\mathbf{z}) e^{-2\pi i \mathbf{n}\cdot\mathbf{z}}\, d\mathbf{z}, \qquad (24.5)$$

where of course $a_{\mathbf{0}} = \int_{I^3} f\, d\mathrm{V}$, and if f is replaced with a translation $f_{\mathbf{w}}$ (see (2.30)) then the corresponding Fourier coefficient changes as follows:

$$a_{\mathbf{n}}(f_{\mathbf{w}}) = e^{-2\pi i \mathbf{n}\cdot\mathbf{w}} a_{\mathbf{n}}(f). \qquad (24.6)$$

Combining (24.4), (24.5) and (24.6), we have

$$N(\mathcal{Y};\text{Energy};\mathcal{R};\omega;f_{\mathbf{w}};t) - \text{Energy}(t)\int_{I^3} f\, d\mathrm{V}$$

$$= \sum_{\substack{\mathbf{n}\in\mathbb{Z}^3:\\ \mathbf{n}\neq\mathbf{0}}} e^{-2\pi i \mathbf{n}\cdot\mathbf{w}} a_{\mathbf{n}} \sum_{k=1}^{N} \mu_k e^{2\pi i \mathbf{n}\cdot(\mathbf{y}_k+\vartheta_k(\rho_k\mathbf{r}_k(t)))} m_k\rho_k^2 \left|\mathbf{r}_k'(t)\right|^2/2. \qquad (24.7)$$

By (24.7) we have

$$\mathbf{D}^2(\mathcal{Y};\text{Energy};\mathcal{R};\omega;f_{\mathbf{w}};t) = \left| N(\mathcal{Y};\text{Energy};\mathcal{R};\omega;f_{\mathbf{w}};t) - \text{Energy}(t)\int_{I^3} f\, d\mathrm{V}\right|^2$$

$$= \sum_{k=1}^{N} m_k^2\rho_k^4 \left|\mathbf{r}_k'(t)\right|^4/4 \sum_{\substack{\mathbf{n}\in\mathbb{Z}^3:\\ \mathbf{n}\neq\mathbf{0}}} |a_{\mathbf{n}}|^2$$

$$+ \sum_{\substack{\mathbf{n}\in\mathbb{Z}^3:\\ \mathbf{n}\neq\mathbf{0}}} |a_{\mathbf{n}}|^2 \sum_{1\le k,\ell\le N: k\neq\ell} m_k\rho_k^2 \left|\mathbf{r}_k'(t)\right|^2/2 \cdot m_\ell\rho_\ell^2 \left|\mathbf{r}_\ell'(t)\right|^2/2 \cdot e^{2\pi i \mathbf{n}\cdot(\mathbf{y}_k-\mathbf{y}_\ell)}$$

$$\cdot e^{2\pi i \mathbf{n}\cdot(\vartheta_k(\rho_k\mathbf{r}_k(t))-\vartheta_\ell(\rho_\ell\mathbf{r}_\ell(t)))} + \sum_{\substack{\mathbf{n}_j\in\mathbb{Z}^3: j=1,2\\ \mathbf{0}\neq\mathbf{n}_1\neq\mathbf{n}_2\neq\mathbf{0}}} e^{-2\pi i(\mathbf{n}_1-\mathbf{n}_2)\cdot\mathbf{w}} \cdot a_{\mathbf{n}_1}\overline{a_{\mathbf{n}_2}}$$

$$\cdot \sum_{1\le k,\ell\le N} m_k\rho_k^2 \left|\mathbf{r}_k'(t)\right|^2/2 \cdot m_\ell\rho_\ell^2 \left|\mathbf{r}_\ell'(t)\right|^2/2 \cdot e^{2\pi i(\mathbf{n}_1\cdot\mathbf{y}_k-\mathbf{n}_2\cdot\mathbf{y}_\ell)}$$

$$e^{2\pi i(\vartheta_k(\rho_k\mathbf{r}_k(t))\cdot\mathbf{n}_1-\vartheta_\ell(\rho_\ell\mathbf{r}_\ell(t))\cdot\mathbf{n}_2)}. \qquad (24.8)$$

Since

$$\int_{I^3} e^{-2\pi i \mathbf{n}\cdot\mathbf{w}}\, d\mathbf{w} = 0 \qquad (24.9)$$

for all $\mathbf{n} \in \mathbb{Z}^3$ except the trivial case $\mathbf{n} = \mathbf{0}$, by (24.8) and (24.9) we have

$$\overline{\mathbf{D}^2}(\mathcal{Y}; \text{Energy}; \mathcal{R}; \omega; f; t) = \int_{I^3} \mathbf{D}^2(\mathcal{Y}; \text{Energy}; \mathcal{R}; \omega; f_{\mathbf{w}}; t)\, d\mathbf{w}$$

$$= \sum_{k=1}^{N} m_k^2 \rho_k^4 \left|\mathbf{r}_k'(t)\right|^4 /4 \cdot \sum_{\substack{\mathbf{n}\in\mathbb{Z}^3:\\ \mathbf{n}\neq\mathbf{0}}} |a_{\mathbf{n}}|^2 + \sum_{\substack{\mathbf{n}\in\mathbb{Z}^3:\\ \mathbf{n}\neq\mathbf{0}}} |a_{\mathbf{n}}|^2$$

$$\cdot \sum_{1\le k,\ell\le N: k\neq\ell} m_k \rho_k^2 \left|\mathbf{r}_k'(t)\right|^2 /2 \cdot m_\ell \rho_\ell^2 \left|\mathbf{r}_\ell'(t)\right|^2 /2 \cdot e^{2\pi i \mathbf{n}\cdot(\mathbf{y}_k - \mathbf{y}_\ell)}$$

$$e^{2\pi i \mathbf{n}\cdot(\vartheta_k(\rho_k \mathbf{r}_k(t)) - \vartheta_\ell(\rho_\ell \mathbf{r}_\ell(t)))}. \tag{24.10}$$

Next we integrate (24.10) over the big initial velocity space $\Omega_g = ([0,\infty) \times SO(3))^N$:

$$\overline{\mathbf{D}^2}_{\Omega_g}(\mathcal{Y}; \text{Energy}; \mathcal{R}; f; t) = \int_{\Omega_g} \overline{\mathbf{D}^2}(\mathcal{Y}; \text{Energy}; \mathcal{R}; \omega; f; t)\, d\omega$$

$$= \int_{\rho_1=0}^{\infty} \int_{\vartheta_1 \in SO(3)} \cdots \int_{\rho_N=0}^{\infty} \int_{\vartheta_N \in SO(3)} \mathbf{D}^2(\mathcal{Y}; \text{Energy}; \mathcal{R}; \omega; f; t) \rho_1^2$$

$$\cdot g(\rho_1)\, d\rho_1\, d\vartheta_1 \cdots \rho_N^2 g(\rho_N)\, d\rho_N\, d\vartheta_N$$

$$= \sum_{k=1}^{N} m_k^2 \rho_k^4 \left|\mathbf{r}_k'(t)\right|^4 /4 \cdot \sum_{\substack{\mathbf{n}\in\mathbb{Z}^3:\\ \mathbf{n}\neq\mathbf{0}}} |a_{\mathbf{n}}|^2$$

$$+ \sum_{\substack{\mathbf{n}\in\mathbb{Z}^3:\\ \mathbf{n}\neq\mathbf{0}}} |a_{\mathbf{n}}|^2 \sum_{1\le k,\ell\le N: k\neq\ell} m_k \left|\mathbf{r}_k'(t)\right|^2 /2 \cdot m_\ell \left|\mathbf{r}_\ell'(t)\right|^2 /2 \cdot e^{2\pi i \mathbf{n}\cdot(\mathbf{y}_k - \mathbf{y}_\ell)}$$

$$\cdot \int_{\rho_1=0}^{\infty} \int_{\vartheta_1 \in SO(3)} \cdots \int_{\rho_N=0}^{\infty} \int_{\vartheta_N \in SO(3)} e^{2\pi i \mathbf{n}\cdot(\vartheta_k(\rho_k \mathbf{r}_k(t)) - \vartheta_\ell(\rho_\ell \mathbf{r}_\ell(t)))} \rho_1^2$$

$$\cdot g(\rho_1)\, d\rho_1\, d\vartheta_1 \cdots \rho_N^2 g(\rho_N)\, d\rho_N\, d\vartheta_N$$

$$= \sum_{k=1}^{N} m_k^2 \rho_k^4 \left|\mathbf{r}_k'(t)\right|^4 /4 \cdot \sum_{\substack{\mathbf{n}\in\mathbb{Z}^3:\\ \mathbf{n}\neq\mathbf{0}}} |a_{\mathbf{n}}|^2$$

$$+ \sum_{\substack{\mathbf{n}\in\mathbb{Z}^3:\\ \mathbf{n}\neq\mathbf{0}}} |a_{\mathbf{n}}|^2 \sum_{1\le k,\ell\le N: k\neq\ell} m_k \left|\mathbf{r}_k'(t)\right|^2/2 \cdot m_\ell \left|\mathbf{r}_\ell'(t)\right|^2/2 \cdot e^{2\pi i\mathbf{n}\cdot(\mathbf{y}_k-\mathbf{y}_\ell)}$$

$$\cdot \int_0^\infty \int_{\vartheta\in SO(3)} e^{2\pi i\vartheta(\rho\mathbf{r}_k(t))\cdot\mathbf{n}} \rho^2 g(\rho)\, d\rho\, d\vartheta$$

$$\cdot \int_0^\infty \int_{\vartheta\in SO(3)} e^{-2\pi i\vartheta(\rho\mathbf{r}_\ell(t))\cdot\mathbf{n}} \rho^2 g(\rho)\, d\rho\, d\vartheta, \tag{24.11}$$

where $d\vartheta$ means the normed Haar measure on $SO(3)$: $\int_{SO(3)} 1\, d\vartheta = 1$, i.e., it is a probability measure.

By using the symmetry of the sphere, we clearly have ($\mathbf{e}\in\mathbf{S}^2$)

$$\int_0^\infty \int_{\vartheta\in SO(3)} e^{-2\pi i\vartheta(\rho\mathbf{e})\cdot\mathbf{w}} \rho^2 g(\rho)\, d\rho\, d\vartheta$$

$$= \int_0^\infty \frac{1}{4\pi}\int_{\mathbf{u}\in\mathbf{S}^2} e^{-2\pi i\rho\mathbf{u}\cdot\mathbf{w}} \rho^2 g(\rho)\, d\rho\, d\mathrm{SA}(\mathbf{u}). \tag{24.12}$$

We define the *isotropic Fourier transform* of a function $h(u)$ that is integrable on the half-line $0\le u<\infty$ as follows (see Theorem 4.2(b) for the special where h is a probability density function on the set of positive real numbers): let

$$\mathbf{F}_h(w) = \mathbf{F}_h(\mathbf{w}) = \int_0^\infty \frac{1}{4\pi}\int_{\mathbf{v}\in\mathbf{S}^2} e^{iu\mathbf{v}\cdot\mathbf{w}} h(u)\, du\, d\mathrm{SA}(\mathbf{v}), \tag{24.13}$$

where $\mathbf{w}$ is an arbitrary 3-dimensional vector, $w = |\mathbf{w}|$ is the length, and $d\mathrm{SA}(\mathbf{v})$ represents the surface area.

Combining (24.12) and (24.13), we have ($\mathbf{e}\in\mathbf{S}^2$)

$$\int_0^\infty \int_{\vartheta\in SO(3)} e^{-2\pi i\vartheta(\rho\mathbf{e})\cdot\mathbf{w}} \rho^2 g(\rho)\, d\rho\, d\vartheta = \mathbf{F}_h(2\pi|\mathbf{w}|) \text{ with } h(u) = u^2 g(u). \tag{24.14}$$

Next we use (24.14) in (24.11):

$$\overline{\mathbf{D}^2}_{\Omega_g}(\mathcal{Y}; \text{Energy}; \mathcal{R}; f; t) = \int_{\Omega_g} \overline{\mathbf{D}^2}(\mathcal{Y}; \text{Energy}; \mathcal{R}; \omega; f; t)\, d\omega$$

$$= \sum_{k=1}^N m_k^2 \rho_k^4 \left|\mathbf{r}_k'(t)\right|^4/4 \cdot \sum_{\substack{\mathbf{n}\in\mathbb{Z}^3:\\ \mathbf{n}\neq\mathbf{0}}} |a_{\mathbf{n}}|^2$$

$$+\sum_{\substack{\mathbf{n}\in\mathbb{Z}^3:\\ \mathbf{n}\neq\mathbf{0}}} |a_{\mathbf{n}}|^2 \sum_{1\le k,\ell\le N:k\neq\ell} m_k\,|\mathbf{r}'_k(t)|^2/2\cdot m_\ell\,|\mathbf{r}'_\ell(t)|^2/2\cdot e^{2\pi i\mathbf{n}\cdot(\mathbf{y}_k-\mathbf{y}_\ell)}$$

$$\cdot\,\mathbf{F}_h(2\pi|\mathbf{r}_k(t)|\;|\mathbf{n}|)\mathbf{F}_h(2\pi|\mathbf{r}_\ell(t)|\;|\mathbf{n}|), \tag{24.15}$$

where $h(u)=u^2g(u)$.

We can rewrite (24.15) in the alternative form

$$\overline{\mathbf{D}^2}_{\Omega_g}(\mathcal{Y};\text{Energy};\mathcal{R};f;t)$$

$$=\sum_{k=1}^{N} m_k^2\,|\mathbf{r}'_k(t)|^4/4\sum_{\substack{\mathbf{n}\in\mathbb{Z}^3:\\ \mathbf{n}\neq\mathbf{0}}}|a_{\mathbf{n}}|^2\left(1-\mathbf{F}_h^2(2\pi|\mathbf{r}_k(t)|\;|\mathbf{n}|)\right)$$

$$+\sum_{\substack{\mathbf{n}\in\mathbb{Z}^3:\\ \mathbf{n}\neq\mathbf{0}}}|a_{\mathbf{n}}|^2\left|\sum_{k=1}^{N} m_k\,|\mathbf{r}'_k(t)|^2/2\cdot e^{2\pi i\mathbf{n}\cdot\mathbf{y}_k}\cdot\mathbf{F}_h(2\pi|\mathbf{r}_k(t)|\;|\mathbf{n}|)\right|^2, \tag{24.16}$$

where $h(u)=u^2g(u)$.

Consider now the special case where the (parametrized) curves are the same $\mathbf{r}_1(t)=\mathbf{r}_2(t)=\ldots=\mathbf{r}_N(t)=\mathbf{r}(t)$. Then using the notation

$$\Phi(\mathcal{Y};\text{Mass};\mathbf{n})=\sum_{k=1}^{N} m_k e^{2\pi i\mathbf{n}\cdot\mathbf{y}_k}, \tag{24.17}$$

and also $\mathcal{R}=\mathbf{r}$ (because $\mathcal{R}(t)=|\mathbf{r}(t)|$), we can rewrite (24.16) as follows:

$$\overline{\mathbf{D}^2}_{\Omega_g}(\mathcal{Y};\text{Energy};\mathcal{R}=\mathbf{r};f;t)=\int_{\Omega_g}\overline{\mathbf{D}^2}(\mathcal{Y};\text{Energy};\mathcal{R}=\mathbf{r};\omega;f;t)\,d\omega$$

$$=\frac{1}{4}|\mathbf{r}'(t)|^4\sum_{k=1}^{N} m_k^2\sum_{\substack{\mathbf{n}\in\mathbb{Z}^3:\\ \mathbf{n}\neq\mathbf{0}}}|a_{\mathbf{n}}|^2\left(1-\mathbf{F}_h^2(2\pi|\mathbf{r}(t)|\;|\mathbf{n}|)\right)$$

$$+\frac{1}{4}|\mathbf{r}'(t)|^4\sum_{\substack{\mathbf{n}\in\mathbb{Z}^3:\\ \mathbf{n}\neq\mathbf{0}}}|a_{\mathbf{n}}|^2\,|\Phi(\mathcal{Y};\text{Mass};\mathbf{n})|^2\,\mathbf{F}_h^2(2\pi|\mathbf{r}(t)|\;|\mathbf{n}|), \tag{24.18}$$

where $h(u)=u^2g(u)$.

Notice that (24.18) is the perfect analog of (4.19') in Theorem 4.2(b) with the natural modifications that the mass μ_k is replaced by the kinetic energy $m_k\,|\mathbf{r}'(t)|^2/2$ of the k-th particle, and in the isotropic Fourier transform we replace $g(u)$ by $h(u)=u^2g(u)$.

In particular, the last term in (24.18)

$$\frac{1}{4}\left|\mathbf{r}'(t)\right|^4 \sum_{\substack{\mathbf{n}\in\mathbb{Z}^3:\\ \mathbf{n}\neq\mathbf{0}}} |a_{\mathbf{n}}|^2 \left|\sum_{k=1}^{N} m_k e^{2\pi i\mathbf{n}\cdot\mathbf{y}_k}\right|^2 \mathbf{F}_h^2(2\pi|\mathbf{r}(t)|\,|\mathbf{n}|) \tag{24.19}$$

(with $h(u) = u^2 g(u)$) is the perfect analog of the last term in (4.19′)

$$\sum_{\substack{\mathbf{n}\in\mathbb{Z}^3:\\ \mathbf{n}\neq\mathbf{0}}} |a_{\mathbf{n}}|^2 \left|\sum_{k=1}^{N} \mu_k e^{2\pi i\mathbf{n}\cdot\mathbf{y}_k}\right|^2 \cdot \mathbf{F}_g^2(2\pi|\mathbf{r}(t)|\,|\mathbf{n}|). \tag{24.20}$$

We recall that (24.20) is exactly the critical "unevenness part" in Theorem 5.1:

$$\overline{\mathbf{D}^2}_{F_g,uneven}(t) = \overline{\mathbf{D}^2}_{F_g,uneven}(\mathcal{Y},\mathcal{M};\mathbf{r};f;t)$$

$$= \sum_{\substack{\mathbf{n}\in\mathbb{Z}^3:\\ \mathbf{n}\neq\mathbf{0}}} |a_{\mathbf{n}}|^2 \left|\sum_{k=1}^{N} \mu_k e^{2\pi i\mathbf{n}\cdot\mathbf{y}_k}\right|^2 \mathbf{F}_g^2(2\pi|\mathbf{r}(t)|\,|\mathbf{n}|). \tag{24.21}$$

So, by (24.19) the perfect analog of the "unevenness part" (24.21) goes as follows:

$$\overline{\mathbf{D}^2}_{\mathbf{F}_g,uneven}(\text{Energy};t) = \overline{\mathbf{D}^2}_{\mathbf{F}_g,uneven}(\mathcal{Y};\text{Energy};\mathbf{r};f;t)$$

$$= \frac{1}{4}\left|\mathbf{r}'(t)\right|^4 \sum_{\substack{\mathbf{n}\in\mathbb{Z}^3:\\ \mathbf{n}\neq\mathbf{0}}} |a_{\mathbf{n}}|^2 \left|\sum_{k=1}^{N} m_k e^{2\pi i\mathbf{n}\cdot\mathbf{y}_k}\right|^2 \mathbf{F}_h^2(2\pi|\mathbf{r}(t)|\,|\mathbf{n}|). \tag{24.22}$$

Similarly, the perfect analog of the equilibrium part in Theorem 5.1

$$\text{Eqlbrm} = \text{Eqlbrm}(\mathcal{M};f) = \sum_{\substack{\mathbf{n}\in\mathbb{Z}^3:\\ \mathbf{n}\neq\mathbf{0}}} |a_{\mathbf{n}}|^2 \sum_{k=1}^{N} \mu_k^2 = \sigma_0^2(f)\sum_{k=1}^{N}\mu_k^2 \tag{24.23}$$

is the following sum:

$$\text{Eqlbrm(Energy)} = \text{Eqlbrm(Energy};f) = \sigma_0^2(f)\frac{1}{4}\left|\mathbf{r}'(t)\right|^4 \sum_{k=1}^{N} m_k^2. \tag{24.24}$$

So, repeating the proof of Theorem 5.1 by replacing (24.21) and (24.23) with the corresponding terms (24.22) and (24.24), we obtain the following analog of Theorem 5.1:

$$\text{Variance}_{\omega\in\Omega_g}\overline{\mathbf{D}^2}(\mathcal{Y};\text{Energy};\mathbf{r};\omega;f;t)$$

$$\le 14\text{Eqlbrm(Energy)} \cdot \overline{\mathbf{D}^2}_{\mathbf{F}_g,uneven}(\text{Energy}; t)$$

$$+ 14\,(\text{Eqlbrm(Energy)})^{3/2}\left(\overline{\mathbf{D}^2}_{\mathbf{F}_g,uneven}(\text{Energy}; t)\right)^{1/2}$$

$$+ 14\,(\text{Eqlbrm(Energy)})^2. \tag{24.25}$$

The proofs of our Second Law results are all based on (4.19′) and Theorem 5.1. Since (24.18) is the analog of (4.19′), and (24.25) is the analog of Theorem 5.1, it is straightforward to extend our Second Law results for the case of *energy balance*. The only difference is that in the isotropic Fourier transform we replace $g(u)$ by $h(u) = u^2 g(u)$.

This leads to the question: How does it affect the isotropic Fourier transform to replace $g(u)$ by $h(u) = u^2 g(u)$? To answer this question, first we study some special cases.

Let us start with the first kind (Gaussian distribution)

$$g(u) = \sqrt{\frac{2}{\pi}}u^2 e^{-u^2/2}, \tag{24.26}$$

and then

$$h(u) = \sqrt{\frac{2}{\pi}}u^4 e^{-u^2/2}. \tag{24.26′}$$

Note that in this special case we already know the isotropic Fourier transform $\mathbf{F}_h$. Indeed, in Section 18 we studied the "fifth kind" with speed density function (see (18.8))

$$g_5(u) = \frac{1}{3}\sqrt{\frac{2}{\pi}}u^4 e^{-u^2/2}, \tag{24.27}$$

and succeeded to determine its isotropic Fourier transform $\mathbf{F}_{g_5}$ (see (18.13)):

$$\mathbf{F}_{g_5}(w) = \frac{1}{3}\left(3 - w^2\right)e^{-w^2/2}. \tag{24.27′}$$

Multiplying both (24.27) and (24.27′) by 3, we obtain

$$\mathbf{F}_h(w) = \left(3 - w^2\right)e^{-w^2/2}, \text{ where } h(u) = \sqrt{\frac{2}{\pi}}u^4 e^{-u^2/2}. \tag{24.28}$$

On the other hand, for the first kind we have

$$\mathbf{F}_g(w) = e^{-w^2/2}, \text{ where } g(u) = \sqrt{\frac{2}{\pi}}u^2 e^{-u^2/2}. \tag{24.29}$$

Notice that $\mathbf{F}_h(w)$ and $\mathbf{F}_g(w)$ have the same superexponentially fast decay as $w \to \infty$. As we explained in Section 3 (see (3.13)), the torus model of the first kind reaches square-root equilibrium if and only if

$$e^{4\pi^2\mathcal{R}^2(t)} \geq N \iff \mathcal{R}(t) \geq \frac{\sqrt{\log N}}{2\pi}. \tag{24.30}$$

Here we assume, for simplicity, that every particle has unit mass.

In the case of *energy balance*, comparing (24.28) and (24.29), we obtain that the analog equation of (24.30) is the following:

$$\left(3 - 4\pi^2\mathcal{R}^2(t)\right)^2 e^{4\pi^2\mathcal{R}^2(t)} \geq N \iff \mathcal{R}(t) \geq \frac{\sqrt{\log N}}{2\pi} + o(1). \tag{24.31}$$

Notice that the two thresholds in (24.30) and (24.31)

$$\mathcal{R}(t) \geq \frac{\sqrt{\log N}}{2\pi} \text{ and } \mathcal{R}(t) \geq \frac{\sqrt{\log N}}{2\pi} + o(1)$$

are the *same* apart from a negligible additive constant that tends to zero. It proves, in a quantitative way, that the typical time evolution of the first kind reaches square-root equilibrium in two different ways **exactly the same time** — the two ways are (1) in terms of the *mass balance*, and (2) in terms of the *energy balance*.

Note that the case of the second kind is trivial, since then $g = h = 1$.

We consider next the third kind $g(u) = 3u^2$, $0 \leq u \leq 1$, and zero outside; then $h(u) = 3u^4$, $0 \leq u \leq 1$, and zero outside. We can easily determine the corresponding isotropic Fourier transform $\mathbf{F}_h$ by using integration by parts (similarly to the proof of Lemma 8.1). Indeed, by using the first part (3.35) of Lemma 3.1, we have

$$\mathbf{F}_h(w) = \frac{3}{4\pi}\int_{u=0}^{1} u^4 \left(\int_{\mathbf{v}\in\mathbf{S}^2} e^{i\mathbf{w}\cdot u\mathbf{v}}\, d\mathrm{SA}(\mathbf{v})\right) du = 3\int_{u=0}^{1} u^4 \left(\frac{\sin(uw)}{uw}\right) du$$

$$= \frac{3}{w}\int_{u=0}^{1} u^3 \sin(uw)\, du, \tag{24.32}$$

where $w = |\mathbf{w}|$ and $d\mathrm{SA}(\mathbf{v})$ indicates the surface area. Using integration by parts in the last line of (24.32), we have

$$\int_{u=0}^{1} u^3 \sin(uw)\, du = -\frac{\cos w}{w} + \frac{3\sin w}{w^2} + \frac{6\cos w}{w^3} - \frac{6\sin w}{w^4},$$

and using this fact in (24.32), we obtain

$$\mathbf{F}_h(w) = -\frac{3\cos w}{w^2} + \frac{9\sin w}{w^3} + \frac{18\cos w}{w^4} - \frac{18\sin w}{w^5}. \tag{24.33}$$

On the other hand, for the third kind we have

$$\mathbf{F}_g(w) = -\frac{3\cos w}{w^2} + \frac{3\sin w}{w^3}, \text{ where } g(u) = 3u^2, 0 \le u \le 1 \text{ and zero outside.} \tag{24.34}$$

Notice that $\mathbf{F}_h(w)$ and $\mathbf{F}_g(w)$ have the same polynomial decay as $w \to \infty$, due to the same dominating term $3w^{-2}\cos w$. So, similarly to the case of the first kind, the typical time evolution of the third kind reaches square-root equilibrium in two different ways **exactly the same time**. Again the two ways are (1) in terms of the *mass balance*, and (2) in terms of the *energy balance*.

In general, if the symmetric extension of $g(u)$ over the whole real line is very smooth, then the same holds for $h(u) = u^2 g(u)$, and as we explained in Section 19 (see Lemma 19.1), it follows that the typical time evolution reaches square-root equilibrium very quickly in both ways: (1) in terms of the *mass balance*, and (2) in terms of the *energy balance*.

It does not mean, however, that the typical time evolution always reaches square-root equilibrium exactly the same time in both ways. Consider, for example, the "sixth kind" (see (19.1)) $g = g_6(u) = \frac{1}{2}u^2 e^{-u}$. Then by (19.5)

$$\mathbf{F}_g(w) = \mathbf{F}_{g_6}(w) = \frac{1}{(1+w^2)^2},$$

which implies a polynomial decay of size w^{-4} as $w \to \infty$.

On the other hand, the study of the "seventh kind" (see (19.6)–(19.7)) shows that $\mathbf{F}_h(w)$ (where $h(u) = \frac{1}{2}u^4 e^{-u}$) exhibits a *faster* polynomial decay of size w^{-6} as $w \to \infty$. The faster decay implies that the typical time evolution reaches square-root equilibrium in terms of the energy balance *faster* than in terms of the mass balance. A reasonable explanation for this difference is that the extra factor u^2 in $h(u) = u^2 g(u)$ makes $h(u)$ "smoother" at $u = 0$.

By using the results of Sections 18–19 one can study many more examples.

Next we switch from LA-equilibrium to L-equilibrium.

2. Approach to L-equilibrium and a corresponding Second Law. Our Second Law that we used above was based on the concept of LA-equilibrium (see Definition 2.4). Here we explain how to modify the argument to obtain an analog Second Law based on the concept of L-equilibrium (see Definition 2.3).

Our Second Law described the typical increase of the Micro-Entropy (that we may call an LA-entropy). The key step in the definition of our Micro-Entropy was the definition of the Unevennes Function in (19.12), and the latter was motivated partly by formula (4.19′) about $\overline{\mathbf{D}^2}_{\Omega_g}(\mathcal{Y},\mathcal{M};\mathbf{r};f;t)$, and partly by the idea of taking maximum over the whole space $f \in L^2$ (as we explained it in (12.12)–(12.16) in the special case of the first kind).

The first step toward a Second Law for L-equilibrium is to find the analog of (4.19′) for $\mathbf{D}^2_{\Omega_g}(\mathcal{Y},\mathcal{M};\mathbf{r};f;t)$. This involves a little bit longer calculations. Repeating the argument of (7.9) in the general case, we have (note that (7.9) is about the special case of the first kind, but the argument extends to the general case without any modification; in fact, we just need $\mathcal{R} = \mathbf{r}$)

$$\mathbf{D}^2_{\Omega_g}(\mathcal{Y},\mathcal{M};\mathbf{r};f;t) = \int_{\Omega_g} \mathbf{D}^2(\mathcal{Y},\mathcal{M};\mathbf{r};\omega;f;t)\,d\omega$$

$$= \sum_{k=1}^{N}\mu_k^2 \sum_{\mathbf{n}\in\mathbb{Z}^3\setminus\mathbf{0}} |a_\mathbf{n}|^2 + \sum_{\mathbf{n}\in\mathbb{Z}^3\setminus\mathbf{0}} |a_\mathbf{n}|^2\,\mathbf{F}_g^2(2\pi|\mathbf{r}(t)|\;|\mathbf{n}|)$$

$$\cdot \sum_{1\le k,\ell\le N:k\ne\ell} \mu_k\mu_\ell e^{2\pi \mathrm{i}\mathbf{n}\cdot(\mathbf{y}_k-\mathbf{y}_\ell)} + \sum_{\substack{\mathbf{n}_j\in\mathbb{Z}^3:j=1,2\\ \mathbf{0}\ne\mathbf{n}_1\ne\mathbf{n}_2\ne\mathbf{0}}} a_{\mathbf{n}_1}$$

$$\cdot\,\overline{a_{\mathbf{n}_2}}\,\mathbf{F}_g(2\pi|\mathbf{r}(t)|\;|\mathbf{n}_1|)\mathbf{F}_g(2\pi|\mathbf{r}(t)|\;|\mathbf{n}_2|) \sum_{1\le k,\ell\le N:k\ne\ell} \mu_k\mu_\ell e^{2\pi\mathrm{i}(\mathbf{n}_1\cdot\mathbf{y}_k-\mathbf{n}_2\cdot\mathbf{y}_\ell)}$$

$$+ \sum_{\substack{\mathbf{n}_j\in\mathbb{Z}^3:j=1,2\\ \mathbf{0}\ne\mathbf{n}_1\ne\mathbf{n}_2\ne\mathbf{0}}} a_{\mathbf{n}_1}\overline{a_{\mathbf{n}_2}}\,\mathbf{F}_g(2\pi|\mathbf{r}(t)|\;|\mathbf{n}_1-\mathbf{n}_2|) \sum_{1\le k\le N} \mu_k^2 e^{2\pi\mathrm{i}(\mathbf{n}_1-\mathbf{n}_2)\cdot\mathbf{y}_k}$$

$$= \left(\sum_{\mathbf{n}_1\in\mathbb{Z}^3\setminus\mathbf{0}} a_{\mathbf{n}_1}\mathbf{F}_g(2\pi|\mathbf{r}(t)|\;|\mathbf{n}_1|) \sum_{1\le k\le N} \mu_k e^{2\pi\mathrm{i}\mathbf{n}_1\cdot\mathbf{y}_k}\right)$$

$$\cdot\left(\sum_{\mathbf{n}_2\in\mathbb{Z}^3\setminus\mathbf{0}} \overline{a_{\mathbf{n}_2}}\mathbf{F}_g(2\pi|\mathbf{r}(t)|\;|\mathbf{n}_2|) \sum_{1\le \ell\le N} \mu_\ell e^{-2\pi\mathrm{i}\mathbf{n}_2\cdot\mathbf{y}_\ell}\right)$$

$$+\sum_{k=1}^{N}\mu_k^2 \sum_{\mathbf{n}\in\mathbb{Z}^3\setminus\mathbf{0}} |a_\mathbf{n}|^2\left(1-\mathbf{F}_g^2(2\pi|\mathbf{r}(t)|\;|\mathbf{n}|)\right)$$

$$+\sum_{1\le k\le N}\mu_k^2 \sum_{\substack{\mathbf{n}_j\in\mathbb{Z}^3:j=1,2\\ \mathbf{0}\ne\mathbf{n}_1\ne\mathbf{n}_2\ne\mathbf{0}}} a_{\mathbf{n}_1}\overline{a_{\mathbf{n}_2}}$$

$$\cdot\Big(\mathbf{F}_g(2\pi|\mathbf{r}(t)|\,|\mathbf{n}_1-\mathbf{n}_2|)-\mathbf{F}_g(2\pi|\mathbf{r}(t)|\,|\mathbf{n}_1|)\mathbf{F}_g(2\pi|\mathbf{r}(t)|\,|\mathbf{n}_2|)\Big)e^{2\pi i(\mathbf{n}_1-\mathbf{n}_2)\cdot\mathbf{y}_k}. \tag{24.35}$$

By using the notation introduced in (3.6) and (7.17)

$$\text{Eqlbrm}(\mathcal{M};f)=\sigma_0^2(f)\sum_{k=1}^N\mu_k^2$$

$$=\Bigg(\sum_{\mathbf{n}\in\mathbb{Z}^3\setminus\mathbf{0}}|a_{\mathbf{n}}|^2\Bigg)\Bigg(\sum_{k=1}^N\mu_k^2\Bigg)\quad(\text{“equilibrium part”}) \tag{24.36}$$

and

$$\Phi(\mathcal{Y},\mathcal{M};\mathbf{n})=\sum_{k=1}^N\mu_k e^{2\pi i\mathbf{n}\cdot\mathbf{y}_k}, \tag{24.37}$$

we can rewrite (24.35) in the form

$$\mathbf{D}^2_{\Omega_g}(\mathcal{Y},\mathcal{M};\mathbf{r};f;t)=\text{Eqlbrm}(\mathcal{M};f)$$

$$+\Bigg|\sum_{\mathbf{n}\in\mathbb{Z}^3\setminus\mathbf{0}}a_{\mathbf{n}}\mathbf{F}_g(2\pi|\mathbf{r}(t)|\,|\mathbf{n}|)\Phi(\mathcal{Y},\mathcal{M};\mathbf{n})\Bigg|^2$$

$$-\sum_{k=1}^N\mu_k^2\sum_{\mathbf{n}\in\mathbb{Z}^3\setminus\mathbf{0}}|a_{\mathbf{n}}|^2\,\mathbf{F}_g^2(2\pi|\mathbf{r}(t)|\,|\mathbf{n}|)+\sum_{1\le k\le N}\mu_k^2\sum_{\substack{\mathbf{n}_j\in\mathbb{Z}^3:j=1,2\\ \mathbf{0}\ne\mathbf{n}_1\ne\mathbf{n}_2\ne\mathbf{0}}}a_{\mathbf{n}_1}\overline{a_{\mathbf{n}_2}}$$

$$\cdot\Big(\mathbf{F}_g(2\pi|\mathbf{r}(t)|\,|\mathbf{n}_1-\mathbf{n}_2|)-\mathbf{F}_g(2\pi|\mathbf{r}(t)|\,|\mathbf{n}_1|)\mathbf{F}_g(2\pi|\mathbf{r}(t)|\,|\mathbf{n}_2|)\Big)e^{2\pi i(\mathbf{n}_1-\mathbf{n}_2)\cdot\mathbf{y}_k}. \tag{24.38}$$

To avoid possible divergence in (24.38), we assume that the isotropic Fourier transform satisfies the decay condition

$$|\mathbf{F}_g(w)|\le\frac{C}{w^\alpha}\text{ with }\alpha>3=\text{dimension for all }w\ge 1, \tag{24.39}$$

where α and $C>0$ are some fixed constants.

Under the condition of (24.39), and also assuming $|\mathbf{r}(t)| \geq 1$, we can easily estimate the last sum in (24.38):

$$\left| \sum_{1\leq k\leq N} \mu_k^2 \sum_{\substack{\mathbf{n}_j\in\mathbb{Z}^3:j=1,2\\ \mathbf{0}\neq\mathbf{n}_1\neq\mathbf{n}_2\neq\mathbf{0}}} a_{\mathbf{n}_1}\overline{a_{\mathbf{n}_2}} \right.$$

$$\left. \cdot\Big(\mathbf{F}_g(2\pi|\mathbf{r}(t)|\,|\mathbf{n}_1-\mathbf{n}_2|)-\mathbf{F}_g(2\pi|\mathbf{r}(t)|\,|\mathbf{n}_1|)\mathbf{F}_g(2\pi|\mathbf{r}(t)|\,|\mathbf{n}_2|)\Big)e^{2\pi \mathrm{i}(\mathbf{n}_1-\mathbf{n}_2)\cdot\mathbf{y}_k}\right|$$

$$\leq \left(\sum_{1\leq k\leq N}\mu_k^2\right)\sum_{\substack{\mathbf{n}_j\in\mathbb{Z}^3:j=1,2\\ \mathbf{0}\neq\mathbf{n}_1\neq\mathbf{n}_2\neq\mathbf{0}}} |a_{\mathbf{n}_1}|\,|a_{\mathbf{n}_2}|$$

$$\cdot\big(C(|\mathbf{r}(t)|\,|\mathbf{n}_1-\mathbf{n}_2|)^{-\alpha}+C(|\mathbf{r}(t)|\,|\mathbf{n}_1|)^{-\alpha}C(|\mathbf{r}(t)|\,|\mathbf{n}_2|)^{-\alpha}\big)\,. \tag{24.40}$$

To estimate (24.40), first we focus on the subsum

$$\sum_{\substack{\mathbf{n}_j\in\mathbb{Z}^3:j=1,2\\ \mathbf{0}\neq\mathbf{n}_1\neq\mathbf{n}_2\neq\mathbf{0}}} |a_{\mathbf{n}_1}|\,|a_{\mathbf{n}_2}|\,|\mathbf{n}_1-\mathbf{n}_2|^{-\alpha}, \tag{24.41}$$

and applying the simple inequality

$$|a_{\mathbf{n}_1}a_{\mathbf{n}_2}| \leq \frac{|a_{\mathbf{n}_1}|^2+|a_{\mathbf{n}_2}|^2}{2}$$

in (24.41), we have

$$\sum_{\substack{\mathbf{n}_j\in\mathbb{Z}^3:j=1,2\\ \mathbf{0}\neq\mathbf{n}_1\neq\mathbf{n}_2\neq\mathbf{0}}} |a_{\mathbf{n}_1}|\,|a_{\mathbf{n}_2}|\,|\mathbf{n}_1-\mathbf{n}_2|^{-\alpha}$$

$$\leq \sum_{\mathbf{n}_1\in\mathbb{Z}^3\setminus\mathbf{0}} |a_{\mathbf{n}_1}|^2 \sum_{\substack{\mathbf{n}_2\in\mathbb{Z}^3\setminus\mathbf{0}\\ \mathbf{n}_2\neq\mathbf{n}_1}} |\mathbf{n}_1-\mathbf{n}_2|^{-\alpha}. \tag{24.42}$$

Since $\alpha > 3 =$ dimension,

$$\sum_{\substack{\mathbf{n}_2\in\mathbb{Z}^3\setminus\mathbf{0}\\ \mathbf{n}_2\neq\mathbf{n}_1}} |\mathbf{n}_1-\mathbf{n}_2|^{-\alpha} \leq \sum_{\mathbf{n}\in\mathbb{Z}^3\setminus\mathbf{0}} |\mathbf{n}|^{-\alpha}$$

$$=\sum_{\ell=1}^{\infty}\sum_{\substack{\mathbf{n}\in\mathbb{Z}^3\setminus\mathbf{0}\\ 2^{\ell-1}\leq|\mathbf{n}|<2^{\ell}}} |\mathbf{n}|^{-\alpha} \leq \sum_{\ell=1}^{\infty}\sum_{\substack{\mathbf{n}\in\mathbb{Z}^3\setminus\mathbf{0}\\ 2^{\ell-1}\leq|\mathbf{n}|<2^{\ell}}} 2^{-\alpha(\ell-1)}$$

$$\leq \sum_{\ell=1}^{\infty} 2^{3\ell+3} 2^{-\alpha(\ell-1)} = 2^6 \sum_{\ell=1}^{\infty} 2^{-(\alpha-3)(\ell-1)} = \frac{64}{1-2^{-(\alpha-3)}}, \tag{24.43}$$

where the factor $2^{3\ell+3}$ is an upper bound on the number of lattice points in the ball-minus-ball type set $2^{\ell-1} \leq |\mathbf{n}| < 2^{\ell}$ ("the number of lattice points is close to the volume, which is a cubic function"). By using (24.43) in (24.42), we have

$$\sum_{\substack{\mathbf{n}_j \in \mathbb{Z}^3 : j=1,2 \\ \mathbf{0} \neq \mathbf{n}_1 \neq \mathbf{n}_2 \neq \mathbf{0}}} |a_{\mathbf{n}_1}|\,|a_{\mathbf{n}_2}|\,|\mathbf{n}_1 - \mathbf{n}_2|^{-\alpha} \leq \frac{64}{1-2^{-(\alpha-3)}} \sigma_0^2(f). \tag{24.44}$$

Moreover, by the Cauchy–Schwarz inequality

$$\left(\sum_{\mathbf{n} \in \mathbb{Z}^3 \setminus \mathbf{0}} |a_{\mathbf{n}}|\,|\mathbf{n}|^{-\alpha} \right)^2 \leq \sum_{\mathbf{n} \in \mathbb{Z}^3 \setminus \mathbf{0}} |a_{\mathbf{n}}|^2 \cdot \sum_{\mathbf{n} \in \mathbb{Z}^3 \setminus \mathbf{0}} |\mathbf{n}|^{-2\alpha}$$

$$\leq \sigma_0^2(f) \cdot \frac{64}{1-2^{-(2\alpha-3)}}, \tag{24.45}$$

since the upper bound

$$\sum_{\mathbf{n} \in \mathbb{Z}^3 \setminus \mathbf{0}} |\mathbf{n}|^{-2\alpha} \leq \frac{64}{1-2^{-(2\alpha-3)}}$$

follows from (24.43) replacing α with 2α.

Using (24.44) and (24.45) in (24.40), we have

$$\Bigg| \sum_{1 \leq k \leq N} \mu_k^2 \sum_{\substack{\mathbf{n}_j \in \mathbb{Z}^3 : j=1,2 \\ \mathbf{0} \neq \mathbf{n}_1 \neq \mathbf{n}_2 \neq \mathbf{0}}} a_{\mathbf{n}_1} \overline{a_{\mathbf{n}_2}}$$

$$\cdot \Big(\mathbf{F}_g(2\pi |\mathbf{r}(t)|\,|\mathbf{n}_1 - \mathbf{n}_2|) - \mathbf{F}_g(2\pi |\mathbf{r}(t)|\,|\mathbf{n}_1|) \mathbf{F}_g(2\pi |\mathbf{r}(t)|\,|\mathbf{n}_2|) \Big) e^{2\pi i (\mathbf{n}_1 - \mathbf{n}_2) \cdot \mathbf{y}_k} \Bigg|$$

$$\leq \left(\sum_{1 \leq k \leq N} \mu_k^2 \right) \left(C |\mathbf{r}(t)|^{-\alpha} \frac{64}{1-2^{-(\alpha-3)}} \sigma_0^2(f) + C^2 |\mathbf{r}(t)|^{-2\alpha} \frac{64}{1-2^{-(2\alpha-3)}} \sigma_0^2(f) \right)$$

$$= \sigma_0^2(f) \left(\sum_{1 \leq k \leq N} \mu_k^2 \right) \left(C |\mathbf{r}(t)|^{-\alpha} \frac{64}{1-2^{-(\alpha-3)}} + C^2 |\mathbf{r}(t)|^{-2\alpha} \frac{64}{1-2^{-(2\alpha-3)}} \right). \tag{24.46}$$

Using (24.46) in (24.38), we have

$$\left|\mathbf{D}^2_{\Omega_g}(\mathcal{Y},\mathcal{M};\mathbf{r};f;t)-\text{Eqlbrm}(\mathcal{M};f)-\left|\sum_{\mathbf{n}\in\mathbb{Z}^3\setminus\mathbf{0}} a_{\mathbf{n}}\mathbf{F}_g(2\pi|\mathbf{r}(t)|\,|\mathbf{n}|)\Phi(\mathcal{Y},\mathcal{M};\mathbf{n})\right|^2\right|$$

$$\leq \sum_{k=1}^{N}\mu_k^2 \sum_{\mathbf{n}\in\mathbb{Z}^3\setminus\mathbf{0}} |a_{\mathbf{n}}|^2\,\mathbf{F}_g^2(2\pi|\mathbf{r}(t)|\,|\mathbf{n}|)$$

$$+\sigma_0^2(f)\left(\sum_{1\leq k\leq N}\mu_k^2\right)\left(C|\mathbf{r}(t)|^{-\alpha}\frac{64}{1-2^{-(\alpha-3)}}+C^2|\mathbf{r}(t)|^{-2\alpha}\frac{64}{1-2^{-(2\alpha-3)}}\right)$$

$$= o\left(\sigma_0^2(f)\sum_{1\leq k\leq N}\mu_k^2\right) \quad \text{as} \quad |\mathbf{r}(t)|\to\infty. \tag{24.47}$$

We prefer to rewrite (24.47) in the form

$$\mathbf{D}^2_{\Omega_g}(\mathcal{Y},\mathcal{M};\mathbf{r};f;t)=\left|\sum_{\mathbf{n}\in\mathbb{Z}^3\setminus\mathbf{0}} a_{\mathbf{n}}\mathbf{F}_g(2\pi|\mathbf{r}(t)|\,|\mathbf{n}|)\Phi(\mathcal{Y},\mathcal{M};\mathbf{n})\right|^2$$

$$+\,\text{Eqlbrm}(\mathcal{M};f)+\text{negligible}$$

as long as $|\mathbf{r}(t)|$ exceeds a sufficiently large absolute constant. (24.48)

This is an analog of (4.19′), which has the following equivalent form:

$$\overline{\mathbf{D}^2}_{\Omega_g}(\mathcal{Y},\mathcal{M};\mathbf{r};f;t)$$

$$=\sum_{\substack{\mathbf{n}\in\mathbb{Z}^3:\\ \mathbf{n}\neq\mathbf{0}}} |a_{\mathbf{n}}|^2\,|\Phi(\mathcal{Y},\mathcal{M};\mathbf{n})|^2\,\mathbf{F}_g^2(2\pi|\mathbf{r}(t)|\,|\mathbf{n}|)$$

$$+\,\text{Eqlbrm}(\mathcal{M};f)+\text{negligible}$$

as long as $|\mathbf{r}(t)|$ exceeds a sufficiently large absolute constant.

The next step from (24.48) to the corresponding Unevenness Function is to take maximum over the whole space $f\in L^2$. We repeat the argument of (12.12)–(12.16). The only change is an application of the Cauchy–Schwarz inequality, including the extra fact that the inequality becomes an equality under the "constant ratio" condition:

$$\sup_{\tau\geq t}\ \sup_{f\in L^2:\ \sigma_0^2(f)>0}\frac{1}{\sigma_0^2(f)}\left|\sum_{\mathbf{n}\in\mathbb{Z}^3\setminus\mathbf{0}}a_{\mathbf{n}}\mathbf{F}_g(2\pi|\mathbf{r}(\tau)|\ |\mathbf{n}|)\Phi(\mathcal{Y},\mathcal{M};\mathbf{n})\right|^2$$

$$=\sup_{\tau\geq t}\ \sup_{f\in L^2:\ \sigma_0^2(f)>0}\frac{1}{\sigma_0^2(f)}\left(\sum_{\mathbf{n}\in\mathbb{Z}^3\setminus\mathbf{0}}|a_{\mathbf{n}}|^2\right)$$

$$\left(\sum_{\mathbf{n}\in\mathbb{Z}^3\setminus\mathbf{0}}\mathbf{F}_g^2(2\pi|\mathbf{r}(\tau)|\ |\mathbf{n}|)\,|\Phi(\mathcal{Y},\mathcal{M};\mathbf{n})|^2\right)$$

$$=\sup_{\tau\geq t}\ \sup_{f\in L^2:\ \sigma_0^2(f)>0}\frac{1}{\sigma_0^2(f)}\sigma_0^2(f)\left(\sum_{\mathbf{n}\in\mathbb{Z}^3\setminus\mathbf{0}}\mathbf{F}_g^2(2\pi|\mathbf{r}(\tau)|\ |\mathbf{n}|)\,|\Phi(\mathcal{Y},\mathcal{M};\mathbf{n})|^2\right)$$

$$=\sup_{\tau\geq t}\sum_{\mathbf{n}\in\mathbb{Z}^3\setminus\mathbf{0}}\mathbf{F}_g^2(2\pi|\mathbf{r}(\tau)|\ |\mathbf{n}|)\left|\sum_{k=1}^{N}\mu_k e^{2\pi i\mathbf{n}\cdot\mathbf{y}_k}\right|^2, \tag{24.49}$$

where in the last step we used (24.37). The (absolute) convergence of (24.49) is guaranteed by the decay condition (24.39), assuming $t\geq 1$ (note that (24.49) is divergent for $t=0$).

By using (24.48) and (24.49) we can define the corresponding Unevenness Function $\text{UnevenFun}_{g,L}(\mathcal{Y},\mathcal{M};\mathbf{r};t)$ as an analog of $\text{UnevenFun}_g(\mathcal{Y},\mathcal{M};\mathbf{r};t)$, where the extra index L indicates that the underlying equilibrium concept is L-equilibrium (see Definition 2.3) instead of LA-equilibrium (see Definition 2.4):

$$\text{UnevenFun}_{g,L}(\mathcal{Y},\mathcal{M};\mathbf{r};t)$$

$$=\max\left\{\sup_{\tau\geq t}\sum_{\mathbf{n}\in\mathbb{Z}^3\setminus\mathbf{0}}\mathbf{F}_g^2(2\pi|\mathbf{r}(\tau)|\ |\mathbf{n}|)\left|\sum_{k=1}^{N}\mu_k e^{2\pi i\mathbf{n}\cdot\mathbf{y}_k}\right|^2,\sum_{k=1}^{N}\mu_k^2\right\}. \tag{24.50}$$

Similarly to (19.12), the right-hand side of (24.50) describes the typical time evolution of the system toward L-equilibrium for the majority of $\omega\in\Omega_g$ (the proof is similar).

It follows from (24.50) that the corresponding torus model $(\mathcal{Y},\mathcal{M};\mathbf{r};\Omega_g)$ reaches L-equilibrium at time $t=t_0$ when

$$\sup_{\tau\geq t_0}\sum_{\mathbf{n}\in\mathbb{Z}^3\setminus\mathbf{0}}\mathbf{F}_g^2(2\pi|\mathbf{r}(\tau)|\ |\mathbf{n}|)\left|\sum_{k=1}^N\mu_k e^{2\pi i\mathbf{n}\cdot\mathbf{y}_k}\right|^2=\sum_{k=1}^N\mu_k^2. \tag{24.51}$$

The decay condition (24.39) guarantees the existence of t_0.

For illustration, consider the class of probability density functions g satisfying the decay condition (19.11) with $k\geq 4$ (which clearly implies (24.39)).

Using (19.11) in (24.51), we have

$$\text{UnevenFun}_{g,L}(\mathcal{Y},\mathcal{M};\mathbf{r};t)$$

$$=\max\left\{\sum_{\mathbf{n}\in\mathbb{Z}^3\setminus\mathbf{0}}\left|\sum_{k=1}^N\mu_k e^{2\pi i\mathbf{n}\cdot\mathbf{y}_k}\right|^2\min\left\{\sup_{\tau\geq t}\frac{C^2}{(2\pi|\mathbf{r}(\tau)|\ |\mathbf{n}|)^{2k}},1\right\},\sum_{k=1}^N\mu_k^2\right\}$$

$$+\text{negligible}. \tag{24.52}$$

It follows from (24.52) that the corresponding torus model reaches L-equilibrium at time t_0 when

$$2\pi\inf_{\tau\geq t_0}|\mathbf{r}(\tau)|=\left(\left(\sum_{\mathbf{n}\in\mathbb{Z}^3\setminus\mathbf{0}}\frac{1}{|\mathbf{n}|^{2k}}\frac{C^2\left|\sum_{k=1}^N\mu_k e^{2\pi i\mathbf{n}\cdot\mathbf{y}_k}\right|^2}{\sum_{k=1}^N\mu_k^2}\right)^{\frac{1}{2k}}\ \text{or}\ 1\right)_{\max}. \tag{24.53}$$

We refer to the time t_0 and the euclidean distance $\inf_{\tau\geq t_0}|\mathbf{r}(\tau)|$ defined by the equality (24.53) as the L-Relaxation Time and the L-Relaxation Drifting Distance, respectively, of the system $(\mathcal{Y},\mathcal{M};\mathbf{r};\Omega_g)$.

Note that in the interesting special case $|\mathbf{r}(t)|=t$ (which includes, among others, the Bernoulli billiard/torus model)

$$\left(\left(\sum_{\mathbf{n}\in\mathbb{Z}^3\setminus\mathbf{0}}\frac{1}{|\mathbf{n}|^{2k}}\frac{C^2\left|\sum_{k=1}^N\mu_k e^{2\pi i\mathbf{n}\cdot\mathbf{y}_k}\right|^2}{\sum_{k=1}^N\mu_k^2}\right)^{\frac{1}{2k}}\ \text{or}\ 1\right)_{\max}$$

represents the true order of magnitude of the L-Relaxation Time (i.e., the time to reach L-equilibrium) for the typical time evolution of the system $(\mathcal{Y},\mathcal{M};\mathbf{r};\Omega_g)$.

In (24.53) we basically used the function $\left(C^2x\right)^{1/2k}$. To define the L-Imbalance, we apply the inverse function $C^{-2}x^{2k}$ in (24.53): we have

$$\text{Imbalance}_{g,L}(\mathcal{Y},\mathcal{M})=\left(\left(\left(\sum_{\mathbf{n}\in\mathbb{Z}^3\setminus\mathbf{0}}\frac{1}{|\mathbf{n}|^{2k}}\frac{\left|\sum_{k=1}^{N}\mu_k e^{2\pi i\mathbf{n}\cdot\mathbf{y}_k}\right|^2}{\sum_{k=1}^{N}\mu_k^2}\right)^{\frac{1}{2k}}\right)^{2k}\text{ or }1\right)_{\max}$$

$$=\left(\left(\sum_{\mathbf{n}\in\mathbb{Z}^3\setminus\mathbf{0}}\frac{1}{|\mathbf{n}|^{2k}}\frac{\left|\sum_{k=1}^{N}\mu_k e^{2\pi i\mathbf{n}\cdot\mathbf{y}_k}\right|^2}{\sum_{k=1}^{N}\mu_k^2}\right)\text{ or }1\right)_{\max}. \tag{24.54}$$

We may say, therefore, that $\text{Imbalance}_{g,L}(\mathcal{Y},\mathcal{M})$ represents the $2k$-th power of the L-Relaxation Drifting Distance for the typical time evolution of $(\mathcal{Y},\mathcal{M};\mathbf{r};\Omega_g)$.

The corresponding entropy has the usual complementary relation with the Imbalance:

$$\text{MicroEntropy}_{g,L}(\mathcal{Y},\mathcal{M})=\frac{N^2}{\sum_{k=1}^{N}\mu_k^2}-\text{Imbalance}_{g,L}(\mathcal{Y},\mathcal{M}). \tag{24.55}$$

We are now ready to prove an L-analog of Theorem 19.1 for the whole class (19.11) of density functions g.

Again we work with the usual notation

$$\mathcal{Y}_{\mathbf{r};\omega;t}=\{\mathbf{z}_k=\mathbf{y}_k+\vartheta_k(\rho_k\mathbf{r}(t))\text{ modulo one}:\ 1\le k\le N\},$$

where

$$\omega=(\rho_1,\vartheta_1,\rho_2,\vartheta_2,\ldots,\rho_N,\vartheta_N)\in\Omega_g=([0,\infty)\times SO(3))^N$$

with

$$\Pr[\rho_k\le u]=\int_0^u g(y)\,dy.$$

Because of the complementary relation between our L-entropy and the L-Imbalance, it suffices to study

$$\text{Imba}_{g,L}(\mathcal{Y}_{\mathbf{r};\omega;t},\mathcal{M})=\text{Imbalance}_{g,L}(\mathcal{Y}_{\mathbf{r};\omega;t},\mathcal{M}).$$

Again we want to find the relation between $\text{Imba}_{g,L}(\mathcal{Y}_{\mathbf{r};\omega;t},\mathcal{M})$ for the majority of $\omega\in\Omega_g$ and a corresponding ω-independent function. This ω-independent function, called L-Time Evolution Function, is defined as follows (t is of course the time variable)

$$\text{TiEvFun}_{g,L}(t)=\text{TiEvFun}_{g,L}(\mathcal{Y},\mathcal{M};\mathbf{r};t)$$

$$= \left(\left(\sum_{\mathbf{n}\in\mathbb{Z}^3\setminus\mathbf{0}} \frac{1}{|\mathbf{n}|^{2k}} \frac{\left|\sum_{k=1}^N \mu_k e^{2\pi i \mathbf{n}\cdot\mathbf{y}_k}\right|^2 \mathbf{F}_g^2\left(2\pi|\mathbf{r}(t)|\ |\mathbf{n}|\right)}{\sum_{k=1}^N \mu_k^2} \right) \text{ or } 1 \right)_{\max} . \tag{24.56}$$

Notice that the L-Time Evolution Function (24.56) is an analog of the (ordinary) Time Evolution Function (19.17); the only change is that $\max_{\mathbf{n}\in\mathbb{Z}^3\setminus\mathbf{0}}$ is replaced by $\sum_{\mathbf{n}\in\mathbb{Z}^3\setminus\mathbf{0}}$. The same applies for the L-Imbalance and the L-entropy.

So far we assumed that $k \geq 4$ in (19.11), but the series

$$\sum_{\mathbf{n}\in\mathbb{Z}^3\setminus\mathbf{0}} \frac{1}{|\mathbf{n}|^{2k}}$$

in (24.54) and (24.56) are convergent for $k \geq 2$. So, it is meaningful to define the L-Imbalance, the L-entropy and the L-Time Evolution Function for $k \geq 2$.

By repeating the proof of Theorem 19.1 — with the necessary minor changes — we obtain an analog of Theorem 19.1. This analog is a Second Law for the class (19.11) describing the typical time evolution to L-equilibrium, and goes as follows.

Theorem 24.1. *Let $g(u)$, $0 \leq u < \infty$ be a probability density function such that the isotropic Fourier transform $\mathbf{F}_g$ satisfies (19.11) with $k \geq 2$. Consider the corresponding 3-dimensional isotropic torus model $(\mathcal{Y}, \mathcal{M}; \mathbf{r}; \Omega_g)$. Given a time $t > 0$ and $0 < \varepsilon < 10^{-2}$, there exists a measurable subset $\Omega_g^* = \Omega_g^*(\varepsilon; t) \subset \Omega_g$ of the corresponding initial velocity space such that*

$$\mathrm{ProdMeas}_g(\Omega_g^*) > 1 - 10^2\varepsilon$$

and

$$\left|\mathrm{Imbalance}_{g,L}(\mathcal{Y}_{\mathbf{r};\omega;t}, \mathcal{M}) - \mathrm{TiEvFun}_{g,L}(\mathcal{Y}, \mathcal{M}; \mathbf{r}; t)\right|$$

$$\leq \frac{4}{\sqrt{\varepsilon}\sqrt{\mathrm{TiEvFun}_{g,L}(t)}} \left(1 + \frac{2}{\sqrt{\mathrm{TiEvFun}_{g,L}(t)}}\right) \cdot \mathrm{TiEvFun}_{g,L}(t)$$

holds for all $\omega \in \Omega_g^*$. □

Note that the recipe "replace $\max_{\mathbf{n}\in\mathbb{Z}^3\setminus\mathbf{0}}$ by $\sum_{\mathbf{n}\in\mathbb{Z}^3\setminus\mathbf{0}}$" works far beyond the class of probability density functions g defined in (19.11).

For example, consider our basic example: the first kind. Then the L-Imbalance is

$$\text{Imbalance}_{1,L}(\mathcal{Y},\mathcal{M}) = \left(\sum_{\mathbf{n}\in\mathbb{Z}^3\setminus\mathbf{0}} \frac{\left|\sum_{k=1}^{N}\mu_k e^{2\pi i\mathbf{n}\cdot\mathbf{y}_k}\right|^2}{\sum_{k=1}^{N}\mu_k^2} \text{ or } 1 \right)_{\max}^{1/|\mathbf{n}|^2}, \tag{24.57}$$

and the L-Time Evolution Function is

$$\text{TiEvFun}_{1,L}(\mathcal{Y},\mathcal{M};\mathbf{r};t)$$

$$= \left(e^{-(2\pi|\mathbf{r}(t)|)^2} \sum_{\mathbf{n}\in\mathbb{Z}^3\setminus\mathbf{0}} \left(\frac{\left|\sum_{k=1}^{N}\mu_k e^{2\pi i\mathbf{n}\cdot\mathbf{y}_k}\right|^2}{\sum_{k=1}^{N}\mu_k^2} \right)^{1/|\mathbf{n}|^2} \text{ or } 1 \right)_{\max}. \tag{24.58}$$

As always, the corresponding L-entropy comes from the complementary relation (24.55).

Using (24.57) and (24.58) instead of the corresponding LA-versions, we obtain a perfect analog of Theorem 13.1 (the same proof works).

Similarly, consider the third kind. Then the L-Imbalance is

$$\text{Imbalance}_{3,L}(\mathcal{Y},\mathcal{M}) = \left(\sum_{\mathbf{n}\in\mathbb{Z}^3\setminus\mathbf{0}} \frac{1}{|\mathbf{n}|^4} \frac{\left|\sum_{k=1}^{N}\mu_k e^{2\pi i\mathbf{n}\cdot\mathbf{y}_k}\right|^2}{\sum_{k=1}^{N}\mu_k^2} \text{ or } 1 \right)_{\max}, \tag{24.59}$$

and the L-Time Evolution Function is

$$\text{TiEvFun}_{3,L}(\mathcal{Y},\mathcal{M};\mathbf{r};t) = \left(\frac{9}{\sum_{k=1}^{N}\mu_k^2} \sum_{\mathbf{n}\in\mathbb{Z}^3\setminus\mathbf{0}} \left|\sum_{k=1}^{N}\mu_k e^{2\pi i\mathbf{n}\cdot\mathbf{y}_k}\right|^2 \right.$$

$$\left. \left(\frac{\cos(2\pi|\mathbf{r}(t)|\ |\mathbf{n}|)}{4\pi^2|\mathbf{r}(t)|^2|\mathbf{n}|^4} - \frac{\sin(2\pi|\mathbf{r}(t)|\ |\mathbf{n}|)}{8\pi^3|\mathbf{r}(t)|^3|\mathbf{n}|^5} \right)^2 \text{ or } 1 \right)_{\max}. \tag{24.60}$$

Using (24.59) and (24.60) instead of the corresponding LA-versions, we obtain a perfect analog of Theorem 13.2 (the same proof works).

We skipped the second kind, because then a formal application of the recipe "replace $\max_{\mathbf{n}\in\mathbb{Z}^3\setminus\mathbf{0}}$ by $\sum_{\mathbf{n}\in\mathbb{Z}^3\setminus\mathbf{0}}$" leads to divergence. Indeed, then the recipe gives the L-Imbalance

$$\text{Imbalance}_{2,L}(\mathcal{Y},\mathcal{M}) = \left(\sum_{\mathbf{n}\in\mathbb{Z}^3\setminus\mathbf{0}} \frac{1}{|\mathbf{n}|^2} \frac{\left|\sum_{k=1}^N \mu_k e^{2\pi i \mathbf{n}\cdot\mathbf{y}_k}\right|^2}{\sum_{k=1}^N \mu_k^2} \quad \text{or} \quad 1 \right)_{\max} . \tag{24.61}$$

Unfortunately we cannot guarantee the convergence of (24.61) due to the fact

$$\sum_{\mathbf{n}\in\mathbb{Z}^3\setminus\mathbf{0}} \frac{1}{|\mathbf{n}|^2} = \infty.$$

Chapter 5

Long-Term Stability of Equilibrium

25. Simultaneous box equilibrium

What is in Chapter 5? *Here we give a time-quantitative treatment of the long-term stability of equilibrium (in our implicit interaction models).*

1. Equilibrium for all nice sets. Our equilibrium concepts, LA-equilibrium and L-equilibrium, are two forms of "square-root equilibrium". (More precisely, L-equilibrium is a "square-root equilibrium" and LA-equilibrium is an "average square-root equilibrium".)

Definitions 2.3 and 2.4 have the following common message: once the system reaches equilibrium at time t, it stays in equilibrium forever ("for every time τ with $\tau \geq t$"). This characterizes the *collective* behavior of the system. But this message — once the system reaches equilibrium, it stays in equilibrium forever — is certainly false for the *individual* behavior of the system, i.e., for the time evolution of a typical trajectory of the system.

Why does it fail? (A physicist would probably mention the well-known Zermelo–Poincare paradox, which is a good intuition, but it is formally incorrect, since the initial condition is a zero measure set, so Poincare's recurrence theorem does not apply.) To give a rigorous explanation, we consider the simplest Bernoulli model, and apply Lemma 1.2 (instead of Poincare's recurrence theorem). Lemma 1.2 tells us that, given an arbitrary initial configuration of starting points, if the initial velocity coordinates are linearly independent over the rationals (representing "typical" directions), then the time evolution of this individual trajectory of the system violates square-root equilibrium in the worst possible ways infinitely many times. In this section we focus on the following natural question: after reaching square-root equilibrium, *how long* does the time evolution of a typical individual trajectory (of the system) stay in square-root equilibrium?

We start with the simplest model: N point billiards in the unit cube (see (1.10)–(1.13)), the initial velocity distribution is Gaussian (see (1.9′)), and as usual,

$$\mathcal{Y} = \{\mathbf{y}_1, \mathbf{y}_2, \ldots, \mathbf{y}_N\}$$

denotes an arbitrary but fixed initial configuration.

Of course we cannot expect that, starting from $\mathcal{Y}$, and after reaching equilibrium, the typical time evolution of the system stays in L-equilibrium with respect to *all measurable* test sets simultaneously. (Indeed, this follows from a trivial measure-theoretic argument: the union of the N trajectories of the N particles has volume zero, so the system does not visit the complement of the union set at all, despite the fact that the complement has measure one, so the majority of the particles should be in the complement set in the majority of the time.)

What we may expect, however, is that, starting from $\mathcal{Y}$, and after reaching equilibrium, the typical time evolution of the system stays in L-equilibrium with respect to *all nice* test sets simultaneously for a very, very long time. And indeed, we are going to prove such a result.

Of course first we have to clarify what *nice test sets* mean. Aligned boxes (= Cartesian products of intervals) are the simplest test sets, so they are certainly *nice*. Since we study isotropic (= rotation-invariant) velocity distributions, it is natural to consider a rotation-invariant family of test sets. We consider, therefore, the family of all rotated (= aligned and tilted) boxes *contained* in the 3-dimensional unit torus $[0,1)^3$ in the sense that the diameter of a box is $\le \sqrt{3}$. The set of rotated boxes is a 9-parameter family in the following sense: we pick two vertices, P_1 and P_2 (each has 3 coordinates), pick a perpendicular direction to the P_1P_2-line (determined by an angle between 0 and 2π), pick a length (between 0 and $\sqrt{3}$) to determine a third vertex P_3, and pick another length (between 0 and $\sqrt{3}$) to determine a fourth vertex P_4 (P_1P_2, P_1P_3 and P_1P_4 are three pairwise orthogonal directions). Notice that the four vertices P_1, P_2, P_3, P_4 determine the rotated box, and altogether we picked $3+3+1+1+1 = 9$ real numbers (each one is between 0 and 2π).

Let $\varepsilon > 0$ be a small number to be specified later; we want to define some kind of ε-approximations of a given rotated box. We start with the case of aligned boxes $B = [a_1, b_1] \times [a_2, b_2] \times [a_3, b_3]$: for $j = -2, -1, 1, 2$ let

$$B(j\varepsilon) = [a_1 - j\varepsilon, b_1 + j\varepsilon] \times [a_2 - j\varepsilon, b_2 + j\varepsilon] \times [a_3 - j\varepsilon, b_3 + j\varepsilon].$$

Clearly

$$B(-2\varepsilon) \subset B(-\varepsilon) \subset B \subset B(\varepsilon) \subset B(2\varepsilon). \tag{25.1}$$

Next comes the case of tilted boxes. Every tilted box has the form ϑB where $\vartheta \in SO(3)$ is a rotation of the 3-space and B is an aligned box. The straightforward analog of (25.1) is the rotated version

$$\vartheta B(-2\varepsilon) \subset \vartheta B(-\varepsilon) \subset \vartheta B \subset \vartheta B(\varepsilon) \subset \vartheta B(2\varepsilon), \tag{25.2}$$

representing the ε-approximations of the rotated box ϑB. For convenience we assume that $\vartheta B(2\varepsilon)$ is a subset of the unit torus $[0,1)^3$ and $\vartheta B(-2\varepsilon)$ is nonempty; if this is not the case then we just need a trivial modification of the argument below.

Let $K \geq 100/\varepsilon$ be an integer. It is easy to see that, given any rotated box $A = \vartheta B$, there exist two (rotated) boxes $A_{K,-\varepsilon}$ and $A_{K,\varepsilon}$ such that

$$\vartheta B(-2\varepsilon) \subset A_{K,-\varepsilon} \subset \vartheta B(-\varepsilon), \tag{25.3}$$

$$\vartheta B(\varepsilon) \subset A_{K,\varepsilon} \subset \vartheta B(2\varepsilon), \tag{25.4}$$

and the coordinates of the vertices of the boxes $A_{K,-\varepsilon}$ and $A_{K,\varepsilon}$ are all rational numbers with denominator K. Let $A^c_{K,\varepsilon}$ denote the complement of the set $A_{K,\varepsilon}$ with respect to the unit torus $[0,1)^3$. We call the sets $A_{K,\varepsilon}$ and $A^c_{K,\varepsilon}$ K-special. The total number of K-special sets is clearly less than $(2\pi K)^9$ (since K is a common denominator, and the set of rotated boxes is a 9-parameter family).

We need to use the point-counting function $N_S(\mathcal{Y};\omega;t)$: given a concrete initial velocity distribution $\omega \in \Omega_1$ (= Gaussian case), a measurable subset $S \subset [0,1)^3$ and a time $t > 0$, let $N_S(\mathcal{Y};\omega;t)$ denote the number of point billiards in the set S at time t, assuming the system started at $t = 0$ from the initial configuration $\mathcal{Y}$ with initial velocities ω.

We recall that $\Omega_1 = \Omega_1(\textit{Bernoulli})$ (see the "Gaussian case" after (1.9′)), and the corresponding probability measure is denoted by ProdMeas_1.

We study the consequences of the following (at first sight artificially complicated)

Hypothesis A: For every member t_i, $0 \leq i \leq L$ of the arithmetic progression of time points

$$t_0,\ t_1 = t_0 + \gamma,\ t_2 = t_0 + 2\gamma,\ t_3 = t_0 + 3\gamma, \ldots,\ t_L = t_0 + L\gamma,$$

and for every K-special set $S = A_{K,-\varepsilon}$ or $A^c_{K,\varepsilon}$, where ε is some fixed real number in the interval $0 < \varepsilon \leq \frac{1}{24}N^{-1/2}$ and $K \geq 100/\varepsilon$ is a fixed integer,

$$\text{ProdMeas}_1\left(|N_S(\mathcal{Y};\omega;t_i) - \text{vol}(S)N| \geq 29\sqrt{N}\right) < \frac{\eta}{2(L+1)\cdot(2\pi K)^9}, \tag{25.5}$$

and also,

$$\text{ProdMeas}_1\left(\text{there are more than } \sqrt{N}/2 \text{ fast point billiards}\right) < \eta/2, \tag{25.6}$$

where a point billiard is called *fast* if its speed is $\geq \varepsilon/\gamma$.

If a point billiard is not fast, we call it *slow*. The crucial fact here is that, in a time interval of length γ, a slow point-billiard *cannot* cover a distance ε.

We show that Hypothesis A has the following interesting corollary.

Corollary of Hypothesis A: Under the condition of Hypothesis A, for every time t in the interval $t_0 \leq t \leq t_L = t_0 + L\gamma$, and for every rotated box ϑB contained in the unit torus $[0,1)^3$,

$$|N_{\vartheta B}(\mathcal{Y};\omega;t) - \text{vol}(B)N| < 30\sqrt{N} \tag{25.7}$$

holds for at least $1-\eta$ part of the initial velocities $\omega \in \Omega_1$.

Remark. If $\eta > 0$ is very small and $L\gamma$ is very, very large, then the Corollary says exactly what we want: starting from $\mathcal{Y}$, for $t \geq t_0$ the typical time evolution of the system stays in L-equilibrium (with constant factor $C = 30$) with respect to all "nice" test sets (= rotated boxes) for a very, very long time.

Deduction of the Corollary. Let

$$\Omega_1(bad;1) = \Big\{\omega \in \Omega_1 : \text{ there exist a time } t_i = t_0 + i\gamma \ (0 \leq i \leq L \text{ integer}) \\ \text{and a } K\text{–special set } S = A_{K,-\varepsilon} \text{ or } A^c_{K,\varepsilon} \\ \text{with } |N_S(\mathcal{Y};\omega;t_i) - \text{vol}(S)N| \geq 29\sqrt{N}\Big\}, \tag{25.8}$$

and let

$$\Omega_1(bad;2) = \Big\{\omega \in \Omega_1 : \text{ there are more than } \sqrt{N}/2 \text{ fast point billiards}\Big\}. \tag{25.9}$$

By (25.8)–(25.9) and Hypothesis A,

$$\text{ProdMeas}_1\left(\Omega_1(bad;1) \cup \Omega_1(bad;2)\right) \\ \leq (L+1)\cdot(2\pi K)^9 \cdot \frac{\eta}{2(L+1)\cdot(2\pi K)^9} + \frac{\eta}{2} = \eta. \tag{25.10}$$

Let

$$\omega \in \Omega_1 \setminus (\Omega_1(bad;1) \cup \Omega_1(bad;2)), \tag{25.11}$$

let t be a time in the interval $t_0 \le t \le t_L = t_0 + L\gamma$, and finally let ϑB be a rotated box contained in the unit torus $[0,1)^3$.

Let $t_i = t_0 + i\gamma \le t < t_{i+1} = t_0 + (i+1)\gamma$ be the two "neighbors" of t in the arithmetic progression in Hypothesis A.

Let $A_{K,-\varepsilon}$ and $A_{K,\varepsilon}$ be the two boxes (defined in (25.3)–(25.4)) that correspond to ϑB:

$$\vartheta B(-2\varepsilon) \subset A_{K,-\varepsilon} \subset \vartheta B(-\varepsilon) \subset \vartheta B \subset \vartheta B(\varepsilon) \subset A_{K,\varepsilon} \subset \vartheta B(2\varepsilon), \tag{25.12}$$

and the coordinates of the vertices of the boxes $A_{K,-\varepsilon}$ and $A_{K,\varepsilon}$ are all rational numbers with denominator K.

By Hypothesis A, for both K-special sets $S = A_{K,-\varepsilon}$ and $A^c_{K,\varepsilon}$, we have

$$|N_S(\mathcal{Y};\omega;t_i) - \text{vol}(S)N| < 29\sqrt{N}. \tag{25.13}$$

We recall the crucial fact that, in a time interval of length γ, a slow point-billiard *cannot* cover a distance ε. Moreover, by (25.9) and (25.11), there are at most $\sqrt{N}/2$ fast point billiards. Combining these facts with (25.12), we conclude that the point-billiards that are in $A_{K,-\varepsilon}$ at time t_i are still in the larger set ϑB at time t, except possibly some fast billiards. Similarly, the point-billiards that are in $A^c_{K,\varepsilon}$ at time t_i still cannot reach the set ϑB at time t, except possibly some fast billiards. Formally,

$$N_{A_{K,-\varepsilon}}(\mathcal{Y};\omega;t_i) - \text{Fast}_1 \le N_{\vartheta B}(\mathcal{Y};\omega;t) \le N - N_{A^c_{K,\varepsilon}}(\mathcal{Y};\omega;t_i) + \text{Fast}_2, \tag{25.14}$$

where $\text{Fast}_1 \ge 0$ and $\text{Fast}_2 \ge 0$ with

$$\text{Fast}_1 + \text{Fast}_2 \le \sqrt{N}/2. \tag{25.15}$$

Moreover, we use the trivial inequalities that follow from (25.12):

$$\text{vol}(B) \ge \text{vol}(A_{K,-\varepsilon}) \ge \text{vol}(B) - 6\cdot 2\varepsilon, \tag{25.16}$$

and

$$1 - \text{vol}(B) \ge \text{vol}(A^c_{K,\varepsilon}) \ge 1 - \text{vol}(B) - 6\cdot 2\varepsilon, \tag{25.17}$$

where the factor 6 in (25.16)–(25.17) comes from the fact that a box has 6 faces.

Combining (25.13)–(25.16),

$$N_{\vartheta B}(\mathcal{Y};\omega;t) - \text{vol}(B)N \ge N_{A_{K,-\varepsilon}}(\mathcal{Y};\omega;t_i) - \text{vol}(B)N - \text{Fast}_1$$

$$= \left(N_{A_{K,-\varepsilon}}(\mathcal{Y};\omega;t_i) - \text{vol}(A_{K,-\varepsilon})N\right) - N\left(\text{vol}(A_{K,-\varepsilon}) - \text{vol}(B)\right) - \text{Fast}_1$$

$$\geq -29\sqrt{N} - 12\varepsilon N - \sqrt{N}/2 \geq -30\sqrt{N}, \tag{25.18}$$

since by Hypothesis A, $0 < \varepsilon \leq \frac{1}{24}N^{-1/2}$.

Similarly, by (25.13)–(25.15) and (25.17),

$$N_{\vartheta B}(\mathcal{Y};\omega;t) - \text{vol}(B)N \leq N - N_{A^c_{K,\varepsilon}}(\mathcal{Y};\omega;t_i) - \text{vol}(B)N + \text{Fast}_2$$

$$= -\left(N_{A^c_{K,\varepsilon}}(\mathcal{Y};\omega;t_i) - \text{vol}(A^c_{K,\varepsilon})N\right) + N\left(1 - \text{vol}(B) - \text{vol}(A^c_{K,\varepsilon})\right) + \text{Fast}_2$$

$$\leq 29\sqrt{N} + 12\varepsilon N + \sqrt{N}/2 \leq 30\sqrt{N}. \tag{25.19}$$

Inequalities (25.18)–(25.19) combined with (25.10)–(25.11) complete the deduction of the Corollary from Hypothesis A. □

2. An intuitive proof of Hypothesis A. Assume that, starting from the initial configuration $\mathcal{Y}$ at $t = 0$, the point billiard system becomes "totally random" for $t \geq t_0$. Total randomness intuitively means that the N point billiards behave like N independent random variables, uniformly distributed in the unit cube $[0,1]^3$ each. According to this intuition, the large deviation type probability in (25.5)

$$\text{ProdMeas}_1\left(|N_S(\mathcal{Y};\omega;t_i) - \text{vol}(S)N| \geq 29\sqrt{N}\right)$$

is well approximated with the tail of the corresponding binomial distribution

$$\sum_{\substack{0\leq k\leq N:\\ |k-pN|\geq 29\sqrt{N}}} \binom{N}{k} p^k q^{N-k} \quad \text{where } p = \text{vol}(S) \text{ and } q = 1-p. \tag{25.20}$$

By the well-known Bernstein inequality in probability theory ("large deviation" inequality; see e.g. in [Fel71]),

$$\sum_{\substack{0\leq k\leq N:\\ |k-pN|\geq 29\sqrt{N}}} \binom{N}{k} p^k q^{N-k} = \sum_{\substack{0\leq k\leq N:\\ |k-pN|\geq \frac{29}{\sqrt{pq}}\sqrt{pqN}}} \binom{N}{k} p^k q^{N-k}$$

$$= \sum_{\substack{0\leq k\leq N:\\ |k-pN|\geq \lambda\cdot\text{StandardDeviation}}} \binom{N}{k} p^k q^{N-k} \leq e^{-\lambda^2/2}, \text{ where } \lambda = \frac{29}{\sqrt{pq}}, \tag{25.21}$$

and of course $\sqrt{pqN}$ is the standard deviation of the binomial distribution.

Since $pq = p(1-p) \leq 1/4$, combining the arguments above, we obtain

$$\text{ProdMeas}_1\left(|N_S(\mathcal{Y};\omega;t_i) - \text{vol}(S)N| \geq 29\sqrt{N}\right)$$

$$\approx \sum_{\substack{0\leq k\leq N:\\ |k-pN|\geq 29\sqrt{N}}} \binom{N}{k} p^k q^{N-k} = \sum_{\substack{0\leq k\leq N:\\ |k-pN|\geq \lambda\cdot\text{StandardDeviation}}} \binom{N}{k} p^k q^{N-k}$$

$$\leq e^{-\lambda^2/2} = e^{-29^2/(2pq)} \leq e^{-2\cdot 29^2} = e^{-1682} < 10^{-730} < 10^{-700}. \qquad (25.22)$$

We take advantage of the fact that 10^{-700} is a very, very small probability. Let $N = 10^{27}$ (it is a realistic number for the number of gas particles in a cube-shaped container of volume 1 cubic meter), and choose $\varepsilon = 10^{-15}$, $K = 10^{17}$, $\gamma = 10^{-17}$, $L = 10^{240}$ and $\eta = 10^{-200}$. Thus the conditions

$$\varepsilon \leq \frac{1}{24\sqrt{N}} \quad \text{and} \quad K \leq 100/\varepsilon$$

are clearly satisfied. Moreover, by (25.22),

$$\text{ProdMeas}_1\left(|N_S(\mathcal{Y};\omega;t_i) - \text{vol}(S)N| \geq 29\sqrt{N}\right)$$

$$< 10^{-700} < \frac{\eta}{2(L+1)\cdot(2\pi K)^9},$$

which proves (25.5).

Finally, the verification of (25.6) is almost trivial. Indeed, $\varepsilon/\eta = 100$, and by(1.9′) and the definition of *fast billiard*,

$$p_1 = \sqrt{\frac{2}{\pi}} \int_{100}^{\infty} y^2 e^{-y^2/2}\, dy < 10^4 e^{-10^4/2} < 10^{-1000}.$$

Thus we have

$$\text{ProdMeas}_1\left(\text{there are more than } \sqrt{N}/2 \text{ fast point billiards}\right)$$

$$= \sum_{\substack{0\leq k\leq N:\\ k>\sqrt{N}/2}} \binom{N}{k} p_1^k (1-p_1)^{N-k} < \sum_{k>\sqrt{N}/2} (Np_1)^k$$

$$< \sum_{k>\sqrt{N}/2} \left(10^{27-1000}\right)^k < 10^{27-1000} = 10^{-973},$$

which is of course *much* smaller than the requested upper bound $\eta/2 = 10^{-240}/2$. This completes the *intuitive* proof of Hypothesis A in this particular case.

Note that the average speed of a gas molecule in room temperature is in the range of 10^3 meter per second, that is, one meter per millisecond. Since we measure distance in meter, it is reasonable to use millisecond as our time unit. Since Hypothesis A applies, in view of the Corollary, with probability $> 1 - \eta = 1 - 10^{-200}$ (where of course probability means ProdMeas_1), after reaching "total randomness", the system stays in L-equilibrium for a very, very long time. More precisely, the system stays in Square-Root Spatial Equilibrium (with constant factor $C = 30$) with respect to all "nice" sets (= rotated boxes) simultaneously for a time interval of length $L\gamma = 10^{240-17} = 10^{223}$ milliseconds. Note that 10^{223} milliseconds is more than 10^{211} years, which is more than 10^{200} times the estimated age of the Universe (since the Big Bang). This is a truly long time! Similarly, probability $> 1 - 10^{-200}$ means a truly overwhelming majority of the initial velocities.

Notice that L-equilibrium with constant factor $C = 30$ guarantees that the density of every "nice" set remains practically constant. Indeed, consider a "nice" test set of volume $1/2$ (say); then the expected number of point billiards in the set is $N/2$, and the actual number of point billiards remains between

$$\frac{1}{2}N - 30\sqrt{N} \quad \text{and} \quad \frac{1}{2}N + 30\sqrt{N}.$$

So the relative error is at most

$$\frac{30\sqrt{N}}{N/2} = \frac{60}{\sqrt{N}} < 2 \cdot 10^{-12},$$

which is practically negligible, and implies practically constant density for an extremely long time.

The goal of Sections 26–28 is to replace the intuitive proof of (25.22) with a precise one. We are going to give a rigorous proof of the key upper bound

$$\text{ProdMeas}_1\left(|N_S(\mathcal{Y};\omega;t_i) - \text{vol}(S)N| \geq 29\sqrt{N}\right) < 10^{-700}, \qquad (25.23)$$

see Theorem 25.1 below. Before formulating Theorem 25.1, we need to discuss the technical issue of *extra condition on the Fourier coefficients.*

Extra condition on the Fourier coefficients. Consider first the special case of an aligned box

$$B = [-b_1/2, b_1/2] \times [-b_2/2, b_2/2] \times [-b_3/2, b_3/2]$$

centered at the origin and contained in the unit cube $[-1/2, 1/2]^3$. Let $f = \chi_B$ denote the 0-1 valued characteristic function of the box. A trivial calculation shows that the (3-dimensional) Fourier coefficients $a_{\mathbf{n}}$, $\mathbf{n} = (n_1, n_2, n_3) \in \mathbb{Z}^3$ of f have the explicit form

$$a_{\mathbf{n}} = \frac{\sin(\pi n_1 b_1)}{\pi n_1} \cdot \frac{\sin(\pi n_2 b_2)}{\pi n_2} \cdot \frac{\sin(\pi n_3 b_3)}{\pi n_3}, \tag{25.24}$$

noting that in the case of a zero coordinate n_j the corresponding factor is b_j, $1 \le j \le 3$.

If we translate a set, the absolute value of a Fourier coefficient of the characteristic function does not change. Therefore, if f is the characteristic function of an aligned box contained in the unit cube, then the Fourier coefficients $a_{\mathbf{n}}$, $\mathbf{n} = (n_1, n_2, n_3) \in \mathbb{Z}^3$ of f satisfy the upper bound (see (25.24))

$$|a_{\mathbf{n}}| \le \frac{1}{\max\{\pi|n_1|, 1\}} \cdot \frac{1}{\max\{\pi|n_2|, 1\}} \cdot \frac{1}{\max\{\pi|n_3|, 1\}}. \tag{25.25}$$

Moreover, if f is the characteristic function of a rotated box ϑB (where B is an aligned box and $\vartheta \in SO(3)$) contained in the unit torus $I^3 = [0,1)^3$, then the Fourier coefficients $a_{\mathbf{n}}$, $\mathbf{n} = (n_1, n_2, n_3) \in \mathbb{Z}^3$ of f satisfy the analog upper bound

$$|a_{\mathbf{n}}| \le \frac{1}{\max\{\pi|\vartheta^{-1}\mathbf{n}(1)|, 1\}} \cdot \frac{1}{\max\{\pi|\vartheta^{-1}\mathbf{n}(2)|, 1\}} \cdot \frac{1}{\max\{\pi|\vartheta^{-1}\mathbf{n}(3)|, 1\}}, \tag{25.26}$$

where $\vartheta^{-1}\mathbf{n}(j)$, $j = 1, 2, 3$ are the 3 coordinates of the rotated lattice point $\vartheta^{-1}\mathbf{n}$ (of course ϑ^{-1} is the inverse of ϑ). Moreover, we have the trivial upper bound $|a_{\mathbf{n}}| \le 1$, due to the fact $|f| \le 1$.

Note that (25.26) easily follows from (25.25) by using a trivial equality of dot products:

$$\vartheta\mathbf{x} \cdot \mathbf{n} = \mathbf{x} \cdot \vartheta^{-1}\mathbf{n};$$

this dot product shows up in the definition of the Fourier coefficients.

Finally, note that the switch from the billiard model in the cube to the torus model requires the geometric operation of *unfolding*, which generates $2^3 = 8$ half-copies of the original test set. So the "billiard version" of (25.26) is

$$|a_{\mathbf{n}}| \le \min\left\{\frac{1}{\max\{\pi|\vartheta\mathbf{n}(1)|/2, 1\} \cdot \max\{\pi|\vartheta\mathbf{n}(2)|/2, 1\} \cdot \max\{\pi|\vartheta\mathbf{n}(3)|/2, 1\}}, 1\right\}. \tag{25.26'}$$

Upper bound (25.26′) explains/motivates the following *extra condition.* Beside the usual condition $f \in L^2(I^3)$, we also assume that the Fourier coefficients $a_{\mathbf{n}}$, $\mathbf{n} \in \mathbb{Z}^3$ of f satisfy the upper bound

$$|a_{\mathbf{n}}| \le \frac{1}{\max\{|\vartheta\mathbf{n}(1)|, 1\} \cdot \max\{|\vartheta\mathbf{n}(2)|, 1\} \cdot \max\{|\vartheta\mathbf{n}(3)|, 1\}}, \tag{25.27}$$

where $\vartheta = \vartheta(f) \in SO(3)$ is some rotation depending only on f, and $\vartheta\mathbf{n}(j)$, $j = 1, 2, 3$ denote the 3 coordinates of the rotated lattice point $\vartheta\mathbf{n}$. Of course (25.27) is a weaker requirement than (25.26′) (since $\pi/2 > 1$), and, for notational convenience, we replaced ϑ^{-1} with ϑ in (25.26′).

Now we are ready to formulate a theorem.

Theorem 25.1. *Consider a general N-particle torus model with Gaussian initial velocity distribution (see (1.9′)) in the 3-dimensional unit torus I^3, equipped with a family of N continuous rectifiable parametrized space curves*

$$\mathcal{R} = (\mathbf{r}_1(t), \mathbf{r}_2(t), \ldots, \mathbf{r}_N(t))$$

such that $\mathbf{r}_k(0) = \mathbf{0}$. Let

$$\mathcal{Y} = \{\mathbf{y}_1, \mathbf{y}_2, \ldots, \mathbf{y}_N\}$$

be an arbitrary initial point configuration in the unit torus I^3 (i.e., the position of the kth particle at the start $t = 0$ is $\mathbf{y}_k \in I^3$, $1 \le k \le N$). So the trajectory of the kth particle is

$$\mathbf{x}_k(t) = \mathbf{y}_k + \vartheta_k(\rho_k \mathbf{r}_k(t)) \quad \text{modulo one,}$$

which describes the time evolution of the whole N-point system (note that $\vartheta_k \in SO(3)$ and $0 < \rho_k < \infty$, $1 \le k \le N$).

Let f be an arbitrary 0-1 valued measurable function defined on the torus I^3 such that the Fourier coefficients $a_{\mathbf{n}}$, $\mathbf{n} \in \mathbb{Z}^3$ of f satisfy the upper bound (25.27).

Assume that $N \ge 10^{27}$, and at time $t > 0$ the drifting distance

$$\mathcal{R}(t) = \min_{1 \le k \le N} |\mathbf{r}_k(t)|$$

satisfies the inequality

$$\mathcal{R}(t) \ge \frac{\sqrt{\log N + 4\log 10}}{2\pi} = \frac{\sqrt{\log N}}{2\pi} + \frac{\log 10}{\pi\sqrt{\log N}} + \cdots.$$

Then with

$$D_f(\omega; t) = N(\mathcal{Y}; \mathcal{R}; \omega; f; t) - N \int_{I^3} f \, dV$$

we have

$$\text{ProdMeas}_1\left(|D_f(\omega;t)| \geq 29\sqrt{N}\right) < 10^{-700}.$$

Notice that Theorem 25.1 proves (25.23). We discuss the proof of Theorem 25.1 in Sections 26–28.

Note that Theorem 25.1 is about the *general* torus model (far beyond constant speed motions on torus lines). So, by using Theorem 25.1, we can easily prove *approach to equilibrium and long term stability of equilibrium* for the general case of "possibly different types of parametrized curves for different particles", far beyond the special case of constant speed motion on torus lines in (25.23), and beyond the class

$$\mathbf{r}(t) = (\mathbf{r}_1(t) = \mathbf{r}_2(t) = \ldots = \mathbf{r}_N(t))$$

that we studied in Chapters 3 and 4.

We conclude Section 25 with an interesting detour.

3. On the alternation of equilibrium and off-equilibrium states. We may say that Theorem 25.1 justifies the probabilistic intuition in the form of a Large Deviation type inequality for many-particle dynamical systems. Consider the simplest Bernoulli model: N point billiards in the unit cube (see (1.10)–(1.13)), the initial velocity distribution is Gaussian ("first kind"), and as usual,

$$\mathcal{Y} = \{\mathbf{y}_1, \mathbf{y}_2, \ldots, \mathbf{y}_N\}$$

denotes an arbitrary but fixed initial configuration. By using the trick of *unfolding*, we switch from the Bernoulli model to the simplest torus model; formally, $\mathbf{r}_1(t) = \ldots = \mathbf{r}_N(t) = (t, 0, 0)$, see Section 1. Let $\varepsilon > 0$ be a fixed small positive constant (e.g. $\varepsilon = 1/10$). We say that at time t the system is in $(\frac{1}{2} + \varepsilon)$-*power-equilibrium* if for every rotated box in the unit cube, the number of particles in the box is very close to the expected value ($= N$ times the volume of the box) in the quantitative sense that the difference has absolute value less than $N^{(1+\varepsilon)/2}$. The probabilistic intuition suggests that, starting from an arbitrary initial configuration $\mathcal{Y} = \{\mathbf{y}_1, \mathbf{y}_2, \ldots, \mathbf{y}_N\}$, the typical individual time evolution of the system reaches the state of $(\frac{1}{2} + \varepsilon)$-*power-equilibrium* extremely fast in time less than $\sqrt{\log N}$, and the system stays in the state of $(\frac{1}{2} + \varepsilon)$-*power-equilibrium* for a very long time time in the range of

$$e^{N^\varepsilon/2},$$

which is the reciprocal of the tail probability in the normal distribution. (25.28)

We refer to (25.28) as an *exponentially long time*. In view of Lemma 1.2, the individual time evolution of a system eventually leaves the state of $(\frac{1}{2}+\varepsilon)$-*power-equilibrium*, but a typical time evolution goes back to the state of $(\frac{1}{2}+\varepsilon)$-*power-equilibrium* extremely fast in time less than $\sqrt{\log N}$, and again the system stays in the state of $(\frac{1}{2}+\varepsilon)$-*power-equilibrium* for a very long time time in the range of (25.28), i.e., for an exponentially long time. And so on, for a typical time evolution we have these alternating time intervals of $(\frac{1}{2}+\varepsilon)$-*power-equilibrium* and off-equilibrium. The equilibrium blocks are exponentially long in the range of (25.28), and the off-equilibrium blocks are extremely short: less than $\sqrt{\log N}$. By using an adaptation of the proof technique of Theorem 25.1 we can easily prove that this alternation between exponentially long equilibrium blocks and extremely short off-equilibrium blocks will happen *many-many times in a row*, at least for the typical individual time evolution of the system. This raises the following natural question: can we replace *many-many times in a row* with *always*? The answer is no: if we wait long enough, there will be *long* off-equilibrium blocks, much longer than $\sqrt{\log N}$. This follows from a combination of Lemma 1.2 ("superdiscrepancy") and the following Theorem 25.2. For the details of the deduction, see the Remark below.

Theorem 25.2. *Consider the simplest torus model of the first kind* $(\mathcal{Y};\Omega_1)$ *where* $\mathbf{r}_1(t) = \ldots = \mathbf{r}_N(t) = (t,0,0)$ *and every mass is one (i.e., each particle moves on a straight line with constant speed, and of course different particles have different speed). Let* $1 > \varepsilon > 0$ *be arbitrary. Then there is a subset* $\Omega_1^* \subset \Omega_1$ *such that* $\text{ProdMeas}_1(\Omega_1^*) > 9/10$, *and for every*

$$\omega = (\rho_1, \vartheta_1, \rho_2, \vartheta_2, \ldots, \rho_N, \vartheta_N) \in \Omega_1^*$$

there is an initial configuration $\mathcal{Z} = \mathcal{Z}(\varepsilon;\omega) = \{\mathbf{z}_1, \ldots, \mathbf{z}_N\}$ *such that, starting from* $(\mathcal{Z},\omega)$ *at* $t = 0$, *in the interval*

$$0 \le t \le 2^{-5} N^{(1-\varepsilon)/2} \tag{25.29}$$

the time evolution of the system does not reach any $(\frac{1}{2}+\varepsilon)$-*power-equilibrium state, assuming* N *is large enough in terms of* ε. *What is more, if* t *is in the interval (25.29), then the number of particles in the left half* $[0,1/2] \times [0,1] \times [0,1]$ *of the unit cube is more than* $N/2 + N^{(1+\varepsilon)/2}$ *(assuming* N *is large enough in terms of* ε*).*

Remark. The existence of long off-equilibrium blocks can be deduced from Lemma 1.2 and Theorem 25.2 as follows. We can assume that in $\omega \in \Omega_1^*$ the $3N$ initial velocity coordinates are linearly independent over the rationals (since the complement set has measure zero, where the relevant measure is ProdMeas_1). Therefore, starting from $(\mathcal{Y}, \omega)$, if we wait long enough, the system will get arbitrarily close to $(\mathcal{Z}, \omega)$ with $\mathcal{Z} = \mathcal{Z}(\varepsilon; \omega) = \{\mathbf{z}_1, \ldots, \mathbf{z}_N\}$. More precisely, given any $\delta > 0$, it will happen that the image of $\mathbf{y}_i$ gets δ-close to $\mathbf{z}_i$, $1 \le i \le N$. Since $\delta > 0$ can be arbitrarily small, Theorem 25.2 applies, and guarantees the existence of an off-equilibrium block much longer than $\sqrt{\log N}$; namely, the length is polynomial in N, see (25.29).

Proof of Theorem 25.2. Let

$$\Omega_1^* = \Big\{ \omega \in \Omega_1 : \text{ there are } \ge 2N^{(1+\varepsilon)/2} \text{ initial velocities } \mathbf{v}_i(\omega)$$

$$= (v_{i,1}, v_{i,2}, v_{i,3}) \text{ such}$$

$$\text{that } |v_{i,1}| < 8N^{-(1-\varepsilon)/2}, \text{ and } |\mathbf{v}_j(\omega)| < N^{1/4},\ 1 \le j \le N \Big\}. \tag{25.30}$$

It is trivial from the definition of the Gaussian velocity distribution that

$$\text{ProdMeas}_1(\Omega_1^*) > 9/10,$$

assuming N is large enough in terms of $\varepsilon > 0$.

Let $\omega \in \Omega_1^*$ be arbitrary. By rearranging the indices, we can assume that

$$|v_{i,1}| < 8N^{-(1+\varepsilon)/2} \text{ holds for all } N_0 = N - 2N^{(1+\varepsilon)/2} < i \le N. \tag{25.31}$$

For every $1 \le j \le N_0 = N - 2N^{(1+\varepsilon)/2}$, let ξ_j be a random variable uniformly distributed in the unit interval, and assume that ξ_j, $1 \le j \le N_0$ are independent. Since the torus is translation invariant, for notational convenience we can switch from the interval $[0, 1/2)$ to $[1/4, 3/4)$, i.e., we translate the left half of the unit cube to the right by $1/4$. The product form of the test set explains why it suffices to study the first coordinates of the particles.

Let $\eta = N^{-3/4}$, and write

$$I_1 = [1/4 + \eta, 3/4 - \eta] \subset I_2 = [1/4, 3/4).$$

For $1 \le j \le N_0$, let

$$\zeta_j(t) = 1 \text{ if } \{\xi_j + t v_{j,i}\} \in I_1, \text{ and zero otherwise.} \tag{25.32}$$

Note that in (25.32) $\{y\}$ stands for the fractional part of a real number y. Notice that for a fixed t, $\zeta_j(t)$, $1 \le j \le N_0$ are independent 0-1 valued

Bernoulli random variables with $p = \text{length}(I_1) = 1/2 - 2\eta$. Using the simplest form of the classical Bernstein's large deviation inequality,

$$\Pr\left[\sum_{j=1}^{N_0} \zeta_j(t) \le \left(\frac{1}{2} - 2\eta\right) N_0 - \lambda\sqrt{N_0}\right] \le \exp\left(\frac{-\lambda^2}{2(1-p)}\right) \le e^{-\lambda^2/2} \tag{25.33}$$

holds for every fixed time t.

Let

$$0 < T < \frac{e^{N^\varepsilon/8}}{N} \tag{25.34}$$

be fixed. We apply (25.33) with $\lambda = \frac{1}{2}N^{\varepsilon/2}$ for every t of the form $0 < t = k/N \le T$, k integer (which represents TN time points). By (25.34)

$$e^{-\lambda^2/2} = e^{-N^\varepsilon/8} < \frac{1}{TN},$$

so by (25.33), for every t of the form $0 < t = k/N \le T$, k integer, there exist real numbers $\xi_j(0) \in [0,1)$, $1 \le j \le N_0$ such that

$$\left\{\xi_j(0) + \frac{k}{N} v_{j,i}\right\} \in I_1 = [1/4 + \eta, 3/4 - \eta]$$

holds for at least

$$\left(\frac{1}{2} - 2\eta\right) N_0 - \lambda\sqrt{N_0} \ge \left(\frac{1}{2} - 2\eta\right)\left(N - 2N^{(1+\varepsilon)/2}\right) - \lambda\sqrt{N}$$

$$\ge \frac{1}{2}N - 2\eta N - \frac{1}{2}N^{(1+\varepsilon)/2} > \frac{1}{2}N - N^{(1+\varepsilon)/2} \tag{25.35}$$

indices j in the interval $1 \le j \le N_0$ (in the last step of (25.35) we used that N is large enough). Since no particle has speed $\ge N^{1/4}$, if a particle was in the box $B_1 = I_1 \times [0,1) \times [0,1)$ at time $t = k/N$, then this particle stays in the larger box $B_2 = I_2 \times [0,1) \times [0,1)$ in the whole time interval $k/N \le t < (k+1)/N$.

Now we are ready to specify the desired point configuration $\mathcal{Z} = \{\mathbf{z}_1, \ldots, \mathbf{z}_N\}$ for the box B_2. Let $\xi_j(0)$ be the first coordinate of $\mathbf{z}_j$ for $1 \le j \le N_0$, and let $1/2$ be the first coordinate of $\mathbf{z}_j$ for $N_0 < j \le N$ (the second and third coordinates can be chosen freely).

Then it follows from (25.35) that the box $B_2 = [1/4, 3/4) \times [0,1) \times [0,1)$ contains more than

$$\frac{1}{2}N - N^{(1+\varepsilon)/2}$$

particles with index j in $1 \leq j \leq N_0$ during $0 \leq t \leq T$, assuming (25.34) holds.

On the other hand, by (25.31),

$$|v_{i,1}| < 8N^{-(1-\varepsilon)/2} \text{ holds for all } N_0 = N - 2N^{(1+\varepsilon)/2} < i \leq N. \quad (25.36)$$

Choosing

$$T_0 = N^{(1-\varepsilon)/2}/32, \quad (25.37)$$

and combining the fact

$$T_0 \cdot 8N^{-(1-\varepsilon)/2} = \frac{1}{4} = \text{radius of the interval } I_2 = [1/4, 3/4]$$

with (25.36), it follows that the box $B_2 = [1/4, 3/4) \times [0, 1) \times [0, 1)$ contains every particle with index j in $N_0 < j \leq N$ during $0 \leq t \leq T_0$ (see (25.37), and use the fact that $1/2$ is the first coordinate of the starting point $\mathbf{z}_j$ for $N_0 < j \leq N$).

Comparing (25.34) and (25.37), we see that T_0 satisfies (25.34) assuming N is large enough in terms of $\varepsilon > 0$.

Summarizing, the box $B_2 = [1/4, 3/4) \times [0, 1) \times [0, 1)$ contains more than

$$\frac{1}{2}N - N^{(1+\varepsilon)/2} + (N - N_0) = \frac{1}{2}N - N^{(1+\varepsilon)/2} + 2N^{(1+\varepsilon)/2} = \frac{1}{2}N + N^{(1+\varepsilon)/2}$$

particles during $0 \leq t \leq T_0$, which completes the proof of Theorem 25.2. □

26. Starting the proof of Theorem 25.1

We formulate another result — Theorem 26.1 — that implies Theorem 25.1.

Proving a Large Deviation type inequality via higher moments. Let f be a real valued test function defined on the unit torus $I^3 = [0, 1)^3$. Assume $f \in L^2(I^3)$; we need the Fourier expansion

$$f(\mathbf{z}) = \sum_{\mathbf{n} \in \mathbb{Z}^3} a_{\mathbf{n}} e^{2\pi \mathrm{i} \mathbf{n} \cdot \mathbf{z}} \text{ with } a_{\mathbf{n}} = a_{\mathbf{n}}(f) = \int_{I^3} f(\mathbf{z}) e^{-2\pi \mathrm{i} \mathbf{n} \cdot \mathbf{z}} \, d\mathbf{z}, \quad (26.1)$$

where, as usual, $\mathbf{n} \cdot \mathbf{z} = n_1 z_1 + n_2 z_2 + n_3 z_3$ and $\mathrm{i} = \sqrt{-1}$. Clearly $a_{\mathbf{0}} = \int_{I^3} f \, d\mathrm{V}$, $a_{\mathbf{n}}$ and $a_{-\mathbf{n}}$ are complex conjugates (since f is real-valued), and by Parseval's formula,

$$\sigma_0^2(f) = \int_{I^3} (f(\mathbf{z}) - a_{\mathbf{0}})^2 \, d\mathbf{z} = \sum_{\substack{\mathbf{n} \in \mathbb{Z}^3: \\ \mathbf{n} \neq \mathbf{0}}} |a_{\mathbf{n}}|^2 . \quad (26.2)$$

By (26.1),

$$D_f(\omega;t) = N(\mathcal{Y};\mathcal{R};\omega;f;t) - N\int_{I^3} f\,dV$$

$$= \sum_{k=1}^{N}\left(f(\mathbf{x}_k(t)) - \int_{I^3} f\,dV\right) = \sum_{\substack{\mathbf{n}\in\mathbb{Z}^3:\\ \mathbf{n}\neq\mathbf{0}}} a_{\mathbf{n}} \sum_{k=1}^{N} e^{2\pi i\mathbf{n}\cdot\mathbf{x}_k(t)}$$

$$= \sum_{\substack{\mathbf{n}\in\mathbb{Z}^3:\\ \mathbf{n}\neq\mathbf{0}}} a_{\mathbf{n}} \sum_{k=1}^{N} e^{2\pi i\mathbf{n}\cdot(\mathbf{y}_k+\vartheta_k(\rho_k\mathbf{r}_k(t)))}. \tag{26.3}$$

Next we integrate the $2m$-th power of (26.3) over the big "initial velocity space" $\omega = (\vartheta_1,\rho_1,\ldots,\vartheta_N,\rho_N) \in \Omega_1 = ([0,\infty)\times SO(3))^N$ (torus model of the first kind):

$$\mathbf{Moment}_{2m}(\mathcal{Y};\mathcal{R};f;t) = \int_{\Omega_1} (D_f(\omega;t))^{2m}\, d\text{ProdMeas}_1(\omega)$$

$$= \int_{\Omega_1}\left(\sum_{\substack{\mathbf{n}\in\mathbb{Z}^3:\\ \mathbf{n}\neq\mathbf{0}}} a_{\mathbf{n}} \sum_{k=1}^{N} e^{2\pi i\mathbf{n}\cdot(\mathbf{y}_k+\vartheta_k(\rho_k\mathbf{r}_k(t)))}\right)^{2m} d\text{ProdMeas}_1(\omega)$$

$$= \sum_{\substack{\mathbf{n}_1\in\mathbb{Z}^3:\\ \mathbf{n}_1\neq\mathbf{0}}} \sum_{k_1=1}^{N} \sum_{\substack{\mathbf{n}_2\in\mathbb{Z}^3:\\ \mathbf{n}_2\neq\mathbf{0}}} \sum_{k_2=1}^{N} \sum_{\substack{\mathbf{n}_3\in\mathbb{Z}^3:\\ \mathbf{n}_3\neq\mathbf{0}}} \sum_{k_3=1}^{N} \cdots \sum_{\substack{\mathbf{n}_{2m}\in\mathbb{Z}^3:\\ \mathbf{n}_{2m}\neq\mathbf{0}}} \sum_{k_{2m}=1}^{N} a_{\mathbf{n}_1}$$

$$\cdot\, a_{\mathbf{n}_2}a_{\mathbf{n}_3}\cdots a_{\mathbf{n}_{2m}} \int_{\Omega_1} \prod_{j=1}^{2m} e^{2\pi i\mathbf{n}_j\cdot(\mathbf{y}_{k_j}+\vartheta_{k_j}(\rho_{k_j}\mathbf{r}_{k_j}(t)))}\, d\text{ProdMeas}_1(\omega). \tag{26.4}$$

Given a sequence $\mathbf{n}_j, k_j$, $j = 1,2,3,\ldots,2m$, for every integer s in $1\le s\le N$ we define its multiplicity and the corresponding vector-sum as follows. Suppose that s shows up ν times in the sequence $k_1,k_2,k_3,\ldots,k_{2m}$: $s = k_{j_i}$ for $1\le i\le\nu$ with $1\le j_1<\ldots<j_\nu\le 2m$, and $s\neq k_j$ for the rest of the j's; we call ν the *multiplicity* of s, and denote it with

$$\text{mult}(s) = \text{mult}(\mathbf{n}_j,k_j,1\le j\le 2m;s). \tag{26.5}$$

If $\text{mult}(s)\ge 1$, we call $\mathbf{n}_{j_1}+\ldots+\mathbf{n}_{j_\nu}$ the corresponding *vector sum*, and denote it with

$$\vec{v}(s) = \vec{v}(\mathbf{n}_j,k_j,1\le j\le 2m;s) = \sum_{i=1}^{\nu}\mathbf{n}_{j_i}. \tag{26.6}$$

If $\text{mult}(s) = 0$ then let $\vec{v}(s) = \mathbf{0}$.

Using (26.5), (26.6) and Lemma 7.1, we can rewrite (26.4):

$$\mathbf{Moment}_{2m}(\mathcal{Y};\mathcal{R};f;t) = \int_{\Omega_1} D_f^{2m}(\omega;t)\,d\mathrm{ProdMeas}_1(\omega)$$

$$= \sum_{\substack{\mathbf{n}_1\in\mathbb{Z}^3:\\ \mathbf{n}_1\neq\mathbf{0}}} \sum_{k_1=1}^{N} \sum_{\substack{\mathbf{n}_2\in\mathbb{Z}^3:\\ \mathbf{n}_2\neq\mathbf{0}}} \sum_{k_2=1}^{N} \sum_{\substack{\mathbf{n}_3\in\mathbb{Z}^3:\\ \mathbf{n}_3\neq\mathbf{0}}} \sum_{k_3=1}^{N} \cdots \sum_{\substack{\mathbf{n}_{2m}\in\mathbb{Z}^3:\\ \mathbf{n}_{2m}\neq\mathbf{0}}} \sum_{k_{2m}=1}^{N} a_{\mathbf{n}_1}$$

$$\cdot a_{\mathbf{n}_2} a_{\mathbf{n}_3} \cdots a_{\mathbf{n}_{2m}} \prod_{s=1}^{N} e^{2\pi i \vec{v}(s)\cdot\mathbf{y}_s} \cdot e^{-2\pi^2|\vec{v}(s)|^2|\mathbf{r}_s(t)|^2}. \tag{26.7}$$

We divide $\mathbf{Moment}_{2m}(\mathcal{Y};\mathcal{R};f;t)$ into two parts:

$$\mathbf{Moment}_{2m}(\mathcal{Y};\mathcal{R};f;t) = \mathbf{Moment}_{2m}(\mathcal{Y};\mathcal{R};f;t;\mathrm{PartOne})$$

$$+\mathbf{Moment}_{2m}(\mathcal{Y};\mathcal{R};f;t;\mathrm{PartTwo}), \tag{26.8}$$

where $\mathbf{Moment}_{2m}(\mathcal{Y};\mathcal{R};f;t;\mathrm{PartOne})$ means the total contribution of the terms in the big sum in (26.7) for which every nonzero multiplicity is 2 and every vector sum is zero:

$$\vec{v}(\mathbf{n}_j, k_j, 1\le j\le 2m; s) = \mathbf{0}, \tag{26.9}$$

and $\mathbf{Moment}_{2m}(\mathcal{Y};\mathcal{R};f;t;\mathrm{PartTwo})$ means the rest, i.e., it means the total contribution of the terms in the big sum in (26.7) for which at least one of the following two requirements hold:
Requirement One:

$$\text{there is a multiplicity} \notin \{0,2\}; \tag{26.10}$$

Requirement Two:
there is a nonzero multiplicity $\mathrm{mult}(\mathbf{n}_j, k_j, 1\le j\le 2m; s)\ge 1$ which is paired with a nonzero vector sum

$$\vec{v}(\mathbf{n}_j, k_j, 1\le j\le 2m; s) \neq \mathbf{0}. \tag{26.11}$$

We start with the estimation of $\mathbf{Moment}_{2m}(\mathcal{Y};\mathcal{R};f;t;\mathrm{PartOne})$. By definition (see (26.7))

$$\mathbf{Moment}_{2m}(\mathcal{Y};\mathcal{R};f;t;\mathrm{PartOne})$$

$$= \binom{N}{m}\binom{2m}{2}\binom{2m-2}{2}\binom{2m-4}{2}\cdots\binom{4}{2}\binom{2}{2}\left(\sum_{\substack{\mathbf{n}\in\mathbb{Z}^3:\\ \mathbf{n}\neq\mathbf{0}}} |a_{\mathbf{n}}|^2\right)^m$$

$$= \binom{N}{m}(2m)!2^{-m}\left(\sigma_0^2(f)\right)^m, \tag{26.12}$$

where in the last step we used (26.2).

Guiding Intuition. Assume for simplicity that

$$\mathbf{Moment}_{2m}(\mathcal{Y};\mathcal{R};f;t)$$

$$= \mathbf{Moment}_{2m}(\mathcal{Y};\mathcal{R};f;t;\text{PartOne}) + \text{ negligible}$$

$$= (1+o(1))\binom{N}{m}(2m)!2^{-m}\left(\sigma_0^2(f)\right)^m. \tag{26.13}$$

By (26.4) and the "simplicity hypothesis" (26.13),

$$\text{ProdMeas}_1\left(|D_f(\omega;t)| \geq \lambda\sqrt{N}\sigma_0(f)\right)\cdot\left(\lambda\sqrt{N}\sigma_0(f)\right)^{2m}$$

$$\leq \int_{\Omega_1} D_f^{2m}(\omega;t)\,d\text{ProdMeas}_1(\omega) = \mathbf{Moment}_{2m}(\mathcal{Y};\mathcal{R};f;t)$$

$$= (1+o(1))\binom{N}{m}(2m)!2^{-m}\left(\sigma_0^2(f)\right)^m \leq (1+o(1))\frac{N^m(2m)!}{m!2^m}\left(\sigma_0^2(f)\right)^m, \tag{26.14}$$

which holds for every $\lambda > 0$.

After simplification, we can rewrite inequality (26.14) as follows:

$$\text{ProdMeas}_1\left(|D_f(\omega;t)| \geq \lambda\sqrt{N}\sigma_0(f)\right) \leq (1+o(1))\frac{(2m)!}{m!2^m\lambda^{2m}}. \tag{26.15}$$

Given a fixed $\lambda > 0$, we want to minimize

$$\frac{(2m)!}{m!2^m\lambda^{2m}} \quad \text{as an integer valued function of } m. \tag{26.16}$$

By using Stirling's formula $m! = (1+o(1))(m/e)^m\sqrt{2\pi m}$, we have

$$\frac{(2m)!}{m!2^m\lambda^{2m}} = (1+o(1))\frac{(2m/e)^{2m}\sqrt{4\pi m}}{2^m(m/e)^m\sqrt{2\pi m}\lambda^{2m}}$$

$$= (1+o(1))\sqrt{2}\left(\frac{2m}{e\lambda^2}\right)^m. \tag{26.17}$$

So to solve the optimization problem raised in (26.16), for a given $\lambda > 0$ we need to minimize

$$\left(\frac{2m}{e\lambda^2}\right)^m, \tag{26.18}$$

where $m \geq 1$ is an integer variable.

Switching to real variables, the analog problem is to minimize the function

$$f(x) = (x/c)^x = e^{g(x)} \quad \text{where} \quad g(x) = x\log x - x\log c$$

and $c > 0$ is a given constant. Since $g'(x) = \log x + 1 - \log c = 0$ implies $x = c/e$, we conclude that $\min_x f(x) = e^{-c/e}$. Using this fact in (26.18), we obtain that

$$\min_m \left(\frac{2m}{e\lambda^2}\right)^m = e^{-\lambda^2/2} \text{ with the choice } m = \lambda^2/2, \tag{26.19}$$

that is, the minimum is attained at $m = \lambda^2/2$ (for simplicity assume that $m = \lambda^2/2$ is an integer).

Combining (26.15), (26.17) and (26.19), we obtain

$$\text{ProdMeas}_1\left(|D_f(\omega;t)| \geq \lambda\sqrt{N}\sigma_0(f)\right) \leq (1+o(1))\sqrt{2}e^{-\lambda^2/2}, \tag{26.20}$$

which is exactly the superexponentially small upper bound — we ignore the negligible factor $(1+o(1))\sqrt{2}$ — that we used in the *intuitive* proof of Hypothesis A; see (25.22).

This explains our approach: in order to replace the intuitive proof of Hypothesis A with a precise proof, we actually prove the following somewhat modified form of the "simplicity hypothesis" (26.13):

$$\begin{aligned}&\mathbf{Moment}_{2m}(\mathcal{Y};\mathcal{R};f;t)\\ &= \mathbf{Moment}_{2m}(\mathcal{Y};\mathcal{R};f;t;\text{PartTwo}: \text{ every nonzero multiplicity is 1})\\ &\quad + \mathbf{Moment}_{2m}(\mathcal{Y};\mathcal{R};f;t;\text{PartOne}) + \text{negligible}.\end{aligned} \tag{26.21}$$

Clearly (see (26.5)–(26.7))

$$\begin{aligned}&|\mathbf{Moment}_{2m}(\mathcal{Y};\mathcal{R};f;t;\text{PartTwo}: \text{ every nonzero multiplicity is 1})|\\ &\leq \left(\sum_{\mathbf{n}\in\mathbb{Z}^3\setminus\mathbf{0}}\sum_{k=1}^{N}|a_{\mathbf{n}}|\,e^{-2\pi^2|\mathbf{n}|^2|\mathbf{r}_k(t)|^2}\right)^{2m}\\ &\leq \left(\sum_{\mathbf{n}\in\mathbb{Z}^3\setminus\mathbf{0}} Ne^{-2\pi^2|\mathbf{n}|^2(\mathcal{R}(t))^2}\right)^{2m},\end{aligned}$$

since $|f| \leq 1$ implies $|a_{\mathbf{n}}| \leq 1$, and

$$\mathcal{R}(t) = \min_{1\leq k\leq N}|\mathbf{r}_k(t)|$$

is the "drifting distance" introduced in (2.20).

Since

$$\sum_{\mathbf{n}\in\mathbb{Z}^3\setminus\mathbf{0}} e^{-2\pi^2|\mathbf{n}|^2(\mathcal{R}(t))^2} \approx e^{-2\pi^2(\mathcal{R}(t))^2},$$

the choice

$$Ne^{-2\pi^2(\mathcal{R}(t))^2} \le \sqrt{N} \iff \mathcal{R}(t) \ge \frac{\sqrt{\log N}}{2\pi} \tag{26.22}$$

guarantees the upper bound

$$|\mathbf{Moment}_{2m}(\mathcal{Y};\mathcal{R};f;t;\text{PartTwo}: \text{ every nonzero multiplicity is } 1)|$$

$$\le \left(\sum_{\mathbf{n}\in\mathbb{Z}^3\setminus\mathbf{0}} Ne^{-2\pi^2|\mathbf{n}|^2(\mathcal{R}(t))^2}\right)^{2m} \le \left(\sqrt{N}\right)^{2m} = N^m,$$

which basically means that in (26.21) the first term on the right-hand side

$$|\mathbf{Moment}_{2m}(\mathcal{Y};\mathcal{R};f;t;\text{PartTwo}: \text{ every nonzero multiplicity is } 1)|$$

becomes negligible compared to the second term on the right-hand side

$$\mathbf{Moment}_{2m}(\mathcal{Y};\mathcal{R};f;t;\text{PartOne}),$$

justifying the "simplicity hypothesis" (26.13).

We emphasize the fact that (26.22) is exactly the same threshold as in Corollary 3.1. This has the following message: once the system reaches square-root equilibrium, it stays in square-root equilibrium for a very, very long time.

We carry out this plan as follows (unfortunately, the technical details are somewhat unpleasant, requiring long, ugly estimations). In view of (26.8) and (26.12), it remains to estimate

$$\mathbf{Moment}_{2m}(\mathcal{Y};\mathcal{R};f;t;\text{PartTwo}).$$

The estimation of $\mathbf{Moment}_{2m}(\mathcal{Y};\mathcal{R};f;t;\text{PartOne})$ was very easy, because by Parseval's formula we could evaluate the relevant sum exactly:

$$S_2(\mathbf{0};f) = \sum_{\substack{(\mathbf{n}_1,\mathbf{n}_2):\ \mathbf{n}_i\in\mathbb{Z}^3\setminus\mathbf{0}\\ \mathbf{n}_1+\mathbf{n}_2=\mathbf{0}}} a_{\mathbf{n}_1}a_{\mathbf{n}_2} = \sum_{\substack{\mathbf{n}\in\mathbb{Z}^3:\\ \mathbf{n}\neq\mathbf{0}}} |a_{\mathbf{n}}|^2 = \sigma_0^2(f). \tag{26.23}$$

The analog problem here requires (among others) the estimation of the sum of the triple products of the Fourier coefficients; as an upper bound, we take the sum of the absolute values of the products:

$$S_3(\mathbf{0};f) = \sum_{\substack{(\mathbf{n}_1,\mathbf{n}_2,\mathbf{n}_3):\ \mathbf{n}_i\in\mathbb{Z}^3\setminus\mathbf{0}\\ \mathbf{n}_1+\mathbf{n}_2+\mathbf{n}_3=\mathbf{0}}} |a_{\mathbf{n}_1}a_{\mathbf{n}_2}a_{\mathbf{n}_3}|. \tag{26.24}$$

And in the general case we need to estimate the sum of the k-tuple products of the Fourier coefficients

$$S_k(\mathbf{0}; f) = \sum_{\substack{(\mathbf{n}_1,\ldots,\mathbf{n}_k):\ \mathbf{n}_i \in \mathbb{Z}^3\setminus \mathbf{0} \\ \mathbf{n}_1+\ldots+\mathbf{n}_k=\mathbf{0}}} |a_{\mathbf{n}_1} \cdots a_{\mathbf{n}_k}| \tag{26.25}$$

where $k = 3, 4, 5, \ldots$.

Unfortunately the Parseval formula does not generalize for triple products (or other longer products), and the condition $f \in L^2(I^3)$ does not guarantee the convergence of the infinite series (26.25). Therefore, to estimate (25.34), and (25.35) in general, we need some extra information on the Fourier coefficients. This extra information is exactly condition (25.27), the condition of Theorem 25.1.

The following result immediately implies Theorem 25.1.

Theorem 26.1. *Under the condition of Theorem 25.1, the high-power integral*

$$\mathbf{Moment}_{2m}(\mathcal{Y}; \mathcal{R}; f; t) = \int_{\Omega_1} (D_f(\omega; t))^{2m}\, d\text{ProdMeas}_1(\omega)$$

of the discrepancy function

$$D_f(\omega; t) = N(\mathcal{Y}; \mathcal{R}; \omega; f; t) - N \int_{I^3} f\, dV$$

satisfies the inequality

$$\mathbf{Moment}_{2m}(\mathcal{Y}; \mathcal{R}; f; t) < 10^{10} \left(\frac{mN}{2e}\right)^m$$

with the choice $m = 2 \cdot 29^2 = 1682$.

27. Proof of Theorem 26.1 (I)

The deduction of Theorem 25.1 from Theorem 26.1 is easy, and goes as follows. With $\mu = 29$ we have

$$\text{ProdMeas}_1\left(|D_f(\omega; t)| \geq \mu\sqrt{N}\right) \cdot \left(\mu\sqrt{N}\right)^{2m}$$

$$\leq \int_{\Omega_1} (D_f(\omega; t))^{2m}\, d\text{ProdMeas}_1(\omega)$$

$$= \mathbf{Moment}_{2m}(\mathcal{Y}; \mathcal{R}; f; t) \leq \left(\frac{mN}{2e}\right)^m,$$

which implies (with $m = 2\mu^2 = 2 \cdot 29^2 = 1682$)

$$\text{ProdMeas}_1\left(|D_f(\omega;t)| \geq \mu\sqrt{N}\right) < 10^{10}\left(\frac{m}{2e\mu^2}\right)^m = 10^{10}e^{-m}$$

$$= 10^{10}e^{-1682} < 10^{10} \cdot 10^{-730} < 10^{-700},$$

proving Theorem 25.1.

By using Theorem 25.1 it is trivial to replace the intuitive proof of Hypothesis A with a precise proof: we just repeat the argument of Section 25.2 (see (25.20)–(25.23)). Let us return to the concrete model in Section 25.2: we study an N-element point billiard system with $N = 10^{27}$ in a cube container of side length 1 meter. We study the time evolution: the system starts from an arbitrary but fixed initial configuration $\mathcal{Y}$ at $t = 0$, and the initial velocity distribution is Gaussian. Assume that the average speed of a point billiard is 10^3 meter per second (which is realistic for gas molecules in room temperature). The "time threshold" t_0 is defined by the equation about the drifting distance

$$\mathcal{R}(t_0) = \frac{\sqrt{\log N + 4\log 10}}{2\pi} = \frac{\sqrt{\log N}}{2\pi} + \frac{\log 10}{\pi\sqrt{\log N}} + \cdots.$$

In the billiard model it takes about 4 bounces to reach the threshold t_0 (2 jumps in the torus model); with the average speed 1000 m/sec it is just a few milliseconds. The hypothesis of Theorem 25.1 is satisfied for every time $t \geq t_0$. Then for more than $1 - 10^{-200}$ part of the initial velocities (a totally overwhelming majority!), the system stays in Square-Root Spatial Equilibrium with constant factor 30 with respect to all "nice" sets (= rotated boxes) simultaneously for a time interval $t_0 \leq t \leq T$ with $T = 10^{211}$ years. Note that 10^{211} years is more than 10^{200} times the estimated age of the Universe — an incredibly long time! Staying in this kind of Square-Root Spatial Equilibrium with constant factor 30 implies that the density of every "nice" set remains practically constant for such an extremely long time.

The factor 10^{10} in Theorem 26.1 is "accidental", and it can be easily reduced to $1+\varepsilon$ by replacing the threshold $N \geq 10^{27}$ with a larger threshold. Here we just show an illustration where 10^{10} is reduced to 2.

Assume that $N \geq 10^{39}$, and at time $t > 0$ the drifting distance

$$\mathcal{R}(t) = \min_{1 \leq k \leq N} |\mathbf{r}_k(t)|$$

satisfies the inequality

$$\mathcal{R}(t) \geq \frac{\sqrt{\log N + 12\log 10}}{2\pi} = \frac{\sqrt{\log N}}{2\pi} + \frac{3\log 10}{\pi\sqrt{\log N}} + \cdots;$$

then

$$\mathbf{Moment}_{2m}(\mathcal{Y};\mathcal{R};f;t) < 2\left(\frac{mN}{2e}\right)^m \quad \text{with } m = 2\cdot 29^2 = 1682. \quad (27.1)$$

Note that the factor $(mN/2e)^m$ is best possible.

We prove (27.1) at the end of Section 28; see the Remarks.

It remains to give a

Proof of Theorem 26.1. Let $\mathbf{x} = (x_1, x_2, x_3) \in \mathbf{R}^3$ be a 3-dimensional vector of length ≥ 1, and assume that length$(\mathbf{x}) \approx 2^h$ for some integer $h \geq 1$ in the precise sense

$$2^{h-1} \leq |\mathbf{x}| = \sqrt{x_1^2 + x_2^2 + x_3^2} < 2^h, \quad h \geq 1 \text{ integer.}$$

Given $\delta_i = \pm 1$ and $h_i \geq 2$, $i = 1, 2, 3$ integers, we define the aligned boxes

$$B(\delta_1 h_1, \delta_2 h_2, \delta_3 h_3) = I_{\delta_1}(h_1) \times I_{\delta_2}(h_2) \times I_{\delta_3}(h_3) \quad (27.2)$$

as a Cartesian product of the special binary intervals

$$I_1(\ell) = [2^{\ell-1}, 2^\ell), \; I_{-1}(\ell) = (-2^\ell, -2^{\ell-1}], \quad \text{for } \ell \geq 3, \quad (27.3)$$

and $I_1(2) = [0, 4)$, $I_{-1}(2) = (-4, 0]$.

It is easy to see that the "ball minus ball" type region in the 3-space $2^{h-1} \leq |\mathbf{x}| < 2^h$ with $h \geq 2$ integer is contained in the union of the aligned boxes

$$B(\delta_1 h_1, \delta_2 h_2, \delta_3 h_3), \; \delta_i = \pm 1, \; h_i \geq 2, \; (i = 1, 2, 3), \; \max_{1\leq i\leq 3} h_i = h - 1 \text{ or } h. \quad (27.4)$$

Let $\vartheta \in SO(3)$ be an arbitrary but fixed rotation, and consider the rotated lattice $\vartheta\mathbb{Z}^3$. We are going to estimate from above the number of rotated lattice points in a tilted box (27.2)–(27.3); formally, we estimate

$$Z_\vartheta(\delta_1 h_1, \delta_2 h_2, \delta_3 h_3) = \left|\vartheta\mathbb{Z}^3 \cap B(\delta_1 h_1, \delta_2 h_2, \delta_3 h_3)\right|. \quad (27.5)$$

We use a simple volume-comparison argument. For every $\vartheta\mathbf{n} \in B(\delta_1 h_1, \delta_2 h_2, \delta_3 h_3)$ take a copy of the rotated unit cube $\vartheta[-\frac{1}{2}, \frac{1}{2})^3$ centered at $\vartheta\mathbf{n}$, and consider their union. This union set is clearly contained in the $\sqrt{3}$-extension of the aligned box $B(\delta_1 h_1, \delta_2 h_2, \delta_3 h_3)$. Note that $\sqrt{3}$ is the diameter of the unit cube in the 3-space, and the $\sqrt{3}$-extension of the box (27.2)–(27.3) means that for $\delta = 1$ a factor interval (see (27.3))

$$I_1(\ell) = [2^{\ell-1}, 2^\ell) \text{ is replaced by } [-\sqrt{3} + 2^{\ell-1}, 2^\ell + \sqrt{3}) \text{ for } \ell \geq 3, \quad (27.6)$$

$I_1(2) = [0, 4)$ is replaced by $[-\sqrt{3}, 4 + \sqrt{3})$, and the usual mirror image for $\delta = -1$.

The $\sqrt{3}$-extension of the aligned box $B(\delta_1 h_1, \delta_2 h_2, \delta_3 h_3)$ is a larger box of volume

$$\prod_{i=1}^{3} \max\left\{2^{h_i-1} + 2\sqrt{3}, 4 + 2\sqrt{3}\right\}. \tag{27.7}$$

Thus we have

$$\text{volume}(\sqrt{3}\text{–extension}) = \prod_{i=1}^{3} \max\left\{2^{h_i-1} + 2\sqrt{3}, 4 + 2\sqrt{3}\right\}. \tag{27.8}$$

Comparing the volumes, we obtain

$$Z_\vartheta(\delta_1 h_1, \delta_2 h_2, \delta_3 h_3) = \left|\vartheta\mathbb{Z}^3 \cap B(\delta_1 h_1, \delta_2 h_2, \delta_3 h_3)\right|$$

$$= \text{volume(union set)} \le \text{volume}(\sqrt{3}\text{–extension})$$

$$= \prod_{i=1}^{3} \max\left\{2^{h_i-1} + 2\sqrt{3}, 4 + 2\sqrt{3}\right\}, \tag{27.9}$$

where we combined (27.5) and (27.8).

We summarize these simple facts in a lemma.

Lemma 27.1. *For every integer $h \ge 2$ the "ball minus ball" type region in the 3-space $2^{h-1} \le |\mathbf{x}| < 2^h$ is contained in the union of the aligned boxes (see (27.4))*

$$B(\delta_1 h_1, \delta_2 h_2, \delta_3 h_3),\ \delta_i = \pm 1,\ h_i \ge 2,\ (i = 1, 2, 3),\ \max_{1 \le i \le 3} h_i = h - 1 \text{ or } h.$$

Moreover, for every rotation $\vartheta \in SO(3)$ and every aligned box $B(\delta_1 h_1, \delta_2 h_2, \delta_3 h_3)$,

$$Z_\vartheta(\delta_1 h_1, \delta_2 h_2, \delta_3 h_3) = \left|\vartheta\mathbb{Z}^3 \cap B(\delta_1 h_1, \delta_2 h_2, \delta_3 h_3)\right|$$

$$\le \prod_{i=1}^{3} \max\left\{2^{h_i-1} + 2\sqrt{3}, 4 + 2\sqrt{3}\right\} \tag{27.10}$$

□

Now we are ready to estimate the sum (26.24) under the condition (25.27). Let $\mathbf{n}_1, \mathbf{n}_2, \mathbf{n}_3 \in \mathbb{Z}^3 \setminus \mathbf{0}$, then there is an integer $h \ge 1$ such that

$$2^{h-1} \le \max_{1 \le i \le 3} |\mathbf{n}_i| < 2^h. \tag{27.11}$$

Without loss of generality we can assume that $\mathbf{n}_3$ is the longest one. By (25.27) and (27.11),

$$|a_{\mathbf{n}_3}| \le \frac{1}{\max\{|\vartheta\mathbf{n}_3(1)|, 1\} \cdot \max\{|\vartheta\mathbf{n}_3(2)|, 1\} \cdot \max\{|\vartheta\mathbf{n}_3(3)|, 1\}}$$

$$\leq \min\left\{\frac{1}{2^{h-1}-2},1\right\}. \tag{27.12}$$

Combining these simple facts with the first part of Lemma 27.1, we can estimate (27.24) as follows:

$$S_3(\mathbf{0};f) = \sum_{\substack{(\mathbf{n}_1,\mathbf{n}_2,\mathbf{n}_3):\ \mathbf{n}_i\in\mathbb{Z}^3\setminus\mathbf{0}\\ \mathbf{n}_1+\mathbf{n}_2+\mathbf{n}_3=\mathbf{0}}} |a_{\mathbf{n}_1}|\,|a_{\mathbf{n}_2}|\,|a_{\mathbf{n}_3}|$$

$$\leq 3\sum_{h=2}^{\infty}\left(\sum_{\substack{\delta_i=\pm1\\ 1\leq i\leq 3}}\ \sum_{\substack{(h_1,h_2,h_3):h_i\geq 2\\ \max_i h_i\leq h}}\ \sum_{\vartheta\mathbf{n}\in B(\delta_1h_1,\delta_2h_2,\delta_3h_3)} |a_{\mathbf{n}}|\right)^2 \cdot\min\left\{\frac{1}{2^{h-1}-2},1\right\}, \tag{27.13}$$

where the factor 3 comes from the 3 choices for the longest vector among $\mathbf{n}_1$, $\mathbf{n}_2$ and $\mathbf{n}_3$.

By using the definition of the box $B(\delta_1h_1,\delta_2h_2,\delta_3h_3)$ and (25.27),

$$\sum_{\vartheta\mathbf{n}\in B(\delta_1h_1,\delta_2h_2,\delta_3h_3)} |a_{\mathbf{n}}| \leq \left|\vartheta\mathbb{Z}^3\cap B(\delta_1h_1,\delta_2h_2,\delta_3h_3)\right|$$

$$\cdot\frac{1}{\max\{|\vartheta\mathbf{n}(1)|,1\}\cdot\max\{|\vartheta\mathbf{n}(2)|,1\}\cdot\max\{|\vartheta\mathbf{n}(3)|,1\}}$$

$$\leq\left(\prod_{i=1}^{3}\max\left\{2^{h_i-1}+2\sqrt{3},4+2\sqrt{3}\right\}\right)\cdot\frac{1}{2^{h_1-1}\cdot 2^{h_2-1}\cdot 2^{h_3-1}}$$

$$=\prod_{i=1}^{3}\frac{\max\left\{2^{h_i-1}+2\sqrt{3},4+2\sqrt{3}\right\}}{2^{h_i-1}}\leq\prod_{i=1}^{3}\frac{\max\left\{2^{h_i-1}+4,8\right\}}{2^{h_i-1}}$$

$$=\prod_{i=1}^{3}\min\left\{1+2^{-h_i+3},2^{-h_i+4}\right\}, \tag{27.14}$$

where in the last step we used the second part of Lemma 27.1.

Note that

$$\sum_{j=2}^{h}\min\left\{1+2^{-j+3},2^{-j+4}\right\} = 4+2+\left(1+\frac{1}{2}\right)+\left(1+\frac{1}{4}\right)$$

$$+\ldots+\left(1+\frac{1}{2^{h-3}}\right) < 4+2+(h-3)+1 = h+4. \tag{27.15}$$

Using these simple facts and (27.14)–(27.15) in (27.13), we obtain

$$S_3(\mathbf{0};f) \le 3\sum_{h=2}^{\infty}\left(2^3(h+4)^3\right)^2 \cdot \min\left\{\frac{1}{2^{h-1}-2},1\right\}$$

$$= 3\cdot 2^6 \sum_{h=2}^{\infty} \min\left\{\frac{(h+4)^6}{2^{h-1}-2},(h+4)^6\right\} < 10^8, \tag{27.16}$$

where the last step is a routine numerical calculation estimating a convergent series.

The case $k \ge 4$ in (26.25) can be estimated similarly. A straightforward analog of (27.14) goes as follows:

$$S_k(\mathbf{0};f) = \sum_{\substack{(\mathbf{n}_1,\mathbf{n}_2,\ldots,\mathbf{n}_k):\ \mathbf{n}_i\in\mathbb{Z}^3\setminus\mathbf{0}\\ \mathbf{n}_1+\mathbf{n}_2+\ldots+\mathbf{n}_k=\mathbf{0}}} |a_{\mathbf{n}_1}|\,|a_{\mathbf{n}_2}|\cdots|a_{\mathbf{n}_k}|$$

$$\le k\sum_{h=2}^{\infty}\left(\sum_{\substack{\delta_i=\pm1\\ 1\le i\le k}}\ \sum_{\substack{(h_1,h_2,\ldots,h_k):h_i\ge2\\ \max_i h_i\le h}}\ \sum_{\vartheta\mathbf{n}\in B(\delta_1h_1,\delta_2h_2,\delta_3h_3)} |a_{\mathbf{n}}|\right)^{k-1}$$

$$\cdot\min\left\{\frac{1}{|2^{h-1}-k+1|},1\right\}. \tag{27.17}$$

The same way as we derived (27.16) from (27.14) and (27.15), we obtain from (27.17) and (27.15) the following:

$$S_k(\mathbf{0};f) \le k\sum_{h=2}^{\infty}\left(2^3(h+4)^3\right)^{k-1} \cdot \min\left\{\frac{1}{|2^{h-1}-k+1|},1\right\}$$

$$= k\cdot 2^{3(k-1)} \sum_{h=2}^{\infty} \min\left\{\frac{(h+4)^{3(k-1)}}{|2^{h-1}-k+1|},(h+4)^{3(k-1)}\right\}. \tag{27.18}$$

We estimate the end-sum in (27.18):

$$\sum_{h=2}^{\infty} \min\left\{\frac{(h+4)^{3(k-1)}}{|2^{h-1}-k+1|},(h+4)^{3(k-1)}\right\} \le \sum_{\substack{h\ge2:\\ 2^{h-2}\le k}} (h+4)^{3(k-1)}$$

$$+4\sum_{\substack{h\ge2:\\ 2^h>4k}} \frac{(h+4)^{3(k-1)}}{2^h}. \tag{27.19}$$

The last sum in (27.19) motivates the study of the function

$$f(x) = (x+4)^{3(k-1)}2^{-x} \quad \text{in the interval } x > 2 + \log_2 k, \tag{27.20}$$

where $\log_2$ denotes the binary logarithm. The derivative

$$f'(x) = (3(k-1) - (x+4)\log 2)\,(x+4)^{3k-4}2^{-x} \tag{27.21}$$

shows that the function $f(x)$ is monotone increasing if

$$3(k-1) \geq (x+4)\log 2 \iff x \leq c_0 = \frac{3(k-1)}{\log 2} - 4,$$

the function attains its maximum at $x = c_0$, and becomes monotone decreasing if

$$3(k-1) < (x+4)\log 2 \iff x > c_0 = \frac{3(k-1)}{\log 2} - 4.$$

We have

$$\max_x f(x) = f(c_0) = (c_0+4)^{3(k-1)}2^{-c_0} = 2^4\left(\frac{3(k-1)}{e\log 2}\right)^{3(k-1)}. \tag{27.22}$$

We can rewrite (27.21) in the form

$$f'(x) = \left(\frac{3(k-1)}{x+4} - \log 2\right) f(x),$$

so using the logarithmic integral, for every $v > u \geq c_0 = \frac{3(k-1)}{\log 2} - 4$ we have

$$\log\frac{f(v)}{f(u)} = \log f(v) - \log f(u) = \int_u^v \frac{f'(x)}{f(x)}\,dx = \int_u^v \left(\frac{3(k-1)}{x+4} - \log 2\right) dx. \tag{27.23}$$

Clearly

$$\frac{3(k-1)}{x+4} - \log 2 = \frac{3(k-1)}{(x-c_0)+(c_0+4)} - \log 2 = \frac{3(k-1)}{(x-c_0)+\frac{3(k-1)}{\log 2}} - \log 2$$

$$= \log 2\left(\frac{3(k-1)}{(x-c_0)\log 2 + 3(k-1)} - 1\right) = -\log 2\frac{x-c_0}{(x-c_0)+(c_0+4)},$$

and using it in (27.23), we obtain

$$\log\frac{f(v)}{f(u)} = \int_u^v \left(\frac{3(k-1)}{x+4} - \log 2\right) dx = -\log 2\int_u^v \frac{x-c_0}{(x-c_0)+(c_0+4)}\,dx. \tag{27.24}$$

In the special case $v = 2c_0 + 5$ and $u = 2c_0 + 4$ equation (27.24) yields

$$\log \frac{f(v)}{f(u)} \leq -\log 2 \int_u^v \frac{c_0 + 4}{(c_0 + 4) + (c_0 + 4)}\, dx = -\log \sqrt{2}, \tag{27.25}$$

and so

$$f(2c_0 + 5) \leq \frac{1}{\sqrt{2}} f(2c_0 + 4).$$

Similarly, by repeating the argument of (27.25), we obtain

$$f(2c_0 + y + 1) \leq \frac{1}{\sqrt{2}} f(2c_0 + y) \quad \text{for every real } y \geq 4. \tag{27.26}$$

Combining the arguments in (27.20)–(27.26), we can easily estimate the last sum in (27.19):

$$\begin{aligned}
\sum_{\substack{h \geq 2:\\ 2^{h-2} > k}} \frac{(h+4)^{3(k-1)}}{2^h} &\leq (2c_0 + 5) \max_x f(x) + \left(\sum_{j=1}^{\infty} 2^{-j/2} \right) \max_x f(x) \\
&= (2c_0 + 5) f(c_0) + \frac{2^{-1/2}}{1 - 2^{-1/2}} f(c_0) < (2c_0 + 8) f(c_0) \\
&= \frac{6(k-1)}{\log 2} \cdot 2^4 \left(\frac{3(k-1)}{e \log 2} \right)^{3(k-1)}.
\end{aligned} \tag{27.27}$$

Furthermore, we have the trivial upper bound for the other sum in (27.19):

$$\sum_{\substack{h \geq 2:\\ 2^{h-2} \leq k}} (h+4)^{3(k-1)} \leq (6 + \log_2 k)^{3(k-1)+1}. \tag{27.28}$$

Combining (27.19), (27.27) and (27.28),

$$\begin{aligned}
\sum_{h=2}^{\infty} \min \left\{ \frac{(h+4)^{3(k-1)}}{|2^{h-1} - k + 1|}, (h+4)^{3(k-1)} \right\} &\leq \sum_{\substack{h \geq 2:\\ 2^{h-2} \leq k}} (h+4)^{3(k-1)} \\
&\quad + \sum_{\substack{h \geq 2:\\ 2^{h-2} > k}} \frac{(h+4)^{3(k-1)}}{2^{h-2}} \\
&\leq \frac{6(k-1)}{\log 2} \cdot 2^6 \left(\frac{3(k-1)}{e \log 2} \right)^{3(k-1)} + (6 + \log_2 k)^{3(k-1)+1} < (3k)^{3k-2}
\end{aligned}$$

holds for every integer $k \geq 4$.

Using this in (27.18), we conclude

$$S_k(\mathbf{0};f) \le k\cdot 2^{3(k-1)} \sum_{h=2}^{\infty} \min\left\{ \frac{(h+4)^{3(k-1)}}{|2^{h-1}-k+1|}, (h+4)^{3(k-1)} \right\}$$

$$\le k\cdot 2^{3(k-1)}\cdot (3k)^{3k-2} = \frac{k}{2}(6k)^{3k-2}. \tag{27.29}$$

Combining (27.16) and (27.29), we conclude Section 27 with the following lemma.

Lemma 27.2. *Assume that the Fourier coefficients $a_{\mathbf{n}}$, $\mathbf{n}\in\mathbb{Z}^3$ of $f\in L^2(I^3)$ satisfy the upper bound (25.27). Then the sum*

$$S_3(\mathbf{0};f) = \sum_{\substack{(\mathbf{n}_1,\mathbf{n}_2,\mathbf{n}_3):\ \mathbf{n}_i\in\mathbb{Z}^3\setminus\mathbf{0}\\ \mathbf{n}_1+\mathbf{n}_2+\mathbf{n}_3=\mathbf{0}}} |a_{\mathbf{n}_1}a_{\mathbf{n}_2}a_{\mathbf{n}_3}|$$

has the upper bound

$$S_3(\mathbf{0};f)\le 10^8.$$

In the general case, for every $k\ge 4$ integer, the sum

$$S_k(\mathbf{0};f) = \sum_{\substack{(\mathbf{n}_1,\ldots,\mathbf{n}_k):\ \mathbf{n}_i\in\mathbb{Z}^3\setminus\mathbf{0}\\ \mathbf{n}_1+\ldots+\mathbf{n}_k=\mathbf{0}}} |a_{\mathbf{n}_1}\cdots a_{\mathbf{n}_k}|$$

has the upper bound

$$S_k(\mathbf{0};f)\le \frac{k}{2}\cdot(6k)^{3k-2}.$$

□

Given a nonzero lattice point $\mathbf{w}\in\mathbb{Z}^3\setminus\mathbf{0}$, we can generalize the sums (26.25) as follows: for $k\ge 2$ let

$$S_k(\mathbf{w};f) = \sum_{\substack{(\mathbf{n}_1,\ldots,\mathbf{n}_k):\ \mathbf{n}_i\in\mathbb{Z}^3\setminus\mathbf{0}\\ \mathbf{n}_1+\ldots+\mathbf{n}_k=\mathbf{w}}} |a_{\mathbf{n}_1}\cdots a_{\mathbf{n}_k}|.$$

For $k=2$ the Cauchy–Schwarz inequality gives the upper bound

$$S_2(\mathbf{w};f) = \sum_{\substack{(\mathbf{n}_1,\mathbf{n}_2):\ \mathbf{n}_i\in\mathbb{Z}^3\setminus\mathbf{0}\\ \mathbf{n}_1+\mathbf{n}_2=\mathbf{w}}} |a_{\mathbf{n}_1}\cdots a_{\mathbf{n}_k}|$$

$$\le \left(\sum_{\mathbf{n}_1\in\mathbb{Z}^3\setminus\mathbf{0}} |a_{\mathbf{n}_1}|^2\right)^{1/2}\cdot\left(\sum_{\substack{\mathbf{n}_1\in\mathbb{Z}^3\setminus\mathbf{0}:\\ \mathbf{w}-\mathbf{n}_1\ne\mathbf{0}}} |a_{\mathbf{w}-\mathbf{n}_1}|^2\right)^{1/2}$$

$$\leq \sum_{\mathbf{n} \in \mathbb{Z}^3 \setminus \mathbf{0}} |a_{\mathbf{n}}|^2 = \sigma_0^2(f) = S_2(\mathbf{0}; f). \tag{27.30}$$

For $k \geq 3$, we can repeat the proof of Lemma 27.2, and obtain the analog results. Under the condition of Lemma 27.2, for every nonzero lattice point $\mathbf{w} \in \mathbb{Z}^3 \setminus \mathbf{0}$,

$$S_3(\mathbf{w}; f) \leq 10^8 \text{ and } S_k(\mathbf{w}; f) \leq \frac{k}{2} \cdot (6k)^{3k-2} \text{ for every } k \geq 4. \tag{27.31}$$

Indeed, the proof of Lemma 27.2 is based on the simple observation that, the longest lattice point among $\mathbf{n}_1, \ldots, \mathbf{n}_k$ is uniquely determined by the rest, and of course this holds in both cases $\mathbf{n}_1 + \ldots + \mathbf{n}_k = \mathbf{0}$ and $\mathbf{n}_1 + \ldots + \mathbf{n}_k = \mathbf{w}$ (with a fixed $\mathbf{w}$).

We complete the proof of Theorem 26.1 in the next section.

28. Proof of Theorem 26.1 (II)

Now we are ready to estimate $\mathbf{Moment}_{2m}(\mathcal{Y}; \mathcal{R}; f; t; \text{PartTwo})$; see (27.10)–(27.11). Given a corresponding sequence $\mathbf{n}_j, k_j, 1 \leq j \leq 2m$ in (27.7), we study the nonzero multiplicities

$$\text{mult}(s) = \text{mult}(\mathbf{n}_j, k_j, 1 \leq j \leq 2m; s)$$

where $1 \leq s \leq N$. Let $\mu_1 > \ldots > \mu_\ell$ denote the multiplicities ≥ 4; possibly $\ell = 0$. If $\ell \geq 1$, suppose that $\text{mult}(s) = \mu_1$ for $h_1 \geq 1$ values of s, $\text{mult}(s) = \mu_2$ for $h_2 \geq 1$ values of s, and so on, finally suppose that $\text{mult}(s) = \mu_\ell$ for $h_\ell \geq 1$ values of s. Then

$$\mu_1 > \ldots > \mu_\ell \geq 4 \text{ and } \sum_{i=1}^{\ell} h_i \mu_i \leq 2m. \tag{28.1}$$

If $\sum_{i=1}^{\ell} h_i \mu_i < 2m$ then of course there must exist a multiplicity between 1 and 3. Suppose that $\text{mult}(s) = 1$ for α values of s, $\text{mult}(s) = 2$ for β values of s, and finally, $\text{mult}(s) = 3$ for γ values of s. Clearly

$$\sum_{i=1}^{\ell} h_i \mu_i + 3\gamma + 2\beta + \alpha = 2m. \tag{28.1'}$$

Note that the sum $\mathbf{Moment}_{2m}(\mathcal{Y}; \mathcal{R}; f; t; \text{PartTwo})$ is characterized by the property $\beta < m$.

By (26.7), (26.10) and (26.11),

$$\mathbf{Moment}_{2m}(\mathcal{Y}; \mathcal{R}; f; t; \text{PartTwo})$$

$$\leq \sum_{\ell\geq 0} \sum_{\substack{(h_1,\mu_1;\ldots;h_\ell,\mu_\ell;\gamma;\beta;\alpha):h_i\geq 1,\alpha,\gamma\geq 0,0\leq\beta<m\\ \mu_1>\ldots>\mu_\ell\geq 4,\sum_{i=1}^{\ell} h_i\mu_i+3\gamma+2\beta+\alpha=2m}} \binom{N}{h_1}\binom{2m}{\mu_1}\cdots\binom{2m-(h_1-1)\mu_1}{\mu_1}$$

$$\cdot\binom{N-h_1}{h_2}\binom{2m-h_1\mu_1}{\mu_2}\cdots\binom{2m-h_1\mu_1-(h_2-1)\mu_2}{\mu_2}$$

$$\cdots\binom{2m-\sum_{i=1}^{\ell} h_i\mu_i+\mu_\ell}{\mu_\ell}\binom{N-h_1-\ldots-h_\ell}{\gamma}\binom{N-h_1-\ldots-h_\ell-\gamma}{\beta}$$

$$\cdot\binom{N-h_1-\ldots-h_\ell-\gamma-\beta}{\alpha}\frac{\left(2m-\sum_{i=1}^{\ell} h_i\mu_i\right)!}{(3!)^{\gamma}2^{\beta}}\cdot\left(S_{\mu_1}(f;\tau)\right)^{h_1}$$

$$\left(S_{\mu_2}(f;\tau)\right)^{h_2}\cdots\left(S_{\mu_\ell}(f;\tau)\right)^{h_\ell}\left(S_3(f;\tau)\right)^{\gamma}\left(S_2(f;\tau)\right)^{\beta}\left(S_1(f;\tau)\right)^{\alpha}, \quad (28.2)$$

where

$$S_k(f;x) = \sum_{\substack{(\mathbf{n}_1,\ldots,\mathbf{n}_k):\\ \mathbf{n}_i\in\mathbb{Z}^3\setminus\mathbf{0},1\leq i\leq k}} \left|a_{\mathbf{n}_1}\cdots a_{\mathbf{n}_k}\right| e^{-|\mathbf{n}_1+\ldots+\mathbf{n}_k|^2x^2/2} \quad (28.3)$$

for every integer $k\geq 1$, and

$$\tau = 2\pi\mathcal{R}(t) \ \text{ with } \ \mathcal{R}(t) = \min_{1\leq k\leq N} |\mathbf{r}_k(t)|, \quad (28.4)$$

where $\mathcal{R}(t)$ is the "drifting distance".

Note that for $k\geq 2$ (see the end of Section 27 after Lemma 27.2)

$$S_k(f;x) = S_k(\mathbf{0};f) + \sum_{\mathbf{w}\in\mathbb{Z}^3\setminus\mathbf{0}} S_k(\mathbf{w};f)e^{-|\mathbf{w}|^2x^2/2}, \quad (28.5)$$

and for $k=1$ we simply have

$$S_1(f;x) = \sum_{\mathbf{n}\in\mathbb{Z}^3\setminus\mathbf{0}} |a_{\mathbf{n}}|\, e^{-|\mathbf{n}|^2x^2/2}. \quad (28.6)$$

By using (27.31) and Lemma 27.2 in (28.5), for all $k\geq 4$,

$$S_k(f;x) = S_k(\mathbf{0};f) + \sum_{\mathbf{w}\in\mathbb{Z}^3\setminus\mathbf{0}} S_k(\mathbf{w};f)e^{-|\mathbf{w}|^2x^2/2}$$

$$\leq \frac{k}{2}(6k)^{3k-2}\cdot B(x), \ \text{ where } \ B(x) = 1 + \sum_{\mathbf{w}\in\mathbb{Z}^3\setminus\mathbf{0}} e^{-|\mathbf{w}|^2x^2/2}. \quad (28.7)$$

Similarly, by (27.31) and Lemma 27.2,

$$S_3(f;x) \leq 10^8 B(x), \quad (28.8)$$

and by (27.30),

$$S_2(f;x) \le \sigma_0^2(f)B(x). \tag{28.9}$$

In (28.6) we use the Cauchy–Schwarz inequality:

$$S_1(f;x) = \sum_{\mathbf{n}\in\mathbb{Z}^3\setminus\mathbf{0}} |a_{\mathbf{n}}|\, e^{-|\mathbf{n}|^2x^2/2}$$

$$\le \left(\sum_{\mathbf{n}\in\mathbb{Z}^3\setminus\mathbf{0}} |a_{\mathbf{n}}|^2\right)^{1/2} \cdot \left(\sum_{\mathbf{n}\in\mathbb{Z}^3\setminus\mathbf{0}} e^{-x^2|\mathbf{n}|^2}\right)^{1/2}$$

$$= \sigma_0(f)A(x), \quad \text{where} \quad A(x) = \left(\sum_{\mathbf{n}\in\mathbb{Z}^3\setminus\mathbf{0}} e^{-x^2|\mathbf{n}|^2}\right)^{1/2}. \tag{28.10}$$

Let us return now to (28.2): by using the trivial inequality $\binom{M}{h} \le M^h/h!$ and the fact $\sum_{i=1}^{\ell} h_i\mu_i + 3\gamma + 2\beta + \alpha = 2m$, we have

$$\mathbf{Moment}_{2m}(\mathcal{Y};\mathcal{R};f;t;\text{PartTwo})$$

$$\le \sum_{\ell\ge 0} \sum_{\substack{(h_1,\mu_1;\ldots;h_\ell,\mu_\ell;\gamma;\beta;\alpha):h_i\ge 1,\alpha,\gamma\ge 0,0\le\beta<m \\ \mu_1>\ldots>\mu_\ell\ge 4,\sum_{i=1}^{\ell} h_i\mu_i+3\gamma+2\beta+\alpha=2m}} \frac{N^{2m}}{h_1!h_2!\cdots h_\ell!\gamma!\beta!\alpha!}$$

$$\cdot \frac{(2m)!}{(\mu_1!)^{h_1}(\mu_2!)^{h_2}\cdots(\mu_\ell!)^{h_\ell}(3!)^\gamma 2^\beta} \cdot N^m$$

$$\cdot \left(\frac{S_{\mu_1}(f;\tau)}{N^{\mu_1/2}}\right)^{h_1} \left(\frac{S_{\mu_2}(f;\tau)}{N^{\mu_2/2}}\right)^{h_2} \cdots \left(\frac{S_{\mu_\ell}(f;\tau)}{N^{\mu_\ell/2}}\right)^{h_\ell}$$

$$\cdot \left(\frac{S_3(f;\tau)}{N^{3/2}}\right)^{\gamma} \left(\frac{S_2(f;\tau)}{N}\right)^{\beta} \left(\frac{S_1(f;\tau)}{N^{1/2}}\right)^{\alpha}$$

$$= \frac{(2m)!}{2^m m!} N^m \sum_{\ell\ge 0} \sum_{\substack{(h_1,\mu_1;\ldots;h_\ell,\mu_\ell;\gamma;\beta;\alpha):h_i\ge 1,\alpha,\gamma\ge 0,0\le\beta<m \\ \mu_1>\ldots>\mu_\ell\ge 4,\sum_{i=1}^{\ell} h_i\mu_i+3\gamma+2\beta+\alpha=2m}} \frac{2^m m!}{h_1!h_2!\cdots h_\ell!\gamma!\beta!\alpha!}$$

$$\cdot \left(S_3(f;\tau)N^{-1/2}/3!\right)^{\gamma} \left(S_2(f;\tau)/2\right)^{\beta} \left(S_1(f;\tau)N^{1/2}\right)^{\alpha}$$

$$\cdot \prod_{i=1}^{\ell} \left(S_{\mu_i}(f;\tau)N^{1-(\mu_i/2)}/\mu_i!\right)^{h_i}. \tag{28.11}$$

Note that

$$\sum_{(h,\mu):h\geq 1,\mu\geq 4}\frac{1}{h!\mu^2}=\sum_{h=1}^{\infty}\frac{1}{h!}\sum_{\mu=4}^{\infty}\frac{1}{\mu^2}=(e-1)\left(\frac{\pi^2}{6}-1-\frac{1}{4}-\frac{1}{9}\right)<\frac{1}{2},$$

which implies that

$$\sum_{k\geq 0}\sum_{\substack{(h_1,\mu_1;\ldots;h_k,\mu_k):h_i\geq 1\\ \mu_1>\ldots>\mu_k\geq 4,\sum_{i=1}^k h_i\mu_i\leq 2m}}\frac{1}{h_1!\mu_1^2h_2!\mu_2^2\cdots h_k!\mu_k^2}$$

$$<\sum_{k\geq 0}\sum_{\substack{(h_1,\mu_1;\ldots;h_k,\mu_k):\\ h_i\geq 1,\mu_k\geq 4}}\frac{1}{h_1!\mu_1^2h_2!\mu_2^2\cdots h_k!\mu_k^2}$$

$$=\sum_{k\geq 0}\left(\sum_{(h,\mu):h\geq 1,\mu\geq 4}\frac{1}{h!\mu^2}\right)^k\leq\sum_{k\geq 0}(1/2)^k=2. \tag{28.12}$$

Using (28.12) in (28.11), we have

$$\mathbf{Moment}_{2m}(\mathcal{Y};\mathcal{R};f;t;\text{PartTwo})$$

$$\leq\frac{(2m)!}{2^m m!}N^m\cdot 2^m m!\sum_{\gamma\geq 0}\frac{\left(S_3(f;\tau)N^{-1/2}/3!\right)^{\gamma}}{\gamma!}$$

$$\cdot\sum_{0\leq\beta<m}\frac{(S_2(f;\tau)/2)^{\beta}}{\beta!}\sum_{\substack{\alpha\geq 0:\\ 3\gamma+2\beta+\alpha\leq 2m}}\frac{\left(S_1(f;\tau)N^{1/2}\right)^{\alpha}}{\alpha!}$$

$$\cdot 2\max_{\substack{(h_1,\mu_1;\ldots;h_\ell,\mu_\ell):\ell\geq 0,h_i\geq 1\\ \mu_1>\ldots>\mu_\ell\geq 4,\sum_{i=1}^{\ell}h_i\mu_i=2m-3\gamma-2\beta-\alpha}}\prod_{i=1}^{\ell}\mu_i^2\left(S_{\mu_i}(f;\tau)N^{1-(\mu_i/2)}/\mu_i!\right)^{h_i}. \tag{28.13}$$

Assume that

$$e^{2\pi^2(\mathcal{R}(t))^2}\geq 10^2N^{1/2}\iff\mathcal{R}(t)\geq\frac{\sqrt{\log N+4\log 10}}{2\pi}. \tag{28.14}$$

Clearly

$$\frac{\sqrt{\log N+4\log 10}}{2\pi}=\frac{\sqrt{\log N}}{2\pi}+\frac{\log 10}{\pi\sqrt{\log N}}+\ldots=\frac{\sqrt{\log N}}{2\pi}+o(1),$$

which shows that (28.14) is the same as (26.22) or the threshold in Corollary 3.1 (here we ignore the negligible term $o(1)$).

By using (28.4), (28.14) and some trivial lattice point counting in the 3-space, we have

$$A(\tau) = \left(\sum_{\mathbf{n}\in\mathbb{Z}^3\backslash\mathbf{0}} e^{-4\pi^2(\mathcal{R}(t))^2|\mathbf{n}|^2} \right)^{1/2} \leq \left(\sum_{\mathbf{n}\in\mathbb{Z}^3\backslash\mathbf{0}} \left(10^4N\right)^{-|\mathbf{n}|^2} \right)^{1/2}$$

$$\leq \left(6\left(10^4N\right)^{-1} + 12\left(10^4N\right)^{-2} + 8\left(10^4N\right)^{-3} + 6\left(10^4N\right)^{-4} + \dots\right)^{1/2}$$

$$\leq \left(6.25\left(10^4N\right)^{-1}\right)^{1/2} = 2.5\cdot 10^2 N^{-1/2},$$

since $N \geq 10^{27}$ is very large. By using this in (28.10),

$$S_1(f;\tau)N^{1/2} \leq 2.5\cdot 10^{-2}\sigma_0(f) = 2.5\cdot 10^{-2}\sqrt{pq} \leq 1.25\cdot 10^{-2}, \quad (28.15)$$

since f is the characteristic function of a set (the volume of this set is p, and $pq = p(1-p) \leq 1/4$).

By using (28.14) in (28.7) we have $B(\tau)$ is very close to 1; indeed,

$$B(\tau) - 1 \leq \sum_{\mathbf{w}\in\mathbb{Z}^3\backslash\mathbf{0}} \left(10^4N\right)^{-|\mathbf{w}|^2/2} = 6\left(10^4N\right)^{-1/2} + 12\left(10^4N\right)^{-1}$$

$$+ 8\left(10^4N\right)^{-3/2} + 6\left(10^4N\right)^{-2} + \ldots \leq 7\cdot 10^{-2}N^{-1/2},$$

since $N \geq 10^{27}$ is very large. Using this upper bound in (28.9), we have

$$S_2(f;\tau) \leq \left(1 + 7\cdot 10^{-2}N^{-1/2}\right)\sigma_0^2(f)$$

$$= \left(1 + 7\cdot 10^{-2}N^{-1/2}\right)pq \leq \left(1 + 7\cdot 10^{-2}N^{-1/2}\right)/4. \quad (28.16)$$

By using the notation

$$\xi = \left(1 + 7\cdot 10^{-2}N^{-1/2}\right), \quad (28.16')$$

we can rewrite (28.16) in the form

$$S_2(f;\tau) \leq \xi/4. \quad (28.16'')$$

Similarly, by (28.8) and $B(\tau) \leq \frac{6}{5}$,

$$S_3(f;\tau)N^{1-(3/2)}/3! \leq \frac{6}{5}\cdot 10^8 N^{-1/2}/6 = 10^8 N^{-1/2}/5. \quad (28.17)$$

To estimate the cases $k \geq 4$ we use the following trivial elementary lemma.

Lemma 28.1. *If* $N \geq 10^{27}$ *then*

$$\frac{k}{2}(6k)^{3k-2}N^{1-(k/2)}/k! \leq \frac{1}{2}10^{-3k} \quad \text{for all } 4 \leq k \leq 3400.$$

Proof. The cases $k = 4$ and 5 are trivial by direct inspection: for $k = 4$,

$$\frac{4}{2}(6 \cdot 4)^{10}N^{-1}/24 < \frac{1}{2}10^{-12},$$

and for $k = 5$,

$$\frac{5}{2}(6 \cdot 5)^{13}N^{-3/2}/120 < \frac{1}{2}10^{-15},$$

since $N \geq 10^{27}$.

For $6 \leq k \leq 3400$, we have

$$\frac{k}{2}(6k)^{3k-2}N^{1-(k/2)}/k! \leq \frac{e(6k)^{3k-2}}{2(k/e)^{k-1}\sqrt{N}^{k-2}}$$

$$= \frac{e(6k)^4}{2(k/e)}\left(\frac{(6k)^3}{(k/e)\sqrt{N}}\right)^{k-2} = \frac{e^2 6^4 k^3}{2}\left(\frac{e \cdot 6^3 k^2}{\sqrt{N}}\right)^{k-2}. \tag{28.18}$$

First assume that $6 \leq k \leq 100$. Then $e \cdot 6^3k^2 < 10^7$, and using it in (28.18) we have

$$\frac{k}{2}(6k)^{3k-2}N^{1-(k/2)}/k! \leq \frac{e^2 6^4 k^3}{2}\left(\frac{10^7}{\sqrt{N}}\right)^{k-2} < \frac{1}{2}10^{-3k},$$

since $N \geq 10^{27}$.

Finally, consider the case $100 < k \leq 3400$. Then

$$\frac{e \cdot 6^3 k^2}{\sqrt{N}} \leq e \cdot 6^3(3400)^2 10^{-13.5} \leq 10^{-3.5},$$

and using it in (28.8) we have

$$\frac{k}{2}(6k)^{3k-2}N^{1-(k/2)}/k! \leq \frac{e^2 6^4 k^3}{2} \cdot 10^{-3.5(k-2)} < \frac{1}{2}10^{-3k},$$

where in the last step we used that $k > 100$ is "large". This completes the proof of Lemma 28.1. □

Combining (28.7), (28.14), Lemma 28.1 and $B(\tau) \leq 2$, we have

$$S_k(f;\tau) \leq 10^{-3k} \quad \text{for all } 4 \leq k \leq 3400. \tag{28.19}$$

We choose $m = 2 \cdot 29^2 = 1684$ (notice that $2m < 3400$).

By using (28.14)–(28.17) and (28.19) in (28.13),

$$\mathbf{Moment}_{2m}(\mathcal{Y};\mathcal{R};f;t;\text{PartTwo})$$

$$\leq \frac{(2m)!}{2^m m!}N^m\cdot 2^m m!\sum_{\gamma\geq 0}\frac{\left(10^8 N^{-1/2}/5\right)^\gamma}{\gamma!}$$

$$\cdot\sum_{0\leq\beta<m}\frac{1}{(8/\xi)^\beta\beta!}\sum_{\substack{\alpha\geq 0:\\ 3\gamma+2\beta+\alpha\leq 2m}}\frac{\left(1.25\cdot 10^{-2}\right)^\alpha}{\alpha!}$$

$$\cdot 2\max_{\substack{(h_1,\mu_1;\ldots;h_\ell,\mu_\ell):\ell\geq 0,h_i\geq 1\\ \mu_1>\ldots>\mu_\ell\geq 4,\sum_{i=1}^\ell h_i\mu_i=2m-3\gamma-2\beta-\alpha}}\left(\prod_{i=1}^{\ell}\mu_i^2\right)10^{-3(2m-\alpha-2\beta-3\gamma)}$$

$$\leq \frac{(2m)!}{2^m m!}N^m\cdot 2^m m!\sum_{\gamma\geq 0}\frac{\left(10^8 N^{-1/2}/5\right)^\gamma}{\gamma!}$$

$$\cdot\sum_{0\leq\beta<m}\frac{1}{(8/\xi)^\beta\beta!}\sum_{\substack{\alpha\geq 0:\\ 3\gamma+2\beta+\alpha\leq 2m}}\frac{\left(1.25\cdot 10^{-2}\right)^\alpha}{\alpha!}$$

$$\cdot 2\left((2m-\alpha-2\beta-3\gamma)/4\right)^{\sqrt{2(2m-\alpha-2\beta-3\gamma)}}\cdot 10^{-3(2m-\alpha-2\beta-3\gamma)}, \quad (28.20)$$

where the factor

$$\left((2m-\alpha-2\beta-3\gamma)/4\right)^{\sqrt{2(2m-\alpha-2\beta-3\gamma)}}$$

comes from the facts $\mu_1>\ldots>\mu_\ell\geq 4$ and $\mu_1+\ldots+\mu_\ell\leq(2m-\alpha-2\beta-3\gamma)/4$ that trivially imply

$$\prod_{i=1}^{k}\mu_i^2\leq\left((2m-\alpha-2\beta-3\gamma)/4\right)^{\sqrt{2(2m-\alpha-2\beta-3\gamma)}}.$$

We simplify (28.20) to the following more transparent form

$$\mathbf{Moment}_{2m}(\mathcal{Y};\mathcal{R};f;t;\text{PartTwo})$$

$$\leq \frac{(2m)!}{2^m m!}N^m\cdot 2\sum_{0\leq\beta<m}\frac{1}{(4/\xi)^\beta}\cdot\frac{2^{m-\beta}\cdot\frac{m!}{\beta!}\cdot\left((m-\beta)/2\right)^{2\sqrt{m-\beta}}}{10^{6(m-\beta)}}$$

$$\cdot \sum_{\substack{\alpha\geq 0:\\ \alpha+2\beta\leq 2m}} \frac{\left(1.25\cdot 10^{3-2}\right)^{\alpha}}{\alpha!} \sum_{\substack{\gamma\geq 0:\\ \alpha+2\beta+3\gamma\leq 2m}} \frac{\left(10^{8+9}N^{-1/2}/5\right)^{\gamma}}{\gamma!}. \tag{28.21}$$

Clearly $m!/\beta! \leq m^{m-\beta}$; using this in (28.21), and applying the substitution $k = m - \beta$, we obtain

$$\mathbf{Moment}_{2m}(\mathcal{Y};\mathcal{R};f;t;\text{PartTwo})$$

$$\leq \frac{(2m)!}{2^m m!}N^m \cdot 2 \sum_{1\leq k\leq m} \frac{1}{(4/\xi)^{m-k}} \cdot \frac{2^k \cdot m^k \cdot (k/2)^{2\sqrt{k}}}{10^{6k}}$$

$$\cdot \sum_{\substack{\alpha\geq 0:\\ \alpha\leq 2k}} \frac{(1.25\cdot 10)^{\alpha}}{\alpha!} \sum_{\substack{\gamma\geq 0:\\ \alpha+3\gamma\leq 2k}} \frac{\left(10^{17}N^{-1/2}/5\right)^{\gamma}}{\gamma!}. \tag{28.22}$$

Note that

$$\frac{2^k}{(4/\xi)^{m-k}} = \frac{8^k\xi^{-k}}{(4/\xi)^m} \leq \frac{8^k}{(4/\xi)^m},$$

since $\xi > 1$ (see (28.16′)). Using this in (28.22), we have

$$\mathbf{Moment}_{2m}(\mathcal{Y};\mathcal{R};f;t;\text{PartTwo})$$

$$\leq \frac{(2m)!}{2^m m!}N^m \cdot 2 \sum_{1\leq k\leq m} \frac{1}{(4/\xi)^{m}} \cdot \left(\frac{8m\cdot (k/2)^{2/\sqrt{k}}}{10^6}\right)^k$$

$$\cdot \sum_{\substack{\alpha\geq 0:\\ \alpha\leq 2k}} \frac{(1.25\cdot 10)^{\alpha}}{\alpha!} \sum_{\substack{\gamma\geq 0:\\ \alpha+3\gamma\leq 2k}} \frac{10^{3\gamma}}{\gamma!}, \tag{28.22′}$$

since $10^{17}N^{-1/2}/5 \leq 10^{17-13.5}/5 < 10^3$.

Elementary calculus gives that

$$\max_{x>0}(x/2)^{2/\sqrt{x}} < 3,$$

and using this, and also $\alpha + 3\gamma \leq 2k$, we have

$$\left(\frac{8m\cdot (k/2)^{2/\sqrt{k}}}{10^6}\right)^k \leq \left(\frac{\sqrt{24m}}{10^3}\right)^{2k}$$

$$\leq \left(\frac{\sqrt{24m}}{10^3}\right)^{\alpha} \cdot \left(\frac{\sqrt{24m}}{10^3}\right)^{3\gamma} \cdot \left(\frac{\sqrt{24m}}{10^3}\right)^{2k-\alpha-3\gamma}. \tag{28.23}$$

Using (28.23) in (28.22′), we obtain

$$\mathbf{Moment}_{2m}(\mathcal{Y};\mathcal{R};f;t;\text{PartTwo})$$

$$\leq \frac{(2m)!}{2^m m!}N^m \cdot \frac{2}{(4/\xi)^m}\sum_{\alpha\geq 0}\frac{\left(1.25\sqrt{24m}10^{1-3}\right)^\alpha}{\alpha!}$$

$$\cdot\sum_{\gamma\geq 0}\frac{\left(\sqrt{24m}10^{1-3}\right)^{3\gamma}}{\gamma!}\sum_{h\geq 0}\left(\sqrt{24m}10^{-3}\right)^h, \tag{28.24}$$

noting that h plays the role of $2k-\alpha-3\gamma$ in (28.23).

Of course

$$\sum_{\alpha\geq 0}\frac{\left(1.25\sqrt{24m}10^{-4}\right)^\alpha}{\alpha!} = \exp\left(1.25\sqrt{24m}10^{-2}\right) < e^3, \tag{28.25}$$

and similarly,

$$\sum_{\gamma\geq 0}\frac{\left(\sqrt{24m}10^{-2}\right)^{3\gamma}}{\gamma!} = \exp\left(\left(\sqrt{24m}10^{-2}\right)^3\right) < e^{10}, \tag{28.26}$$

since

$$\sqrt{24m} = \sqrt{24\cdot 2\cdot 29^2} = 29\sqrt{48} < 210. \tag{28.27}$$

Finally, by using (28.27) one more time,

$$\sum_{h\geq 0}\left(\sqrt{24m}10^{-3}\right)^h \leq \sum_{h\geq 0}2^{-h} = 2. \tag{28.28}$$

Combining (28.24)–(28.28), we obtain

$$\mathbf{Moment}_{2m}(\mathcal{Y};\mathcal{R};f;t;\text{PartTwo})$$

$$\leq \frac{(2m)!}{2^m m!}N^m \cdot \frac{2}{(4/\xi)^m}\cdot e^3\cdot e^{10}\cdot 2 = \frac{(2m)!}{8^m m!}N^m\cdot \xi^m \cdot 4e^{13}. \tag{28.29}$$

By (28.16″),

$$\xi^m = \left(1+7\cdot 10^{-2}N^{-1/2}\right)^m \leq \left(1+7\cdot 10^{-2-13.5}\right)^{1682} < 5/4,$$

and using it in (28.29), we have

$$\mathbf{Moment}_{2m}(\mathcal{Y};\mathcal{R};f;t;\text{PartTwo})$$

$$\leq \frac{(2m)!}{8^m m!}N^m\cdot 5e^{13}. \tag{28.30}$$

By (26.8), (26.12) and (28.30),

$$\textbf{Moment}_{2m}(\mathcal{Y};\mathcal{R};f;t)$$

$$= \textbf{Moment}_{2m}(\mathcal{Y};\mathcal{R};f;t;\text{PartOne}) + \textbf{Moment}_{2m}(\mathcal{Y};\mathcal{R};f;t;\text{PartTwo})$$

$$\leq \binom{N}{m}(2m)!2^{-m}\left(\sigma_0^2(f)\right)^m + \frac{(2m)!}{8^m m!}N^m \cdot 5e^{13}$$

$$\leq \frac{(2m)!}{8^m m!}N^m\left(\left(4\sigma_0^2(f)\right)^m + 5e^{13}\right)$$

$$\leq \frac{(2m)!}{8^m m!}N^m\left(1+5e^{13}\right), \tag{28.31}$$

because $4\sigma_0^2(f) = 4pq = 4p(1-p) \leq 1$.

By using Stirling's formula and $m = 2\cdot 29^2 = 1682$ in (28.31),

$$\textbf{Moment}_{2m}(\mathcal{Y};\mathcal{R};f;t) \leq \frac{(2m)!}{8^m m!}N^m\left(1+5e^{13}\right)$$

$$\leq \frac{\sqrt{2}\cdot(2m/e)^{2m}\sqrt{4\pi m}}{8^m(m/e)^m\sqrt{2\pi m}}N^m\left(1+5e^{13}\right)$$

$$= \left(\frac{mN}{2e}\right)^m \cdot 2\left(1+5e^{13}\right) < \left(\frac{mN}{2e}\right)^m \cdot 10^{10},$$

completing the proof of Theorem 26.1. □

Remarks. Here we explain how to modify the proof of Theorem 26.1 to obtain (27.1). The hypothesis

$$\mathcal{R}(t) \geq \frac{\sqrt{\log N + 12\log 10}}{2\pi} \iff e^{2\pi^2(\mathcal{R}(t))^2} \geq 10^6 N^{1/2}$$

implies the following analog of (28.15):

$$S_1(f;\tau)N^{1/2} \leq 1.25\cdot 10^{-6}, \tag{28.32}$$

and similarly,

$$B(\tau) - 1 \leq \sum_{\mathbf{w}\in\mathbb{Z}^3\setminus\mathbf{0}} \left(10^{12}N\right)^{-|\mathbf{w}|^2/2} \leq 7\cdot 10^{-6}N^{-1/2},$$

since $N \geq 10^{39}$ is very large. Using this upper bound in (28.9), we have

$$S_2(f;\tau) \leq \left(1 + 7\cdot 10^{-6}N^{-1/2}\right)\sigma_0^2(f)$$

$$= \left(1 + 7 \cdot 10^{-6} N^{-1/2}\right) pq \le \left(1 + 7 \cdot 10^{-6} N^{-1/2}\right) /4. \tag{28.33}$$

By using the notation

$$\xi = \left(1 + 7 \cdot 10^{-6} N^{-1/2}\right), \tag{28.33$'$}$$

we can rewrite (28.33) in the form

$$S_2(f;\tau) \le \xi/4, \tag{28.33$''$}$$

which is an analog of (28.16″).

By repeating the proof of Lemma 28.1 with the other hypothesis $N \ge 10^{39}$ instead of $N \ge 10^{27}$, we obtain the following inequality.

First Analog of Lemma 28.1: If $N \ge 10^{39}$ then

$$\frac{k}{2}(6k)^{3k-2} N^{1-(k/2)}/k! \le \frac{1}{2} 10^{-4k} \quad \text{for all } 4 \le k \le 3400.$$

This inequality implies the following analog of (28.19):

$$S_k(f;\tau) \le 10^{-4k} \quad \text{for all } 4 \le k \le 3400. \tag{28.34}$$

Next we obtain the following analog of (28.20) (of course $m = 2{\cdot}29^2 = 1684$; note that $2m < 3400$):

$$\mathbf{Moment}_{2m}(\mathcal{Y}; \mathcal{R}; f; t; \text{PartTwo})$$

$$\le \frac{(2m)!}{2^m m!} N^m \cdot 2^m m! \sum_{\gamma \ge 0} \frac{\left(10^8 N^{-1/2}/5\right)^\gamma}{\gamma!}$$

$$\cdot \sum_{0 \le \beta < m} \frac{1}{(8/\xi)^\beta \beta!} \sum_{\substack{\alpha \ge 0: \\ 3\gamma + 2\beta + \alpha \le 2m}} \frac{\left(1.25 \cdot 10^{-6}\right)^\alpha}{\alpha!}$$

$$\cdot 2\left((2m - \alpha - 2\beta - 3\gamma)/4\right)^{\sqrt{2(2m-\alpha-2\beta-3\gamma)}} \cdot 10^{-4(2m-\alpha-2\beta-3\gamma)}. \tag{28.35}$$

We simplify (28.35) to the following more transparent form (an analog of (28.21)):

$$\mathbf{Moment}_{2m}(\mathcal{Y}; \mathcal{R}; f; t; \text{PartTwo})$$

$$\le \frac{(2m)!}{2^m m!} N^m \cdot 2 \sum_{0 \le \beta < m} \frac{1}{(4/\xi)^\beta} \cdot \frac{2^{m-\beta} \cdot \frac{m!}{\beta!} \cdot ((m-\beta)/2)^{2\sqrt{m-\beta}}}{10^{8(m-\beta)}}$$

$$\cdot \sum_{\substack{\alpha \geq 0: \\ \alpha+2\beta \leq 2m}} \frac{\left(1.25 \cdot 10^{4-6}\right)^{\alpha}}{\alpha!} \sum_{\substack{\gamma \geq 0: \\ \alpha+2\beta+3\gamma \leq 2m}} \frac{\left(10^{8+12} N^{-1/2}/5\right)^{\gamma}}{\gamma!}. \tag{28.36}$$

Using the inequality $m!/\beta! \leq m^{m-\beta}$, and applying the substitution $k = m - \beta$ in (28.36), we obtain

$$\mathbf{Moment}_{2m}(\mathcal{Y}; \mathcal{R}; f; t; \text{PartTwo})$$

$$\leq \frac{(2m)!}{2^m m!} N^m \cdot 2 \sum_{1 \leq k \leq m} \frac{1}{(4/\xi)^m} \cdot \left(\frac{2m \cdot (k/2)^{2/\sqrt{k}}}{10^8}\right)^k$$

$$\cdot \sum_{\substack{\alpha \geq 0: \\ \alpha \leq 2k}} \frac{\left(1.25 \cdot 10^{-2}\right)^{\alpha}}{\alpha!} \sum_{\substack{\gamma \geq 0: \\ \alpha+3\gamma \leq 2k}} \frac{1}{\gamma!}, \tag{28.37}$$

since $10^{20} N^{-1/2}/5 \leq 10^{20-19.5}/5 < 1$.

Using the elementary inequality

$$\max_{x>0} (x/2)^{2/\sqrt{x}} < 3$$

and $\alpha + 3\gamma \leq 2k$, we have

$$\left(\frac{2m \cdot (k/2)^{2/\sqrt{k}}}{10^8}\right)^k \leq \left(\frac{\sqrt{24m}}{10^4}\right)^{2k}$$

$$\leq \left(\frac{\sqrt{24m}}{10^4}\right)^{\alpha} \cdot \left(\frac{\sqrt{24m}}{10^4}\right)^{3\gamma} \cdot \left(\frac{\sqrt{24m}}{10^4}\right)^{2k-\alpha-3\gamma}, \tag{28.38}$$

which is the analog of (28.23). Using (28.38) in (28.37), we obtain

$$\mathbf{Moment}_{2m}(\mathcal{Y}; \mathcal{R}; f; t; \text{PartTwo})$$

$$\leq \frac{(2m)!}{2^m m!} N^m \cdot \frac{2}{(4/\xi)^m} \sum_{\alpha \geq 0} \frac{\left(1.25\sqrt{24m}10^{-2-4}\right)^{\alpha}}{\alpha!}$$

$$\cdot \sum_{\gamma \geq 0} \frac{\left(\sqrt{24m}10^{-4}\right)^{3\gamma}}{\gamma!} \sum_{\substack{h \geq 0: \\ \alpha+\gamma+h \geq 1}} \left(\sqrt{24m}10^{-4}\right)^{h}, \tag{28.39}$$

noting that h plays the role of $2k - \alpha - 3\gamma$ in (28.38) (notice that (28.39) is the analog of (28.24)).

Of course

$$\sum_{\alpha\geq 0}\frac{\left(1.25\sqrt{24m}10^{-6}\right)^{\alpha}}{\alpha!}=\exp\left(1.25\sqrt{24m}10^{-6}\right)<e^{10^{-3}},\tag{28.40}$$

and similarly,

$$\sum_{\gamma\geq 0}\frac{\left(\sqrt{24m}10^{-4}\right)^{3\gamma}}{\gamma!}=\exp\left(\left(\sqrt{24m}10^{-4}\right)^{3}\right)<e^{10^{-3}},\tag{28.41}$$

since

$$\sqrt{24m}=\sqrt{24\cdot 2\cdot 29^2}=29\sqrt{24\cdot 2}<29\cdot 7<210.\tag{28.42}$$

Finally, by using (28.42) one more time,

$$\sum_{h\geq 0}\left(\sqrt{24m}10^{-4}\right)^{h}\leq\sum_{h\geq 0}40^{-h}=\frac{40}{39}.\tag{28.43}$$

Combining (28.39)–(28.43), we obtain

$$\begin{aligned}&\mathbf{Moment}_{2m}(\mathcal{Y};\mathcal{R};f;t;\text{PartTwo})\\ &\leq\frac{(2m)!}{2^m m!}N^m\cdot\frac{2}{(4/\xi)^m}\cdot\left(e^{10^{-3}}\cdot e^{10^{-3}}\cdot\frac{40}{39}-1\right)\\ &<\frac{(2m)!}{8^m m!}N^m\cdot\xi^m\cdot\frac{1}{10}.\end{aligned}\tag{28.44}$$

By (28.33′),

$$\xi^m=\left(1+7\cdot 10^{-6}N^{-1/2}\right)^m\leq\left(1+7\cdot 10^{-6-19.5}\right)^{1682}<10/9,$$

and using it in (28.44), we have

$$\mathbf{Moment}_{2m}(\mathcal{Y};\mathcal{R};f;t;\text{PartTwo})\leq\frac{(2m)!}{8^m m!}N^m\cdot\frac{1}{9}.\tag{28.45}$$

By (26.8), (26.12) and (28.45),

$$\begin{aligned}&\mathbf{Moment}_{2m}(\mathcal{Y};\mathcal{R};f;t)\\ &=\mathbf{Moment}_{2m}(\mathcal{Y};\mathcal{R};f;t;\text{PartOne})+\mathbf{Moment}_{2m}(\mathcal{Y};\mathcal{R};f;t;\text{PartTwo})\\ &\leq\binom{N}{m}(2m)!2^{-m}\left(\sigma_0^2(f)\right)^m+\frac{(2m)!}{8^m m!}N^m\cdot\frac{1}{9}\\ &\leq\frac{(2m)!}{8^m m!}N^m\left(\left(4\sigma_0^2(f)\right)^m+\frac{1}{9}\right)\end{aligned}$$

$$\leq \frac{(2m)!}{8^m m!} N^m \left(1 + \frac{1}{9}\right), \tag{28.46}$$

because $4\sigma_0^2(f) = 4pq = 4p(1-p) \leq 1$.

By using Stirling's formula and $m = 2 \cdot 29^2 = 1682$ in (28.46),

$$\mathbf{Moment}_{2m}(\mathcal{Y}; \mathcal{R}; f; t) \leq \frac{(2m)!}{8^m m!} N^m \left(1 + \frac{1}{9}\right)$$

$$\leq \frac{\frac{6}{5} \cdot (2m/e)^{2m} \sqrt{4\pi m}}{8^m (m/e)^m \sqrt{2\pi m}} N^m \left(1 + \frac{1}{9}\right)$$

$$= \left(\frac{mN}{2e}\right)^m \cdot \frac{6\sqrt{2}}{5} \left(1 + \frac{1}{9}\right) < \left(\frac{mN}{2e}\right)^m \cdot 2,$$

completing the proof of (27.1). □

29. Stability beyond the Gaussian case (I)

Theorem 25.1 is about the general torus model of the first kind (Gaussian initial velocity distribution). Formally the basic difference between the first kind and the second kind is that the key superexponential function e^{-y^2} is replaced by y^{-1}, which has a completely different asymptotic behavior as $y \to \infty$. In particular, the series

$$\sum_{\mathbf{n} \in \mathbb{Z}^3 \setminus \mathbf{0}} e^{-|\mathbf{n}|^2} \quad \text{and} \quad \sum_{\mathbf{n} \in \mathbb{Z}^3 \setminus \mathbf{0}} |\mathbf{n}|^{-1}$$

are completely different: the first one is rapidly convergent and the second one is divergent. Note that the series

$$\sum_{\mathbf{n} \in \mathbb{Z}^3 \setminus \mathbf{0}} |\mathbf{n}|^{-2}$$

that corresponds to the torus model of the third kind is still divergent (power -2 is not enough, for convergence we need a power less than -3).

Despite this technical difficulty, we are able to prove an analog of Theorem 25.1 for the second kind; see Theorem 29.1 below. Note that there is also an analog of Theorem 29.1 for the third kind — the proof is a straightforward adaptation of the case of the second kind; see the Remarks at the end of Section 30.

Theorem 29.1. *Consider a general N-particle torus model of the second kind (see (4.1)–(4.2)) in the 3-dimensional unit torus I^3, equipped with a family of N continuous rectifiable parametrized space curves*

$$\mathcal{R} = (\mathbf{r}_1(t), \mathbf{r}_2(t), \ldots, \mathbf{r}_N(t))$$

such that $\mathbf{r}_k(0) = \mathbf{0}$. *Let*

$$\mathcal{Y} = \{\mathbf{y}_1, \mathbf{y}_2, \ldots, \mathbf{y}_N\}$$

be an arbitrary initial point configuration in the unit torus I^3 *(i.e., the position of the k-th particle at the start* $t = 0$ *is* $\mathbf{y}_k \in I^3$, $1 \le k \le N$*). So the trajectory of the kth particle is*

$$\mathbf{x}_k(t) = \mathbf{y}_k + \vartheta_k(\mathbf{r}_k(t)) \quad \text{modulo one,}$$

which describes the time evolution of the whole N-point system (note that $\vartheta_k \in SO(3)$, $1 \le k \le N$*).*

Let f be an arbitrary 0-1 valued measurable function defined on the torus I^3 *such that the Fourier coefficients* $a_{\mathbf{n}}$, $\mathbf{n} \in \mathbb{Z}^3$ *of f satisfy the upper bound (25.27).*

Assume that $N \ge 10^{39}$, *and at time* $t > 0$ *the drifting distance*

$$\mathcal{R}(t) = \min_{1 \le k \le N} |\mathbf{r}_k(t)|$$

satisfies the inequality

$$\mathcal{R}(t) \ge \frac{10^{13} N^{1/2}}{2\pi}.$$

Then with

$$D_f(\omega; t) = N(\mathcal{Y}; \mathcal{R}; \omega; f; t) - N \int_{I^3} f \, dV$$

we have

$$\text{ProdMeas}_2\left(|D_f(\omega; t)| \ge 29\sqrt{N}\right) < 10^{-700}.$$

Theorem 29.1 is a straightforward corollary of the following result.

Theorem 29.2. *Under the condition of Theorem 29.1, the high-power integral*

$$\mathbf{Moment}_{2m}(2; \mathcal{Y}; \mathcal{R}; f; t) = \int_{\Omega_2} (D_f(\omega; t))^{2m} \, d\text{ProdMeas}_2(\omega)$$

of the discrepancy function

$$D_f(\omega; t) = N(\mathcal{Y}; \mathcal{R}; \omega; f; t) - N \int_{I^3} f \, dV$$

satisfies the inequality

$$\mathbf{Moment}_{2m}(2; \mathcal{Y}; \mathcal{R}; f; t) < 2 \left(\frac{mN}{2e}\right)^m$$

with the choice $m = 2 \cdot 29^2 = 1682$.

Remarks. The deduction of Theorem 29.1 from Theorem 29.2 is exactly the same as the deduction of Theorem 25.1 from Theorem 26.1.

Theorem 29.1 has the same message as that of Theorem 25.1: once the typical time evolution reaches square-root equilibrium, the system stays in square-root equilibrium with factor 30 (say) with respect to all "nice" sets (rotated boxes, say) for an incredibly long time. Indeed, we can simply repeat the argument of Section 25.2 (see (25.20)–(25.23)). The major difference is in the "time threshold" t_0, which is defined by the equation about the drifting distance $\mathcal{R}(t_0) = \text{const}\sqrt{\log N}$ for the first kind and $\mathcal{R}(t_0) = \text{const}\sqrt{N}$ for the second kind. The crucial fact is that $\sqrt{\log N}$ is much, much smaller than $\sqrt{N}$.

Proof of Theorem 29.2. We begin our discussion with the analog of (28.3) for the torus model of the second kind: for $x \geq 1$ we write

$$S_k(2;f;x) = \sum_{\substack{(\mathbf{n}_1,\ldots,\mathbf{n}_k):\mathbf{n}_1+\ldots+\mathbf{n}_k=\mathbf{0}\\ \mathbf{n}_i\in\mathbb{Z}^3\setminus\mathbf{0},1\leq i\leq k}} |a_{\mathbf{n}_1}\cdots a_{\mathbf{n}_k}|$$

$$+ \sum_{\substack{(\mathbf{n}_1,\ldots,\mathbf{n}_k):\mathbf{n}_1+\ldots+\mathbf{n}_k\neq\mathbf{0}\\ \mathbf{n}_i\in\mathbb{Z}^3\setminus\mathbf{0},1\leq i\leq k}} |a_{\mathbf{n}_1}\cdots a_{\mathbf{n}_k}|\,\frac{1}{x|\mathbf{n}_1+\ldots+\mathbf{n}_k|}$$

$$= S_k(\mathbf{0};f) + \frac{1}{x}S_k(2;\neq\mathbf{0};f), \tag{29.1}$$

where

$$S_k(2;\neq\mathbf{0};f) = \sum_{\substack{(\mathbf{n}_1,\ldots,\mathbf{n}_k):\mathbf{n}_1+\ldots+\mathbf{n}_k\neq\mathbf{0}\\ \mathbf{n}_i\in\mathbb{Z}^3\setminus\mathbf{0},1\leq i\leq k}} |a_{\mathbf{n}_1}\cdots a_{\mathbf{n}_k}|\,\frac{1}{|\mathbf{n}_1+\ldots+\mathbf{n}_k|}, \tag{29.2}$$

and of course $a_{\mathbf{n}}$ are the Fourier coefficients of the function $f \in L^2(I^3)$ defined on the 3-dimensional unit torus I^3.

The case $k=1$ is particularly simple:

$$S_1(2;f;x) = \frac{1}{x}S_1(2;\neq\mathbf{0};f) = \frac{1}{x}\sum_{\mathbf{n}\in\mathbb{Z}^3\setminus\mathbf{0}} |a_{\mathbf{n}}|\,\frac{1}{|\mathbf{n}|}.$$

By using Lemma 27.1,

$$S_1(2;\neq\mathbf{0};f) = \sum_{\mathbf{n}\in\mathbb{Z}^3\setminus\mathbf{0}} \frac{|a_{\mathbf{n}}|}{|\mathbf{n}|}$$

$$\leq \sum_{h=2}^{\infty} \left(\sum_{\substack{\delta_i=\pm1 \\ 1\leq i\leq 3}} \sum_{\substack{(h_1,h_2,h_3):h_i\geq 2 \\ \max_i h_i \leq h}} \sum_{\vartheta\mathbf{n}\in B(\delta_1 h_1,\delta_2 h_2,\delta_3 h_3)} |a_{\mathbf{n}}| \right) \cdot \left\{2^{-h+1} \text{ or } 1\right\}, \tag{29.3}$$

where "2^{-h+1} or 1" means 2^{-h+1} for $h \geq 3$ and 1 for $h = 2$. We recall (27.14) and (27.15):

$$\sum_{\vartheta\mathbf{n}\in B(\delta_1 h_1,\delta_2 h_2,\delta_3 h_3)} |a_{\mathbf{n}}| \leq \prod_{i=1}^{3} \min\left\{1 + 2^{-h_i+3}, 2^{-h_i+4}\right\},$$

and

$$\sum_{j=2}^{h} \min\left\{1 + 2^{-j+3}, 2^{-j+4}\right\} < h + 4.$$

Using these simple facts in (29.3), we obtain

$$S_1(2; \neq \mathbf{0}; f) \leq \sum_{h=2}^{\infty} 2^3 (h+4)^3 \cdot \left\{2^{-h+1} \text{ or } 1\right\}$$

$$= 8\left(6^3 + \sum_{h=3}^{\infty} \frac{(h+4)^3}{2^{h-1}}\right) < 10^4, \tag{29.4}$$

where the last step is a routine numerical calculation estimating a convergent series.

Next we estimate $S_k(2; \neq \mathbf{0}; f)$ for $k \geq 2$:

$$S_k(2; \neq \mathbf{0}; f) = \text{Part}_1(2; f; k) + \text{Part}_2(2; f; k), \tag{29.5}$$

where

$$\text{Part}_1(2; f; k) = \sum_{\substack{(\mathbf{n}_1,\ldots,\mathbf{n}_k):\mathbf{n}_i\in\mathbb{Z}^3\setminus\mathbf{0} \\ 1\leq|\mathbf{n}_1+\ldots+\mathbf{n}_k|^3\leq\max_i|\mathbf{n}_i|/4}} |a_{\mathbf{n}_1}\cdots a_{\mathbf{n}_k}| \frac{1}{|\mathbf{n}_1+\ldots+\mathbf{n}_k|}, \tag{29.6}$$

and

$$\text{Part}_2(2; f; k) = \sum_{\substack{(\mathbf{n}_1,\ldots,\mathbf{n}_k):\mathbf{n}_i\in\mathbb{Z}^3\setminus\mathbf{0} \\ |\mathbf{n}_1+\ldots+\mathbf{n}_k|^3>\max_i|\mathbf{n}_i|/4}} |a_{\mathbf{n}_1}\cdots a_{\mathbf{n}_k}| \frac{1}{|\mathbf{n}_1+\ldots+\mathbf{n}_k|}. \tag{29.7}$$

To estimate $\text{Part}_1(2; f; k)$, we use $\mathbf{n}_0 = -(\mathbf{n}_1 + \ldots + \mathbf{n}_k)$ to denote the negative sum, and we use $\mathbf{n}_{max}$ to denote the longest vector among $\mathbf{n}_1, \ldots, \mathbf{n}_k$. In $\text{Part}_1(2; f; k)$ we have

$$|\mathbf{n}_0|^2 \leq (|\mathbf{n}_{max}|/4)^{2/3},$$

moreover, we have the general inequality

$$\prod_{j=1}^{3} \max\{|\vartheta \mathbf{n}(j)|, 1\} \geq \max\{|\mathbf{n}| - 2, 1\}$$

holds for every $\mathbf{n} \in \mathbb{Z}^3 \setminus \mathbf{0}$, and using these two facts (and always using (25.27)) in (29.6) we obtain the upper bound

$$\mathrm{Part}_1(2; f; k) = \sum_{\substack{(\mathbf{n}_1,\ldots,\mathbf{n}_k):\mathbf{n}_i\in\mathbb{Z}^3\setminus\mathbf{0} \\ 1\leq|\mathbf{n}_1+\ldots+\mathbf{n}_k|^3\leq|\mathbf{n}_{max}|/4}} |a_{\mathbf{n}_1}\cdots a_{\mathbf{n}_k}| \frac{1}{|\mathbf{n}_1 + \ldots + \mathbf{n}_k|}$$

$$\leq \sum_{\substack{(\mathbf{n}_1,\ldots,\mathbf{n}_k):\mathbf{n}_i\in\mathbb{Z}^3\setminus\mathbf{0} \\ 1\leq|\mathbf{n}_1+\ldots+\mathbf{n}_k|^3\leq|\mathbf{n}_{max}|/4}} \frac{1}{|\mathbf{n}_1 + \ldots + \mathbf{n}_k|} \prod_{i=1}^{k}\prod_{j=1}^{3} \frac{1}{\max\{|\vartheta \mathbf{n}_i(j)|, 1\}}$$

$$= \sum_{\substack{(\mathbf{n}_0,\mathbf{n}_1,\ldots,\mathbf{n}_k):\mathbf{n}_i\in\mathbb{Z}^3\setminus\mathbf{0} \\ \sum_{i=0}^k \mathbf{n}_i=\mathbf{0},1\leq|\mathbf{n}_0|^3\leq|\mathbf{n}_{max}|/4}} \frac{1}{|\mathbf{n}_0|} \prod_{j=1}^{3} \frac{1}{\max\{|\vartheta \mathbf{n}_{max}(j)|, 1\}}$$

$$\cdot \prod_{1\leq i\leq k:\mathbf{n}_i\neq\mathbf{n}_{max}} \prod_{j=1}^{3} \frac{1}{\max\{|\vartheta \mathbf{n}_i(j)|, 1\}}$$

$$\leq \sum_{\substack{(\mathbf{n}_0,\mathbf{n}_1,\ldots,\mathbf{n}_k):\mathbf{n}_i\in\mathbb{Z}^3\setminus\mathbf{0} \\ \sum_{i=0}^k \mathbf{n}_i=\mathbf{0},1\leq|\mathbf{n}_0|^3\leq|\mathbf{n}_{max}|/4}} \frac{1}{|\mathbf{n}_0|} \cdot \frac{1}{|\mathbf{n}_0|^2} \cdot \frac{(|\mathbf{n}_{max}|/4)^{2/3}}{\prod_{j=1}^{3} \max\{|\vartheta \mathbf{n}_{max}(j)|, 1\}}$$

$$\cdot \prod_{1\leq i\leq k:\mathbf{n}_i\neq\mathbf{n}_{max}} \prod_{j=1}^{3} \frac{1}{\max\{|\vartheta \mathbf{n}_i(j)|, 1\}}$$

$$\leq \sum_{\substack{(\mathbf{n}_0,\mathbf{n}_1,\ldots,\mathbf{n}_k):\mathbf{n}_i\in\mathbb{Z}^3\setminus\mathbf{0} \\ \sum_{i=0}^k \mathbf{n}_i=\mathbf{0},1\leq|\mathbf{n}_0|^3\leq|\mathbf{n}_{max}|/4}} \frac{(|\mathbf{n}_{max}|/4)^{2/3}}{\max\{|\mathbf{n}_{max}| - 2, 1\}}$$

$$\cdot \prod_{0\leq i\leq k:\mathbf{n}_i\neq\mathbf{n}_{max}} \prod_{j=1}^{3} \frac{1}{\max\{|\vartheta \mathbf{n}_i(j)|, 1\}}$$

$$\le \sum_{\substack{(\mathbf{n}_0,\mathbf{n}_1,\ldots,\mathbf{n}_k):\mathbf{n}_i\in\mathbb{Z}^3\setminus\mathbf{0}\\ \sum_{i=0}^k \mathbf{n}_i=\mathbf{0},1\le|\mathbf{n}_0|^3\le|\mathbf{n}_{max}|/4}} \frac{(|\mathbf{n}_{max}|/4)^{2/3}}{|\mathbf{n}_{max}|/2} \prod_{0\le i\le k:\mathbf{n}_i\ne\mathbf{n}_{max}} \prod_{j=1}^{3} \frac{1}{\max\{|\vartheta\mathbf{n}_i(j)|,1\}}$$

$$< \sum_{\substack{(\mathbf{n}_0,\mathbf{n}_1,\ldots,\mathbf{n}_k):\mathbf{n}_i\in\mathbb{Z}^3\setminus\mathbf{0}\\ \sum_{i=0}^k \mathbf{n}_i=\mathbf{0},1\le|\mathbf{n}_0|^3\le|\mathbf{n}_{max}|/4}} |\mathbf{n}_{max}|^{-1/3} \prod_{0\le i\le k:\mathbf{n}_i\ne\mathbf{n}_{max}} \prod_{j=1}^{3} \frac{1}{\max\{|\vartheta\mathbf{n}_i(j)|,1\}}, \tag{29.8}$$

where

$$|\mathbf{n}_{max}| = \max_{1\le i\le k} |\mathbf{n}_i| = \max_{0\le i\le k} |\mathbf{n}_i| \ge 4. \tag{29.9}$$

To estimate $\text{Part}_2(2; f; k)$ (see (29.7)) we use again $\mathbf{n}_{max}$ to denote the longest vector among $\mathbf{n}_1, \ldots, \mathbf{n}_k$, and of course we always use (25.27):

$$\text{Part}_2(2; f; k) = \sum_{\substack{(\mathbf{n}_1,\ldots,\mathbf{n}_k):\mathbf{n}_i\in\mathbb{Z}^3\setminus\mathbf{0}\\ |\mathbf{n}_1+\ldots+\mathbf{n}_k|^3>\max_i |\mathbf{n}_i|/4}} |a_{\mathbf{n}_1}\cdots a_{\mathbf{n}_k}| \frac{1}{|\mathbf{n}_1+\ldots+\mathbf{n}_k|}$$

$$\le \sum_{\substack{(\mathbf{n}_1,\ldots,\mathbf{n}_k):\mathbf{n}_i\in\mathbb{Z}^3\setminus\mathbf{0}\\ |\mathbf{n}_1+\ldots+\mathbf{n}_k|^3>|\mathbf{n}_{max}|/4}} \frac{1}{|\mathbf{n}_1+\ldots+\mathbf{n}_k|} \prod_{i=1}^{k}\prod_{j=1}^{3} \frac{1}{\max\{|\vartheta\mathbf{n}_i(j)|,1\}}$$

$$\le \sum_{(\mathbf{n}_1,\ldots,\mathbf{n}_k):\mathbf{n}_i\in\mathbb{Z}^3\setminus\mathbf{0}} \min\left\{(|\mathbf{n}_{max}|/4)^{-1/3},1\right\} \prod_{i=1}^{k}\prod_{j=1}^{3} \frac{1}{\max\{|\vartheta\mathbf{n}_i(j)|,1\}}. \tag{29.10}$$

By repeating the arguments of (27.14) and (27.15) in (29.10), we obtain the following analog of (27.18)–(27.19):

$$\text{Part}_2(2; f; k)$$

$$\le \sum_{(\mathbf{n}_1,\ldots,\mathbf{n}_k):\mathbf{n}_i\in\mathbb{Z}^3\setminus\mathbf{0}} \min\left\{(|\mathbf{n}_{max}|/4)^{-1/3},1\right\} \prod_{i=1}^{k}\prod_{j=1}^{3} \frac{1}{\max\{|\vartheta\mathbf{n}_i(j)|,1\}}$$

$$\le k\sum_{h=2}^{\infty} \left(2^3(h+4)^3\right)^{k-1} \cdot \min\left\{\frac{1}{2^{(h-1-2)/3}},1\right\}$$

$$< k\cdot 2^{3(k-1)} \sum_{h=2}^{\infty} \frac{(h+4)^{3(k-1)}}{\frac{1}{2}2^{h/3}} = k\cdot 2^{3(k-1)}\cdot 2\sum_{h=2}^{\infty} \frac{(h+4)^{3(k-1)}}{2^{h/3}}. \tag{29.11}$$

Repeating the same argument for (29.8), we have

$$\text{Part}_1(2; f; k)$$

$$\leq \sum_{\substack{(\mathbf{n}_0,\mathbf{n}_1,\ldots,\mathbf{n}_k):\mathbf{n}_i\in\mathbb{Z}^3\setminus\mathbf{0} \\ \sum_{i=0}^k \mathbf{n}_i=\mathbf{0}}} |\mathbf{n}_{max}|^{-1/3} \prod_{0\leq i\leq k:\mathbf{n}_i\neq\mathbf{n}_{max}} \prod_{j=1}^{3} \frac{1}{\max\{|\vartheta\mathbf{n}_i(j)|,1\}}$$

$$\leq k\sum_{h=2}^{\infty}\left(2^3(h+4)^3\right)^{k-1}\cdot\frac{1}{2^{(h-1)/3}}$$

$$= k\cdot 2^{3(k-1)}\cdot 2^{1/3}\sum_{h=2}^{\infty}\frac{(h+4)^{3(k-1)}}{2^{h/3}}.$$

Combining this with (29.5) and (29.11), we obtain for $k\geq 2$:

$$S_k(2;\neq\mathbf{0};f) = \text{Part}_1(2;f;k)+\text{Part}_2(2;f;k)$$

$$\leq k\cdot 2^{3(k-1)}\left(2^{1/3}+2\right)\sum_{h=2}^{\infty}\frac{(h+4)^{3(k-1)}}{2^{h/3}}. \tag{29.12}$$

The last sums in (29.12) motivates the study of the function

$$g(x) = (x+4)^{3(k-1)}2^{-x/3} \quad \text{in the interval } x>0. \tag{29.13}$$

Because of the similarity of (27.20) and (29.13), we basically repeat the arguments of (27.20)–(27.27). The derivative

$$g'(x) = \left(3(k-1)-(x+4)\frac{1}{3}\log 2\right)(x+4)^{3k-4}2^{-x/3} \tag{29.14}$$

shows that the function $g(x)$ is monotone increasing if

$$3(k-1)\geq(x+4)\frac{1}{3}\log 2 \iff x\leq c_0=\frac{9(k-1)}{\log 2}-4,$$

the function attains its maximum at $x=c_0$, and becomes monotone decreasing if

$$3(k-1)<(x+4)\frac{1}{3}\log 2 \iff x>c_0=\frac{9(k-1)}{\log 2}-4.$$

We have

$$\max_x g(x) = g(c_0) = (c_0+4)^{3(k-1)}2^{-c_0} = 2^4\left(\frac{9(k-1)}{e\log 2}\right)^{3(k-1)}, \tag{29.15}$$

which is the analog of (27.22). We can rewrite (29.14) in the form

$$g'(x) = \left(\frac{3(k-1)}{x+4} - \frac{1}{3}\log 2\right) g(x),$$

so using the logarithmic integral, for every $v > u \geq c_0 = \frac{9(k-1)}{\log 2} - 4$ we have

$$\log\frac{g(v)}{g(u)} = \log g(v) - \log g(u) = \int_u^v \frac{g'(x)}{g(x)}\,dx = \frac{1}{3}\int_u^v \left(\frac{9(k-1)}{x+4} - \log 2\right) dx. \tag{29.16}$$

Clearly

$$\frac{9(k-1)}{x+4} - \log 2 = \frac{9(k-1)}{(x-c_0)+(c_0+4)} - \log 2 = \frac{9(k-1)}{(x-c_0)+\frac{9(k-1)}{\log 2}} - \log 2$$

$$= \log 2\left(\frac{9(k-1)}{(x-c_0)\log 2 + 9(k-1)} - 1\right) = -\log 2 \frac{x-c_0}{(x-c_0)+(c_0+4)},$$

and using it in (29.16), we obtain

$$\log\frac{g(v)}{g(u)} = \frac{1}{3}\int_u^v \left(\frac{9(k-1)}{x+4} - \log 2\right) dx = -\frac{1}{3}\log 2\int_u^v \frac{x-c_0}{(x-c_0)+(c_0+4)}\,dx. \tag{29.17}$$

In the special case $v = 2c_0 + 5$ and $u = 2c_0 + 4$ equation (29.17) yields

$$\log\frac{g(v)}{g(u)} \leq -\frac{1}{3}\log 2\int_u^v \frac{c_0+4}{(c_0+4)+(c_0+4)}\,dx = -\log 2^{1/6}, \tag{29.18}$$

and so

$$g(2c_0+5) \leq \frac{1}{2^{1/6}} g(2c_0+4).$$

Similarly, by repeating the argument of (29.18), we obtain

$$g(2c_0+y+1) \leq \frac{1}{2^{1/6}} g(2c_0+y) \quad \text{for every real } y \geq 4. \tag{29.19}$$

Combining the arguments in(29.13)–(29.19), we can easily estimate the last sum in (29.12):

$$\sum_{h=2}^{\infty} \frac{(h+4)^{3(k-1)}}{2^{h/3}} \leq (2c_0+5)\max_x g(x) + \left(\sum_{j=1}^{\infty} 2^{-j/6}\right) \max_x g(x)$$

$$= (2c_0+5)g(c_0) + \frac{2^{-1/6}}{1-2^{-1/6}} g(c_0) < (2c_0+15)g(c_0)$$

$$= \left(\frac{18(k-1)}{\log 2}+7\right)\cdot 2^4\left(\frac{9(k-1)}{e\log 2}\right)^{3(k-1)}. \tag{29.20}$$

Using (29.20) in (29.12), for $k \geq 2$ we have

$$S_k(2;\neq \mathbf{0};f) = \mathrm{Part}_1(2;f;k)+\mathrm{Part}_2(2;f;k) \leq k\cdot 2^{3(k-1)+2}\sum_{h=2}^{\infty}\frac{(h+4)^{3(k-1)}}{2^{h/3}}$$

$$\leq k\cdot 2^{3(k-1)+2}\left(\frac{18(k-1)}{\log 2}+7\right)\cdot 2^4\left(\frac{9(k-1)}{e\log 2}\right)^{3(k-1)} < (18k)^{3k-1}. \tag{29.21}$$

Combining (29.2)–(29.4) and (29.21), we obtain the following analog of Lemma 27.2.

Lemma 29.1. *Assume that the Fourier coefficients $a_{\mathbf{n}}$, $\mathbf{n}\in \mathbb{Z}^3$ of $f\in L^2(I^3)$ satisfy the upper bound (25.27). Then*

$$S_1(2;\neq \mathbf{0};f) = \sum_{\mathbf{n}\in\mathbb{Z}^3\setminus\mathbf{0}}\frac{|a_{\mathbf{n}}|}{|\mathbf{n}|} < 10^{-4},$$

and for $k\geq 2$,

$$S_k(2;\neq \mathbf{0};f) = \sum_{\substack{(\mathbf{n}_1,\ldots,\mathbf{n}_k):\mathbf{n}_1+\ldots+\mathbf{n}_k\neq\mathbf{0}\\ \mathbf{n}_i\in\mathbb{Z}^3\setminus\mathbf{0},1\leq i\leq k}}\frac{|a_{\mathbf{n}_1}\cdots a_{\mathbf{n}_k}|}{|\mathbf{n}_1+\ldots+\mathbf{n}_k|} < (18k)^{3k-1}.$$

□

We complete the proof of Theorem 29.2 in the next section.

30. Stability beyond the Gaussian case (II)

For the torus model of the second kind we have the following analogs of (26.3) and (26.4):

$$D_f(\omega;t) = N(\mathcal{Y};\mathcal{R};\omega;f;t) - N\int_{I^3} f\,dV$$

$$= \sum_{\substack{\mathbf{n}\in\mathbb{Z}^3:\\ \mathbf{n}\neq\mathbf{0}}} a_{\mathbf{n}}\sum_{k=1}^{N} e^{2\pi i\mathbf{n}\cdot(\mathbf{y}_k+\vartheta_k(\mathbf{r}_k(t)))}, \tag{30.1}$$

and we integrate the $2m$-th power of the discrepancy function (30.1) over the "initial velocity space" $\omega=(\vartheta_1,\ldots,\vartheta_N)\in\Omega_2=(SO(3))^N$:

$$\mathbf{Moment}_{2m}(2;\mathcal{Y};\mathcal{R};f;t) = \int_{\Omega_2}\left(D_f(\omega;t)\right)^{2m}\,d\mathrm{ProdMeas}_2(\omega)$$

$$= \int_{\Omega_2} \left(\sum_{\substack{\mathbf{n}\in\mathbb{Z}^3:\\ \mathbf{n}\neq\mathbf{0}}} a_{\mathbf{n}} \sum_{k=1}^{N} e^{2\pi i\mathbf{n}\cdot(\mathbf{y}_k+\vartheta_k(\mathbf{r}_k(t)))} \right)^{2m} d\mathrm{ProdMeas}_2(\omega)$$

$$= \sum_{\substack{\mathbf{n}_1\in\mathbb{Z}^3:\\ \mathbf{n}_1\neq\mathbf{0}}} \sum_{k_1=1}^{N} \sum_{\substack{\mathbf{n}_2\in\mathbb{Z}^3:\\ \mathbf{n}_2\neq\mathbf{0}}} \sum_{k_2=1}^{N} \sum_{\substack{\mathbf{n}_3\in\mathbb{Z}^3:\\ \mathbf{n}_3\neq\mathbf{0}}} \sum_{k_3=1}^{N} \cdots \sum_{\substack{\mathbf{n}_{2m}\in\mathbb{Z}^3:\\ \mathbf{n}_{2m}\neq\mathbf{0}}} \sum_{k_{2m}=1}^{N} a_{\mathbf{n}_1}$$

$$\cdot a_{\mathbf{n}_2} a_{\mathbf{n}_3} \cdots a_{\mathbf{n}_{2m}} \int_{\Omega_2} \prod_{j=1}^{2m} e^{2\pi i\mathbf{n}_j\cdot(\mathbf{y}_{k_j}+\vartheta_{k_j}(\mathbf{r}_{k_j}(t)))}\, d\mathrm{ProdMeas}_2(\omega). \tag{30.2}$$

Similarly to the proof of Theorem 26.1, we make use of the concepts of *multiplicity* and *vector sum*; see (26.5) and (26.6). The only natural modification — coming from the switch from the *first kind* to the *second kind* — is that we use the fact

$$\frac{1}{4\pi} \int_{\mathbf{v}\in\mathbf{S}^2} e^{i\mathbf{w}\cdot\mathbf{v}}\, d\mathbf{v} = \frac{\sin(|\mathbf{w}|)}{|\mathbf{w}|} \tag{30.3}$$

that holds for every 3-dimensional vector $\mathbf{w}$.

Using (26.5), (26.6) and (30.3), we can rewrite (30.2):

$$\mathbf{Moment}_{2m}(2;\mathcal{Y};\mathcal{R};f;t) = \int_{\Omega_2} D_f^{2m}(\omega;t)\, d\mathrm{ProdMeas}_2(\omega)$$

$$= \sum_{\substack{\mathbf{n}_1\in\mathbb{Z}^3:\\ \mathbf{n}_1\neq\mathbf{0}}} \sum_{k_1=1}^{N} \sum_{\substack{\mathbf{n}_2\in\mathbb{Z}^3:\\ \mathbf{n}_2\neq\mathbf{0}}} \sum_{k_2=1}^{N} \sum_{\substack{\mathbf{n}_3\in\mathbb{Z}^3:\\ \mathbf{n}_3\neq\mathbf{0}}} \sum_{k_3=1}^{N} \cdots \sum_{\substack{\mathbf{n}_{2m}\in\mathbb{Z}^3:\\ \mathbf{n}_{2m}\neq\mathbf{0}}} \sum_{k_{2m}=1}^{N} a_{\mathbf{n}_1}$$

$$\cdot a_{\mathbf{n}_2} a_{\mathbf{n}_3} \cdots a_{\mathbf{n}_{2m}} \prod_{s=1}^{N} e^{2\pi i\vec{v}(s)\cdot\mathbf{y}_s} \cdot \frac{\sin(2\pi|\vec{v}(s)||\mathbf{r}_s(t)|)}{2\pi|\vec{v}(s)||\mathbf{r}_s(t)|}. \tag{30.4}$$

Similarly to Section 26, we divide $\mathbf{Moment}_{2m}(2;\mathcal{Y};\mathcal{R};f;t)$ into two parts:

$$\mathbf{Moment}_{2m}(2;\mathcal{Y};\mathcal{R};f;t) = \mathbf{Moment}_{2m}(2;\mathcal{Y};\mathcal{R};f;t;\mathrm{PartOne})$$

$$+\mathbf{Moment}_{2m}(2;\mathcal{Y};\mathcal{R};f;t;\mathrm{PartTwo}), \tag{30.5}$$

where $\mathbf{Moment}_{2m}(2;\mathcal{Y};\mathcal{R};f;t;\mathrm{PartOne})$ means the total contribution of the terms in the big sum in (30.4) for which every nonzero multiplicity is 2 and every vector sum is zero:

$$\vec{v}(\mathbf{n}_j, k_j, 1\le j\le 2m; s) = \mathbf{0}, \tag{30.6}$$

and $\mathbf{Moment}_{2m}(2;\mathcal{Y};\mathcal{R};f;t;\text{PartTwo})$ means the rest, i.e., it means the total contribution of the terms in the big sum in (30.4) for which at least one of the following two requirements hold:
Requirement One:

$$\text{there is a multiplicity} \notin \{0,2\};$$

Requirement Two:
there is a nonzero multiplicity $\text{mult}(\mathbf{n}_j, k_j, 1 \le j \le 2m; s) \ge 1$ which is paired with a nonzero vector sum

$$\vec{v}(\mathbf{n}_j, k_j, 1 \le j \le 2m; s) \ne \mathbf{0}.$$

By (30.4)–(30.6) we clearly have

$$\mathbf{Moment}_{2m}(2;\mathcal{Y};\mathcal{R};f;t;\text{PartOne})$$

$$= \binom{N}{m}\binom{2m}{2}\binom{2m-2}{2}\binom{2m-4}{2}\cdots\binom{4}{2}\binom{2}{2}\left(\sum_{\substack{\mathbf{n}\in\mathbb{Z}^3:\\ \mathbf{n}\ne\mathbf{0}}}|a_{\mathbf{n}}|^2\right)^m$$

$$= \binom{N}{m}(2m)!2^{-m}\left(\sigma_0^2(f)\right)^m, \tag{30.7}$$

which is the perfect analog of (26.12).

Next we estimate $\mathbf{Moment}_{2m}(2;\mathcal{Y};\mathcal{R};f;t;\text{PartTwo})$. Given a corresponding sequence $\mathbf{n}_j, k_j, 1 \le j \le 2m$ in (30.4), we study the nonzero multiplicities

$$\text{mult}(s) = \text{mult}(\mathbf{n}_j, k_j, 1 \le j \le 2m; s)$$

where $1 \le s \le N$. Let $\mu_1 > \ldots > \mu_\ell$ denote the multiplicities ≥ 3; possibly $\ell = 0$. If $\ell \ge 1$, suppose that $\text{mult}(s) = \mu_1$ for $h_1 \ge 1$ values of s, $\text{mult}(s) = \mu_2$ for $h_2 \ge 1$ values of s, and so on, finally suppose that $\text{mult}(s) = \mu_\ell$ for $h_\ell \ge 1$ values of s. Then

$$\mu_1 > \ldots > \mu_\ell \ge 3 \ \text{ and } \sum_{i=1}^{\ell} h_i\mu_i \le 2m. \tag{30.8}$$

If $\sum_{i=1}^{\ell} h_i\mu_i < 2m$ then of course there must exist a multiplicity $\in \{1,2\}$. Suppose that $\text{mult}(s) = 1$ for α values of s and $\text{mult}(s) = 2$ for β values of s. Clearly

$$\sum_{i=1}^{\ell} h_i\mu_i + 2\beta + \alpha = 2m. \tag{30.9}$$

Note that the sum $\mathbf{Moment}_{2m}(2;\mathcal{Y};\mathcal{R};f;t;\text{PartTwo})$ is characterized by the property $\beta < m$.

Let us return to (30.4): we have the following analog of (28.2):

$$\mathbf{Moment}_{2m}(2;\mathcal{Y};\mathcal{R};f;t;\text{PartTwo})$$

$$\leq \sum_{\ell\geq 0} \sum_{\substack{(h_1,\mu_1;\ldots;h_\ell,\mu_\ell;\beta;\alpha):h_i\geq 1,\alpha\geq 0,0\leq\beta<m \\ \mu_1>\ldots>\mu_\ell\geq 3,\sum_{i=1}^{\ell} h_i\mu_i+2\beta+\alpha=2m}} \binom{N}{h_1}\binom{2m}{\mu_1}\cdots\binom{2m-(h_1-1)\mu_1}{\mu_1}$$

$$\cdot\binom{N-h_1}{h_2}\binom{2m-h_1\mu_1}{\mu_2}\cdots\binom{2m-h_1\mu_1-(h_2-1)\mu_2}{\mu_2}$$

$$\cdots\binom{2m-\sum_{i=1}^{\ell} h_i\mu_i+\mu_\ell}{\mu_\ell}\binom{N-h_1-\ldots-h_\ell}{\beta}$$

$$\cdot\binom{N-h_1-\ldots-h_\ell-\beta}{\alpha}\frac{\left(2m-\sum_{i=1}^{\ell} h_i\mu_i\right)!}{2^\beta}\cdot\left(S_{\mu_1}(2;f;\tau)\right)^{h_1}$$

$$\left(S_{\mu_2}(2;f;\tau)\right)^{h_2}\cdots\left(S_{\mu_\ell}(2;f;\tau)\right)^{h_\ell}\left(S_2(2;f;\tau)\right)^{\beta}\left(S_1(2;f;\tau)\right)^{\alpha}, \quad (30.10)$$

where

$$\tau = 2\pi\mathcal{R}(t) \ \text{ with } \ \mathcal{R}(t) = \min_{1\leq k\leq N} |\mathbf{r}_k(t)| \quad (30.11)$$

($\mathcal{R}(t)$ is the "drifting distance"), and $S_k(2;f;x)$ was defined in (29.1):

$$S_k(2;f;x) = \sum_{\substack{(\mathbf{n}_1,\ldots,\mathbf{n}_k):\mathbf{n}_1+\ldots+\mathbf{n}_k=\mathbf{0} \\ \mathbf{n}_i\in\mathbb{Z}^3\setminus\mathbf{0},1\leq i\leq k}} |a_{\mathbf{n}_1}\cdots a_{\mathbf{n}_k}|$$

$$+ \sum_{\substack{(\mathbf{n}_1,\ldots,\mathbf{n}_k):\mathbf{n}_1+\ldots+\mathbf{n}_k\neq\mathbf{0} \\ \mathbf{n}_i\in\mathbb{Z}^3\setminus\mathbf{0},1\leq i\leq k}} |a_{\mathbf{n}_1}\cdots a_{\mathbf{n}_k}|\frac{1}{x|\mathbf{n}_1+\ldots+\mathbf{n}_k|}$$

$$= S_k(\mathbf{0};f) + \frac{1}{x}S_k(2;\neq\mathbf{0};f) \quad (30.12)$$

for $k\geq 2$, and

$$S_1(2;f;x) = \frac{1}{x}S_1(2;\neq\mathbf{0};f) = \frac{1}{x}\sum_{\mathbf{n}\in\mathbb{Z}^3\setminus\mathbf{0}}\frac{|a_{\mathbf{n}}|}{|\mathbf{n}|}. \quad (30.13)$$

The use of the function $S_k(2; f; x)$ in the upper bound (30.10) is explained by the trivial inequality

$$\frac{\sin x}{x} \le \min\left\{\frac{1}{|x|}, 1\right\} \quad \text{for all real } x,$$

where $\sin x/x$ comes from (30.3)–(30.4).

Applying Lemma 29.1 in (30.13), we have

$$S_1(2; f; x) < \frac{10^4}{x}. \tag{30.14}$$

Next we apply Lemma 29.1 in (30.12) with $k = 2$:

$$S_2(2; f; x) = S_2(\mathbf{0}; f) + \frac{1}{x} S_2(2; \neq \mathbf{0}; f)$$

$$= \sigma_0^2(f) + \frac{1}{x} S_2(2; \neq \mathbf{0}; f) < \sigma_0^2(f) + \frac{36^5}{x}. \tag{30.15}$$

Finally we apply Lemma 27.2 and Lemma 29.1 in (30.12) with $k \ge 3$:

$$S_k(2; f; x) = S_k(\mathbf{0}; f) + \frac{1}{x} S_k(2; \neq \mathbf{0}; f)$$

$$\le \frac{1}{12}(6k)^{3k-1} + \frac{(18k)^{3k-1}}{x}. \tag{30.16}$$

Let us return now to (30.10): by using the trivial inequality $\binom{M}{h} \le M^h/h!$ and the fact $\sum_{i=1}^{\ell} h_i \mu_i + 2\beta + \alpha = 2m$, we have the following analog of (28.11):

$$\mathbf{Moment}_{2m}(2; \mathcal{Y}; \mathcal{R}; f; t; \text{PartTwo})$$

$$\le \frac{(2m)!}{2^m m!} N^m \sum_{\ell \ge 0} \sum_{\substack{(h_1,\mu_1;\ldots;h_\ell,\mu_\ell;\beta;\alpha): h_i \ge 1, \alpha \ge 0, 0 \le \beta < m \\ \mu_1 > \ldots > \mu_\ell \ge 3, \sum_{i=1}^{\ell} h_i \mu_i + 2\beta + \alpha = 2m}} \frac{2^m m!}{h_1! h_2! \cdots h_\ell! \beta! \alpha!}$$

$$\cdot (S_2(f;\tau)/2)^\beta \left(S_1(f;\tau) N^{1/2}\right)^\alpha \cdot \prod_{i=1}^{\ell} \left(S_{\mu_i}(f;\tau) N^{1-(\mu_i/2)}/\mu_i!\right)^{h_i}. \tag{30.17}$$

Note that

$$\sum_{(h,\mu): h \ge 1, \mu \ge 3} \frac{1}{h! \mu^2} = \sum_{h=1}^{\infty} \frac{1}{h!} \sum_{\mu=3}^{\infty} \frac{1}{\mu^2} = (e-1)\left(\frac{\pi^2}{6} - 1 - \frac{1}{4}\right) < \frac{3}{4},$$

which implies that

$$\sum_{k\geq 0} \sum_{\substack{(h_1,\mu_1;\ldots;h_k,\mu_k):h_i\geq 1\\ \mu_1>\ldots>\mu_k\geq 3,\sum_{i=1}^k h_i\mu_i\leq 2m}} \frac{1}{h_1!\mu_1^2 h_2!\mu_2^2\cdots h_k!\mu_k^2}$$

$$<\sum_{k\geq 0} \sum_{\substack{(h_1,\mu_1;\ldots;h_k,\mu_k):\\ h_i\geq 1,\mu_k\geq 3}} \frac{1}{h_1!\mu_1^2 h_2!\mu_2^2\cdots h_k!\mu_k^2}$$

$$=\sum_{k\geq 0}\left(\sum_{(h,\mu):h\geq 1,\mu\geq 3}\frac{1}{h!\mu^2}\right)^k \leq \sum_{k\geq 0}(3/4)^k = 4. \tag{30.18}$$

Using (30.18) in (30.17), we have

$$\mathbf{Moment}_{2m}(2;\mathcal{Y};\mathcal{R};f;t;\text{PartTwo})$$

$$\leq \frac{(2m)!}{2^m m!}N^m \cdot 2^m m! \sum_{0\leq\beta<m}\frac{(S_2(2;f;\tau)/2)^\beta}{\beta!} \sum_{\substack{\alpha\geq 0:\\ 2\beta+\alpha\leq 2m}} \frac{\left(S_1(2;f;\tau)N^{1/2}\right)^\alpha}{\alpha!}$$

$$\cdot 4 \max_{\substack{(h_1,\mu_1;\ldots;h_\ell,\mu_\ell):\ell\geq 0,h_i\geq 1\\ \mu_1>\ldots>\mu_\ell\geq 3,\sum_{i=1}^\ell h_i\mu_i=2m-3\gamma-2\beta-\alpha}} \prod_{i=1}^{\ell}\mu_i^2\left(S_{\mu_i}(2;f;\tau)N^{1-(\mu_i/2)}/\mu_i!\right)^{h_i}. \tag{30.19}$$

Assume that

$$2\pi\mathcal{R}(t)\geq 10^{14}N^{1/2}. \tag{30.20}$$

(30.11), (30.14) and (30.20) imply the following analog of (28.15):

$$S_1(2;f;\tau)N^{1/2}\leq 10^4\cdot 10^{-14}=10^{-10}, \tag{30.21}$$

and similarly, by (30.15),

$$S_2(2;f;\tau)\leq \sigma_0^2(f)+36^5\cdot 10^{-14}$$

$$=pq+36^5\cdot 10^{-14}\leq\frac{1}{4}+10^{-6}=\frac{\xi}{4}\ \text{ with }\ \xi=1+4\cdot 10^{-6}. \tag{30.22}$$

By repeating the proof of Lemma 28.1 with $N\geq 10^{39}$ instead of $N\geq 10^{27}$, we obtain the following inequality.

Second Analog of Lemma 28.1: If $N\geq 10^{39}$ then

$$(18k)^{3k-1}N^{1-(k/2)}/k!\leq\frac{1}{2}10^{-4k}\ \text{ for all } 3\leq k\leq 3400.$$

Combining this inequality with the First Analog of Lemma 28.1 (see the end of Section 28), and using (30.16), we obtain the following analog of (28.19):

$$S_k(2; f; \tau) \leq 10^{-4k} \quad \text{for all } 4 \leq k \leq 3400. \tag{30.23}$$

Using (30.21)–(30.23) in (30.19), we obtain the following analog of (28.20) (note that $m = 2 \cdot 29^2 = 1684$, and so $2m < 3400$):

$$\begin{aligned} &\mathbf{Moment}_{2m}(2; \mathcal{Y}; \mathcal{R}; f; t; \text{PartTwo}) \\ &\leq \frac{(2m)!}{2^m m!} N^m \cdot 2^m m! \sum_{0 \leq \beta < m} \frac{1}{(8/\xi)^\beta \beta!} \sum_{\substack{\alpha \geq 0: \\ 2\beta + \alpha \leq 2m}} \frac{\left(10^{-10}\right)^\alpha}{\alpha!} \\ &\cdot 4 \cdot ((2m - \alpha - 2\beta)/4)^{\sqrt{2(2m-\alpha-2\beta)}} \cdot 10^{-4(2m-\alpha-2\beta)}. \end{aligned} \tag{30.24}$$

We simplify (30.24) to the following more transparent form (an analog of (28.21)):

$$\begin{aligned} &\mathbf{Moment}_{2m}(2; \mathcal{Y}; \mathcal{R}; f; t; \text{PartTwo}) \\ &\leq \frac{(2m)!}{2^m m!} N^m \cdot 4 \sum_{0 \leq \beta < m} \frac{1}{(4/\xi)^\beta} \cdot \frac{2^{m-\beta} \cdot \frac{m!}{\beta!} \cdot ((m-\beta)/2)^{2\sqrt{m-\beta}}}{10^{8(m-\beta)}} \\ &\cdot \sum_{\substack{\alpha \geq 0: \\ \alpha + 2\beta \leq 2m}} \frac{\left(10^{4-10}\right)^\alpha}{\alpha!}. \end{aligned} \tag{30.25}$$

Using the inequality $m!/\beta! \leq m^{m-\beta}$, and applying the substitution $k = m - \beta$ in (30.25), we obtain

$$\begin{aligned} &\mathbf{Moment}_{2m}(2; \mathcal{Y}; \mathcal{R}; f; t; \text{PartTwo}) \\ &\leq \frac{(2m)!}{2^m m!} N^m \cdot 4 \sum_{1 \leq k \leq m} \frac{1}{(4/\xi)^{m-k}} \cdot \frac{2^k \cdot m^k \cdot (k/2)^{2\sqrt{k}}}{10^{8k}} \\ &\cdot \sum_{\substack{\alpha \geq 0: \\ \alpha \leq 2k}} \frac{\left(10^{-10}\right)^\alpha}{\alpha!} \\ &\leq \frac{(2m)!}{2^m m!} N^m \cdot 4 \sum_{1 \leq k \leq m} \frac{1}{(4/\xi)^m} \cdot \left(\frac{8m \cdot (k/2)^{2/\sqrt{k}}}{10^8} \right)^k \cdot \sum_{\substack{\alpha \geq 0: \\ \alpha \leq 2k}} \frac{\left(10^{-2}\right)^\alpha}{\alpha!}, \end{aligned} \tag{30.26}$$

which is an analog of (28.22)–(28.22′).

Using the elementary inequality

$$\max_{x>0}(x/2)^{2/\sqrt{x}} < 3$$

and $\alpha \le 2k$, we have

$$\left(\frac{8m\cdot(k/2)^{2/\sqrt{k}}}{10^8}\right)^k \le \left(\frac{\sqrt{24m}}{10^4}\right)^{2k} \le \left(\frac{\sqrt{24m}}{10^4}\right)^{\alpha}\cdot\left(\frac{\sqrt{24m}}{10^4}\right)^{2k-\alpha}, \tag{30.27}$$

which is the analog of (28.23). Using (30.27) in (30.26), we obtain

$$\mathbf{Moment}_{2m}(2;\mathcal{Y};\mathcal{R};f;t;\mathrm{PartTwo})$$

$$\le \frac{(2m)!}{2^m m!}N^m\cdot\frac{4}{(4/\xi)^m}\sum_{\alpha\ge 0}\frac{\left(\sqrt{24m}10^{-6-4}\right)^{\alpha}}{\alpha!}\sum_{\substack{h\ge 0:\\ \alpha+h\ge 1}}\left(\sqrt{24m}10^{-4}\right)^h, \tag{30.28}$$

noting that h plays the role of $2k-\alpha$ in (30.27) (notice that (30.28) is the analog of (28.24)).

Of course

$$\sum_{\alpha\ge 0}\frac{\left(\sqrt{24m}10^{-10}\right)^{\alpha}}{\alpha!} = \exp\left(\sqrt{24m}10^{-10}\right) < e^{10^{-7}}, \tag{30.29}$$

since

$$\sqrt{24m} = \sqrt{24\cdot 2\cdot 29^2} = 29\sqrt{24\cdot 2} < 29\cdot 7 < 210, \tag{30.30}$$

and by using (30.30) one more time,

$$\sum_{h\ge 0}\left(\sqrt{24m}10^{-4}\right)^h \le \sum_{h\ge 0}40^{-h} = \frac{40}{39}. \tag{30.31}$$

Combining (30.28)–(30.31), we obtain

$$\mathbf{Moment}_{2m}(2;\mathcal{Y};\mathcal{R};f;t;\mathrm{PartTwo})$$

$$\le \frac{(2m)!}{2^m m!}N^m\cdot\frac{4}{(4/\xi)^m}\cdot\left(e^{10^{-7}}\cdot\frac{40}{39}-1\right) < \frac{(2m)!}{8^m m!}N^m\cdot\xi^m\cdot\frac{1}{9}. \tag{30.32}$$

By (30.22),

$$\xi^m = (1+4\cdot 10^{-6})^{1682} < \frac{9}{8},$$

and using it in (30.32), we have

$$\mathbf{Moment}_{2m}(2;\mathcal{Y};\mathcal{R};f;t;\mathrm{PartTwo}) \le \frac{(2m)!}{8^m m!}N^m\cdot\frac{1}{8}. \tag{30.33}$$

By (30.5), (30.7) and (30.33),

$$\mathbf{Moment}_{2m}(2;\mathcal{Y};\mathcal{R};f;t)$$

$$= \mathbf{Moment}_{2m}(2;\mathcal{Y};\mathcal{R};f;t;\text{PartOne}) + \mathbf{Moment}_{2m}(\mathcal{Y};\mathcal{R};f;t;\text{PartTwo})$$

$$\leq \binom{N}{m}(2m)!2^{-m}\left(\sigma_0^2(f)\right)^m + \frac{(2m)!}{8^m m!}N^m \cdot \frac{1}{8}$$

$$\leq \frac{(2m)!}{8^m m!}N^m\left(\left(4\sigma_0^2(f)\right)^m + \frac{1}{8}\right)$$

$$\leq \frac{(2m)!}{8^m m!}N^m\left(1+\frac{1}{8}\right), \tag{30.34}$$

because $4\sigma_0^2(f) = 4pq = 4p(1-p) \leq 1$.

By using Stirling's formula and $m = 2 \cdot 29^2 = 1682$ in (30.34),

$$\mathbf{Moment}_{2m}(2;\mathcal{Y};\mathcal{R};f;t) \leq \frac{(2m)!}{8^m m!}N^m\left(1+\frac{1}{8}\right)$$

$$\leq \frac{\frac{6}{5}\cdot(2m/e)^{2m}\sqrt{4\pi m}}{8^m(m/e)^m\sqrt{2\pi m}}N^m\left(1+\frac{1}{8}\right) = \left(\frac{mN}{2e}\right)^m \cdot \frac{6\sqrt{2}}{5}\left(1+\frac{1}{8}\right)$$

$$< \left(\frac{mN}{2e}\right)^m \cdot 2,$$

completing the proof. □

Remarks. It is easy to see that the proof of Theorem 29.2 works for the torus model of the third kind. Indeed, we know that the switch from the second kind to the third kind means that the key function

$$\frac{\sin x}{x} \quad \text{is replaced by} \quad -3\frac{\cos x}{x^2} + 3\frac{\sin x}{x^3},$$

see Lemma 8.1. In fact, we work with the upper bounds: the inequality

$$\frac{\sin x}{x} \leq \min\left\{\frac{1}{|x|}, 1\right\}$$

is replaced by

$$\left|-3\frac{\cos x}{x^2} + 3\frac{\sin x}{x^3}\right| \leq \min\left\{\frac{3}{x^2} + \frac{3}{|x|^3}, 1\right\} \leq \min\left\{\frac{6}{x^2}, 1\right\}.$$

That means we can simply repeat the proof of Theorem 29.2 by replacing the function $S_k(2; f; x)$ in (29.1)–(29.2) with the following new function: for $x \geq 1$ write

$$S_k(3; f; x) = \sum_{\substack{(\mathbf{n}_1,\ldots,\mathbf{n}_k):\mathbf{n}_1+\ldots+\mathbf{n}_k=\mathbf{0} \\ \mathbf{n}_i\in\mathbb{Z}^3\setminus\mathbf{0}, 1\leq i\leq k}} |a_{\mathbf{n}_1} \cdots a_{\mathbf{n}_k}|$$

$$+ \sum_{\substack{(\mathbf{n}_1,\ldots,\mathbf{n}_k):\mathbf{n}_1+\ldots+\mathbf{n}_k\neq\mathbf{0} \\ \mathbf{n}_i\in\mathbb{Z}^3\setminus\mathbf{0}, 1\leq i\leq k}} |a_{\mathbf{n}_1} \cdots a_{\mathbf{n}_k}| \frac{6}{x^2|\mathbf{n}_1 + \ldots + \mathbf{n}_k|^2}$$

$$= S_k(\mathbf{0}; f) + \frac{6}{x^2} S_k(3; \neq \mathbf{0}; f), \tag{30.35}$$

where

$$S_k(3; \neq \mathbf{0}; f) = \sum_{\substack{(\mathbf{n}_1,\ldots,\mathbf{n}_k):\mathbf{n}_1+\ldots+\mathbf{n}_k\neq\mathbf{0} \\ \mathbf{n}_i\in\mathbb{Z}^3\setminus\mathbf{0}, 1\leq i\leq k}} |a_{\mathbf{n}_1} \cdots a_{\mathbf{n}_k}| \frac{1}{|\mathbf{n}_1 + \ldots + \mathbf{n}_k|^2}$$

for $k \geq 2$, and

$$S_1(3; f; x) = \frac{1}{x^2} S_1(3; \neq \mathbf{0}; f) = \frac{6}{x^2} \sum_{\mathbf{n}\in\mathbb{Z}^3\setminus\mathbf{0}} |a_{\mathbf{n}}| \frac{1}{|\mathbf{n}|^2}.$$

Comparing (28.1) with (30.35) we see that x is replaced by x^2 (more precisely, by $x^2/6$), and this leads to a "square-root change in the time threshold". More precisely, the "time threshold" t_0 is defined by the equation about the drifting distance $\mathcal{R}(t_0) = \text{const}\sqrt{N}$ for the second kind and $\mathcal{R}(t_0) = \text{const}\sqrt{\sqrt{N}} = \text{const}N^{1/4}$ for the third kind. It follows that Theorem 29.2 (and also Theorem 29.1) has a straightforward generalization for the torus model of the third kind, and the only relevant change is that the "time threshold" t_0 is defined by the equation $\mathcal{R}(t_0) = \text{const}N^{1/4}$ (instead of $\mathcal{R}(t_0) = \text{const}\sqrt{N}$). The crucial fact here is that $N^{1/4}$ is much smaller than $\sqrt{N}$.

Appendix 1

The Second Law in Physics

The Second Law defines the thermodynamic arrow of time, which leads to the irreversibility paradox. The latter can be best summarized by a well known quote due to Lord Kelvin: "The equations of motion in (Newtonian) dynamics are perfectly reversible; any solution of these equations remain valid when the time variable t is replaced by $-t$. Physical processes, on the other hand, are irreversible: for example, the friction of solids, conduction of heat, and diffusion." Physicists explain the observed irreversibility of macroscopic systems with the Second Law of Thermodynamics, which is considered an axiom of physics. (Boltzmann claimed a "mathematical proof" of the Second Law via his H-theorem; we will return to this soon.) The Second Law says that in an isolated large (= many degrees of freedom) system, entropy (which is, very roughly speaking, a measure of microscopic "randomness") tends to increase with time. It is the increasing entropy that explains irreversibility.

Physics books usually rely on very convincing real life examples like "we never see a scrambled egg turn back to a whole egg" or "we never see a glass of water at uniform cool temperature turn back to a mix of ice cubes and warm water". Another typical textbook illustration goes as follows (see e.g. a truly great book: The Feynmann Lectures on Physics): suppose we have a box with two kinds of gases separated by a barrier in the middle: on the left side is neon (we may call them "red" molecules), and on the right side is argon (we may call them "blue" molecules). It is clear that if we pull the barrier out, everything will get mixed up after a long time. This change from a systematic arrangement ("red" on the left and "blue" on the right) to a mixed-up arrangement (random mix of "red" and "blue") is the reason behind the Second Law, and also explains irreversibility (arrow of time). Textbooks supply some related probabilistic calculation coming from the viewpoint of statistical mechanics — here is an oversimplified illustration. Tossing a fair coin 40 times, the probability of having 40 Heads — representing a "simple rule" — is about *hundred billion times* less likely than to have 20 Heads and 20 Tails — representing mixed-upness. The precise ratio is $\binom{40}{20} = \frac{40!}{20!20!} \approx 10^{11}$; if 40 is replaced with the Avogadro number (around 10^{23}) then of course the result becomes much more striking. This is how physics textbooks illustrate Boltzmann's classical observation that in a gas there are many more states of the system that seem random than seem systematically arranged. Well, this is great, but something is clearly missing from the picture: it is the dynamic aspects.

Other physics books refer to "chaos" (= instability with respect to small changes in the initial condition), or to cosmology, or to unknown features of quantum mechanics.

What physics books do not provide is a precise quantitative mathematical theorem, based on Newtonian mechanics, focusing on the individual trajectory of the system in the phase space, which supports the Second-Law/Arrow-of-Time.

The Second Law of thermodynamics was formulated more than hundred years ago, but the basic reasons for its validity remain mysterious, and the surrounding mathematics remains controversial. Perhaps the educated reader is wandering: wait a minute, what about Boltzmann's H-theorem? Or, what about the application of ergodic theory? Or what about the rigorous mathematics of the Ehrenfest's urn model, or the Kac's ring model, or Lanford's result on the hard core gas model?

Let me start with Boltzmann's H-theorem. First, the H-theorem is a statistical statement about the distribution function of particle velocity $\mathbf{v}$ in position $\mathbf{x}$ (infinitesimal element in the 6-dimensional phase space) at time t. That is, the H-theorem is *not* about the individual trajectory of the whole system (note that we *do* study the individual trajectory of the whole system). Second, and this is the main issue, there is a notorious problem with mathematical rigor. Boltzmann's H-theorem claims that particle-particle collision increases the entropy. Unfortunately, from the viewpoint of rigorous mathematics, Boltzmann's argument is controversial, because he (indirectly) assumed that molecules are uncorrelated before collision, but not after the collision, so he introduced a fundamental asymmetry in time. This is a logical fallacy. This is why many mathematicians are sceptical about the H-theorem: they believe that the arrow of time in the H-theorem is *not* in fact purely mechanical (due to the particle-particle collision), but really a consequence of assumptions about initial conditions ("Stosszahlensatz").

Let me elaborate on this a little bit. Boltzmann's H-theorem is a simple consequence of the so-called Boltzmann transport equation, a nonlinear integro-differential equation designed to describe the time evolution of rarefied gases. Boltzmann assumed "molecular chaos" (Stosszahlansatz), which means that there is no correlation between the particles. However, this is very "tricky" from a conceptual point of view, because when a collision occurs, then the involved particles *become correlated.* So correlations are continuously created, but, by the assumption of "molecular chaos", they are continuously forgotten. This has triggered a lot of discussions, endless debates, hundred years of controversy involving dozens of physicists and mathematicians. There is (basically) one rigorous result — on the hard core gas model — due to Lanford [L75]. It gives a description (under an *ad hoc* initial condition) of the time evolution of the system in a very short time interval, when most particles did not even have one collision yet. This is rather disappointing, since the real difficulty begins after every particle is already participated in at least a few dozen collisions. (Another problematic part in Lanford's work is the *ad hoc* initial condition involving the Boltzmann–Grad limit, but it is far too technical to be discussed here.) We conclude that Boltzmann's assumption of "molecular chaos" remains unproven, and may or may not be true. Thus, the whole issue of the justification of the Boltzmann equation on the basis of classical mechanics remains a notorious open problem.

Physicists in general have an unconditional love affair with Boltzmann's H-theorem. One likely reason is that the majority of physics books present Boltzmann's kinetic argument (particle-particle collision leads to entropy increase) as a rigorous mathematical proof, more-or-less hiding the fact that it is a conditional result based on the unproved hypothesis of "molecular chaos".

Of course it is perfectly natural to have a strong faith in the Second Law — as most physicists do — based on overwhelming experimental evidence. But one should honestly admit the lack of mathematical proof. Many physicists refuse to admit that, claiming the Boltzmann equation and the H-theorem settled the issue a long time ago.

Mathematicians, on the other hand, regard Boltzmann's argument a very exciting and challenging hypothesis/intuition that deserves a precise mathematical discussion/clarification, hopefully some kind of a mathematical proof. It is fair to say that present-day mathematics is not ready to clarify/prove the Boltzmann equation rigorously. This is rather disappointing, since the Boltzmann equation is widely used in nonequilibrium thermodynamics (and statistical mechanics). (There is a rather large literature of the Boltzmann equation, and its simplified versions. This line of research basically started with the great Hilbert. We cannot solve the Boltzmann equation in general, but there are some very interesting partial results; see e.g. Villani's work in [Vi2010].)

A well known old criticism of Boltzmann's H-theorem is due to Loschmidt, a teacher of Boltzmann, who replaced Kelvin's time switch $t \to -t$ with velocity reverse. For every motion of a large system $t > t_0$, we can associate with the reverse motion where every velocity is reversed at t_0, and the time evolution of the reverse motion contradicts the law of increasing entropy (Second Law). This is called the Loschmidt paradox.

Boltzmann gave several answers to the paradox. The shortest answer was: "Go on, reverse them!". This is partly a joke, and partly very serious, emphasizing the difficulty of carrying out a concrete experiment which reverses the velocity of every single gas particle.

Boltzmann's second answer to the Loschmidt-paradox/irreversibility-paradox was *statistical*, and it can be summarized as follows. Given the energy of a many-particle system, the overwhelming majority of its phase points on the corresponding energy surface area "equilibrium points", all of which look macroscopically the same. Since there are many more "equilibrium points" than "nonequilibrium points" in the phase space, it is plausible to expect that the time evolution of a typical trajectory (of the system) in the phase space exhibits a rapid transition from "nonequilibrium" to "equilibrium" as the time $t \to \infty$.

To make this qualitative picture more convincing, books usually supply some quantitative calculations motivated by the Boltzmann entropy. Consider a gas of N particles (with fixed energy E) initially confined by a barrier to the right half of a box container (the left half is empty), and suppose that this barrier is removed at time t_0. The phase space volume — we are talking about the $6N$-dimensional volume — available to this N-particle system for times $t > t_0$ is then enlarged compared to what it was initially by a factor of (roughly) 2^N. Since N is already in the range of the huge Avogadro number, 2^N is just "ridiculously large", beyond comprehension.

Well, this is a great insight, a very convincing intuition, we may call it a framework, but it is nowhere near to a rigorous mathematical proof. A big hole in the argument is how to switch from the statistical statement to the time-evolution of a typical individual trajectory (of the system) in the phase space. One may call it the dilemma of statistical mechanics versus the Newtonian deterministic viewpoint.

Can ergodic theory help? At first sight this seems extremely promising, since Birkhoff's well known ergodic theorem is exactly about the equality of the phase space average and the asymptotic time average ($t \to \infty$). (For example, the distinguished mathematician A.I. Khinchin included Birkhoff's ergodic theorem with proof in his book *Mathematical Foundations of Statistical Mechanics*, and declared it the status of being one of the three main mathematical ingredients of the theory; the other two were Liouville theorem and the central limit theorem.) Note that the ergodic theorem makes some of the early intuitions of Boltzmann and Gibbs precise. Boltzmann and Gibbs introduced the *ergodic hypothesis*, which roughly says that the curve representing the motion of the system in the phase space "passes through every point of the energy surface". In this rigid form the ergodic hypothesis is false, but later a meaningful weaker form has been worked out in terms of ergodic theory and the ergodic theorem.

Well, the first problem with the application of traditional measure-theoretic ergodic theory is that it ignores sets of volume zero in the phase space, and nonequilibrium usually represents a zero measure set; for example, starting from Big Bang. Another example: when all N particles are in the right half of the box, then the corresponding volume in the phase space is the ridiculously small 2^{-N}, which is practically zero.

The second problem is that Birkhoff's individual ergodic theorem — what we need to study individual orbits of systems — does not provide any error term. Even if we can prove that a given dynamical system is ergodic (which is already hard!), Birkhoff's ergodic theorem itself cannot answer the key question: how long does it take for the dynamical system to reach equilibrium. Our main interest here is exactly the time scale; what happens in realistic short time (from a split second to less than a day, say).

We briefly elaborate on the issue of time scale versus ergodic theory (see e.g. the books of Thompson [Th72] and Farquhar [Fa64]). By Liouville's well known theorem the physical system has a natural time-invariant measure in its phase space, and ergodic theory says that for a typical initial state the time evolution of the system will be described by that invariant measure in the limit when the time t grows to infinity; see Birkhoff's ergodic theorem. That invariant measure is a probability distribution that corresponds to the principles of equilibrium statistical mechanics. In other words, the laws of equilibrium statistical mechanics must "set in" eventually, in a very, very distant future ... — due to the unfortunate fact that Birkhoff's ergodic theorem does not give any error term. In other words, we have to assume infinite time. Note that there are a few finite quantitative result, but, unfortunately, the time scale seems to increase exponentially with the number of particles (Avogadro number), which is practically the same as infinite time — completely unrealistic.

Realistic short time is absolutely crucial here, since if we allow very, very long time — e.g., exponential in terms of the number of particles — then it is inevitable to have even a most absurd phenomenon such as the so-called Poincare recurrence — this is the subject of the well known Zermelo paradox. This famous paradox led to some confusion in early interpretations of the Second Law. It is based on Poincare's recurrence theorem (a special case of the ergodic theorem, discovered well before Birkhoff's ergodic theorem), which seemingly contradicts the Second Law/Arrow-of-Time. The recurrence theorem says that an arbitrary closed mechanical system *eventually* gets at least arbitrarily close to repeating itself, i.e., starting from "nonequilibrium", it eventually goes back to "nonequilibrium". (In sharp contrast to the Loschmidt paradox and the controversy around the Second Law, the solution of the Zermelo–Poincare recurrence paradox is quite easy. *Eventually* means such an enormous time — around 2^N — that the recurrence theorem has no practical relevance in large physical systems, where the particle-number N is in the range of the Avogadro number.)

Birkhoff's ergodic theorem — a beautiful qualitative result — gives a deep insight (and of course ergodic theory is an increasingly important branch of mathematics), but it has limited use in the foundations of statistical mechanics, unless it can say something quantitative about the time scale. Large systems reach equilibrium in realistic short time; Zermelo–Poincare recurrence happens in unrealistic time. A relevant mathematical theory should reflect the fundamental difference between the two cases; i.e., the theory should focus on realistic time scale.

It is quite surprising that even today, hundred years after Boltzmann and Gibbs, there is still an ongoing debate (often a shouting match!) about the Second Law. There is no consensus even about the foundations of Statistical Mechanics — not even in the equilibrium case, and especially not in the nonequilibrium case. The situation may be best characterized as a battlefield between many different schools; see e.g. the book of Sklar [S93]. The "ergodic theory school" is just one of them, and certainly not the most influential among physicists. There is of course the traditional school of "a priory probabilities", which in fact splits into two main groups: those who champion Boltzmann, and those who prefer Gibbs; we also have the Markovians — all of them use coarse-graining. Moreover, we have Jaynes, Prigogine, interventionism, and so on. The fact that such disagreement persists is a strong indication that the basic questions has not been completely answered yet.

The reason behind this diversity of opinions is that each approach has a weakness. As we said above, it is a weakness of the ergodic theory approach how to explain the (highly unlikely) nonequilibrium case. A weakness of the traditional approach is how to justify a priory probabilities in Newtonian mechanics, especially far from equilibrium. Similarly, a weakness of the Markovians is how to justify the Markovian postulate, which says that the successive observations of states of a macroscopic physical system constitute a Markov chain. A common weakness of these schools is that coarse-graining is "subjective" and "anthropomorphic".

Let us return to Boltzmann's statistical resolution of the Loschmidt paradox and the irreversibility paradox. Needless to say, it is a brilliant intuition, but it is quite vague in the sense that it lacks the dynamic aspects: it lacks the quantitative description of the mechanism of the time-evolution of a typical individual trajectory (of the system) in the phase space. To make such a quantitative analysis, books have to rely on non-mechanical toy models (since the realistic many-particle mechanical models are hopelessly difficult, partly due to the effect of particle-particle collisions). The most popular deterministic toy model is the Kac ring model, which is basically combinatorics. We quote a few sentences from the book of Mark Kac [Ka59], where this toy model was introduced. Kac wrote: "The objections of Loschmidt and Zermelo made it clear that the naive formulation of the H-theorem is untenable. Boltzman himself proposed that the H-theorem be interpreted statistically. In spite of the brilliant analysis of P. and T. Ehrenfest [Eh11] which elucidated Boltzmann's not always clear ideas and made them highly plausible, the question even now (45 years after Ehrenfest's work!) is not fully settled. There are broadly (and somewhat vaguely) speaking two problems.

Problem I: Is it possible to reconcile both time reversibility and recurrence with "observable" irreversibility?

Problem II: Is it possible to achieve such a reconciliation in the realm of classical mechanics?"

Kac then points out that the well known Ehrenfest urn model ("dog-flea model") is an attempt to solve Problem I. This urn model was introduced in 1907 in order to illustrate/clarify Boltzmann's somewhat vague ideas by working out the details on a rigorous mathematical model (P. Ehrenfest was a student of Boltzmann). The Ehrenfest urn model has an elegant complete solution (in fact, the first complete solution was found by Kac himself in

1947), and since then it is a standard part of the theory of finite Markov chains (almost every textbook in probability theory gives a more-or-less detailed description of the model). Unfortunately, the Ehrenfest model has an obvious shortcoming: it is a probabilistic model — a Markov chain, which initiated the Markovian school — that does not help to clarify the fundamental problem of how probability enters Newtonian mechanics. It is reasonable to guess that this shortcoming inspired Kac to introduce his deterministic ring model, which is closer to "reality". Kac's ring model is another contribution to Problem I.

Many physicists believe that Newtonian Mechanics and Statistical Mechanics (explaining thermodynamics) represent two different axiom systems. To quote an authority, in their well known book on statistical mechanics Landau and Lifshitz wrote (see [La-Li40]): "It is not clear whether the law of increasing entropy can be in principle derived from Newtonian mechanics." As far as we know every such attempt failed so far. The relevant kinetic equations — describing the time evolution to equilibrium — have all been obtained by some approximation, which introduces a non-deterministic element (like Boltzmann's "molecular chaos") that is absent in the original mechanical system. Therefore, a missing link has remained between mechanics and thermodynamics.

This is why Boltzmann's resolution of the irreversibility problem is not fully satisfying. It is a brilliant insight, but a proper resolution of the irreversibility problem should be carried out strictly inside Newtonian mechanics, describing the time evolution of a typical individual trajectory (of the system) in the phase space. This is the main motivation for Problem II.

Not much changed in the last 50 years since Kac formulated Problems I and II. There is still an ongoing debate about reversibility versus irreversibility. Most physics books give "soft" qualitative arguments. For example, they often use some information-theoretic interpretation referring to Shannon's entropy. A new trend is to refer to "chaos". The reason is that the Gibbs entropy (which, by the way, is preferred by the majority of physics textbooks) remains constant over time. So the "Gibbs school" has to find a way to go around this "paradox", and to explain the Second Law they usually rely on the vague concept of "chaos", supplying impressive pictures of fractals, but never provide any concrete relevant mathematical theorem.

In the rare cases when the books or lectures feel the need to include a convincing quantitative mathematical proof, they go back to the good old toy models, the Ehrenfest urn model and/or the Kac ring model, to explain the Second Law and irreversibility.

The two problems of Kac reflect the viewpoint of a mathematician (in his later years Kac became increasingly fascinated and involved in mathematical physics). We call them *mathematical* problems because the lack of rigorous mathematical foundations is not a major headache for the physicists — the majority of them are pragmatists anyway. Most physicists are perfectly satisfied with the fact that the Second Law "works": it can correctly predict the outcomes of experiments. Agreement with experiment is the best substitute for a rigorous mathematical proof.

Similarly to Kac, we share the viewpoint of a mathematician. With all due respect (and admiration!) to the physicists, a mathematician by training is obliged to point out the characteristic fallacy: "inductive experience that the Second Law works is not a rigorous mathematical proof".

Even if the majority of physicists are happy with the current explanations, there are plenty of others who nevertheless feel that these questions have not been completely answered yet. For illustration, here is what the physicist Gerald Feinberg wrote about it in his book [Fei85]:

"Why does time flow in only one way in the world of everyday life, when it appears to flow in both directions in the microworld? Of those questions in physics that have not yet been answered to everyone's satisfaction, this is one of the oldest. It was first raised in the latter part of the nineteenth century. While various answers have been proposed over the years, there is still a sense among many physicists, including myself, that the last word has not been said on the subject. The usual answer is that the notion that time has only one direction in the world of large-scale objects is only approximately true. It is overwhelmingly probable that things will proceed this way, but not absolutely certain.

It is argued that equally likely situations on the microlevel correspond to more *disorderly* situations on the level of familiar objects than to *orderly* ones. Therefore, when changes in a system take place, such as the adjustments in temperature when two objects of unequal temperature are placed in contact, the changes are more likely to be in the direction of *decreasing order* (in a very broad sense). This *decrease in order* is what we mean by entropy increase. It is the cornerstone of the physicists concept of time moving forward in the macroworld.

This approach to the *decrease of order* has been generally accepted for about a century, since it was recognized by Austrian physicist Ludwig Boltzmann and others. However, there are problems connected with the theory that leave many physicists uneasy. One question involves whether "order" really does decrease. There is little doubt that order decreases on the macroscopic level. When you pour a quart of colt water into a quart of hot water, the resulting lukewarm water represents a decrease in the order because if the atoms had been randomly assigned to one or another of the original quarts, it is highly unlikely that all the "hot" ones, those with higher energy, would have ended up in one of the quarts. But it has been suggested that on the microlevel, the amount of order always remains the same (since the Gibbs entropy = fine-grained entropy is constant by Liouville's theorem). We may think we are observing a process in which order decreases, but what is actually happening is that order being transferred from one form, which is apparent to us, into another forms that are not so apparent. In other words, the "decrease" is only in our information, what we know about the system (= the increase in coarse-grained entropy), and not an objective change in the system itself.

While this idea appeals to some physicists, perhaps on the grounds that it is better if nothing is ever lost, other physicists resist it. They consider it an unnecessary introduction of subjectivity into physical laws. These points of view are strongly held by one or another group of physicists."

Next we quote from the well known mathematical physicist and cosmologist Roger Penrose from his book [PeR2004]]:

"Boltzmann entropy involves the division of the phase space into subregions ("boxes") — called a coarse-graining of the phase space — where the points of a given box represent physical states that are macroscopically indistinguishable. Boltzmann's definition of entropy of a state in a box is the logarithm of the volume of the box in the phase space.

I have said that Boltzmann's definition makes the notion of entropy "clear". But for the logarithmic formula of Boltzmann's entropy to represent something physically precise, it would be necessary to have a clear-cut prescription for the coarse graining that our family of "boxes" is supposed to represent. There is undoubtedly something "arbitrary" in the particular division into boxes that one might happen to select. The definition seems to depend upon how closely one chooses to examine the system. Two states that are "macroscopically indistinguishable" to one experimenter might be distinguishable to another. Moreover, exactly where the boundary between two boxes happens to be drawn is again very arbitrary, since two neighboring points of the coarse-graining, with one on each side of a boundary, might be assigned quite different entropies, despite their being virtually identical. There is still something very subjective about the definition of Boltzmann's entropy, despite its being an undoubted improvement on the idea of just a measure of the "randomness" in a system.

My own position concerning the physical status of entropy is that I do not see it as an "absolute" notion in present day physical theory, although it is certainly a very useful one.

In my view, entropy has the status of "convenience", in present day theory, rather than being "fundamental"."

It is interesting to point out that entropy is a central concept in cosmology. For example, a key message of the two books [PeR94] and [PeR2004] of Roger Penrose is based on this notion of "convenience": it is what Penrose calls "our extraordinarily special Big Bang with its absurdly low entropy". It comes from the Bekenstein–Hawking black hole entropy, and Penrose's estimation of the entropy of the universe. We have to point out, however, that this is rather controversial, since the "entropy of the universe" is not a well-established concept, and similarly, the black hole entropy is far from being fully understood.

This example nicely illustrates how entropy is widely used by contemporary physicists far beyond classical thermodynamics, which makes it even more desirable to have a solid, mathematically rigorous conceptual foundation.

Our goal in this book is to work out a rigorous time-quantitative approach, which explains how, starting from off-equilibrium, large dynamical systems reach equilibrium and exhibit entropy increase quickly, in realistic short time, for more-or-less realistic systems. We feel our idealized mechanical models are much more convincing than the non-mechanical Ehrenfest urn model or the Kac ring model. We do not use ergodic theory; our approach is based on Fourier analysis and combinatorics.

Appendix 2

Proving Lemmas 1.1–2 and Beyond

We start with the proof of (1.7), and explain how it implies both Lemma 1.1 and Lemma 1.2.

More precisely, we study N point billiards in the unit cube $[0,1]^3$, and the motion of the k-th point billiard in $[0,1]^3$ is described by $\mathbf{x}_k(t)$, $0 \le t < \infty$. The time-evolution of the system is represented by the following parametrized curve in the $3N$-dimensional cube $[0,1]^d = [0,1]^{3N}$ (= configuration space) that we call the trajectory of the system

$$\Gamma(t) = (\mathbf{x}_1(t), \mathbf{x}_2(t), \ldots, \mathbf{x}_N(t))$$

$$= \Big(2\|(y_{1,1}+v_{1,1}t)/2\|, 2\|(y_{1,2}+v_{1,2}t)/2\|, 2\|(y_{1,3}+v_{1,3}t)/2\|, \ldots$$

$$\ldots, 2\|(y_{N,1}+v_{N,1}t)/2\|, 2\|(y_{N,2}+v_{N,2}t)/2\|, 2\|(y_{N,3}+v_{1,3}t)/2\|\Big). \quad \text{(A1)}$$

Note that (A1) corresponds to (1.19).

Assume that the $3N$ initial velocity coordinates

$$\{v_{k,1}, v_{k,2}, v_{k,3} : \ 1 \le k \le N\}$$

are linearly independent over the rationals, and let $S \subset [0,1]^3$ be an arbitrary but fixed Jordan measurable subset of the unit cube with volume $0 < \mathrm{vol}(S) < 1$. We use notation (1.5): $N_S(t)$ denotes the point-counting function

$$N_S(t) = \sum_{\substack{1 \le k \le N: \\ \mathbf{x}_k(t) \in S}} 1.$$

We show that the Kronecker–Weyl theorem, applied in the configuration space, and combined with unfolding, implies (1.7)

$$\lim_{T\to\infty} \frac{1}{T} \text{length}\,\{t \in [0,T] : \ N_S(t) = n\} = \binom{N}{n} p^n q^{N-n}, \qquad \text{(A2)}$$

where $p = \text{vol}(S)$, $q = 1 - p$, and *length* simply means the 1-dimensional Lebesgue measure.

To prove (A2), we use the following notation: let $S^{(1)} = S \subset [0,1]^3$, and let $S^{(0)} = [0,1]^3 \setminus S$ be the complement of S, and for every N-dimensional vector $\Delta = (\delta_1, \delta_2, \ldots, \delta_N)$ with coordinates $\delta_i = 0$ or 1, let

$$S_{\mathcal{C}}(\Delta) = S^{(\delta_1)} \times S^{(\delta_2)} \times \cdots \times S^{(\delta_N)} \tag{A3}$$

be the Cartesian product of the N sets $S^{(\delta_i)}$, $1 \leq i \leq N$ in this order (the index $\mathcal{C}$ indicates the configuration space). If the vector Δ has n coordinates with 1 and $N - n$ coordinates with 0, then

$$\text{vol}_d(S_{\mathcal{C}}(\Delta)) = p^n q^{N-n}, \tag{A4}$$

where vol_d stands for the $d = 3N$-dimensional Lebesgue measure.

We know that the distribution of the billiard path $\Gamma(t)$, $0 \leq t \leq T$ (see (A1) with respect to a test set $S_{\mathcal{C}}(\Delta)$ (see (A3)) is equivalent — via *unfolding*, see Section 1 — to the distribution of the corresponding torus line (see (1.16))

$$\Gamma^*(t) = \Big(\{y_{1,1} + v_{1,1}t\}, \{y_{1,2} + v_{1,2}t\}, \{y_{1,3} + v_{1,3}t\}, \ldots$$

$$\ldots, \{y_{N,1} + v_{N,1}t\}, \{y_{N,2} + v_{N,2}t\}, \{y_{N,3} + v_{1,3}t\}\Big)$$

$$= \Gamma(0) + \Gamma'(0)t \;\text{ modulo one} \tag{A5}$$

with starting point $\Gamma(0)$ and direction vector $\Gamma'(0)$ with respect to the union

$$\bigcup_{i=1}^{2^{3N}} S_{\mathcal{C},i}(\Delta)$$

of the 2^{3N} "reflected halves" of $S_{\mathcal{C}}(\Delta)$. In other words, by using the notation $S^*(\Delta) \subset [0,1]^{3N}$ for this union set, we have

$$\text{length}\,\{t \in [0,T] : \; \Gamma(t) \in S_{\mathcal{C}}(\Delta)\} = \text{length}\,\{t \in [0,T] : \; \Gamma^*(t) \in S^*(\Delta)\}.$$

Clearly (see (A4))

$$\text{vol}_d(S^*(\Delta)) = \text{vol}_d(S_{\mathcal{C}}(\Delta)) = p^n q^{N-n}. \tag{A6}$$

By hypothesis the coordinates $v_{i,j}$, $1 \leq i \leq N$, $1 \leq j \leq 3$ of the direction vector $\Gamma'(0)$ are linearly independent over the rationals, so we can apply the Kronecker–Weyl theorem to the 0-1 valued characteristic function of a set $S^*(\Delta) \subset [0,1]^{3N}$, and obtain

$$\lim_{T\to\infty} \frac{1}{T}\text{length}\,\{t \in [0,T] : \; \Gamma^*(t) \in S^*(\Delta)\} = \text{vol}_d(S^*(\Delta)) = p^n q^{N-n}, \tag{A7}$$

where in the last step we used (A6), and $\Gamma^*(t)$ is defined in (A5). By (A7),

$$\lim_{T\to\infty}\frac{1}{T}\text{length}\,\{t\in[0,T]:\ N_S(t)=n\}$$

$$=\sum_{\substack{\Delta\text{ has }n\\ \text{coordinates }1}}\lim_{T\to\infty}\frac{1}{T}\text{length}\,\{t\in[0,T]:\ \Gamma(t)\in S_C(\Delta)\}$$

$$=\sum_{\substack{\Delta\text{ has }n\\ \text{coordinates }1}}\lim_{T\to\infty}\frac{1}{T}\text{length}\,\{t\in[0,T]:\ \Gamma^*(t)\in S^*(\Delta)\}=\binom{N}{n}p^nq^{N-n},$$

which proves (1.7).

Notice that (A7) also implies Lemma 1.2 ("superdiscrepancy"). Indeed, the Cartesian product of rational boxes $\prod_{i=1}^N B_i$ has positive volume in the $3N$-dimensional configuration space, so, in view of (A7), the torus line will enter this product set infinitely many times. (It is interesting to note that Lemma 1.2 can be derived directly from Kronecker's density theorem in the configuration space, i.e., we do not really need the full power of the Kronecker–Weyl theorem.)

The same argument — applying the Kronecker–Weyl equidistribution theorem in the configuration space via unfolding — proves a far-reaching generalization of (1.7).

Multinomial version of (1.7). *Let $k\ge 2$ be an integer, and let $S_1,S_2,\ldots,S_k$ be arbitrary pairwise disjoint Jordan measurable subsets of the unit cube $[0,1]^3$. Let $0<p_i=\text{vol}(S_i)<1$, $1\le i\le k$. If the $3N$ initial velocity coordinates of the point-billiard system are linearly independent over the rationals, then*

$$\lim_{T\to\infty}\frac{1}{T}\text{length}\,\{t\in[0,T]:\ N_{S_i}(t)=n_i\text{ for all }1\le i\le k\}$$

$$=\frac{N!}{n_1!\cdots n_k!(N-n_1-\ldots-n_k)!}p_1^{n_1}\cdots p_k^{n_k}(1-p_1-\ldots-p_k)^{N-n_1-\ldots-n_k}. \tag{A8}$$

Notice that the right-hand side of (A8) represents the general term of the multinomial distribution (just like as the right-hand side of (A2) represents the general term of the binomial distribution).

Let us return to (1.6)–(1.7) in Section 1. Now the deduction of (1.6) (i.e., Lemma 1.1) from (1.7) is easy. Indeed, by (1.7),

$$\lim_{T\to\infty}\frac{1}{T}\text{length}\left\{t\in[0,T]:\ \frac{N_S(t)-pN}{\sqrt{pqN}}\le y\right\}=\sum_{\substack{0\le n\le N:\\ n\le pN+y\sqrt{pqN}}}\binom{N}{n}p^n q^{N-n},\tag{A9}$$

where $p=\text{vol}(S)$ and $q=1-p$. Finally, we apply the classical Moivre–Laplace theorem, which says that the binomial distribution with parameters N and $0<p<1$ approximates the standard normal distribution with error term $O(1/\sqrt{Npq})$. Combining this fact with (A9), (1.6) follows.

Note that the same argument (applying the Kronecker–Weyl equidistribution theorem in the configuration space via unfolding) proves the generalization of (1.6) where different point billiards may have different masses. Let μ_k, $1\le k\le N$ denote the mass of the k-th point billiard. For simplicity we assume that average mass is one, i.e., $\sum_{k=1}^N\mu_k=N$. Then the point-counting function $N_S(t)$ is replaced by the mass-counting function $N_S(\text{mass};t)$:

$$N_S(\text{mass};t)=\sum_{\substack{1\le k\le N:\\ \mathbf{x}_k(t)\in S}}\mu_k.\tag{A10}$$

It is natural to expect that, when the system reaches "equilibrium", the mass-counting function $N_S(\text{mass};t)$ gets very close to pN, and the discrepancy exhibits "random fluctuation" with "standard deviation" $\sigma\sqrt{pqN}$ as t runs in a long time interval; here

$$\sigma^2=\frac{1}{N}\sum_{k=1}^N\mu_k^2,\ \text{ and for simplicity assume }\frac{1}{N}\sum_{k=1}^N\mu_k=1.\tag{A11}$$

Again the Kronecker–Weyl theorem justifies this intuition; see (A14) below. The first step is the following perfect analog of (A9) — the proof is the same.

Mass-counting version of (A9). *If the $3N$ initial velocity coordinates of the point-billiard system are linearly independent over the rationals, then using the notation (A11), for every real number $-\infty<y<\infty$,*

$$\lim_{T\to\infty}\frac{1}{T}\text{length}\left\{t\in[0,T]:\ \frac{N_S(\text{mass};t)-pN}{\sigma\sqrt{pqN}}\le y\right\}$$
$$=\sum_{\substack{I\subset\{1,2,\ldots,N\}:\\ \sum_{i\in I}\mu_k\le pN+y\sigma\sqrt{pqN}}}p^{|I|}q^{N-|I|},\tag{A12}$$

where $p = \mathrm{vol}(S)$, $q = 1 - p$, *and* $|I|$ *denotes the number of elements of a finite set* I.

The right-hand side of (A9) is the well known binomial distribution; the right-hand side of (A12) is the distribution of the sum $Y = \sum_{k=1}^{N} X_i$ of N independent random variables X_k, $1 \le k \le N$ such that $\Pr[X_k = \mu_k] = p$ and $\Pr[X_k = 0] = q$. Then the expectation of Y is pN, and the variance of Y is $\sigma^2 pqN$. The well known Berry–Esseen inequality in probability theory says that

$$\max_{-\infty<y<\infty} \left| \sum_{\substack{I\subset\{1,2,\dots,N\}:\\ \sum_{i\in I}\mu_k \le pN + y\sigma\sqrt{pqN}}} p^{|I|}q^{N-|I|} - \frac{1}{\sqrt{2\pi}}\int_{-\infty}^{y} e^{-u^2/2}\,du \right|$$

$$\le \frac{\rho^3/\sigma^3}{\sqrt{pqN}}, \quad \text{where } \rho^3 = \frac{1}{N}\sum_{k=1}^{N}\mu_k^3. \tag{A13}$$

Combining (A12) and (A13), we obtain

$$\lim_{T\to\infty}\frac{1}{T}\mathrm{length}\left\{t\in[0,T]: \frac{N_S(\mathrm{mass};t) - pN}{\sqrt{pqN}} \le y\right\}$$

$$= \frac{1}{\sqrt{2\pi}}\int_{-\infty}^{y} e^{-u^2/2}\,du + O\left(\frac{\rho^3/\sigma^3}{\sigma\sqrt{pqN}}\right). \tag{A14}$$

Note that we can easily switch from "mass equilibrium" to "energy equilibrium" by changing the interpretation of μ_i as follows: let $\mu_i = m_i v_i^2/2$, where m_i and v_i denote, respectively, the mass and the speed of the i-th particle.

Let us return to Lemma 1.1; note that in Lemma 1.1 we can assume without loss of generality that $\mathrm{vol}(S) \le 1/2$ (indeed, otherwise we just take the complement of S). If $\mathrm{vol}(S)$ is "not too small" then Lemma 1.1 is basically a central limit theorem. Indeed, since in the applications N is *very large* (around 10^{27}, or larger), the error term $O(1/\sqrt{Npq})$ (with $p = \mathrm{vol}(S)$ and $q = 1 - p$) is still very small, so the distribution on the left-hand side of (1.6) is very close to the standard normal distribution.

On the other hand, if the volume $p = \mathrm{vol}(S)$ is very small — say, it is around $1/N$ — then of course (1.6) becomes useless (since the error term $O(1/\sqrt{Npq})$ is in the range of 1). However, (1.7) still works, and implies the following result.

Poisson Law. *Write* $\lambda = pN = \text{vol}(S)N$. *If the* $3N$ *initial velocity coordinates of the point-billiard system are linearly independent over the rationals, then for every integer* $\ell \geq 0$,

$$\lim_{T\to\infty} \frac{1}{T}\text{length}\,\{t \in [0,T]:\ N_S(t) = \ell\} = \frac{\lambda^\ell e^{-\lambda}}{\ell!} + O(\lambda^2/N). \tag{A15}$$

The probability distribution

$$\frac{\lambda^\ell e^{-\lambda}}{\ell!}, \quad \ell = 0, 1, 2, 3, \ldots$$

on the right-hand side of (A15) is the well known Poisson distribution with parameter $\lambda > 0$.

Notice that (A15) immediately follows from (1.7) combined with the simple approximation

$$\binom{N}{\ell}(\lambda/N)^\ell(1-\lambda/N)^{N-\ell} = \frac{\lambda^\ell e^{-\lambda}}{\ell!} + O(\lambda^2/N).$$

If $\lambda = pN = \text{vol}(S)N$ is in the "constant range" (i.e., the volume of the test set S is very small), then by (A15), the point-counting function $N_S(t)$ has Poisson distribution with negligible error (since the error term $O(\lambda^2/N)$ is very small).

In the general case, where there are several sets $S_1, S_2, \ldots, S_k$, $k \geq 2$, (A8) gives the following.

Multinomial Poisson Law. *Let* $k \geq 2$ *be an integer, and let* $S_1, S_2, \ldots, S_k$ *be* k *small pairwise disjoint Jordan measurable subsets of the unit cube* $[0,1]^3$ *such that* $0 < \lambda_i = p_i N = \text{vol}(S_i)N$, $1 \leq i \leq k$ *are all in the constant range. If the* $3N$ *initial velocity coordinates of the point-billiard system are linearly independent over the rationals, then*

$$\lim_{T\to\infty} \frac{1}{T}\text{length}\,\{t \in [0,T]:\ N_{S_i}(t) = \ell_i \text{ for all } 1 \leq i \leq k\}$$

$$= \frac{N!}{\ell_1!\cdots\ell_k!(N-\ell_1-\ldots-\ell_k)!}(\lambda_1/N)^{\ell_1}$$

$$\cdots(\lambda_k/N)^{\ell_l}(1-(\lambda_1+\ldots+\lambda_k)/N)^{N-\ell_1-\ldots-\ell_k}$$

$$= \prod_{i=1}^{k} \frac{\lambda_i^\ell e^{-\lambda_i}}{\ell!} + O(1/N), \tag{A16}$$

where the implicit constant in the error term $O(1/N)$ depends only on the given constants $\lambda_1, \ldots, \lambda_k$.

The product at the beginning of the last line in (A16) tells us that the long-term time evolution of the system exhibits "asymptotic independence" of "very small" disjoint test sets S_i, $1 \le i \le k$.

Notice that "asymptotic independence" certainly fails for "large" subsets S_i. The simplest example is when we have two sets, and they are complements $S_2 = [0,1]^3 \setminus S_1$. Then we have a deterministic relation between the number of point billiards in S_1 and in S_2 (the sum is constant), which contradicts independence.

What happens if the disjoint test sets S_i, $1 \le i \le k$ are still "small" but not "very small"?

Here is a positive result. Assume that the volumes $p_i = \mathrm{vol}(S_i)$, $1 \le i \le k$ are all in the range of $1/\sqrt{N}$ (i.e., each test set is expected to contain around $\sqrt{N}$ point billiards), then

$$\lim_{T\to\infty} \frac{1}{T}\text{length}\left\{t \in [0,T] : \frac{N_{S_i}(t) - p_i N}{\sqrt{p_i(1-p_i)N}} \le y_i \text{ for all } 1 \le i \le k\right\}$$

$$= \prod_{i=1}^{k} \mathcal{N}(y_i) + O\left(N^{-1/4}\log N\right) \tag{A17}$$

for all fixed constants $-\infty < y_1, y_2 < \infty$, where log stands for the natural (base e) logarithm and $\mathcal{N}(y)$ denotes the normal distribution:

$$\mathcal{N}(y) = \frac{1}{\sqrt{2\pi}} \int_{-\infty}^{y} e^{-u^2/2}\, du.$$

Note that (A17) easily follows from (A8) via routine estimation. We outline this simple — but somewhat long — estimation in the special case of two disjoint midsize sets S_1 and S_2 (i.e., $k = 2$) with $p_i = \mathrm{vol}(S_i) \approx 1/\sqrt{N}$, $i = 1, 2$. In view of (A8) we have to estimate the ratio

$$\frac{\frac{N!}{n_1!n_2!(N-n_1-n_2)!} p_1^{n_1} p_2^{n_2} (1-p_1-p_2)^{N-n_1-n_2}}{\binom{N}{n_1} p_1^{n_1}(1-p_1)^{N-n_1} \cdot \binom{N}{n_2} p_2^{n_2}(1-p_2)^{N-n_2}}, \tag{A18}$$

where the product in the denominator of (A18) represents independence. To prove (A17) (for $k = 2$), it suffices to show that the ratio (A18) is

$$1 + O\left(N^{-1/4}\log N\right) \quad \text{for } p_i N - \log N\sqrt{p_i N} \le n_i \le p_i N + \log N\sqrt{p_i N},$$

$$i = 1, 2. \tag{A19}$$

Indeed, the binomial distribution is close to the normal distribution if Np_i is "large" (see the denominator of (A18)), and $|n_i - p_i N| > \log N \sqrt{p_i N}$ represents a "large deviation" with probability around

$$e^{-(\log N)^2/2} = N^{-\log N/2},$$

which is *superpolynomially small* in terms of N, so it is totally negligible.

We have

$$\text{ratio (A18)} = \frac{(N-n_1)!(N-n_2)!(1-p_1-p_2)^{N-n_1-n_2}}{(N-n_1-n_2)!N!(1-p_1)^{N-n_1}(1-p_2)^{N-n_2}}$$

$$= \frac{(1-p_1-p_2)^{N-n_1-n_2}}{(1-p_1)^{N-n_1}(1-p_2)^{N-n_2}} \prod_{\ell=1}^{n_1}\left(1-\frac{n_2}{N-n_1+\ell}\right)$$

$$= \frac{\left(\frac{1-p_1-p_2}{(1-p_1)(1-p_2)}\right)^{N-n_1-n_2}}{(1-p_1)^{n_2}(1-p_2)^{n_1}} \prod_{\ell=1}^{n_1}\left(1-\frac{n_2}{N-n_1+\ell}\right)$$

$$= \frac{\left(1-\frac{p_1p_2}{(1-p_1)(1-p_2)}\right)^{N-n_1-n_2}}{(1-p_1)^{n_2}(1-p_2)^{n_1}} \prod_{\ell=1}^{n_1}\left(1-\frac{n_2}{N-n_1+\ell}\right)$$

$$= \exp\Bigg((N-n_1-n_2)\log\left(1-\frac{p_1p_2}{(1-p_1)(1-p_2)}\right)$$

$$-n_1\log(1-p_2) - n_2\log(1-p_1) + \sum_{\ell=1}^{n_1}\log\left(1-\frac{n_2}{N-n_1+\ell}\right)\Bigg),$$

where of course $\exp(x) = e^x$. Using the Taylor series

$$\log(1-x) = -(x + x^2/2 + x^3/3 + \ldots) = -x + O(x^2),$$

we have

$$\text{ratio (A18)} = \exp\Bigg(-(N-n_1-n_2)\left(\frac{p_1p_2}{(1-p_1)(1-p_2)} + O(p_1^2p_2^2)\right)$$

$$+n_1p_2 + O(n_1p_2^2) + n_2p_1 + O(n_2p_1^2) - \sum_{\ell=1}^{n_1}\frac{n_2}{N-n_1+\ell} + O(n_1n_2^2N^{-2})\Bigg). \tag{A20}$$

Estimating the Riemann sum with the integral, we have

$$\sum_{\ell=1}^{n_1} \frac{n_2}{N-n_1+\ell} = n_2 \int_0^{n_1} \frac{1}{N-x}\, dx + O(n_2/N) = n_2 \log\left(\frac{N}{N-n_1}\right) + O(n_2/N)$$

$$= -n_2 \log\left(1-\frac{n_1}{N}\right) + O(n_2/N) = \frac{n_1 n_2}{N} + O(n_1^2 n_2 N^{-2}) + O(n_2/N). \quad \text{(A21)}$$

Using (A21) in (A20), we have

$$\text{ratio (A18)} = \exp\Bigg(\left(\frac{Np_1p_2}{(1-p_1)(1-p_2)} - \frac{n_1n_2}{N}\right)$$

$$+ (n_1p_2 - Np_1p_2) + (n_2p_1 - Np_2p_1)$$

$$+ O(Np_1^2p_2^2) + O(n_1p_2^2) + O(n_2p_1^2) + O(n_1n_2^2N^{-2}) + O(n_1^2n_2N^{-2}) + O(n_2/N)\Bigg). \quad \text{(A22)}$$

By using the condition

$$n_i = p_iN \pm \log N\sqrt{p_iN}, \ i = 1,2$$

(see (A19)) in (A22), we conclude

$$\text{ratio (A18)} = \exp\left(O\left(\log N\sqrt{Np_1}p_2\right) + O\left(\log N\sqrt{Np_2}p_1\right)\right)$$

$$= 1 + O\left(N^{-1/4}\log N\right),$$

which is exactly what we wanted to prove (see (A19)).

References

[Be2010] J. Beck, Superuniformity of the typical billiard paths, *An Irregular Mind*, special volume dedicated to the 70th birthday of Endre Szemerédi (Conference in Budapest, August 2010), edited by Imre Bárány, 1–70, Springer 2010.

[Be2015] J. Beck, From Khinchin's conjecture on strong uniformity to superuniform motions, Mathematika **61** issue 3 (2015), 591–707.

[Be2018] J. Beck, *Strong Uniformity and Large Dynamical Systems*, World Scientific Publishing 2018.

[Be-Ch87] J. Beck and W.W.L. Chen, *Irregularities of Distribution*, Cambridge University Press 1987.

[Bo09] P. Bohl, Über ein in der Theorie der säkularen Störungen vorkommendes Problem, J. Reine Angew. Math, **135** (1909), 189–282.

[Ch-M2006] N. Chernov and R. Makarian, *Chaotic Billiards*, Mathematical Surveys and Monographs, vol. 127, American Mathematical Society 2006.

[Dr-Ti97] M. Drmota and R. F. Tichy, *Sequences, Discrepancies and Applications*, Lecture Notes in Math. 1651, Springer 1997.

[Eh11] P. and T. Ehrenfest, *The Conceptual Foundations of the Statistical Approach in Mechanics* (English translation of the original German edition in 1911), Cornell University Press, Ithaca, New York, 1959.

[Fa64] I. Farquhar, *Ergodic Theory in Statistical Mechanics*, Wiley, New York, 1964.

[Fei85] G. Feinberg, *Solid Clues*, Simon& Schuster, Inc., New York, 1985.

[Fel71] W. Feller: *An Introduction to Probability Theory and its Applications*, Vol. 2 (2nd edition), Wiley, New York 1971.

[Go-Le2004] S. Goldstein and J.L. Lebowitz, On the (Boltzmann) entropy of non-equilibrium systems, Physica D **193** (2004), 53–66.

[Ja89] E.T. Jaynes, *Papers on Probability, Statistics and Statistical Physics*, editor: R.D. Rosenkrantz, Kluwer Academic Publishers 1989.

[Ka59] M. Kac, *Probability and Related Topics in Physical Sciences*, Lectures in Applied Math. Series, Vol. 1a, American Mathematical Society 1959.

[Ke-M-S86] S. Kerckhoff, H. Masur and J. Smillie, Ergodicity of billiard flows and quadratic differentials, Annals of Math. **124** (1986), 293–311.

[Kö-Sz13] D. König and A. Szücs, Mouvement d'un point abandonne a l'interier d'un cube, Rend. Circ. Mat. Palermo **36** (1913), 79–90.

[La-Li40] L.D. Landau and E.M. Lifshitz *Statistical Physics*, 1940 (originally published in Russian).

[L75] O.E. Lanford III, Time evolution of large classical systems, pp. 1–111 of *Dynamical systems, theory and applications*, ed. J. Moser, Springer 1975.

[L-M2003] J.L. Lebowitz and Ch. Maes, Entropy: a Dialog; see Chapter 13 in the book *Entropy*, eds. A. Greven, G. Keller, G. Warnecke, Princeton University Press 2003.

[L-Y99] E.H. Lieb and J. Yngvason, The Physics and Mathematics of the Second Law of Thermodynamics, Physics Reports **310** (1999), 1–96.

[Mar70] J. M. Marstrand, On Khinchin's conjecture about strong uniform distribution, Proc. London Math. Soc. **21** (1970), 540–556.

[Mat99] J. Matousek, *Geometric Discrepancy*, Springer 1999 (and corrected softcover printing in 2010).

[PeO70] O. Penrose, *Foundations of Statistical Mechanics, A deductive treatment*, Pergamon Press, Oxford 1970.

[PeR94] R. Penrose, *Shadows of the Mind*, Oxford University Press 1994.

[PeR2004] R. Penrose, *The Road to Reality: A Complete Guide to the Laws of the Universe*, Alfred A. Knopf (publisher) 2004.

[Re62] A. Rényi, *Wahrscheinlichkeitsrechnung*, Deutscher Verlag der Wissenschaften, 1962.

[Ru69] D. Ruelle, *Statistical Mechanics: Rigorous Results*, Benjamin, New York, 1969.

[Ru99] D. Ruelle, Smooth dynamics and new theoretical ideas in non-equilibrium statistical mechanics, J. Statistical Physics **95** (1999), 393–468.

[Si76] Ya.G. Sinai, *Introduction to Ergodic Theory*, Mathematical Notes, Princeton University Press 1976.

[Sk93] L. Sklar, *Physics and Chance: Phylosophical Issues in the Foundations of Statistical Mechanics*, Cambridge University Press 1993.

[Sz2000] *Hard Ball Systems and the Lorentz Gas*, Encyclopaedia of Math. Sci., editor D. Szász, Springer 2000.

[Th72] C. J. Thompson, *Mathematical Statistical Mechanics*, The Macmillan Company 1972.

[U-F63] G.E. Uhlenbeck and G.W. Ford, *Lectures in Statistical Mechanics*, American Mathematical Society 1963.

[Vi2010] C. Villani, (Ir)reversibility and Entropy, pp. 19–79 in the book *Time*, edited by B. Duplantier, Poincare Seminar 2010, Progress in Mathematical Physics vol. 63, Springer Basel 2013 Birkhauser.

[Wa58] G.N. Watson, *A Treatise on the Theory of Bessel Functions*, Cambridge University Press, Cambridge 1958.

[We14] H. Weyl, Sur une application de la théorie des nombres á la mécanique statistique at á la théorie des perturbations, LÉnseignement mathématique 1014.

[We16] H. Weyl, Über die Gleichverteilung von Zahlen mod Eins, Math. Ann. **77** (1916), 313–352.

Index

Lightning Source UK Ltd.
Milton Keynes UK
UKHW020001061020
371064UK00003B/238

9 789811 225550